Based on unrestricted access to papers and personnel at BP, as well as numerous other sources, this second volume of BP's history aims to be an honest and comprehensive examination of the Company in the period 1928 to 1954. Such a history inevitably touches on many different historical interests ranging from international relations to social, economic, political and military topics, primarily in Britain and the Middle East.

The book includes penetrating insights into the direction and management of the Company, the achievements and shortcomings of successive chairmen, and the relationship between the Company and its major shareholder, the British Government. It also deals in detail with matters which have retained an aura of controversy and mystique long after their occurrence, most notably the international petroleum cartel which sought to control world oil markets, and the major international crisis arising from Iran's oil nationalisation.

Concluding his account of these events, Dr Bamberg calls into question some widely held views on the history of BP and the oil industry. Was it in BP's interest to be closely identified with the declining imperial power of Britain during an age of rising nationalism? Did BP, as one of the famous 'Seven Sisters' of the oil industry, control its economic and political environment to its own advantage? Or was it buffeted and thrown off course by events beyond its control?

THE HISTORY OF THE
BRITISH PETROLEUM COMPANY

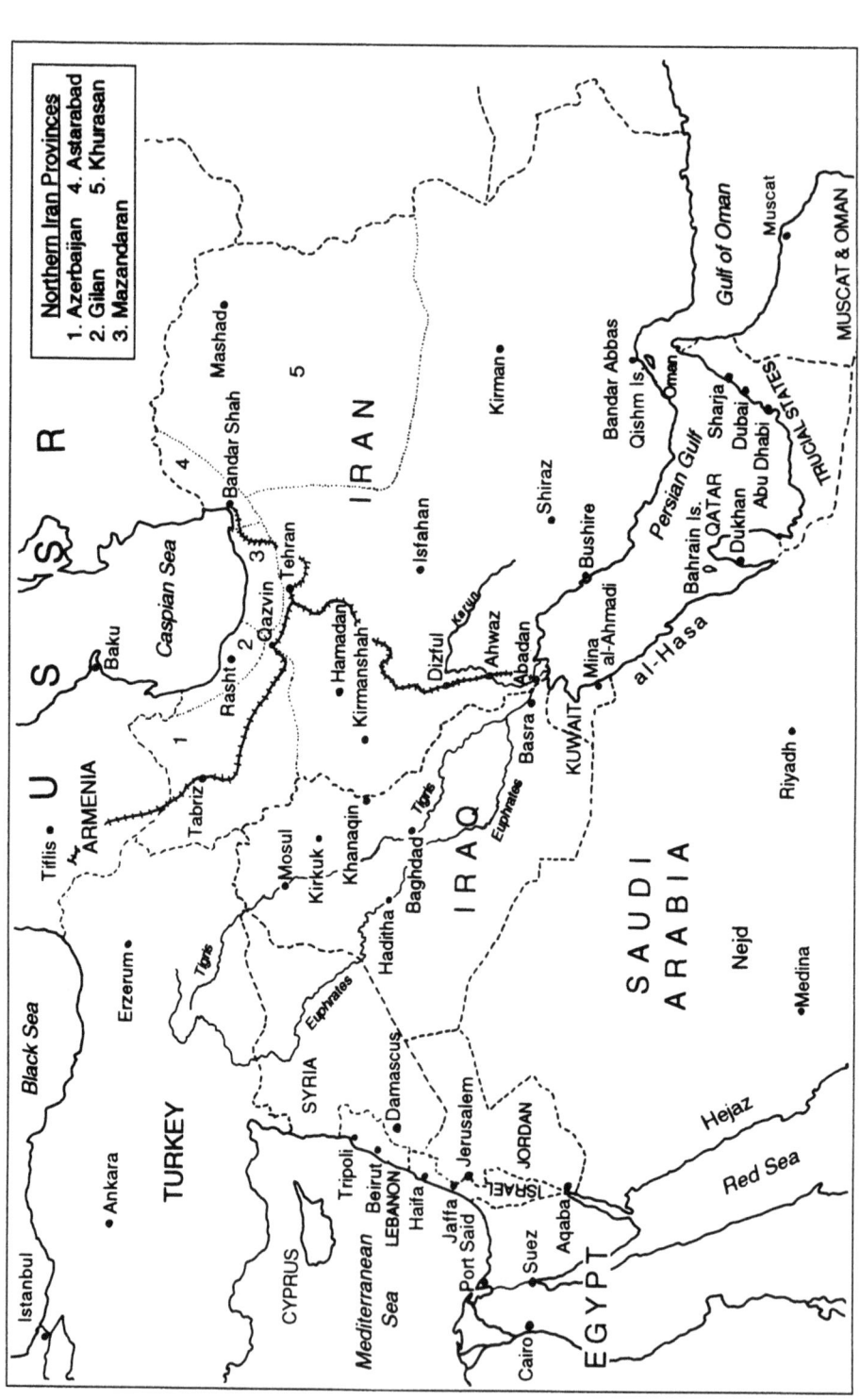

THE HISTORY OF
THE
BRITISH PETROLEUM
COMPANY

Volume 2
The Anglo-Iranian Years,
1928–1954

J. H. BAMBERG

CAMBRIDGE
UNIVERSITY PRESS

CAMBRIDGE UNIVERSITY PRESS
Cambridge, New York, Melbourne, Madrid, Cape Town, Singapore, São Paulo, Delhi

Cambridge University Press
The Edinburgh Building, Cambridge CB2 8RU, UK

Published in the United States of America by Cambridge University Press, New York

www.cambridge.org
Information on this title: www.cambridge.org/9780521117593

First published 1994
Reprinted 1994, 1995, 1996, 2000
This digitally printed version 2009

A catalogue record for this publication is available from the British Library

Library of Congress Cataloguing in Publication data
Ferrier, R. W. (Ronald W.)
The history of The British Petroleum Company.
Vol. 2: by J.H. Bamberg.
Includes bibliographical references and index.
Contents: v. 1. The developing years, 1901–1932–
v. 2 The Anglo-Iranian years, 1928–1954.
1. British Petroleum Company – History.
2. Petroleum industry and trade – Great Britain – History.
I. Bamberg, J.H.   II. Title.
HD9571.9.B73F47   1982   338.7'622338'0941   81-18019
ISBN 0-521-24647-4

ISBN 978-0-521-25950-7 hardback
ISBN 978-0-521-11759-3 paperback

Contents

Tables

Figures

Maps

Graphs and diagrams

Illustrations

Preface

In volume 1 of *The History of The British Petroleum Company* (1982), Ronald Ferrier described the origins and development of BP in approximately the first three decades of the twentieth century. This volume follows on from where volume 1 left off and terminates with the settlement of the Anglo-Iranian oil nationalisation dispute in 1954.

The treatment is somewhat different from the preceding volume. In part this is attributable to the changing nature of the Company as it emerged from the early pioneering years and developed into a more corporate form of enterprise in a changed social, economic and political environment. Mainly, however, differences in style and approach between the two volumes arise from the change in author, which took place after I succeeded Ronald Ferrier as BP historian on his retirement in 1989. By that time he had done much work on the second volume, on which, despite ill health, he continued to work for another year before handing over to me in mid-1990. From that point on, I became responsible for the book.

Writers of commissioned histories tend to be suspected of compromising their impartiality and independence in obeisance to their paymasters, who, it is frequently supposed, seek to be shown in a favourable light. To counteract such suspicions, historians commonly insert disclaimers of bias or censorship in the prefaces of commissioned Company histories. In the case of this history of BP, the scope for doubting its honesty of purpose is if anything magnified by its being written, not by a commissioned outside historian, but by an historian in the employment of the Company. Those who are inclined to doubt the integrity of this history may be reassured to know that no restrictions were placed on access to sources of information in BP, or on the manner in which they were used; that BP's pre-1955 archives, located at the University of Warwick, are now open to outside access, enabling researchers to study the evidence used for this volume and draw their own conclusions; and that the research and writing of this volume was monitored by an Editorial Committee which included, for varying periods, not only successive BP chairmen (Sir David Steel, Sir Peter Walters and Robert Horton) and directors (Lord Robbins and Lord Greenhill), but also eminent external historians (Professor Alfred D. Chandler, Professor Donald Coleman and Dr Peter Mathias), whose combined experience, knowledge and critical acumen could at any time be exercised to uphold the integrity and standards of the history.

Of course, none of that can possibly guarantee that this volume will satisfy all who might be interested in it. Although the aim has been to achieve a full, balanced coverage of the Company's activities, heed has also been taken of Voltaire's cautionary maxim: 'Le secret d'ennuyer est ... de tout dire' ('The way to be a bore [for an author] is to say everything'). For that reason – and for the practical purpose of keeping the book to a manageable length – much detail has been excluded. As a result, there will surely be specialists in particular fields who feel that one aspect or another of BP's history has been underemphasised or overlooked. While some may be disappointed by the treatment or exclusion of their special interests, there are others – particularly those who have made academic business history their speciality – with whom the whole idea of writing individual company histories is out of favour. Adherents to that school of thought hold that the most fruitful approach to expanding knowledge and understanding of the development of business is by the comparative analysis of a number of businesses, not by corporate biographies of single firms – a form of 'life without theory', as Disraeli described biography. It is indeed not difficult to find company histories which conform to that description. Yet there are also pitfalls in going to the opposite extreme and being too doctrinaire in stripping a business of its unique features, reducing enterprise to a mechanistic formula. Although there are repetitions in history, that is not to say that history repeats itself exactly. As Charles Wilson wrote some forty years ago, at the beginning of his classic history of Unilever: 'There is no trick or sequence of tricks that can be learnt to be repeated successfully: too many factors are changing too much of the time'.

For the historian, dealing, like the businessman, with a myriad of variables and subjective factors, there can be no objectively perfect balance between biography and theory, narrative and analysis, qualitative judgement and statistical measurement. Unable to identify an objective optimum, I have tried to interweave the Company's unique characteristics and identity with topics of wider, comparative interest such as finance, marketing, organisation, technology and government relations. Much more space is devoted to international affairs than is usual in business histories, but that, I feel, is an accurate reflection of BP's exceptionally deep involvement in matters of international diplomacy, particularly in relation to Iran. The balance and style will not, of course, suit every purpose or taste. Nor, on matters of interpretation, will my views meet with universal approval – an impossible prospect given the wide range of opinions which are held on some of the more controversial episodes in BP's history.

A work such as this – extending over a number of historical disciplines and drawing on a great volume and range of sources – requires more than a solo effort. A most important contribution was made by the antecedent work of Laurence Lockhart and Rose Greaves, who in 1970 completed a meticulously researched multi-volume record of BP's relations with the Iranian Government up to 1954 for in-house circulation only. Most of the subsequent research was done during Ronald Ferrier's term as BP historian. Among his research assistants, Anne Ewing was concerned with British oil policy; Julian Bowden researched and wrote up the history of the Iraq Petroleum Company; Robert Brown provided information on the Company's employment policies in Iran; Ewen Green produced interesting material on relations between the Company and the British Government after World War II, before moving on to a research fellowship at Brasenose College, Oxford and later to a lectureship at the University of Reading; Pamela Tanzey investigated the Company's marketing activities in continental Europe; and Anne Saunders was resourceful in locating relevant documents at the Public Records Office, Kew. Others who contributed to the history after retiring from their full-time employment with BP were Commander Edward Platt, who did much work on the history of the Company's shipping activities; John Hooper, who was indefatigable in researching the technical side of the Company's history; and Keith Taggart, whose skill in compiling the financial statistics was matched only by the pleasure of his company. In the later stages, after the research phase was largely over, Jenny Ward brought a fresh mind to the challenging task of unravelling and reducing the huge volume of material on the Company's concessionary relations with Iran to manageable proportions. Throughout, the study of documentary evidence was enlivened by discussions with scholars and others – including many past and present BP employees – whose knowledge and recollections helped greatly to clarify subjects and events which might otherwise have remained obscure. They are too numerous for me to acknowledge every one individually, and to mention some would mean excluding others. In this invidious dilemma, I can think of no better course than to thank them all together.

Outside the circle of those involved with the research and writing were others who were concerned with supporting services and administration. The Company's archivists – under first Janette Harley and later Anita Hollier – provided a first-class information retrieval service. Among those who provided secretarial support were, in the early days, Kay Underdown, held in very high esteem by Ronald Ferrier; Sandra Peters, who I shall always remember for showing the strength of mind not to go under as the paper piled higher; and Judy Scherbening, whose tact and efficiency in keeping the office wheels turning when I was preoccupied with the final stages of completing this volume went largely unsung at the time. I have also been extremely fortunate that the senior managers who have been administratively responsible for the history since I became BP historian have, without exception, been people who have combined highly professional managerial skills with a friendly and personal way of going about their business. They include David Walton, Dick Olver, Paul Fowler and, at the time of writing, Robert Pennant Jones.

Finally, I wish to express my gratitude to BP's present board of directors for continuing to support the research and writing of this history – albeit on much reduced resources – during the unusually difficult times through which the Company has passed in recent years. Having spared the means, I hope they will be able to derive some pleasure from the result.

Abbreviations

AGIP	Azienda Generale Italiana Petroli
AIOC	Anglo-Iranian Oil Company
API	American Petroleum Institute
APOC	Anglo-Persian Oil Company
Aramco	Arabian American Oil Company
BOD	British Oil Developments
BPC	Basra Petroleum Company
BTC	British Tanker Company
C	Centigrade
Casoc	California Arabian Standard Oil Company
CFP	Compagnie Française des Pétroles
d	pence
DIB	Di-isobutene
ECA	Economic Co-operation Authority
EGS	Eastern and General Syndicate
ICI	Imperial Chemical Industries
IHP	International Hydrogenation Patents
ILO	International Labour Organisation
IPC	Iraq Petroleum Company
KOC	Kuwait Oil Company
MPC	Mosul Petroleum Company
NEDC	Near East Development Corporation
NIOC	National Iranian Oil Company
OEEC	Organisation for European Economic Co-operation
RAF	Royal Air Force
ROP	Russian Oil Products
s	shillings

SGHP	Société Générale des Huiles de Pétrole
Socal	Standard Oil Company of California
Socony	Standard Oil Company of New York
Standard Oil (NJ)	Standard Oil Company of New Jersey
TEL	Tetraethyl lead
TPC	Turkish Petroleum Company
UOP	Universal Oil Products

A note on the text

Country names

In general, countries are described by the names which were in use at the time. For example, Palestine is used instead of Israel before the state of Israel came into being. However, an exception is made in the case of Persia, whose name was changed to Iran in 1935. In referring to that country in this book, it would be confusing to use the name Persia up to 1935 and then to change to Iran because successive chapters of the book do not always follow one another in strict chronology. The name Iran has therefore been used throughout the text, except in quotations, which are *verbatim* and reflect the common western usage of the name Persia even after 1935.

Company names

The entity which is now known as The British Petroleum Company plc was called the Anglo-Persian Oil Company Limited when it was originally formed in 1909. In 1935 the name was changed to the Anglo-Iranian Oil Company Limited and in 1954 it was changed to The British Petroleum Company Limited. In order to avoid confusion, the entity which underwent these successive changes of name is referred to as the Company throughout the text. That term is not otherwise used as a proper noun, except where it forms part of another company's name, for example in the Burmah Oil Company.

Introduction

At the beginning of the twentieth century, the international oil industry was still a fledgling, undeveloped not only in size, but also in its range of products, technology and organisation. The main centres of oil production, in Russia and the USA, supplied only a very small fraction of the industrialised world's fuel. Energy supply was still dominated by coal, the fuel which more than any other had powered the development of industry, commerce and transport in the nineteenth century.

The oil industry's principal commercial product was kerosene, used mainly for lighting. It was obtained by refining crude oil into its components – mainly fuel oil, kerosene and gasoline – by the process of distillation. That process was, however, subject to the great limitation that the straight distillation of a given crude oil could produce only a fixed ratio of products. As a result, the refiner was unable to adjust the yield of products to meet the pattern of demand. An increase in the output of one product, say kerosene, could only be achieved by expanding the output of other products, which, being unwanted, had to be destroyed.

Not only was the industry limited by the refining technology of the day, but crude oil exploration and production tended to be haphazard and inefficient. The 'wildcatter' remained commonplace in exploration. Under the law of capture which prevailed in the USA, oilfields, once discovered, were prone to be wastefully developed by competing firms scrambling for maximum short-term production, rather than by operators applying scientific principles in the interests of maximising the amount of oil which could be recovered from the field in the long term.

In 1900, the oligopolistic structure which was to characterise the international oil industry for most of the twentieth century had not yet emerged. But the tendency for the industry to become concentrated in the hands of very large firms was already in evidence, especially in the USA where John D. Rockefeller's Standard Oil Company had risen to dominance. The Nobels and the Rothschilds had developed large-scale oil interests in Russia. Other significant firms were the Royal Dutch Company, formed in 1890 to exploit the discovery of oil in Sumatra, and the Shell Transport and Trading Company, formed in 1897 to carry on the oil business which had been built up by Marcus Samuel.

During the next quarter-century the oil industry made great advances. Crude oil production increased about sevenfold, rising from 20 million tons in 1900 to more than 140 million tons in 1925. In Russia, which was the world's largest oil producer in 1900–1, the development of the industry was retarded during the period of communist revolution. Elsewhere, however, new oil producing regions were discovered in, most notably, the Middle East and South America. Meanwhile, as the position of kerosene as a fuel for illumination was challenged by the rise of gas and electricity, other demands for oil expanded rapidly. Fuel oil captured an ever-larger share of coal's traditional markets and the growing use of the motor car led to greatly increased demand for motor spirit.

The industry's ability to supply increased quantities of motor spirit without producing surpluses of other products was enhanced by the new refining technique of 'cracking', whereby large hydrocarbon (oil) molecules were broken down into smaller molecules, with the practical result that relatively low-value heavy fractions of the crude oil could be cracked to produce higher-value motor spirit. In short, the cracking process introduced new flexibility into refining, allowing the product yield to be varied, within limits, so that the output of the more desirable products could be increased without a concomitant surplus of other products. At the same time, exploration techniques were made more scientific by the spread of geological and geophysical methods and by the adoption of rotary drilling instead of the simple pounding action of the percussion system.

These developments were accompanied by the emergence of a few firms which came to dominate the international oil industry. The 'Seven Sisters', as they came to be called, included five US oil companies. They were the Gulf Oil Corporation and the Texas Company (later Texaco), which were formed in the 1900s after the famous oil discovery at Spindletop, Texas in 1901; and the Standard Oil Company of New Jersey (later Exxon), the Standard Oil Company of

New York (later Mobil) and the Standard Oil Company of California (later Chevron) – each of which was part of Rockefeller's Standard Oil Trust before it was dismembered by order of the US Supreme Court (on anti-trust grounds) in 1911. In addition to the five US companies, there was the Anglo-Dutch enterprise, Royal Dutch-Shell, formed by the merger of the Royal Dutch Company and the Shell Transport and Trading Company in 1907. Finally, to complete the rollcall of the Seven Sisters, there was the Anglo-Persian Oil Company (later British Petroleum and hereinafter called the Company), whose origins and rise to an exalted position in the international oil industry were described in volume 1 of this history.[1]

As was seen in that volume, the Company's origins go back to May 1901, when a wealthy Englishman, William Knox D'Arcy, obtained a concession from the Shah of Iran to explore for and exploit the oil resources of that country, excluding the five northern provinces which bordered Russia. Having been granted the concession, D'Arcy employed an engineer, George Reynolds, to undertake the task of exploring for oil in Iran. The inclement weather of that country, the difficult terrain, the lack of infrastructure facilities which would have been taken for granted in Britain and other more developed economies, the shortage of local skilled labour and the problems of dealing with local tribes in the absence of a strong central government – all conspired to make Reynolds' pioneering task an exceptionally arduous one. There was, moreover, no guarantee of success; indeed, growing reason to doubt it as months, then years, passed without oil being discovered in commercial quantities. Meanwhile, the costs mounted, stretching D'Arcy's resources to the point where he sought outside assistance. That came in 1905 from the Burmah Oil Company, a Scottish-registered company with its head office in Glasgow, which had been formed in 1886 to produce, refine and market Burmese oil.[2] More exploration in Iran followed without success until eventually in May 1908 Reynolds and his helpers struck oil in commercial quantities at Masjid i-Suleiman in the province of Khuzistan in south-western Iran. It was the first commercial oil discovery in the Middle East, signalling the emergence of that region as an oil producing area.

After the discovery had been made, the Company was formed in April 1909 to develop the oilfield and work the concession. At the time of its formation, 97 per cent of the ordinary shares were owned by the Burmah Oil Company, which had financed the exploration effort in Iran since May 1905. The remaining 3 per cent were owned by Lord Strathcona, the Company's first chairman, who was then aged 89 and whose interest in Iranian oil affairs stemmed more from imperial than business considerations. Of the eight other directors, three – John Cargill, Charles Wallace (the Company's first managing director) and James Hamilton – were directors of the Burmah Oil Company, the first in a long line of Burmah directors on the Company's board. Although D'Arcy was also appointed a director and remained on the board until his death in 1917, he was not to play a major part in the Company's affairs. His role as the initial risk-taking investor in the uncertain oil prospects of Iran was past and the daunting task of developing the oil discovery into a commercial enterprise passed to others, amongst whom one stands out: Charles Greenway. He had long experience of oil marketing in eastern markets, having joined the mercantile firm of Shaw, Wallace and Co., agents for the Burmah Oil Company in India, in 1893. He became a partner in 1897 and had a longstanding acquaintance with Charles Wallace, who was a co-founder of Shaw, Wallace and Co. As one of the Company's founder-directors, Greenway quickly became a central figure in its affairs, succeeding Wallace as managing director in 1910 and Strathcona as chairman in 1914. Knighted in 1919 and elevated to the peerage in 1927, Greenway was the architect of the Company's early development.

The strategy which he pursued was the same as that followed by other prominent international oil companies: to build up a vertically integrated enterprise, which, in its full form, meant participating in every stage of the flow of oil from the well to the consumer, including crude oil production, transportation, refining and retail marketing. So thoroughly was this strategy executed by the major oil companies that the vertically integrated enterprise became the standard form of organisation in the international oil industry.

Steps in that direction were taken by the Company before World War I, when not only was further drilling carried out at the oilfield, but a pipeline was laid along the 138-mile route from Masjid i-Suleiman to the site of the intended refinery at Abadan, a flat mud island on the Shatt al-Arab, the delta of the rivers Tigris and Euphrates, some thirty miles from the Persian Gulf. Although many difficulties were encountered in laying the pipeline, its construction was a relatively simple operation, completed in July 1911. The erection and commissioning of the Abadan refinery proved to be a challenge of a different order. The properties of Iranian crude oil were then unknown and the Company ran into great problems trying to produce marketable products. As a result, the commissioning of the refinery, whose construction was started in 1909, was subject to long delays.

Meanwhile, in 1911 the Company, expecting that the refinery would soon come onstream, gave consideration to the marketing and branding of its products. However, the failure to master the refining of Iranian crude blighted the Company's early hopes of pursuing a strategy of independence in marketing. As the Company, struggling with its refining problems, came under increasing financial strain, Greenway had no alternative but to fall back on negotiating a ten-year contract to supply crude oil and products to Royal Dutch-Shell, thereby securing a ready-made outlet for the Company's production.

Greenway, anxious to avoid falling under the domination of Royal Dutch-Shell, also turned to another potential source of revenue and capital: the British Government. The basis of an agreement to mutual advantage lay in the Company's desire to find both new capital and an outlet for its fuel oil; and, on the other side, in the Admiralty's desire to obtain secure supplies of fuel oil, which had advantages over coal as a fuel for the ships of the Royal Navy. After lengthy negotiations, the two sides reached agreement in 1914 shortly before the outbreak of World War I, when the Company contracted to supply the Admiralty with fuel oil and the Government injected £2 million of new capital into the Company, receiving in return a majority shareholding and the right to appoint two directors to the Company's board. Although the principle that the Government would not interfere with the Company's normal commercial operations was enshrined in a letter to the Company from Sir John Bradbury, Joint Permanent Secretary of the Treasury, the Government shareholding undeniably brought an unusual political dimension to the Company's affairs. For Greenway, however, the Admiralty contract and the Government's injection of new capital represented a major coup. He had succeeded in securing the Company's survival and independence by finding a new source of finance, while simultaneously diluting the Burmah Oil Company's shareholding and avoiding the seemingly predatory instincts of Royal Dutch-Shell without conceding managerial control to the Government.

One of the advantages of the contract to supply fuel oil to the Admiralty was that it enabled the Company to make bulk supplies of a single product to one customer, obviating the need for a spread of distribution and marketing installations which could not be afforded at the time. However, although the Admiralty and other armed services were by far the most important customers of the Company during World War I, it remained one of the Company's primary objectives to develop its own marketing business. As products of marketable quality began to come into regular production in 1913–14, a network of outlets and agents was built up in the Persian Gulf and Mesopotamia, to which kerosene and benzine were distributed in two-gallon tins. Meanwhile, fuel oil began to be supplied for local steamers, the first contract being with Lynch Bros in December 1912. In 1914 further contracts were made with the Euphrates and Tigris Steam Navigation Company and the Hamburg-Amerika shipping line, followed in 1916 by a contract with the British India Steam Navigation Company. A much larger step was taken in 1917, when the Company acquired the British Petroleum Company, which had developed a substantial marketing presence in Britain after being formed in 1906 as a subsidiary of the German oil company, the Europäische Petroleum Union. Meanwhile, in 1915 the Company had formed a new subsidiary, the British Tanker Company, and built up a significant shipping fleet by the end of the war.

Further moves of vertical integration were made in the decade after the war. In 1919 the Company ventured into marketing on the continent of Europe through its interest in a new marketing associate, L'Alliance in Belgium. Over the next few years further marketing interests were acquired or formed in France, Italy, Austria, Germany, Iceland, Sweden and Holland. At the same time, new marketing methods were introduced, with kerbside pumps replacing two-gallon tins in the distribution of motor spirit. New Company refineries, much smaller than that at Abadan, also came onstream at Llandarcy in South Wales in 1921 and at Grangemouth in Scotland in 1924, added to which the Company's majority-owned French associate had a refinery at Courchelettes, near Douai. On the other side of the world, in Australia the Company and the Australian Government were partners in Commonwealth Oil Refineries Ltd, which commissioned a new refinery at Laverton, near Melbourne, in 1924 and sold oil products in the domestic market. The Company also marketed its products in Iran and Iraq and established an international chain of marine bunkering stations, which provided outlets for fuel oil as naval demand fell after the end of the war.

Thus, by the time that Greenway retired as chairman in March 1927, the Company was established in all phases of the industry. Yet, despite the development of its downstream operations, the Company still held a much smaller share of world oil markets than its main rivals, the Standard Oil Company of New Jersey and Royal Dutch-Shell. The Company's influence in the industry and its competitive strength were based mainly on its upstream position, which, despite widespread exploration activities in the 1920s, continued to be based on its huge oil reserves in the Middle East. Some of those reserves were in Iraq, where the Company discovered oil at Naftkhana in 1923 and connected the field by pipeline to the Alwand refinery, which came onstream in February 1927 to supply products for internal distribution in Iraq. Much more important, however, was the Company's position in Iran, where, as sole concessionaire, it was able to develop the oilfield at Masjid i-Suleiman by applying the best-practice principle of unitisation, by which the field was treated as a single production unit operated exclusively by the Company and not, as happened elsewhere in the industry, as a battleground for competing firms to maximise their individual short-term extraction rates at the expense of longer-term productive efficiency. Moreover, within a month of Greenway's retirement, the discovery of a second Iranian oilfield at Haft Kel gave the Company the added security of no longer being dependent on the one oilfield at Masjid i-Suleiman for its Iranian crude.

The man who had been groomed to succeed Greenway was of a different personality and background. More inclined to co-operate with other firms than to assert the Company's independence, and with an experience of the oil industry which was more technical than commercial, Sir John Cadman had first become interested in petroleum while he was Chief Inspector of Mines in Trinidad in 1904–8. In 1908 he was appointed Professor of Mining and Petroleum Technology at Birmingham University. He continued, however, to maintain close links with the Government and was a member of the Admiralty Commission which visited Iran to inspect and report on its oil prospects in 1913, when discussions on the supply of fuel oil to the Admiralty were in progress. Four years later, Cadman was appointed director of the Petroleum Executive, which was set up by the Government to administer wartime petroleum affairs. He was also a member of the Petroleum Imperial Policy Committee which was established under Lord Harcourt in 1918 – the year in which Cadman was knighted – to formulate a long-term petroleum strategy for the Government. In 1921 Cadman was recruited to the Company, where his advancement to managing director in 1923 and deputy chairman in 1925 was strongly supported by the Government.[3] Like others of his generation who had experienced the unprecedentedly close contact between government and business during World War I, Cadman was a figure who comfortably bridged the traditionally separate worlds of politics and commerce in Britain, bringing more diplomatic skills to the chairmanship than either his predecessor or, as will be seen, his successor.[4]

PART I
DEPRESSION, RECOVERY
AND WAR, 1928–1945

Management and finance,
1928–1939

When Sir John Cadman became chairman in succession to Lord Greenway on 27 March 1927, the Company's position in Iran was fundamental to its existence. Apart from the small Naftkhana oilfield in Iraq, Iran was the Company's single centre of crude oil production until 1934 and remained the predominant source of supply until 1951. The geology and topography of Iran determined the location of oil reservoirs and access to them. The natural composition of Iranian crude oil affected the technology of production, the processes of refining and the yield of products for the markets and stimulated the research required for technical progress, product development and customer satisfaction.

At the same time, the lack of a local industrial and educational infrastructure in the south west of Iran, where the Company's operations were located, necessitated the provision of Company facilities for a whole range of activities from power generation to ice manufacture, from the installation of telecommunications to the institution of medical and educational services, from the supply of accommodation to the erection of workshops. The wide range of requirements necessitated a variety of skills, some of which were to be found only in the expatriate element in the workforce, many of whom were engaged in ancillary occupations rather than purely industrial activities. At a local level, the borderline between municipal and corporate responsibilities was often difficult to define and sometimes caused disagreement between the Company and the Iranian authorities. At the national level, Riza Shah's ambitions for the modernisation of Iran, involving expensive projects which were partly funded by oil revenues, did not always coincide, and sometimes conflicted, with Company objectives. There was, moreover, an international dimension to the Company's relations with the Iranian Government, especially as the British Government's majority shareholding in the Company gave ample scope for confusion over the distinction between affairs of state and matters of commerce.

The Company's dependence on Iran as the main source of its crude oil and refined products put it in the vulnerable position of having most of its eggs in one basket. Not only had the Company to guard against the risk of arbitrary Iranian action, it also had to fend off competition from other oil producers whose expansion might undermine Iran's position in the world oil industry, causing the Iranian Government to become dissatisfied with the Company's performance. The threat of competition was felt most keenly from those who sought to acquire nearby concessions in the Middle East. However, the Company could not ignore more distant sources of new production, particularly those in the oil exporting states of the western hemisphere. These included the USA, where the Company had no producing interests, and the countries of South America where, apart from a minor interest in the oil industry of Argentina, the Company was effectively precluded from participation in oil exploration and production because the British Government's shareholding was legally and intrinsically unacceptable to the principal states concerned, most notably Venezuela.

The handicap of being excluded from South America was greatly magnified by the rapid development of South American oil resources on a massive scale in the 1920s. The resultant surplus of production, already evident in 1927, was exacerbated by the onset of the worldwide economic depression in 1929, which resulted in reduced demand for oil at the very time when supply was greatly augmented by the discovery of the vast US East Texas oilfield, from which production commenced in 1930.

The combination of concessionary problems and excess productive capacity in the late 1920s and early 1930s predisposed the Company to a cautious policy of overall retrenchment and a collaborative effort with two other major oil companies, Standard Oil of New Jersey (NJ) and Royal Dutch-Shell, to stabilise the industry (see chapter 4). Cadman was a leading proponent of rationalisation and co-operation, which satisfied his personal preferences, his scientific principles and the Company's needs. His expressions of concern about the need for each generation to hold in trust the resources of the world for those following were combined with his insistence on the need to bring supply and demand into equilibrium. In 1927 he cautioned that surplus production might, for a time, mean very cheap products for the consumer, but 'it means also industrial disorganisation, the elimination of many of the smaller units in the industry, and the dissipation of one of the world's most precious assets'. Cadman was not an alarmist about the exhaustion of oil reserves, but advised a 'wise and considered policy of conservation'.[1] Publicly and frequently he warned about the problem of uncontrolled production, informing shareholders in 1928 that 'your Board hold strongly to the view that a temperate though steady progressive policy is much to the ultimate advantage of all who look for continuous benefit from the activities of this Company'. Like others of his generation, he maintained:

> It profits nobody, in the long run, that there should be alternate waves of over-production and under-production, of high prices and low prices, of big profits and little or no profits. Least of all does it help the consumer ... to find his fuel bill, over a very brief term of years, looking like a fever chart.

He believed in efficient production, economical distribution and market stability which, 'in no sense ... representing any tendency towards monopoly, should ensure fair and stable prices to the consumer and a more rational and therefore more prosperous future for the oil industry'.[2] He criticised the waste of oil resources arising from poor techniques and irresponsible drilling. In 1930 he praised the technology by which the Company was capable of converting crude oil into those products which the market most needed and of ensuring 'in almost every circumstance a complete producing equilibrium from the field to the market'.[3]

THE BOARD OF DIRECTORS

At the time of Cadman's appointment as chairman, responsibility for directing the Company's affairs was vested in a board whose membership is shown in table 1.1. It was a comparatively large board of seventeen members with a somewhat unusual composition in that there were five non-executive directors who represented the Company's two main shareholders, namely the British Government and the Burmah Oil Company. The Government directors were the former Joint Permanent Secretary of the Treasury, Lord Bradbury (previously Sir John, he had been elevated to the peerage in 1925), and the former Admiralty official, Sir Edward Packe. The Burmah element consisted of Sir John Cargill, Burmah's chairman (knighted in 1920), along with Robert Watson and Gilbert Whigham, who were also on the Burmah

Sir John Cadman
(created Lord Cadman, 1937)

William Fraser
(knighted, 1939)

Arthur Hearn

Hubert Heath Eves

1   The Company's chairman and management directors, 1927–1941

John Lloyd
(knighted, 1928)

Thomas Jacks

James Jameson

Neville Gass

board. In addition, there were five other outside non-executive directors of the Company, namely Admiral Slade, who was vice-chairman; Sir Hugh Barnes and Sir Trevredyn Wynne, both of whom were ageing 'eastern' directors with experience of government and railways respectively in Burma and India; and Frank Tiarks and Frederick Lund, who had City experience in banking and shipping respectively. Finally, so far as the non-executives were concerned, there was Greenway, who remained on the board with the honorific title of president after he resigned as chairman. This long rollcall of non-executives, most of them titled, some of them well advanced in years, and three of them founder-directors of the Company in 1909, added up to a considerable weight of experience on the board. They comfortably outnumbered the much smaller nucleus of Cadman and the management directors (managing directors in modern parlance), including William Fraser, Arthur Hearn, Hubert Heath Eves and John Lloyd, who constituted an unusually youthful team of executive directors, the oldest of whom, Lloyd, was fifty-three years old at the end of 1927. He was also the only executive director who had been on the board for more than four years when Cadman became chairman. Lastly, another young director, Thomas Jacks, had an overseas posting as resident director in Iran. Thus when Cadman succeeded Greenway he had little need to rejuvenate the executive side of the board. As for the non-executive representation, there may have been a case for thinning it out, but a brief examination of its age structure would have assured Cadman that natural wastage could be relied upon to reduce the numbers if such was desired.

That, indeed, was precisely what happened and when Slade died in 1928, followed by Greenway in 1934 and Barnes in 1940, none of them was replaced. Meanwhile, on the resignation of Bradbury in 1927, Sir George Barstow, previously a senior official in the Treasury, was appointed as a Government director in his place. As for the management directors, Hearn retired at the end of 1938 and James Jameson and Neville Gass became directors in February 1939, the last board appointments before Cadman's death in 1941. By that time, the board had been reduced to fourteen members and a more even balance achieved between non-executive and executive directors (moving from a ratio of 11:6 in 1927 to 8:6 in 1941).

As chairman, Cadman dominated the board, loved the limelight, enjoyed his involvement in public affairs and preferred the grand design to the execution of the details. He argued his ideas in public through lectures, writings and addresses in Britain and the USA,

Table 1.1 *The board of directors, 1927–1941*

The board on Cadman's appointment as chairman in March 1927		Date of appointment as director	Age at end of 1927	Date of resignation	Died in office
Chairman and management directors	Sir John Cadman (created Lord Cadman 1937)	1923	50	–	1941
	William Fraser (knighted 1939)	1923	39	–	–
	A. C. Hearn	1927	50	1938	–
	H.B. Heath Eves	1924	44	–	–
	J. B. Lloyd (knighted 1928)	1919	53	–	–
Resident director in Iran	T.L. Jacks	1925	43	1935	–
Government directors	Lord Bradbury	1925	55	1927	–
	Sir Edward Packe	1919	49	–	–
Burmah directors	Sir John Cargill	1909	60	–	–
	R. I. Watson	1918	49	–	–
	G. C. Whigham	1925	50	–	–
Ordinary non-executive directors	Sir Hugh Barnes	1909	74	–	1940
	F. W. Lund	1917	53	–	–
	Admiral Sir Edmond Slade	1914	68	–	1928
	F. C. Tiarks	1917	53	–	–
	Sir Trevredyn Wynne	1915	74	–	–
President	Lord Greenway	1909	70	–	1934

Subsequent appointments:			Age on appointment		
Management directors	J.A. Jameson	1939	54	–	–
	N. A. Gass	1939	46	–	–
Government directors	Sir George Barstow	1927	53	–	–

Note: Slade was a Government director on his appointment to the board in 1914, but became an ordinary director in 1917.

especially in the period from 1928 to 1934. He appealed for the international understanding of oil affairs. He championed the scientific importance of the industry, symbolised by the proceedings of the First World Petroleum Congress in 1933. He offered the Iranian Government a participatory shareholding in the Company, which, had it come about, would have helped to foster a greater sense of partnership. He was actively engaged in the evolving relationship between the British Government and the Company, not only in moments of crisis like the cancellation of the D'Arcy concession in 1932 (see chapter 2), but also in more routine matters, whether in foreign affairs or trade relations, particularly as they affected the Middle East. He made his own contributions to public service by sitting on commissions like those for the Imperial Airways and Broadcasting. However, in the last few years of his life Cadman became less resilient than hitherto. He suffered frequently and severely from ill health and often took the waters at a succession of European spas. He was depressed at the prospects for peace and the state of the world. He was sometimes unhappy in his family life. In alternate moments of pleasure and dejection he was occasionally unappreciative of Fraser's contribution to the management of the Company.[4]

Yet, under Cadman, Fraser was the lynch pin, conversant with technical affairs and possessing commercial and administrative experience, having spent the early part of his business career in the shale oil industry in Scotland, where he became managing director of Scottish Oils Ltd, an amalgamation of the main shale oil producers which was acquired by the Company in 1919. At Cadman's personal and urgent insistence Fraser was persuaded to leave Scotland and settle in England permanently to become his deputy. Fraser, who had been appointed a director at the same time as Cadman, was an all-round oil man, practical, cautious and dedicated to serving the Company loyally. Cadman was grateful, writing in 1935 'You are a wonderful fellow, Willie'.

During the decade from 1929 to 1938 the responsibilities of the directors and the organisation of the Company remained basically unchanged, conforming to a relatively simple line management structure of four main departments, Production, Finance, Distribution and Concessions, under four management directors, namely Fraser, Lloyd, Heath Eves and Hearn respectively.[5] In reality, power at board level rested largely in the hands of a triumvirate consisting of Cadman, Fraser and Lloyd, the last of whom had considerable influence as finance director and exercised tight financial control over the

Company. The power of his position stemmed in part from his role on the Finance Committee, which first met in 1923 and consisted of the chairman, the management directors (of whom only the finance director had voting powers), the Government directors and three other non-executive directors. Essentially, the committee's function was to examine expenditure proposals and to exercise financial supervision, but in 1930 Hearn described it as 'a convenient day-to-day instrument for the regulation of policy and its co-ordination throughout the whole Company'. The Finance Committee therefore played an important role in the workings of the Company.

So too did weekly informal meetings of management directors, at which ideas could be openly discussed outside the more formal atmosphere of full board meetings at which the presence of directors of the Burmah Oil Company, which could be regarded as a competitor, may well have inhibited the disclosure of much detailed Company business. Thus, whilst the full board was kept satisfactorily informed of current developments through the monthly submission of management reports, its proceedings were in all likelihood as bland as the record which was kept in that generally uninformative class of historical records of British companies: the board minutes.

THE MANAGEMENT OF OPERATIONS IN IRAN

Given the importance of Iran in the Company's operations, the lines of control and managerial relationship between Head Office and Iran were of particular significance in the organisational structure of the Company. On a visit to Iran in 1924 Cadman had deplored the lack of a collaborative approach to management among those in charge and the failure of general management to co-ordinate the activities for which they were responsible. He regretted the overriding and insensitive authority exercised by the London management on engineering matters. There was inadequate financial control and no rational sequence in which deliveries were received in relation to the orders placed. Moreover, there was insufficient recognition of the growing central power being exercised from Tehran by Riza Shah. It was for these reasons that Jacks was appointed resident director in Tehran in 1926, reporting to Cadman, with Jameson as general manager in charge of operations in the southern province of Khuzistan, centre of the Company's activities.

Jameson's knowledge of the operations, understanding of the region and its people and qualities of leadership were derived from his experiences over two decades, from being a fitter on the construction of the original pipeline from Masjid i-Suleiman to Abadan to being general manager in Iran from 1926 to 1928. He embodied the pioneering spirit, the practical engineer and the respected manager. Surveyor rather than architect in character, he was the mastermind behind the Company's technical progress in Iran. Jameson was recalled to London in 1928 to be made deputy director of production under Fraser, in which capacity he was responsible for technical affairs in Iran, apart from the refinery at Abadan. He was succeeded as general manager in Khuzistan by Edward Elkington, a cavalier character, an energetic personality, an administrator who ruled from the office. Of a quieter disposition, but with a good technical understanding of the operations, was John Pattinson, who succeeded Elkington as general manager in 1937 and remained in the post until 1945. Meanwhile, the title of resident director was abolished when Jacks retired in 1935. Instead, Leonard Rice was appointed chief representative in Tehran.

These arrangements may have seemed a reasonable form of line management, but in reality they masked a managerial problem which was a serious challenge to the Company. As the geographical spread of operations extended and the technical complexities increased in a variety of dissimilar political and economic conditions the problems of management became more diverse. The Company did not respond to the managerial challenge in the manner of Standard Oil (NJ), which evolved a multi-divisional structure in which operational responsibilities were devolved to autonomous self-contained operating divisions, relieving the main board of day-to-day detail and leaving it free to concentrate on overall appraisal and strategy. One of the features of those developments at Standard Oil (NJ), analysed by the historian A. D. Chandler, was the distinction which was made between the activities of operating companies and the parent company.[6] Such a distinction was accepted by the Company in the manner in which it organised its distribution system through local subsidiaries, whose boards included senior Company representatives and for which Heath Eves had ultimate responsibility as the director in charge. It became the practice of the US oil companies operating overseas to set up local production companies whose executives reported to the parent company. In Kuwait, the Company and Gulf Oil Corporation established a joint company to handle the concession, but the principle of a local operating company was not adopted in Iran, where the Company continued to administer its operations directly, allowing only limited local responsibility.

Given that the concession in Iran was so crucial to the Company, a balanced triangular relationship between London, Tehran and Abadan was essential, but did not always exist. Cadman's ideal management solution for Iran did not always live up to expectations and there was sometimes uncertainty about the balance between local authority and direction from London. There were personal prejudices and rivalries, especially between Jacks and Elkington. As the ambitious plans of Cadman for a closer partnership with the Iranian Government faded with the cancellation of the D'Arcy concession in 1932, so the potential role of the resident director diminished in importance. Following the retirement of Jacks there was no resident director in Iran who might have reduced the need for supervision from London. Former general managers like Jameson and Elkington, who may have resented interference from London whilst in Abadan, returned to London still concerned to assert their authority, still managing rather than directing. At the same time the recall of Gass to London in 1934 to take charge of concessionary affairs, after his having been deputy general manager in Iran, created another focal point of responsibility. While Pattinson was general manager in Abadan from 1937 to 1945 and Rice was chief representative in Tehran, there was a period of close mutual trust and confident association supported by those in charge in London. However, it was the warmth of their personal relations rather than the definition of their duties which cemented their collaborative achievement. Whilst there may have been some corporate indecision, some individual arbitrariness, and later some disorganisation as a result of wartime conditions, there was an unmistakable sense of administrative effectiveness in the conduct of affairs in the operational areas organised by the general management, principally due to the dominating personality of Jameson.

If effective management was to some extent achieved despite, rather than because of, the formal organisational arrangements, it is relevant to consider whether a more self-contained, decentralised management structure in Iran might have been more satisfactory. Cadman could have moved in this direction had he been so inclined, particularly during his management reorganisation in the mid-1920s[7] or after the cancellation and successful renegotiation of the concession in 1932–3. However, far from sympathising with the idea of more decentralisation, he constantly endeavoured to reach a closer association and identification with Iran. His predecessor, Greenway, had also been reluctant to accept the separation of operational responsibilities from the holding company and when such ideas were put forward by Slade in 1924 they were not accepted. Such ideas were precluded by the strong sense of attachment to Iran which pervaded Cadman's thoughts and actions. In 1927 he declared: 'We want Persians to feel that our activities in Persia are not only directed toward extracting oil, but also toward developing a great national industry in the country'.[8] In 1930 he realised 'how closely the welfare of this Company is associated with that of Persia'.[9] In mid-1934 he spoke of 'a spirit of partnership in the development of the oil resources of Persia'.[10]

These were not merely ritual utterings without serious intent. At a meeting in June 1938 the board discussed and approved a policy memorandum which drew attention to the paramount importance of maintaining stable relations with concessionary governments, particularly that of Iran, and made clear that those relations were primarily dependent on the maintenance of production, and consequently royalties, at a high level.[11] That point had been strongly emphasised to Cadman by the Shah and according to the policy memorandum it was 'quite obvious' that concessionary activities outside Iran should not be allowed to prejudice the Company's interests in Iran. The board therefore agreed that the Company should no longer contemplate the acquisition of other concessions outside the British empire and that no new efforts should be made to acquire concessions in such countries as Colombia and Venezuela.[12] The primacy of Iran was confirmed.

PROFITS AND FINANCE

The combination of a world surplus of oil production capacity and the economic depression made trading conditions exceptionally unfavourable in the early 1930s. Falling world oil prices and a decline in the Company's sales tonnage in 1931 (see chapter 4) inevitably put a squeeze on the Company's profits, which fell, before tax, from £6.5 million in 1930 to £3.6 million in 1931, as can be seen from table 1.2. An important casualty of the collapse in profits was the royalty payment to the Iranian Government which, being based on a percentage of profits, was heavily reduced in 1931. The reduction had serious effects on the Company's concessionary position in Iran, where the Shah, annoyed at the diminution in income from oil, cancelled the D'Arcy concession.[13] Some Iranian commentators have described the fall in royalties as 'ridiculous' and 'inexplicable',[14] a view which overlooks the slump in the Company's profits, to which royalty payments were directly coupled under the terms of the concession.

While the renegotiation of the concession was its first concern, the

Table 1.2 *The Company's profitability, 1929–1938*

| | Profits before and after taxes and inflation | | | | | Return on capital | |
	(a) Pre-tax profits at current prices (£millions)	(b) Deduction for tax (£millions)	(c) Post-tax profits at current prices (£millions)	(d) Index of post-tax profits at constant prices (1929 = 100)	(e) Capital employed at current prices (£millions)	(f) Pre-tax return on capital employed: (a) as % of (e) (%)	(g) Post-tax return on capital employed: (c) as % of (e) (%)
1929	6.2	2.2	4.0	100.0	45.4	13.7	8.8
1930	6.5	1.8	4.7	128.9	45.1	14.4	10.4
1931	3.6	1.7	1.9	55.8	45.2	8.0	4.2
1932	3.5	0.8	2.7	79.4	45.0	7.8	6.0
1933	3.1	1.0	2.1	64.3	44.7	6.9	4.7
1934	5.3	1.3	4.0	115.4	43.1	12.3	9.3
1935	5.3	1.4	3.9	109.8	42.0	12.6	9.3
1936	7.8	1.8	6.0	158.5	44.7	17.4	13.4
1937	9.8	2.8	7.0	161.3	47.1	20.8	14.9
1938	8.7	2.3	6.4	146.6	47.5	18.3	13.5

Sources: BP consolidated balance sheets (see Appendix 1). The price index used in calculating profits at constant prices is that for plant and machinery in C. H. Feinstein, *National Income, Expenditure and Output in the United Kingdom, 1855–1965* (Cambridge, 1972), table 63.

Table 1.3 *The company's sources and applications of funds, 1929–1938 (£millions)*

| | Internally generated funds less capital expenditure and dividends | | | | Financial movements | | | |
	(a) Funds generated from operations	(b) Capital expenditure	(c) Dividends	(d) Funds generated or (required) after outflows on (b) and (c)	(e) Shares issued	(f) (Increase) or reduction in loan capital	(g) Increase or (reduction) in liquid resources	(h) Total financial movements: (e) + (f) + (g)
1929	12.7	7.0	3.6	2.1	–	0.8	1.3	2.1
1930	12.5	8.2	2.9	1.4	–	0.1	1.3	1.4
1931	7.7	6.1	1.7	(0.1)	–	1.1	(1.2)	(0.1)
1932	9.1	2.7	2.1	4.3	–	0.5	3.8	4.3
1933	9.8	7.1	2.1	0.6	–	3.0	(2.4)	0.6
1934	11.3	3.8	2.7	4.8	–	5.0	(0.2)	4.8
1935	10.7	4.5	3.1	3.1	–	0.1	3.0	3.1
1936	15.7	8.3	4.4	3.0	–	–	3.0	3.0
1937	16.0	5.0	6.1	4.9	–	–	4.9	4.9
1938	10.8	10.4	5.1	(4.7)	–	(0.1)	(4.6)	(4.7)
Total	116.3	63.1	33.8	19.4	–	10.5	8.9	19.4

Sources: BP consolidated balance sheets (see Appendix 1).

Table 1.4 *The Company's financial gearing, 1929–1938*

	(a) Loan capital at year end (£millions)	(b) Capital employed at year end (£millions)	(c) Gearing: (a) as % of (b) (%)
1929	10.2	44.6	22.9
1930	10.1	45.5	22.2
1931	9.0	44.9	20.0
1932	8.5	45.0	18.9
1933	5.6	44.3	12.6
1934	0.7	41.8	1.7
1935	0.6	42.3	1.4
1936	0.6	47.1	1.3
1937	0.6	47.0	1.3
1938	0.6	47.9	1.3

Note: Rounding to a single decimal point accounts for minor inconsistencies with the debt reduction shown in table 1.3.
Sources: BP consolidated balance sheets (see Appendix 1).

Company also had to contend with the continuing depression in profits in 1932 and 1933. As table 1.3 shows, it responded to its straitened circumstances by cutting dividend payments and drastically reducing capital expenditure so that the funds generated from operations were generally sufficient to cover outflows on those items and a reduction in debt, except in 1931 when the Company drew on its liquid resources, held in cash and marketable securities, to help finance the outflow of funds. With the rise in world oil prices and the rapid expansion of the Company's sales in 1934–7 (see chapter 4), pre-tax profits recovered and in the peak prewar year of 1937 the Company earned a pre-tax return of more than 20 per cent on its capital employed. However, although the earlier reduction of dividends was reversed, a generally cautious financial policy still prevailed. Thus, after allowing for outflows on capital expenditure and dividends, there remained a large surplus of funds generated from operations, which was applied to continued reductions in debt and, in most years, to bolstering liquid resources. The latter were drawn down heavily in 1938 to help finance heavy capital expenditure in a year which saw falls in profits and funds generated from operations.

A summary indication of the Company's financial prudence in the decade 1929–38 is shown in table 1.3, from which it can be seen that debt was reduced by £10.5 million and there was an overall addition of £8.9 million to liquid resources despite their depletion by £4.6 million in 1938. A more accurate measure of financial conservatism is the extent to which a company relies on loans, rather than share capital, to finance its activities. This is generally expressed as the ratio between a company's loan capital and its share capital (known as equity) and is commonly described as 'gearing' or 'leverage'. The higher the ratio or 'gearing', the more a company is exposed to the risks of having to meet fixed interest payments and debt repayments in future trading conditions which are inevitably uncertain. In the case of the Company, table 1.4 shows that in the decade 1929–38 its gearing was reduced to negligible proportions, clear evidence of the cautious, conservative and low-risk financial policy which it pursued during those years.

2

A new concession and a 'fresh start' in Iran, 1932–1939

When the Company was formed in 1909, its licence to operate in Iran was governed by the terms of the oil concession which Muzaffar al-Din Shah had granted to William Knox D'Arcy in 1901. The D'Arcy concession, which was valid for sixty years from the date of its signing, gave the Company the exclusive right to explore for and exploit oil in all of Iran except for the five northern provinces of Azerbaijan, Gilan, Mazandaran, Astarabad and Khurasan. Those provinces were excluded in order to avoid offending Russia, Iran's northern neighbour which looked upon the north of Iran as part of its sphere of influence in the same way that Britain saw southern Iran as falling under its sway.

The most essential provision of the D'Arcy concession was that the Company was to pay the Iranian Government an annual royalty which was defined somewhat vaguely as equal to 16 per cent of the Company's net profits. Formulated before oil had been discovered in commercial quantities in Iran, or anywhere else in the Middle East, and when the future dramatic growth of the Iranian oil industry could not have been foreseen, the imprecise wording of the concession was to result in difficulties in arriving at agreed calculations of the amount of royalty which was due. The heart of the problem was the definition of profits, about which expert opinions, as usual, differed. An attempt to remove some of the ambiguities in the concession by defining the terms more closely was made in the Armitage-Smith Agreement of December 1920. The Agreement, named after Sydney Armitage-Smith, the British Treasury official who was the financial adviser to the Iranian Government in the negotiations, preserved the fundamental principle of the D'Arcy concession, namely that the annual royalty would be equal to 16 per cent of the Company's net profits.[1] It did not, however, prove to be longlasting in the fast-changing Iranian political scene of the 1920s.

FROM QAJAR TO PAHLAVI RULE: THE SCENE IN IRAN

In the nineteen years from the grant of the D'Arcy concession to the Armitage-Smith Agreement, Iran, centre of the ancient and once-glorious Achaemenian Empire, fell to what was, by common consent, a nadir in its history. In the absence of a coherent administrative bureaucracy or an effective army the arbitrary, despotic power of the Qajar dynasty, which had ruled Iran since the late eighteenth century, had for long been maintained not by the science of administration or the force of coercion, but by manipulating rival factions – tribal, religious and racial – against one another. Thus the Qajars retained power not by unifying the country, but by exploiting its fragmentation. Muzaffar al-Din Shah, who acceded to the throne in 1896, was a weak ruler who showed no signs of possessing the qualities of leadership required to instil cohesive purpose into the fractured society of Iran. Rather than reform the underlying economy, he raised and dissipated foreign loans and granted concessions to foreigners, most notably the D'Arcy concession, to help finance his inefficient administration and personal extravagances. To the decadence of Qajar rule was added the humiliation of the country being divided into foreign spheres of interest, the influence of Russia being, as has been mentioned, paramount in northern Iran, while Britain was the dominant foreign power in southern Iran.[2]

The arbitrary despotism of Muzaffar al-Din Shah did not, however, go unchallenged. Within a few years of granting the D'Arcy concession he found himself unable to resist Iranian demands for a new constitution which would replace the absolute power of the Shah with a constitutional monarchy. The Fundamental Laws of the constitution, modelled on the Belgian example, were duly drawn up in 1906–7 and provided, most importantly, for the establishment of a bicameral legislature consisting of a lower, elected house of parliament, known as the Majlis, and a Senate, half of whose members were to be appointed by the Shah. Executive power was vested in the Shah and a Council of Ministers headed by the Prime Minister. For one reason or another, the Senate was not brought into being for many years (until 1949, in fact), but the First Majlis was elected in 1906. One of the more significant provisions of the new constitution was the stipulation, contained in Article 24, that concessions to foreigners were not to be concluded without ratification by the Majlis. This provision was a reaction to the earlier granting of concessions by Muzaffar al-Din Shah, whose authority to bind future generations of Iranians to the D'Arcy concession was later to be denied by the Iranian Government.

The constitutional revolution brought little immediate improvement to Iran's fortunes. Muzaffar al-Din Shah died in 1907, to be succeeded by his son, Muhammad Ali Shah who attempted, without success, to suppress the constitution and stage a counter revolution. He was deposed in 1909, after which his thirteen-year-old son, Ahmad, was put under a Regent until he was crowned Shah just before the outbreak of World War I. The turmoil of revolution and counter revolution brought no relief from foreign domination, which was confirmed in the Anglo-Russian agreement of 1907 in which those two great powers affirmed their spheres of interest in southern and northern Iran respectively. During World War I there was a virtual suspension of the parliamentary process in Iran which, despite her proclaimed neutrality, suffered invasions of British, Russian and Turkish troops. Those events, coupled with tribal uprisings in the south and rebellions in the north, brought the country close to a state of complete chaos and anarchy. After the Russian revolution of 1917 and the defeat of the Central Powers in 1918 Britain emerged as the dominant power in Iran. By an Anglo-Iranian treaty negotiated in 1919 Iran would have become a British protectorate in all but name had the treaty not been rejected by the Majlis. Then, in 1920, Soviet troops entered the northern province of Gilan, which was declared to be a Soviet republic. Iran, in short, was in a state of virtual disintegration. Its national territory was invaded; the central authority of Ahmad Shah carried little weight; society was disunited; and the non-oil economy was locked into a pre-industrial torpor.

Conditions were propitious for a powerful figure to step into the virtual vacuum of authority in the capital city of Tehran. Invited by the conditions, though not by Ahmad Shah, Riza Khan, a military officer, seized power in 1921. After initially taking the post of Minister of War, he consolidated his position and became Prime Minister in 1923. Two years later Ahmad Shah was deposed and in 1926 Riza Khan, having taken the name Pahlavi as his dynastic title, was crowned Riza Shah Pahlavi. The Qajar era was at an end, the Pahlavi era at a beginning.

Combining autocratic rule with vigorous reform and modernisation, Riza Shah has been the subject of contrasting judgements by historians. On the one hand, he has been cast in the heroic mould of Peter the Great of Russia, seeking to revolutionise his country by shaking it out of its previous decadence and lethargy and importing modern western ideas to that end. On the other hand, he has been characterised as a brutal military dictator who eliminated all opposition, abused his power for purposes of self-enrichment and introduced reforms, not for the sake of efficiency and welfare, but to promote his own prestige and status by acquiring the superficial trappings, but not the substance, of a modern state.[3] Between those two extremes of interpretation it may be said, without undue controversy, that Riza Shah was both highly authoritarian and a vigorous reformer of the Iranian economy and society. As the Company was to be directly affected by the style and content of his rule, some further explanation is desirable.

After Riza Khan's military coup of 1921 it cannot have taken the Company long to realise that in its relations with Iran it would no longer be dealing with the traditionally weak governments of the Qajar era, but with a new authoritarian figure. Unlike the Qajars, Riza Khan used the institution he knew and understood best – the army – to establish his authority throughout the country. He swiftly crushed the separatist movement in Gilan and compelled other local rulers and tribal chieftains to submit to his rule, so that by the time he was crowned Shah his power was uncontested. During his reign (1926–41) opposition of all kinds was suppressed. The Majlis remained in being, but wielded no effective power; the Communist Party was outlawed; and opponents of the Shah were arrested, exiled or removed from politics by other measures. For example, in 1926 Sa'id Hussan Mudarris, who enjoyed a reputation as a tribune of the people, was badly wounded in an assassination attempt, widely thought to have been engineered by the police. He recovered, but was barred from standing for the next Majlis and was exiled to Khurasan.[4] Another victim of Riza Shah's dictatorial rule was a figure who was later to play an enormously significant part in relations between the Company and Iran: Dr Muhammad Musaddiq. Elected to the Majlis in 1924, he was an outspoken critic of the Shah in the Fifth and Sixth Majlis (1924–8), before being barred from election to the Seventh Majlis and kept out of active politics for the remainder of Riza Shah's reign.[5]

If Riza Shah's opponents had reason to fear him, so too, in a different way, did his closest ministers and advisers. Exhibiting that common failing of autocrats, an inability to delegate authority with consistency, Riza Shah kept his ministers in a state of insecurity, of never knowing when the Shah's trust would turn to suspicion from which disgrace, exile, imprisonment or death was sure to follow. Forever looking over their shoulders, the Shah's closest associates lived in constant fear of becoming the next fallen favourite as the Shah quickly developed a reputation for irascibility, abruptness in speech and violent invective. Lacking sophistication in manner and invariably wearing military uniform, the Shah was, moreover, socially and culturally at odds with the cultivated and formally educated political notables who acted as his ministers. Early warnings of the speed with which ministers could fall from favour came even before Riza Khan was crowned Shah. In the military coup of 1921 his main political ally was Sa'id Zia al-Din Tabataba'i, a nationalist intellectual with a taste for literature and poetry and a knowledge of history. After the success of the coup he became Prime Minister, but his alliance with Riza Khan proved to be short-lived. Within three months Zia had resigned and fled to Palestine where he remained in opposition to Riza Khan whose rule, he said, represented the negation of his ideals.[6] After Zia's departure, Qavam al-Saltana became Prime Minister, but in 1923 he was accused of plotting against Riza Khan and fled to Europe.[7]

The rise and fall of ministers continued after Riza Khan became Shah and surrounded himself with a small group of associates who, as shall be seen, played an important part in negotiations with the Company. Foremost among them was Abdul Husayn Khan Timurtash, who shared the Shah's nationalism, but was unlike him in other ways. For example, the Shah was largely self-educated and spoke little foreign language apart from some Russian learned during his days as an army officer. Timurtash, on the other hand, spoke Russian and French and understood English. He possessed social grace, with the wit, intelligence and energy, manners and charm, for smooth diplomacy. Riza Shah only once made a state visit outside Iran (to Turkey in 1934), but the cosmopolitan Timurtash visited numerous foreign cities such as Moscow, London, Paris, Rome, Berlin and Brussels. As Minister of Court he was, for a time, virtually the sole route of access to the Shah. He attended the Council of Ministers and was recognised to be more powerful than the Prime Minister until, that is, his disgrace and death in prison.

After his coronation, Riza Shah's first Prime Minister was Mirza Muhammad Ali Khan Furughi, the son of a university professor who had been a Minister of Finance. Furughi spoke English and French and had used his linguistic skills to translate Sir Percy Sykes' *History of Persia* into Persian. His first spell as Prime Minister under Riza Shah lasted only for a short time before he was moved to the Ministry of War in 1926 and then to the Foreign Ministry. In the mid-1930s he again became Prime Minister, from which post he was removed after the guardian of the Shrine of Imam Riza, with whom Furughi had family ties, was court-martialled and hanged.[8]

Another of Riza Shah's early political associates who played a significant part in relations between the Company and the Iranian Government was Sa'id Hassan Khan Taqizadeh, who had been a prominent figure in the constitutional revolution of 1906–7. From Riza Shah's accession until 1930 Taqizadeh was on the fringes of power as Governor of Khurasan and then Minister to London. Then, in 1930, he was appointed Minister of Finance, a post which he held until 1933 when he was sent to France as Minister to Paris. He held that post only briefly before going into self-imposed exile rather than return to Tehran to face Riza Shah's displeasure.

A further member of the group of ministers who gathered round Riza Shah in the early years of his reign was the talented Ali Akbar Davar, of middle-class background and with a law degree from the University of Geneva. In 1927 Riza Shah put Davar in charge of the newly established Ministry of Justice. As Minister of that department Davar was entrusted with the task of drawing up a new legal code. He remained in the post until 1933 when he succeeded Taqizadeh as Minister of Finance. Four years later he committed suicide to avoid imminent disgrace or execution.[9]

As this brief rollcall makes clear, senior Iranian ministers were the creatures of the Shah, insecure in his favour, fearful of his wrath and cautious about taking any decision which might offend him. Such a political system was not, as the Company was to discover, conducive to the exercise of ministerial initiative and responsibility in concessionary negotiations.

As ministers rose and fell, Riza Shah's rapacious appetite for economic and social reforms was seemingly insatiable. He forged Iran's disparate armed forces into a national army and introduced conscription. He set about improving transport facilities both by constructing roads and by building the Trans-Iranian Railway, his most grandiose scheme. Placing great emphasis on the role of the state in economic development, he established state enterprises in a range of industries. He removed education and the judicial system from the traditional control of the religious hierarchy. He introduced the Uniform Dress Law under which Iranians were required to wear western dress. He forbade women to wear the *chador* (veil). He formed a national bank, the Bank Melli, which took over the note-issuing functions previously carried out by the British-owned Imperial Bank of Persia. He brought foreign trade under government control.[10] Most important of all from the Company's point of view, he rejected the validity of the Armitage-Smith Agreement of 1920 on the grounds that Armitage-Smith had exceeded his authority in reaching the agreement. The Company regarded the Agreement as valid, but recognised the desirability of revising the concession, to which end discussions were opened by Cadman and Timurtash in 1928. The subsequent negotiations dragged on through various permutations until mid-1932 when, with agreement apparently in sight, the Company informed the Iranian Government that the estimated royalty due for 1931, in which year profits were badly affected by the worldwide depression, was only £306,872 compared with £1,288,312 for the previous year. Shocked by the precipitate fall in royalties, Timurtash rejected the terms which had been negotiated and proposed to go back to the drawing board. The Iranian Government indicated that it would prepare new proposals, which the Company was awaiting when the Shah intervened in dramatic fashion.[11]

THE CANCELLATION OF THE D'ARCY CONCESSION

The Iranian Government had still not submitted new proposals to the Company by the time of the Council of Ministers' meeting on 26 November 1932. On arriving at the meeting, the Shah, accompanied by Taqizadeh, abused Timurtash for the failure to reach an agreement with the Company. The Shah then dictated a letter cancelling the concession before leaving his surprised ministers. The Prime Minister, Mihdi Quli Hidayat, recollected that in his anger the Shah called for the file on the oil negotiations and had it flung into the stove.[12] In that act of exasperation, the Shah expressed his frustration with the unsuccessful negotiations which had been going on since 1928.

As to the reasons for the failure to reach an agreement, each side, naturally, blamed the other. Jacks, who knew Timurtash well, thought that the delays were caused by enmity and personal intrigues between Timurtash and Taqizadeh. They, on the other hand, argued that it was procrastination on the part of the Company which had prevented a settlement being reached.[13] However the blame is apportioned, the cancellation was an act of national assertion by Riza Shah, entangled in his suspicions of British policy towards Iran.[14] Other motives may have been the need to fend off internal opposition and the hope of extracting higher revenues from the Company to help finance his large expenditures on the army, his new naval fleet, the Trans-Iranian Railway and his ambitious plans for industrialisation. Envy of the

Company's success was also probably a factor. Visiting the south of Iran in October to inspect the navy and review the progress of the railway construction, the Shah had found the contrast between Iranian-run projects and the Company's efficient operations irritating, particularly when Company transport had to rescue him from being bogged down in the autumnal mud of Khuzistan.[15]

No matter the causes, the Iranian press gleefully greeted the cancellation as an act of political emancipation, 'a new page to Persian honour', the restoration of 'national wealth', 'the cleansing of a dirty stain' and the removal of 'a shameful remnant of the past'. The Company was condemned for the 'fabrication of its accounts', its 'failure to negotiate' and its favouritism towards Indian employees. The Soviet paper, *Izvestia*, also attacked the Company and stressed that the cancellation represented 'a serious breach in the colonial policy of England' which, it was said, would lead to a further deepening of the cracks in the decaying edifice of the decrepit British Empire.[16]

The announcement of the cancellation not only excited the press, but also coincided with the anniversary of Muhammad's proclamation of his prophetic mission, the occasion for a public holiday. The police ordered jubilation to take place and Tehran was lit up for two nights of celebration in which entry into cinemas was made free and only a snowstorm dampened the festivities.[17] In this heady atmosphere, Isa Khan, the Imperial Oil Commissioner, was the sole Iranian who openly disapproved of the cancellation. He protested to Taqizadeh that the Iranian Government had been wrong to cancel the concession by unilateral action without first resorting to arbitration.[18] He suggested referring the dispute to the Permanent Court of International Justice at The Hague or the League of Nations, but his advice was rejected and he was dismissed, dying in February 1933 after a short illness.[19]

THE REACTIONS OF THE COMPANY AND THE BRITISH GOVERNMENT

Jacks received the letter of cancellation, signed by Taqizadeh, on 27 November. Complaining that the concession had conflicted with national interests, the Iranian Government claimed that it was not 'legally and logically' bound by concessionary terms which had been granted before the establishment of constitutional government in Iran 'in view of the manner in which such concession was obtained and granted at that time'. However, although it argued that cancellation was the only way to safeguard sovereign rights, the letter showed signs of a willingness to compromise in stating that the Iranian Government would not in principle refuse to grant a new concession.[20]

The Company, feeling that it was impractical to discuss new terms with the concession cancelled, instructed Jacks to ask for the withdrawal of the cancellation notice so that it could enter into friendly negotiations.[21] To that end, Jacks instructed Mustafa Fateh, his Iranian assistant, to call on Taqizadeh in the strictest confidence.[22] Taqizadeh argued that the cancellation was simply meant to expedite negotiations and that it would be political suicide to withdraw it 'for not only this Government but the Shah himself would be in a most precarious position', exposed to tribal scorn, if the decision was reversed. Fateh informed Jacks that since July the Shah had asked Taqizadeh to take over exclusive charge of negotiations with the Company. Although Taqizadeh declared that Iran could not be bound by the acts of pre-constitutional governments which had no connection with the people, it was, apparently, he who had insisted on keeping the door open for the negotiation of a new concession in the letter of cancellation.[23] Jacks felt it would be best to be conciliatory and to support Taqizadeh rather than Timurtash, whose star he rightly felt was on the wane.[24]

In London, Arthur Hearn, the Company director in charge of concessionary affairs, told the Foreign Office that the Company did not want diplomatic assistance until and unless the situation definitely got beyond its own control.[25] The Company hoped to persuade the Iranian Government to negotiate and intended to continue its operations in Iran unless impeded by *force majeure*. That line of action was approved by Sir Lancelot Oliphant, in charge of the Eastern Department of the Foreign Office, whose legal advice was that the cancellation was illegal.[26] Fraser, acting in the absence of Cadman who was visiting the USA, hoped to avoid any action which might aggravate a situation that could be solved 'by fair and just means'.[27] Cadman supported that attitude.[28] Furughi, who was now Foreign Minister, also tried to be conciliatory, arguing that the Iranian Government wanted to conclude a new concession. He considered, however, that the Company was morally responsible for the Iranian action.[29]

On the other hand, Sir Robert Vansittart, Permanent Under-Secretary at the Foreign Office, favoured taking a strong line, thinking 'if we do not make ourselves felt at the outset, we shall have far worse trouble with the Persians later'.[30] Reginald Hoare, British Minister in Tehran, strongly advised against entering into negotiations until the cancellation notice was withdrawn. He asked that contingency plans should be prepared for naval protection and suggested making an appeal to the League of Nations because 'our case is so good and the offence so flagrant'.[31] Sir John Simon, the Foreign Secretary, in briefing the Cabinet, admitted that Cadman had wanted the question left to the Company to deal with, but that provocative press comment, demonstrations and some interference with the Company's property had rendered that difficult.[32]

In early December Hoare, realising that the Shah could not accept the humiliation of formally withdrawing the cancellation, adopted a more moderate approach. 'Our interests', he thought, 'would be best served by helping to build a bridge'.[33] Vansittart was unmoved, feeling that 'we have built enough bridges in Persia'.[34] Fraser, though, seemed to share Hoare's taste for conciliation when, after a visit to the Foreign Office on 3 December, he reflected that it would be necessary to find a 'let out' so that negotiations could be resumed without prejudice to either party's rights or position.[35] After a further meeting at the Foreign Office on 5 December, when Fraser apparently expressed the Company's conciliatory inclinations,[36] there emerges a significant difference between the records of the Company and the Foreign Office regarding Fraser's attitude. According to the Company's record, when Fraser visited the Foreign Office on 6 December, he expounded his view that the Company was anxious to seek a peaceful solution and was ready to act in that direction 'at any moment judged appropriate'. He said nothing stronger than that 'it would be inappropriate to negotiate unless British prestige had been fully maintained and in the face of the blank denial of the existence of concessionary rights'.[37] Yet, according to the Foreign Office, 'the gist of our conversation was that the Company prefer not to do any bridge building at present and to leave matters to be conducted by His Majesty's Government, whose recent strong language showed the intention of not being brow-beaten by the Persians'.[38] Whatever conflicts of understanding can be deduced from this difference of emphasis in the sources, the 'bridge' option was undeniably becoming more difficult. In a formal note dated 5 December Furughi declared that the Iranian Government was within its rights in cancelling the concession. Although he was prepared to negotiate a new concession which would safeguard the rights and interests of Iran, he was not prepared to accept responsibility for any damage suffered by the Company.[39] The Foreign Office regarded that as completely unsatisfactory and was determined to take every legitimate measure at its disposal to challenge the Iranian right to cancel the concession. 'The Persians', noted a Foreign Office official disparagingly, 'have a way of neither meaning what they say nor saying what they mean'.[40] Cadman, who preferred persuasion to compulsion, persisted in looking for the means by which a bridge could be found, but Fraser informed him that the matter had been 'taken so far by the Foreign Office that it was rather out of our hands'.[41]

That was, indeed, an accurate summary of the position which had been reached by the end of the first week of December. Vansittart continued to recommend taking a strong line to obtain the withdrawal of the cancellation, failing which an appeal should, he suggested, be made to the Permanent Court of International Justice. The Cabinet approved and on 7 December the Foreign Office replied to Furughi's note of 5 December, warning that if the Iranian Government did not retract the cancellation the dispute would be urgently referred to the Permanent Court.[42] The next day, Anthony Eden, Under-Secretary of State for Foreign Affairs, read the British reply to Furughi's note in the House of Commons, adding the threat that the Government was prepared to take 'all such measures as the situation may demand'. George Lansbury, for the opposition, asked whether Eden meant armed measures, in response to which Eden produced a sort of British version of neither saying what was meant nor meaning what was said: 'the position', he said, 'is quite clear', by which veiled threat he conveyed the very opposite of clarity.[43]

On 12 December Furughi replied to the British note of 7 December. He denied that the Permanent Court had any competence in the dispute and intimated that the Iranian Government would be within its rights in advising the Council of the League of Nations about the threats and pressure which had been directed against it.[44]

APPEAL TO THE LEAGUE OF NATIONS

The Cabinet, afraid that the Iranian Government would make a complaint to the League of Nations before Britain could make one, authorised an immediate appeal to the Council in Geneva.[45] There was no prior consultation with the Company, though it was required to provide information. These developments coincided with the return of Cadman from the USA on 13 December. Two days later he learnt at the Foreign Office that the Company was not to play a part in the proceedings before the League and that his presence in Geneva was considered inadvisable in view of the publicity it would attract. However, it was felt that a representative of the Company should be available for consultation.[46] Harold Brown, senior partner in Linklaters and Paines, the Company's solicitors, was already in touch with Sir William Malkin, chief legal adviser to the Foreign Office, who thought it would be a mistake to embark on any talks in Tehran until the matter had been raised before the Council.[47] Jacks was therefore advised not to initiate discussions in Tehran.[48] If, after the first meeting in Geneva, the parties came together then there would be no harm in such talks.

As the cancellation escalated into an international issue, Cadman expressed his astonishment at the turn of events. On 16 December, he told the Iranian Minister in London, Abbas Quli Ansari, that the Iranian Government's action had come as a complete bombshell, 'the denunciation of the concession being so utterly at variance with what we had been led to expect'. He found the conduct of the Iranian Government inexplicable: 'It is as though we were about to sit down to a game of chess and the first player swept all the pieces off the board, then asked us to make the next move and was astonished to learn that we did not consider that that was the way chess, or any other game, could possibly be played'.[49] Cadman's objective was to resume discussions, but he did not wish to enter into negotiations in Tehran immediately after the cancellation as this would have amounted to a 'mere dictation of terms'. He told Ansari that the cancellation was an extremely unfriendly act and that the matter was now in the hands of their two Governments. Ansari suggested that the cancellation was only 'theoretical', otherwise the Iranian Government would have seized the Company's property and endeavoured to run its operations.[50] Two days later, he informed Cadman that the Iranian Government would welcome discussions in Tehran, to which Cadman repeated his view that it would be futile for him to go there while the concession remained cancelled, but that once it had been reinstated, a meeting would be possible on neutral ground.[51]

Meanwhile, the British memorandum was presented in Geneva on 19 December. In it, the British Government complained that the cancellation was a unilateral act of confiscation which contravened international law. Expressing the hope that an amicable and equitable settlement could be reached, it requested action to ensure the maintenance of the status quo and to prevent the interests of the Company being prejudiced while proceedings were pending.[52] These preliminary proceedings were adjourned until 23 January, pending a memorandum from the Iranian Government. In his notice of adjournment the President, Sean Lester, pointed out that although the dispute had been brought before the Council this did not mean that the parties could not arrive at an understanding between themselves.[53]

Behind the scenes in the Foreign Office, it was Malkin's opinion which most cogently summarised the situation and received general approval, including that of the Company. He felt that the immediate objective should be to bring about a resumption of negotiations between the Iranian Government and the Company in a way that would not put the Company at a disadvantage: 'on a basis', as he put it, 'of equality'.[54] He thought that the purpose of the Iranian Government in cancelling the concession was not to destroy the Company, but to put it in a position where it would have to accept Iranian terms for a new concession.[55] The intervention of the British Government had prevented that outcome by the referral to the Council, which was not an end in itself, but a means, Malkin argued, towards detailed and fair negotiations between the Company and the Iranian Government.[56] Hoare was accordingly informed that the appeal to the Council was intended to bring the parties together and to facilitate a settlement.[57]

In Tehran the press campaign continued. *Shafaq-i-Surkh* (Red Dawn) condemned the Company as 'the centre for intrigues against the Iranian Government and nation', a 'nest of spies' and 'guilty of acting fraudulently and deceitfully'.[58] It was not only the Company that was under attack; so too was Timurtash, who was dismissed from office on 23 December, the position of Minister of Court being abolished at the same time. His fall was apparently connected, not with the affairs of the Company, but with financial irregularities at the Bank Melli, whose director had committed suicide after being arrested in October. In January 1933 Timurtash was arrested and he was later imprisoned and fined on charges of extortion, bribery and embezzlement. His death followed shortly afterwards.

The new director of the Bank Melli, Husayn Ala, a former Minister in Paris, was one of the Iranian delegates to the League of Nations. The other was Davar, still Minister of Justice. Accompanied by Abdullah Intizam, head of the League of Nations Department in the Foreign Ministry, the delegation left for Geneva on 30 December. The British case was to be put by the Foreign Secretary, Sir John Simon. Although they had no official role, representatives from the Company, including Hearn, L. Lefroy (concessions manager) and Dr V. R. Idelson (legal adviser), arrived in Geneva on 21 January. They were followed a few days later by Cadman, with Dr M. Y. Young as his adviser. The Rapporteur was Dr Edvard Benes, Foreign Secretary of Czechoslovakia.

The burden of the Iranian complaint was that from the beginning of the concession the Company had refused arbitration, made unacceptable claims, falsified its accounts, cheated on its royalty payments, upheld the Armitage-Smith Agreement, discontinued negotiations without justification and prevented a revision of the concession which the Iranian Government had endeavoured to secure. The Company, it was argued, had rendered the concession void and therefore released the Iranian Government from its contractual relationship. The Company was further criticised for extending its operations beyond Iran, employing workmen who were not Iranian and making no attempt to seek any legal remedy for the cancellation by appealing to the Iranian Government or applying to the Iranian courts. The action of the British Government was condemned as contrary to international law. Diplomatic protection was argued to be inapplicable on the grounds that cancellation based on the non-fulfilment of a contract was incontestably not a violation of international law. As the Company had not claimed the legal remedies, the British Government had no right, said the Iranians, to make a diplomatic issue of the case.[59]

Simon, in response, pleaded that by the act of cancellation, subsequently ratified by the Majlis, the Iranian Government had deprived the Company of its right, under Article 17 of the concession, to go to arbitration. No Iranian court, argued Simon, would be able to give any remedy to the Company as that would mean going against the country's own laws. The real motive for cancelling the concession was to dictate new concessionary terms to the Company, having put it in the unfair position of having its concession cancelled. This, claimed Simon, was an indefensible use of sovereign power. The British Government was anxious for the dispute to be settled, but not by condoning an international wrong. He refuted the Iranian allegations. He said that in November 1932 the Company had no reason to suppose that new proposals would not be submitted, as promised, by the Iranian Government which, instead, had abruptly terminated the negotiations. The British Government hoped, concluded Simon, that the Council would be able to bring about an amicable and equitable settlement which would enable the Company to continue its operations in Iran in a harmonious relationship with the Iranian Government.[60]

Davar replied in what Cadman considered a rambling speech,

evading all the points made by Simon and merely repeating the Iranian memorandum. He made no reference to the possibility of amicable settlement in his remarks, which were, Cadman thought, 'altogether truculent in tone and offered no suggestion whatever of yielding an inch of ground'.[61] He complained that the British Government gained more financially than the Iranian Government from the Company's operations. He also contested Simon's opinion that the Company was prevented from seeking redress in the Iranian courts because of the Majlis' ratification of the cancellation, which, argued Davar, was only a motion of confidence and not a law.

After listening to the arguments of both parties Benes ordered an adjournment so that he could study their submissions in detail and make personal contact with both sides before reporting to the Council. The diplomatic trick was to bring them together in direct negotiations. Seeking a formula which would bring about that end, Benes had discussions with Ala and Davar on the one hand, and with Cadman on the other. Nuri Pasha al-Sa'id, the most prominent Iraqi politician of his generation who seemed forever to be forming and dissolving Iraqi governments, was also involved as an intermediary between the two sides in Geneva. On 30 January he managed to bring members of the Company and the Iranian delegation together at an unofficial lunch-eon party. It was a successful manoeuvre and on 3 February Benes submitted a resolution to the Council, proposing that the two Governments should suspend all proceedings before the Council until at least May 1933 while the Company and the Iranian Government conducted negotiations. During the negotations the legal position of each side, as represented to the Council, would be reserved and the Company would continue operations as before. This proposal was approved by the Council and the way was now cleared for direct negotiations to commence.

THE NEW CONCESSION

In Geneva, Davar and Ala told Cadman that they had no authority to hold negotiations, which could only be conducted in Tehran,[62] though after some further discussion Davar disclosed that he had received a telegram from Tehran which gave the Iranian Government's main requirements. These included a 25 per cent shareholding in the Company, a minimum annual payment from the Company of £1 million (gold) on the first 6 million tons of production, 16 per cent of the profits which the Company made on annual oil production above

6 million tons, an agreed basis for the payment of taxes and representation on the Company's board of directors. Cadman protested that he did not propose to take part in discussions based on such excessive demands, and hoped that negotiations would take place in Geneva and Paris, but Davar was not in any hurry. The following day, Company representatives, excluding Cadman who had returned to London the previous evening, met Davar and Ala for a further inconclusive meeting.[63] Davar had no detailed instructions about a new concession, but had authority to submit more points for the attention of the Company in addition to those already given to Cadman. These were concerned with limiting the area of the Company's concession, revoking the Company's exclusive right to construct pipelines in the concessionary area, increasing oil production in Iran, reducing the prices of oil products supplied for Iranian consumption, the refining of oil within Iran and the payment of past and pending claims. It was also made clear that neither Davar nor Ala had authority to agree or disagree with Company counter proposals; all they could do was transmit the Company's views to Tehran. Nevertheless, it was agreed to meet in Paris for further private discussions on 8 February.[64]

The Paris meeting was frustrating because, as was subsequently learnt, the Iranian points had been compiled by Calouste Gulbenkian, and Davar and Ala were not particularly knowledgeable about them. Ala presumed that, as the Company was founded on Iranian oil, the Iranian Government had a right to a 25 per cent shareholding, with which the Company representatives disagreed. Davar was apparently unaware of the Company's internal distribution network in Iran and was surprised to learn that more than 80 per cent of production was refined at Abadan. Moreover, he repeatedly insisted on all serious discussions taking place in Tehran because he was not prepared to risk his own political future by admitting or indeed accepting from the Iranian Government authority to negotiate in Europe. He recalled the Iranian proverb that 'Oil always burns the hands' and stressed the necessity of having the Shah's approval for every word. In the circumstances, Cadman decided that the Company should take the initiative in putting forward proposals and the Iranian delegates were accordingly informed that the Company would propose a new agreement, advantageous to the Iranian Government, consistent with economic possibilities, equitable and practicable.[65] The proceedings at Geneva had drawn the parties back from the brink, but they had yet to make substantive progress.

In Tehran, Taqizadeh informed Jacks that he had been appointed to negotiate with the Company and wished to begin immediately. The negotiations had to be completed by the beginning of May to comply with the League of Nations resolution and he warned that any delay in reaching a settlement would be regarded as the fault of the Company.[66] In London, the board of directors agreed that Cadman himself should travel to Tehran to conduct the negotiations. This, Cadman told Jacks, would prevent the Iranian Government being able to claim that the Company had failed to make a determined attempt to reach a settlement.[67]

Cadman was, however, aware of the tension which would surround his discussions in Tehran. Not only was the Iranian press in full cry against the Company, but some of the members of the Company's board were showing increasing signs of frustration after five years of abortive negotiations on the concession. Lloyd, the finance director, thought that it would be better for the Company to try to uphold the existing concession than to embark on another.[68] Another of the directors, Cargill, referring to the Iranians, felt that it was 'no use wasting time with impossible people like these'.[69] In short, patience was running out when Cadman, in company with Fraser and Young, left London for Iran via Bombay on 2 March. They took with them a draft of a new concession which included the provisions that the royalty should be based on the principle of a fixed rate per ton of oil, rather than a percentage of profits as in the 1901 D'Arcy concession; that the Iranian Government would also receive a sum equivalent to 16 per cent of the Company's dividend payments in excess of £2,013,000 in any year; that the duration of the new concession would be seventy-five years; and that the area covered by the concession would be reduced to 100,000 square miles after allowing a period, yet to be defined, for exploration over double that area. A few days after Cadman's departure, Duncan Anderson, a Company accountant, and Lefroy also left for Tehran by the more direct overland route.

Cadman and his party reached Tehran on 3 April, but found no official welcoming party and an atmosphere in which Iranians were forbidden to associate with Europeans.[70] Their plan was to obtain from the Iranian Government its draft proposals for a new concession before tabling the Company proposals. On 4 April it was agreed, in preliminary discussions with Taqizadeh, that Fraser would lead the negotiations for the Company with Cadman remaining in the background, but being available for consultation or for a direct audience with the Shah.[71] The Iranian negotiating team consisted of Taqizadeh, Davar, Ala and Furughi. When the negotiations began on 5 April,

Taqizadeh provided a summary of proposals on royalty, participation and production, but not a complete draft of the Iranian proposals. Fraser, however, did not want to discuss the terms on a piecemeal basis and asked for the complete Iranian proposals, believing that without them 'it would not be possible to avoid protracted and perhaps fruitless conversations nor to make any counter proposals'. On 7 April Cadman impatiently renewed this request, and regretted that 'everyone seems to keep away from the house and I cannot understand the atmosphere at all'.

Cadman was cordially received by the Shah on 11 April and took him through the history of the negotiations up to the cancellation of the concession. The Shah became excited, disclaimed any desire to go to the League, but seemed baffled by the interference of the British Government. As Cadman recorded in his diary, the Shah claimed that he had only wanted 'to "clean the slate" ... he had no desire for any other party to work the oilfields of Persia; he was satisfied entirely with the Company and his only misgivings were on the monetary terms'. Cadman advised him that the sooner they got to grips the better and complained that after a week in Tehran he had not received the Iranian terms. The Shah also wanted the matter adjusted quickly in a friendly spirit. He was satisfied that the Company's work was 'a model for his people to follow'.

It was probably owing to action by the Shah that the Company representatives received sixteen clauses of an incomplete draft concession very late on 13 April. Fraser, after much difficulty, succeeded in seeing Taqizadeh and reiterated Cadman's view that they had no intention of discussing incomplete proposals and that furthermore there would be no point in staying to discuss them if they were completely unacceptable. Taqizadeh anxiously exclaimed that the proposals were only suggestions and he was ready to consider any amendments. By 17 April Cadman had become very restless. Furughi declared that the cancellation was more a repudiation of Timurtash than anything else and that the Shah never intended it to be taken in the way the British Government had taken it. Cadman was indifferent to such explanations and impressed upon Furughi the need for a rapid settlement.

On 19 April the representatives of the Company and the Iranian Government discussed the Iranian draft. Fraser expressed his disappointment about the lack of detail regarding the reduction of the concessionary area and took exception to the proposal that the duration of the concession should be limited to thirty years. He also disliked the Iranian proposals on a range of issues including minimum production levels; the Company's loss of its exclusive pipeline rights; the requirement that 90 per cent of Iranian production should be refined in Iran; the stipulation that all Company employees should be of Iranian nationality; the Iranian Government's proposed share-holding in the Company; the Iranian power to veto Company deci-sions; and royalties and taxation.[72] Ala was surprised at the dismissive manner in which the Iranian proposals were treated, but Cadman was dismayed at the 'extravagant' Iranian terms. He regretted 'the dreadful atmosphere' in Tehran, where the 'Persian Ministers seem afraid for their necks and are almost terrified to speak to or be seen with me or any of my party'. He doubted 'if reason and equity are included in the Persian mentality of today' and just hoped that 'we may ride in with an agreement on the strength of HIM's [the Shah's] desire not to return to the League for which I should have to thank His Majesty's Govern-ment and its Foreign Secretary'.[73]

On 21 April the Company's counter proposals were communicated to Taqizadeh and discussions ensued on 22 and 23 April.[74] Agreement, however, proved to be impossible. The Iranians, seeking to 'create as much work in Persia as possible and safeguard ourselves against reduction in output', wanted to produce as much oil as possible. Fraser explained to Taqizadeh that increases in production depended upon increases in consumption, the reality of the market. Taqizadeh never-theless insisted on a guaranteed minimum annual production rate and said that if the Company refused, it would have to relinquish territory. Taqizadeh rejected the Company proposal that a new concession should have a duration of seventy-five years and stuck to the Iranian offer of a thirty-year term. His demand for a right of veto to provide against 'any act in the future contrary to the interests of Persia' was refused by the Company delegates. Taqizadeh was convinced that 'our position and interests do not permit us to be merely spectators' and that the protection offered to the shareholders of the Company by English law was irrelevant. Moreover, he argued that Iranian rights far exceeded those of the British Government and that the alleged subser-vience of the Company to the interests of the British Government would have to be prevented in the future. Fraser explained the rights of the British Government as a shareholder and its commitment not to intervene in the commercial affairs of the Company, and claimed that with the right to veto Iran would quickly become the 'cockpit' for political disputes because of the Company's worldwide connections. A possible solution would be to separate Iranian interests from those of the rest of the Company by forming a separate Company-owned subsidiary in Iran. On the question of royalties, Taqizadeh rejected the Company proposals and countered with Iranian demands for a tonnage royalty based on gold and underpinned by a guaranteed minimum annual production. As the talks went on with no appreciable progress, Fraser remarked that they had reached not a convergence of views, but 'a parting of the ways'. On his part, Taqizadeh felt that there was no object in discussing the Company's offer any further as it was 'insufficiently precise'. The Company representatives agreed that they should leave Tehran two days later on 25 April without an agreement to put before the League of Nations in May. The negotiations had reached the point of breakdown. Almost immediately Furughi informed Cadman that he was summoned to attend the Shah.

On the morning of the next day, 24 April, Cadman arrived at the Shah's Palace to find that the whole of the Majlis had assembled there to present greetings on the seventh anniversary of the Shah's coronation. Cadman was called in to see the Shah, who welcomed the proposal that he should chair a meeting of the negotiators. Cadman suggested that the meeting be held straight away, but the Shah, remembering 'the poor members of Parliament waiting outside', indicated that it would be better to hold the meeting in the afternoon. He made it clear that he had no intention of letting Cadman leave Tehran until they had settled the concession which, the Shah told Cadman, 'he was quite prepared and ready to do'.[75]

The afternoon meeting, described by Cadman as 'most historic ... though somewhat Gilbertian', was one of the most important which Cadman ever attended as chairman of the Company. Though quite different in context, it was a summit in international oil diplomacy which at least compared with, and possibly surpassed, his famous meeting at Achnacarry Castle in Scotland some few years earlier. On that occasion, described in chapter 4, Cadman and the heads of the other two largest oil companies in the world, Standard Oil (NJ) and Royal Dutch-Shell, had arrived at what was, in effect, a private international oil treaty to share the world's oil markets. Every bit as significant, though more exotic, was the meeting on 24 April at the Shah's Palace in Tehran. It was a decisive event at which Cadman and the Shah, men of utterly contrasting backgrounds and temperaments, came together with the shared knowledge that each had the undisputed authority and the ultimate responsibility to reach agreement. They achieved a breakthrough.

Cadman brought with him Fraser and Young; the Shah had

Taqizadeh, Furughi and Davar in attendance. Cadman presented the Company's proposals and went over the principal difficulties which had arisen in the negotiations between Fraser and the Shah's ministers. He explained the Company's preference for adopting sterling, rather than gold, as the standard of currency for royalty payments. He told the Shah that the Company could not possibly agree to a minimum annual production of six million tons of oil, particularly in view of the depressed state of world demand. On the financial terms, he said that £750,000 was the most the Company could offer as a minimum annual payment. Other points were also raised, including the area and duration of the concession and the Company's exclusive pipeline rights.

The Shah accepted most of the Company's proposals and said he would not insist on having a representative on the Company's board of directors or on a right to veto its decisions. In a show of his absolute authority he also, Cadman recorded,

> gave a little homily to his Ministers, who, he said, were down on the ground and could not see very far beyond their noses, whilst he was placed on a pinnacle and could see the great world around him ... The Ministers sat in almost subdued silence listening to this lecture like small schoolboys, occasionally putting in a word or two in answering questions when spoken to by HM [the Shah].[76]

It was a scene reminiscent of Thomas Herbert's description in 1634 of the members of the court of Shah Abbas I 'who sat like so many statues rather than living men'.[77]

At the end of the meeting the Company representatives and the Iranian ministers went away to hammer out the details of agreement, an exhausting task which kept Fraser up until three in the morning on 25 and 26 April.[78] When Cadman, accompanied by Fraser and Young, again met the Shah and his ministers later on 26 April two points remained to be settled. On duration, as the Shah would not agree to more than thirty years for the whole area of the former concession, Cadman compromised and agreed to reduce the area to 100,000 square miles, upon which the Shah accepted a sixty-year term for the new concession. The second unresolved problem concerned the oilfield at Naftkhana, straddling the Iran/Iraq border far to the north of the Company's main oilfields in southern Iran. In 1927 the Company, through its wholly owned subsidiary, the Khanaqin Oil Company, had brought a new refinery into operation for the processing of Naftkhana crude at Alwand in Iraq. Cadman explained to the Shah that if Iran and Iraq came to an agreement to allow crude oil and refined products to flow from one country to another across the border, the Company would have no difficulty apportioning the financial proceeds arising from the operation of the field as a whole between the parties concerned. The Shah, however, wanted the Iranian sector of the field, in the province of Kirmanshah, developed independently of Iraq. He particularly requested Cadman to construct a new refinery on the Iranian side so that the dependence of northern Iran on Soviet oil supplies would be reduced. Cadman agreed to erect a new refinery and consented to the inclusion of a clause in the concession whereby the Company would undertake to develop the oil resources of Kirmanshah for the purpose of supplying northern Iran as far as it could economically be done.[79]

The final points having been settled, Cadman's parting audience with the Shah on 28 April was essentially a ceremonial occasion, at which Cadman expressed his 'satisfaction that we had come to an amicable settlement although I felt we had been pretty well plucked'.[80] The Shah also used the occasion to ask for technical assistance with railway engineering and the manufacture of cement, and asked whether Cadman could do anything to improve Iranian contacts with the British Government. A couple of days later, while Cadman and his party were still in Tehran, the matter was raised again by Furughi who talked of the need for a new treaty between Britain and Iran and thought that an understanding between the two countries could best be developed by someone specially charged with the task, by which he can only have meant Cadman.[81] It was, as will be seen, a matter on which Cadman was to take further action after his return to Britain. On 1 May he left Tehran and, after a packed itinerary which took in many visits and meetings including discussions with King Faisal in Baghdad, Cadman arrived back in London on 14 May.[82]

Two weeks later the new concession agreement was ratified by the Majlis without debate and received the royal assent the next day.[83] The Council of the League was officially informed that an agreement had been reached. Under its provisions the area of the concession was to be reduced to 100,000 square miles, which the Company was to select from the area of the former D'Arcy concession by the end of 1938. This would mean the relinquishment of about 80 per cent of the territory covered by the old concession. The duration of the concession was, however, extended to the end of 1993, i.e. it was to last for sixty years. The Company gave up its exclusive rights to construct and own pipelines in its concession area and undertook, as Cadman had promised it would, to proceed forthwith with the production and refining of oil in the province of Kirmanshah. On the matter of

2   Sir John Cadman at Tehran airport after the conclusion of the 1933 concession negotiations

representation, the Iranian Government was to have the right to appoint a representative, with the title of Delegate of the Imperial Government, who would be empowered: to obtain from the Company all the information to which shareholders were entitled; to attend meetings of the board of directors, its committees and meetings of shareholders convened to consider matters arising out of relations between the Company and the Iranian Government; and to request that special meetings of the board of directors be convened to consider any proposal that the Iranian Government wished to present to it. On employment, all of the Company's unskilled employees were to be Iranian nationals and the Company was to recruit its artisans, technical and commercial staff from Iranian nationals to the extent that it could find Iranians who possessed the requisite competence and experience. In addition, the Company and the Iranian Government were to draw up a 'general plan' for the progressive reduction of foreign employees and their replacement by Iranians in the shortest possible time.

In its financial terms, the agreement provided for the Company to make a lump sum payment of £1 million to the Iranian Government in settlement of all past claims. Royalties were to be calculated by a more straightforward method than before, being based not on profits, but on physical volumes of oil and the financial distribution which the Company made to its shareholders. Specifically, beginning on 1 January 1933 the royalty was to be 4s per ton of oil consumed in Iran or exported, plus a sum equal to 20 per cent of the dividends paid to the Company's ordinary shareholders in excess of £671,250. The annual royalty from these sources was guaranteed to be at least £750,000. Moreover, on the expiration or surrender of the concession the Company was also to pay the Iranian Government an amount equal to 20 per cent of the sums added to its general financial reserve since the end of 1932. The royalties for 1931 and 1932 were to be recalculated on the new basis, with the result that £1,339,132 was eventually paid for 1931 (replacing the royalty of £306,872 which had caused so much vexation) and for 1932 the final sum was £1,525,383. Finally, the Company was to be exempt from Iranian taxation for thirty years in consideration of the following payments: during the first fifteen years, 9d per ton on the first 6 million tons of oil and 6d per ton on any additional quantities, with the Company guaranteeing a minimum payment of £225,000; during the next fifteen years, 1s for each of the first 6 million tons and 9d per ton for any additional quantities.[84]

A 'FRESH START'

On arriving back in London on 14 May 1933, Cadman lost little time in acting on the Shah's request for technical assistance on railways and cement manufacture. With regard to railways, he arranged for A. N. Baylis, a Company engineer who had been appointed deputy general manager in charge of technical operations in Iran and Iraq, to be available to give advice to the Iranian Government.[85] In November, he also arranged for F. C. Hall, a mechanical engineer with the Great Western Railway, to go to Iran as technical adviser to the Iranian Government with regard to the best types of locomotive and rolling stock for the Iranian railways. Meanwhile, Cadman wrote to the Shah enclosing technical advice on cement manufacture and in January 1934 instructed G. M. Lees, the Company's chief geologist, to make an unofficial survey into cement production prospects in Iran in the course of his ordinary geological work.[86] Lees later submitted a report on the subject which was forwarded to the Iranian Government.

The Shah had also, it will be recalled, asked Cadman whether he could do anything to improve contacts between the Iranian and British governments. This was a cause to which Cadman devoted considerable effort in 1933–4. Negotiations for a general treaty between Britain and Iran had been begun by Sir Robert Clive in 1927 during his term as Minister in Tehran, but they were impeded by the oil dispute, which inevitably affected the general spirit of Anglo-Iranian relations. However, once the dispute was settled the time was ripe, in the words of Victor Mallet, acting Chargé d'Affaires in Tehran, to 'bury the past and make a fresh start'.[87] Hopes were buoyed by the news from Hoare (by then Sir Reginald) in mid-1933 that the Shah was in a 'most amiable frame of mind'.[88]

Unfortunately, it was not long before the apparently bright outlook was clouded by one of those relatively trivial incidents which is blown up out of all proportion to its real importance. Commander Bayandur, the newly appointed head of the fledgling Iranian navy, in a gesture of enthusiastic but ill-considered defiance, hauled down the British flag at Basidu on the island of Qishm, where there was a British naval cemetery.[89] Much publicity was given to the affair, which became the subject of exchanges between the Foreign Office and the Iranian authorities.[90] Then, on 9 October a British naval guard arrested an Iranian customs officer, also at Basidu. Later that month Cadman telegraphed the Shah to say that, being disturbed by the incidents, he had discussed them confidentially and informally with Ramsay Mac-

Donald, the British Prime Minister, explaining how anxious he was to ensure that there should be the best possible feeling between Britain and Iran.[91] The Shah, in reply, expressed his pleasure and satisfaction with the steps taken by Cadman in the matter.[92]

Cadman, no doubt encouraged by the Shah's approving response, expressed renewed confidence in the prospects of an improvement in Anglo-Iranian relations. In early November he informed MacDonald that he had received from the Iranian Government, at the instigation of the Shah, a request for the purchase of railway rolling stock and to obtain tenders for locomotives. It was, Cadman believed, the first order from the Iranian Government to come to Britain for many years. He saw it as evidence of a desire by the Shah to establish closer relations with Britain and thought that there was now an opportunity for 'making a real friend of this monarch'.[93] A few days later Cadman went further, writing again to MacDonald to say he hoped that in its relations with Iran, Britain would 'hold out the olive branch'. Intimating that he thought he might have a role to play in improving relations between the two countries, Cadman invited MacDonald to 'use me in any way you think fit'.[94] At the beginning of December MacDonald replied that, following the naval incidents in the Gulf, he had been trying to get the situation 'straightened out' and had almost succeeded when the Shah 'developed a bad fit of temper, and things for the moment have taken a backward turn'.[95] It was not, MacDonald added, presumably with reference to Cadman's offer of help, the time to send a special envoy.

Meanwhile, the Shah's bad mood with Britain was worsened by an article which appeared in *The Times* on 28 November, containing statements which were critical of the Iranian Government.[96] Cadman again embarked on a minor mission of unofficial diplomacy to try to smooth things out. Not only did he send a signed letter to *The Times* for publication (it appeared on 7 December), he also made it his business to see Major Astor and Geoffrey Dawson, the proprietor and editor, respectively, of the paper. They undertook to publish a series of articles 'giving prominence to the developments in modern Persia and the aspirations of the Shah for his country': in other words, to try to set things right by flattering the Shah.[97] Elkington, the Company's general manager in Iran, was relieved to hear of Cadman's diplomacy, for the two Basidu incidents and *The Times* article had apparently incensed the Shah to such a degree that he had threatened to break off diplomatic relations with Britain.[98]

In December Cadman, in meetings and correspondence with Simon,

Eden and Vansittart, kept up his pressure on the Foreign Office for a
new initiative to be taken in Anglo-Iranian relations.[99] In discussions
with Simon the possibility of Cadman playing a role in bringing the
two countries together came up in a more definite form than that
which had earlier been tentatively broached with MacDonald. More
specifically, Cadman gained the impression that Simon wanted him to
use the opportunity of a meeting he would in any case probably be
having with the Shah to have a talk on 'general' matters. Cadman felt
quite definitely that this suggestion did not go nearly far enough. He
was, as has been seen, personally acquainted with the Shah with whom
he had held direct and extremely decisive negotiations over the Company's concession. He asked Vansittart to tell Simon:

> from my personal knowledge of the mentality of the Shah that any
> suggestion confined to a mere preliminary discussion, seemed to me
> quite inadequate to the necessities of the case and that I could not
> possibly act in the manner suggested.
> My own view is that the British spokesman, whoever he might be,
> must be fortified with full powers to discuss and settle on the spot. Quite
> frankly I see no prospect of progress if his powers be restricted to
> discussions, appreciations and reports.[100]

Cadman's proposal that he, or someone else, should be endowed
with special powers to settle differences between Britain and Iran was,
it hardly needs to be said, an extremely unconventional idea which cut
across all sorts of demarcation lines in official diplomatic channels.
The matters at issue were much more wide-ranging than concessionary
relations. They included not only the problem of naval tension in the
Gulf, but also other bones of contention: Iranian claims to sovereignty
over Bahrain, with whose Arab rulers Britain had made treaties which
made Bahrain virtually a British protectorate; and the continuing
Iranian disagreement with Iraq over sovereignty of the Shatt al-Arab
waterway. These were matters not of commerce, but of international
diplomacy, conventionally settled not by men of business, but by men
of state. It was a distinction nicely made by the top Treasury official,
Sir Warren Fisher, to whom, as an 'old friend', Cadman explained his
ideas.[101] Fisher replied:

> I think that as far as possible it is desirable to divorce commercial from
> political affairs – at any rate in this case – and I am just afraid that even
> should you be given complete powers to settle all outstandings and
> succeed – as I believe you probably would – that if the present system of
> dealing with Persia is maintained further problems will need settlement
> and there will be a tendency for Persia to look to you or to the

machinery of the APOC [Company] for the settlement of these problems.[102]

To Cadman, however, the complications which might arise from mixing business with diplomacy were evidently less irksome than the complexities of working through official channels. From as early as 1929 he had criticised the lack of coherence in dealing with Iranian affairs arising from the multiple division of responsibilities between government departments.[103] Now, early in January 1934, he was pressing for the control of Iranian and Middle East affairs to be placed in the co-ordinated hands of one department. As he complained to Elkington, on almost every matter which had arisen in the last few years 'the policy and attitude of the several Government Departments which have a finger in the pie always differ and the results we have experienced are as ludicrous as they are disgraceful'.[104]

Cadman thus sought, by a mixture of persuasion and protestation, to bring about a more direct and decisive method of handling Anglo-Iranian relations, in which he believed he might have a useful role to play as a special envoy with, as it were, a mission to settle British differences head-to-head with the Shah. Directness and decisiveness were, as Cadman knew from his experience in renegotiating the concession in Tehran, characteristics which the Shah possessed in abundance. The idea of taking a short cut, by-passing the labyrinth of diplomatic officialdom, did not, however, appeal to the British Government. On 15 January Cadman had lunch with Simon who later that day alerted the Cabinet to Cadman's 'unconventional suggestion that the only way to secure a satisfactory agreement with Persia was by direct negotiation between the Shah himself and some plenipotentiary other than the accredited diplomatic representative'.[105] Simon indicated that the Foreign Office was ready for discussions with the Iranian authorities and on 24 January the Cabinet decided that Hoare should be instructed to resume treaty negotiations.[106]

Despite Cadman's efforts, nothing had really changed and negotiations through the usual channels, which it would be tedious to describe, duly took their laborious course. Iran and Iraq reached a treaty over the Shatt al-Arab waterway in 1937, but no general Anglo-Iranian treaty emerged.

MORE PROBLEMS

In the meantime, Iran was embarked on an impressive but uneven programme of modernisation whose implementation was expensive.

Foreign loans were excluded, for they would have compromised Riza Shah's sense of national independence, so taxation was raised internally on goods such as tea and sugar to fund projects such as that prominent symbol of Riza Shah's ambitions, the Trans-Iranian Railway.[107] At the same time, the extensive state direction of the economy was producing a growing centralised bureaucracy. The contraction of the private sector and the increase of government control was ultimately to have an adverse effect on the Iranian economy and on the political life of the country, where individualism was inhibited by growing regimentation.[108] Even one of the most sympathetic commentators on Riza Shah, Donald Wilber, questioned whether co-operation with the regime compensated for the accompanying loss of political and personal liberties.[109]

In foreign trade, the Soviet Union was traditionally Iran's main partner, providing the only outlet for exports from northern Iran, but times were changing as Iran tried to redress its balance of trade by a boycott of Soviet goods in 1933[110] and by diversifying its trading links. Overtures were made to Japan for the exchange of cotton goods in mid-1933, and also to India.[111] In July 1936 the Ministry of Finance indicated to the British Legation that it would welcome a new initiative on trading relations between the two countries,[112] but despite some British exports of textile machinery, railway equipment, aircraft and chemicals there was not much increase in British trade with Iran.[113]

Germany, on the other hand, was energetic in increasing its share of Iran's foreign trade, seizing the opportunities presented by Iran's rapid industrialisation, which made Tehran an alluring hunting ground for the agents of foreign firms seeking industrial contracts. The growing German presence was noticed by Hughe Knatchbull-Hugessen, who succeeded Hoare as British Minister in Tehran in 1934 and commented in October 1935 on the ubiquity of German agents showing great energy in seeking commercial opportunities.[114] In the same month a Convention for the Regulation of Payments between Germany and Iran was signed, giving Germany a privileged position in Iran. The British initially showed no sign of anxiety, believing that the possibility of German commercial expansion in Iran was limited by the nature of the trade between the two countries and would not be permanent.[115] However, from the end of 1936 it became increasingly obvious that Germany was obtaining a very large share of Iran's foreign trade and was playing an important part in the country's economic life.[116] Some, believing that a close and permanent association was being forged, envisaged Iran growing to economic and industrial strength under the guidance of, and in co-operation with Germany.[117] It was expected that mutual self-interest would bring the two countries closer together because of the tendency for trade to form deepening grooves with time. Germany was replacing Russia as the major foreign force in the Iranian economy.

With the suicide of Davar early in 1937 the restraining brake on Iranian spending was removed and ambition prevailed over prudence. The problem was compounded by the concentration of investment on a few capital-intensive projects with little immediate profitability, most obviously the enormously costly Trans-Iranian Railway, which opened in August 1938. Riza Shah was not, of course, the only national leader of his time to direct funds into activities giving a low economic return, nor was he exceptional in spending a large proportion of his revenues on military items. However, even in favourable circumstances with a satisfactory balance of trade Iran would have been hard-pressed to sustain the expenditure required by the Shah. Moreover, so strong was his determination to assert Iranian independence that there was no prospect of recourse to foreign loans. He could, therefore, ill afford any drop in revenues, a growing proportion of which came from the Company's royalty and tax payments.

For the first few years after the 1933 concession was agreed there was little cause for concern about a fall in Iranian oil revenues. As the world economic depression came to an end there was a revival in demand for oil products and a great expansion of the Company's operations in Iran, as described in more detail in the next chapter. Suffice to say here that the Company's crude oil production in Iran rose from 7,086,706 tons in 1933 to 10,167,795 tons in 1937. Between the same dates, the Company's royalty and tax payments, shown in table 2.1, nearly doubled, rising from £1,812,442 to £3,545,313. To this agreeable state of affairs was added the satisfaction of seeing the completion of the new refinery at Kirmanshah in 1935. Against the background of such favourable economic conditions the period of the mid-1930s was one of cordial relations between the Company and the Iranian Government and Cadman happily told the Company's shareholders at the annual general meeting in June 1936 that the new concession was operating in a way that was 'very gratifying'.[118]

The only discordant note was a splitting of hairs over the unit of measurement to be used in calculating royalty payments. More specifically, was it to be the long ton (2240 lbs) or the slightly smaller metric ton (2205 lbs)? When the matter was raised by the Iranian Ministry of Finance in December 1934, the Company confirmed that

Table 2.1 *The Company's royalty and tax payments to the Iranian Government, 1931–1938*

	£
1931	1,339,132
1932	1,525,383
1933	1,812,442
1934	2,189,853
1935	2,220,648
1936	2,580,205
1937	3,545,313
1938	3,307,478

Notes: 1. Excludes the lump sum payment of £1,000,000 made in settlement of past claims under the 1933 concession agreement.
2. Includes retrospective payments for 1931–2 provided for under the 1933 concession agreement.
Source: BP 4308.

its royalty calculations were based on the long ton, which was its traditional unit of measurement for oil quantities. The Iranians naturally argued in favour of using the metric ton which, being smaller, would give a higher royalty for a given volume of production. Eventually, after the matter had been batted back and forth between the two sides, each claiming that legal opinions supported its case, the Company's board of directors decided that little purpose would be served by jeopardising its generally good relations with the Iranian Government for a relatively small sum of money. In August 1936 the Company accordingly paid a sum of £147,422 in settlement of the arrears that were alleged.[119]

Soon, however, the Company was faced with the familiar concessional dilemma: Iranian pressure for increased oil revenues at a time of reduced oil consumption. The year of 1937 was a bumper one. Rising demand, stimulated by the build-up of reserve stocks as international tension spread and rearmament intensified, enabled the Company to increase its crude production in Iran to the figure, already mentioned, of 10,167,795 tons compared with 8,198,119 tons in 1936, an increase of nearly 25 per cent. The royalty of £3,545,313 for 1937 was nearly 38 per cent higher than that for 1936. Such a spectacular rate of growth could not possibly be sustained, but as in economic cycles through the ages the peak year excited hopes and aroused expectations which stood not the remotest chance of being fulfilled.

In the course of 1938 it became apparent that Iranian oil exports, and consequently royalties, were falling. By the spring, the build-up of reserve stocks was virtually complete and exports for the three months of July, August and September were less than they had been for the same period in 1937. In response to Iranian requests for an explanation, Fraser pointed to market conditions and cautioned that 'phenomenal expansion is not a normal expectation'.[120] In November Cadman, having been elevated to the peerage as Lord Cadman of Silverdale in the Coronation honours of 1937, wrote to the Shah outlining the Company's development plans in Iran.[121] However, the Shah, concerned about his falling revenues, remained dissatisfied. In December he replied to Cadman in a letter from which it is worth quoting at length:

> You will of course agree how unpleasant it is for a progressive country like Iran, which must administer its affairs according to a definite programme, and cannot relegate its business to Chance, nor place its trust upon supernatural assistance, to have to confront such an unexpected issue.
>
> ... Now, can Iran, in need as she is of revenue, and knowing for certain that an important source such as the oil mines under your concession can yield her several times more revenue than what she at present receives from the Company, tolerate such a situation?
>
> It is my desire that your Company should indeed endure and prosper. Unfortunately, however, the policy adopted in its commercial activities is absolutely inconsistent with the needs and expectations of Iran.[122]

Early in January 1939 Cadman responded with a detailed explanation of the reasons for the drop in exports: the fall in demand; competition from sources of oil nearer to the main markets; the growing production of oil from coal and alcohol in countries which pursued policies of economic autarchy; the high tariff barriers in the USA and the prohibition of oil imports into the Soviet Union, which meant that Iranian oil was excluded from the markets which accounted for the bulk of the world's consumption. Cadman pointed out that during the 1930s Iran's share of the world's oil trade had increased substantially, as had the royalties which the Iranian Government received from the Company. He referred to the Company's programme for the expansion and improvement of its facilities in Iran and emphasised that the Company aimed to maximise its sales of Iranian oil and to earn 'reasonable' profits. As the Iranian Government's oil revenues were based on the volume of the Company's sales and a percentage of the profits which were distributed to shareholders, Cadman could argue, with some justification, that the Company's aims were not, as the Shah had claimed, inconsistent with the needs and expectations of Iran. On the contrary, according to Cadman the Company and Iran shared a 'close identity of interests'.[123] Whatever its merits, Cadman's exposition failed to mollify the Shah, who replied on 25 January that the situation was so unsatisfactory that 'it is better that I should abstain from direct discussion of the problem'.[124] It was as if he was breaking off diplomatic relations in the matter.

Cadman, trying to keep the lines of communication open, wrote again to the Shah on 30 January to say that Neville Gass would be visiting Iran in February and he hoped that the Shah would entrust Gass with any messages he wished to convey to Cadman.[125] Gass arrived in Tehran on 11 February to find that the Company was 'definitely in disfavour'. He had interviews with the Prime Minister and Minister of Finance, but was unable to overcome Iranian suspicions of the Company and what he described as the Shah's 'authoritarian disregard of the facts'.[126] After getting back to London in the first half of May, Gass briefed Cadman, who had decided to visit Tehran himself, on the situation there. Gass doubted whether his discussions with ministers carried any weight at all, especially as what he said might not even have been reported accurately to the Shah for fear of incurring his displeasure. However, Gass was convinced that the Iranian need for money was so desperate that the Iranians would go to any lengths to try to get it so that their development projects would not suffer.[127] It seemed clear to Gass that the Shah either held, or was using for negotiating purposes, the belief that the Company had promised a progressive annual increase in production, which had been cut back either for the purpose of storing oil reserves for the future, or as a deliberate attempt to impede the development of Iran's economy by restricting its revenues. Newspaper articles in Iran gave free rein to such allegations and left no doubt, according to Gass, that the inference to be drawn was that the Company's actions were largely dictated by political, not commercial motives. Gass had repeatedly tried to explain to the Shah's ministers the reasons for the fall in consumption of Iranian oil, but he was certain that even if his explanations had been passed on to Riza Shah – which, he repeated, was doubtful – the Shah was not in any case prepared to accept them. 'In simple words', Gass continued, 'he holds the belief that there are enormous reserves of oil within our Concession area, that there is a demand for oil which far exceeds the supply and that oil is money, his sole remaining source of increased revenue'.[128] Moreover, in a country where most industries were controlled by the state it was, Gass thought, probably incomprehensible to the Shah that the British Government, with its large shareholding in the Company, could not arrange to take more Iranian oil and so satisfy Iran's need for additional revenue.[129]

So forewarned of Iranian attitudes, Cadman, who had for several months been suffering from ill health, arrived in Tehran on 27 May. Two days later, in audience with the Shah, he refuted the various allegations made against the Company in the Iranian press and denied that the Company was restricting production on the orders of the British Government.[130] He reiterated the point that the volume of production depended on the level of demand. The Shah was satisfied with Cadman's explanation, but wished to know how the Company could help him raise revenues which he required for 'certain projects he had in view'.[131] It was a question for which Cadman, briefed by Gass, had come prepared. Before leaving London he had investigated the possibility of the Iranian Government obtaining help from the British Exports Credit Guarantee Department. That Department had been set up by the British Government in 1928 with the primary purpose of helping Britain's struggling basic export industries, such as cotton, coal and iron and steel, by giving credit to exporters so that they, in turn, could offer longer credit terms to foreign buyers and so stand a better chance of winning their orders. At his meeting with the Shah on 29 May Cadman suggested that Iran might be able to obtain goods from Britain under the Exports Credit Scheme, in which event the goods could be paid for out of future, rather than present, income. It was an ingenious idea, for it would not involve the Company in a financial outlay; nor would Iran be the direct recipient of a foreign loan to which, as Cadman would have known, the Shah had very strong objections which he confirmed at the meeting 'saying a loan would make his people lazy';[132] and, finally, it would increase British exports and promote closer trading links between Britain and Iran, a cause which, as he had earlier demonstrated in the matter of Iranian railways, was close to Cadman's heart.

After the meeting between Cadman and the Shah, more detailed discussions were held between Mahmud Badr (the Iranian Minister of Finance), Gass and Leonard Rice, who had become the Company's chief representative in Tehran after Jacks returned to London in 1935. Then, on 3 June Cadman had a second meeting with the Shah, at which Cadman promised that if the Shah sent a representative to

Britain the Company would introduce him to the Exports Credit Guarantee Department and do what it could to facilitate the grant of export credits for goods going to Iran. At the end of the meeting Cadman left 'feeling certain that I had regained all the ground we had lost'.[133] Leaving Tehran by air on 5 June, he travelled via Baghdad to London, where he arrived on 7 June.[134] Three days later he wrote to Neville Chamberlain, the British Prime Minister:

> I am glad to say I was able to get HIM [the Shah] to change his attitude towards the Anglo-Iranian Oil Company, and to return to the friendly procedure which has hitherto characterised his dealings with us.
>
> He had it in mind to cancel our Concession out of hand; the Shah will evidently deal with no one but the Chairman.[135]

CONCLUSION

As the international tension hovered ominously over Europe Cadman, at the age of sixty-two and with his health failing, was aware that his working life was coming to an end. He had said as much to the Shah whom he had told, during his visit to Tehran, that 'with the passing of the years the conduct of affairs must ultimately pass into other hands'.[136] If, on returning home he was inclined to reflect on concessionary relations over the previous decade he might have detected signs of a recurring pattern in which certain features were evident. For one thing, concessionary relations had tended to follow the trade cycle. Thus, in years of economic downturn, when demand, profits and royalty payments all fell together, the Iranian Government quickly became dissatisfied and its relations with the Company deteriorated. That was what happened in 1932 when the concession was cancelled and again in 1938–9 when the Shah came close to cancelling it again. The Company, accustomed as it was to operating in world oil markets and to the vagaries of economic conditions, saw economic downturns as a product of an inequation between supply and demand, to be combatted by the traditional remedies of, amongst other things, retrenchment and cutbacks. On the other hand, the Iranians, who had no direct contact with or experience of oil markets and who had for long suffered the oppression less of the trade cycle than of foreign domination, tended to associate falls in oil revenues not with the impersonal laws of economics, but with political interference on the part of Britain, which had for long exercised a powerful influence over Iran. That association was rendered more credible by the speed and vigour with which the Foreign Office came to the Company's defence in 1932; and, of course, by the British Government's majority share-holding in the Company which all too easily gave the impression that the Company was not so much an economic organisation as the political instrument of an imperial power. The point is illustrated by the Iranian allegations that the reductions in Company oil production in 1938–9 were made on the orders of the British Government.

If contemplation of such matters might have helped to identify some of the causes of the concessionary crises of the 1930s, attention might just as fruitfully have been directed to the manner in which differences were resolved. It was not, certainly, by the labyrinthine workings of official diplomacy, with which Cadman displayed almost as much impatience, in his different way, as did the Shah. Nor were solutions worked out and agreed by delegations and committees from the two sides. It was rather in one-to-one negotiations between Cadman and the Shah that concessionary disputes were resolved. Thus, in 1933, after the breakdown of negotiations between the Company representatives led by Fraser and the Shah's ministers led by Taqizadeh, it was Cadman and the Shah who, at their meeting in the Shah's Palace on 24 April, resuscitated the concession. Again, in May–June 1939 it was Cadman's personal visit which persuaded the Shah to draw back from the brink of, possibly, another cancellation. The two men, who on the surface had so little in common, understood the nature of each other's authority and responsibility and were capable of cutting through the knot of negotiations which others were unable to untie. Yet, with so much depending on the chemistry of the Cadman–Shah relationship rather than the more institutional plane of Company–Iranian relations, the question could not forever remain unanswered: what would happen after they were gone?

3

Operations and employment in Iran, 1928–1939

The scale and complexity of the Company's operations in Iran were unparalleled in their time as a single concessionary oil enterprise. There was nothing then comparable to the emergence of the oil industry in the Middle East rising from the dry plains and mountain ranges of south-western Iran. Once the centre of a flourishing civilisation stretching in time from Elam to Islam and scattered with monuments of an eventful antiquity, it had become a neglected nomadic backwater in the nineteenth century, before the discovery of oil and the arrival of the Company. Operating on a large scale in a region of Iran which was largely devoid of the economic infrastructure associated with more highly developed economies, the Company was responsible for a vast network of services including roads, electricity and water supplies, telephone lines, jetties, transportation, accommodation, social amenities, education and security. Such facilities, though major undertakings in themselves, were essentially ancillary to the core operations of exploration, production, transportation and refining, all of which had undergone substantial expansion and improvement in the 1920s. Advances in production methods at the Company's first oilfield at Masjid i-Suleiman had been made by applying the principle of unitisation, whereby the oilfield was developed and operated as a single productive unit and was not, as occurred elsewhere in the oil industry, split between the competing claims of rival firms, a practice which gave rise to waste and inefficiency. In exploration, the discovery of oil at Haft Kel in 1927 enabled the Company to bring a second field into production, reducing dependence on Masjid i-Suleiman as the sole source of Iranian crude. Exploratory drilling was also carried out at Gach Saran, Agha Jari, Pazanun and Naft-i-Shah, the Iranian side of the Naftkhana (Iraq)/Naft-i-Shah (Iran) field which straddled the

Table 3.1 *The Company's capital expenditure in Iran,*
1930–1938 (£thousands)

	Fields	Pipelines	Refineries	Other	Total
1930	198	81	2038	19	2336
1931	33	4	685	15	737
1932	8	2	62	2	74
1933	46	2	143	11	202
1934	76	70	471	28	645
1935	35	181	1709	13	1938
1936	54	12	1120	22	1208
1937	83	260	1385	72	1800
1938	288	366	3499	276	4429

Sources: See Appendix 1.

border between Iraq and Iran. All of these areas were later to be developed as producing fields. In refining, the Abadan refinery, much criticised for its inefficiency by Cadman after his visit there in 1924, was improved in the second half of the 1920s when new fractionating columns for the distillation of crude oil were installed. In 1929, the first thermal cracking units were ordered and by the end of 1931 they were in operation, enabling the refinery to upgrade heavy fractions of the crude to lighter, more valuable products. In transportation, the capacity of the main pipeline from Masjid i-Suleiman to Abadan was increased and new pumping units installed. In addition, a new pipeline was laid to connect the Haft Kel oilfield to the main pipeline at Kut Abdullah. In the same period, shipping access to Abadan was improved by the dredging of the Shatt al-Arab estuary, reducing the cost of transporting products to the Company's markets.

After the onset of the great economic depression, the pace of improvement slowed. Employment was reduced, projects postponed and exploration stopped as the Company pursued a policy of severe retrenchment. Its capital expenditure in Iran, shown in table 3.1, was reduced from £2,336,000 in 1930 to a mere £74,000 in 1932, a year in which the Company suffered great concessionary uncertainty, culminating in the cancellation of the D'Arcy concession. It was not until 1933, when new concessionary terms were agreed and the worst of the depression over, that confidence returned and the Company embarked on an ambitious policy of expansion in Iran, where its capital expenditure climbed to a prewar peak of £4,429,000 in 1938. From the bottom of the depression to the top of the recovery there was, as these figures make clear, an immense expansion in the Company's assets in Iran.

OPERATIONS

It was one of the terms of the 1933 concession that the Company's concessionary area was to be reduced to 200,000 square miles by the end of 1933 and to 100,000 square miles by 1938. The priority in exploration was, therefore, to delineate the most promising area. G. M. Lees, the Company's chief geologist, made the initial selection in consultation with his senior geological staff, choosing a territory which ran roughly from the border with Iraq north-west of Kirmanshah in a south-easterly direction to the border of what is now Pakistan. The second relinquishment required more critical geological information which could only be obtained by detailed surveys with quarter-inch mapping and extensive topographical examination, similar to that carried out earlier in some parts of the country. Over five years most of the possible concessionary area which had not already been investigated was covered with geological reconnaissance surveys in which J. V. Harrison was the driving force, ably assisted by N. L. Falcon, later chief geologist, and other colleagues. Peter Cox, another Company geologist who later rose to the position of chief geologist (and managing director of D'Arcy Exploration Company), described it as 'an achievement without parallel in scope and thoroughness in the history of oil exploration having regard to the time taken and the relatively small number of geologists assigned to it'.[1] In order to assist the Company in its ultimate choice of the most promising concessionary area an advisory panel of prominent geologists was appointed in 1935. The area selected contained most of the oil discovered in Iran to the present day between a line more or less along the western flanks of the Zagros mountains and the Persian Gulf seaboard, shown in map 3.1.

Concurrent with the concessionary surveys, but more specific in purpose, were other geological investigations into individual prospective areas. From 1935, they were supplemented by aerial reconnaissance photography in collaboration with Aerofilms Ltd as contractors. The most important development, however, was the growing use of geophysical methods to detect the nature of subterranean structures by the seismic techniques of refraction and reflection, the principles of which are illustrated in figure 3.1. There were some years of

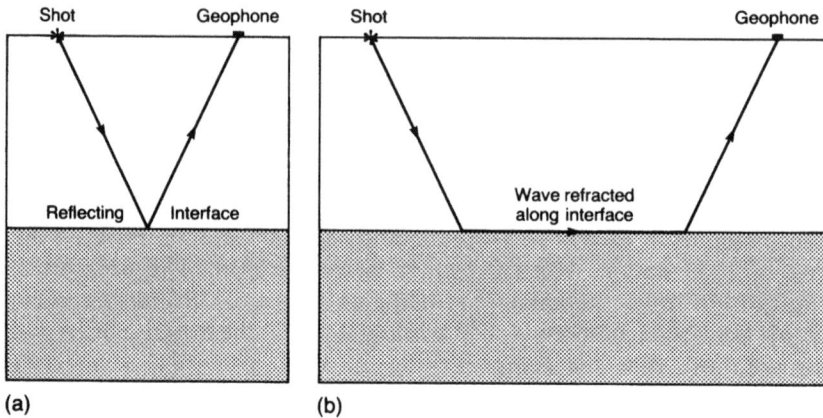

Figure 3.1   Principles of seismic methods. (a) Reflection; (b) Refraction

experimentation during which the chief geophysicist, Dr J. H. Jones, favoured first the refraction method and then the reflection method advocated by another geophysicist, D. T. Germain-Jones. After the reflection technique had been shown to be effective in the siting of a well at Masjid i-Suleiman in 1934, its first great success was in determining the position of the Lali No. 1 well which found the oil-bearing Asmari limestone almost directly beneath a syncline in the Upper Fars structure, as had been predicted from the seismic evidence. After years of doubtful application, geophysical techniques had justified themselves in Iran and more work continued to be undertaken, including the use of gravimeters and newer reflection methods by the French contractors, Générale Géophysique, in 1937–8.

It was, however, only with the drill that the existence of oil could be definitely proven. In the 1930s exploratory wells were drilled at a number of sites, as a result of which oil was discovered at Naft Safid in 1935, Agha Jari in 1937–8 and Lali in 1938. In addition, gas was discovered at Pazanun in 1937. Meanwhile, further drilling was undertaken at Gach Saran and Naft-i-Shah, fields which had been discovered, though not developed, in the 1920s.[2] Of these various fields, the only one to come into commercial production before World War II was the Naft-i-Shah field, being the Iranian side of the reservoir which underlay the Iran/Iraq border. The oilfield, where oil had been discovered on the Iraq side at Naftkhana in 1923 and on the Iranian side in 1928, was geographically isolated from the Company's main fields in south-west Iran, being located in the province of Kirmanshah,

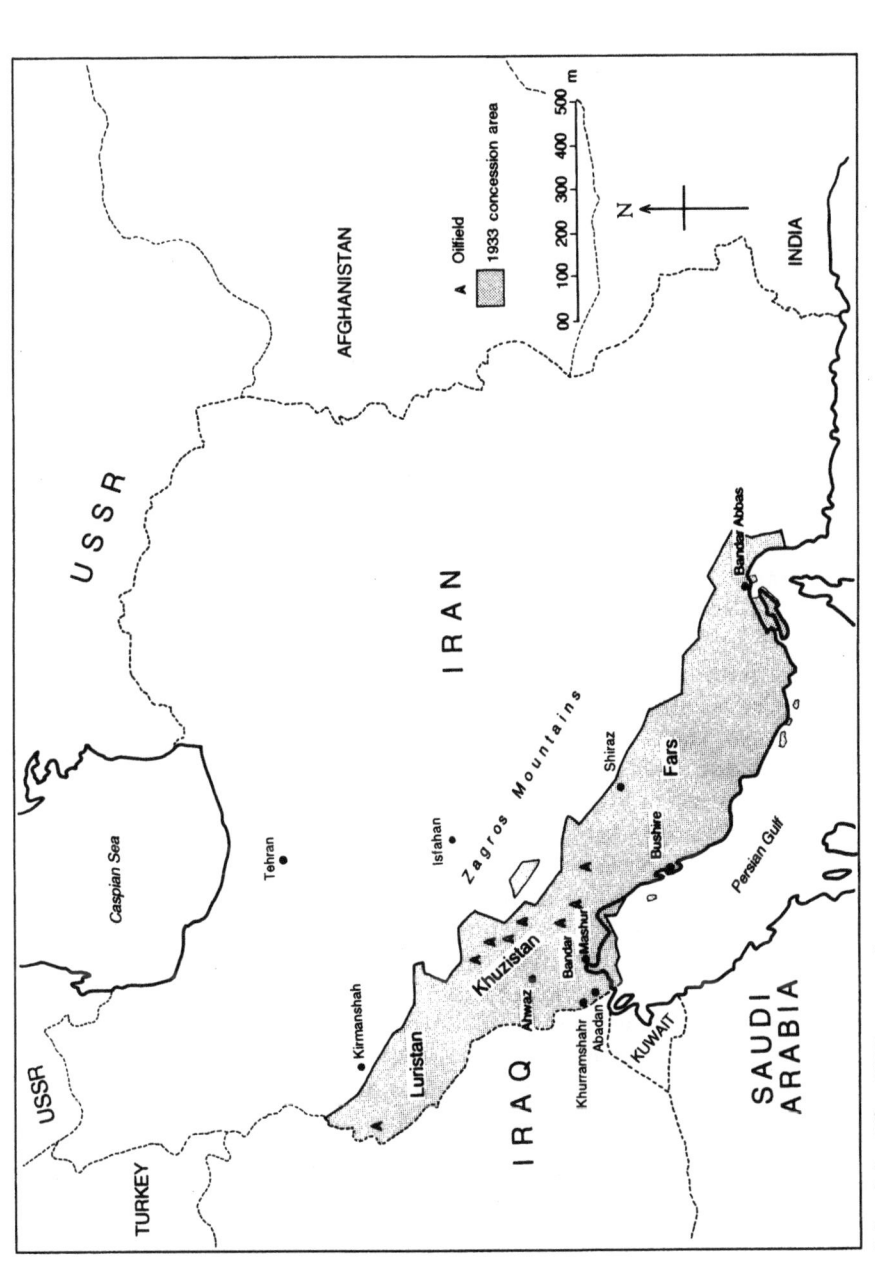

Map 3.1 The 1933 concession area

considerably to the north of the southern fields and the Abadan refinery. In the negotiations for the 1933 concession, Cadman agreed to bring the previously undeveloped Naft-i-Shah side into production separately from Naftkhana and to form a separate company to produce, refine and sell oil from the field for the domestic Iranian market. Accordingly, the Kirmanshah Oil Company was registered in London as a subsidiary in June 1934, drilling at Naft-i-Shah was recommenced and preparations made for the commencement of production. A small 'topping' plant was erected so that fractions of the crude oil not wanted for refining could be separated and recycled into the reservoir. A pipeline was constructed to Kirmanshah, 158 miles distant across mountainous territory. Equipped with four pumping stations, it carried Naft-i-Shah crude to the small Kirmanshah refinery which was erected some three miles from the city and officially opened on 10 October 1935 in the presence of the Prime Minister, Mirza Muhammad Ali Khan Furughi, and other leading Iranian personalities. At the refinery, the distribution organisation was equipped with road vehicles, mostly manufactured by Scammells, in which products were transported in bulk and in tins to depots to service the requirements of local dealers.[3]

While the Naft-i-Shah oilfield and the Kirmanshah refinery met domestic demand for oil in the northern and central areas of Iran, the main oilfields in the south-west of the country were connected into a pipeline and refining system from which the Company made supplies to international markets. The Masjid i-Suleiman and Haft Kel fields were already connected to Abadan by pipeline before 1932, but in the course of the 1930s the pipeline capacity for the two fields was increased to cater for a throughput of 12.5 million tons a year by 1938. The main innovation in pipeline construction during this period was the introduction of welded joints instead of screwed couplings. The enlarged capacity of the pipeline system enabled the Company to increase its production of crude oil in Iran from 6,446,000 tons in 1932 to 10,196,000 tons in 1938. Between those dates the mature Masjid i-Suleiman field went into decline, but the loss of output from that source was more than made up by the very rapid growth in production from Haft Kel, whose output soared from 1,956,000 tons in 1932 to 6,626,000 tons in 1938. Full production figures for the period are shown in table 3.2.

During the same period, the Abadan refinery continued to be expanded, both in the volume of oil which it was capable of processing

Table 3.2 *The Company's crude oil production in Iran,*
1930–1938 (thousand tons)

	Masjid i-Suleiman	Haft Kel	Gach Saran	Naft-i-Shah	Total
1930	4926	1013	–	–	5939
1931	4248	1502	–	–	5750
1932	4490	1956	–	–	6446
1933	5000	2087	–	–	7087
1934	5339	2199	–	–	7538
1935	4228	3231	–	31	7490
1936	3641	4474	–	83	8198
1937	4375	5694	1	98	10,168
1938	3464	6626	10	96	10,196

Sources: See Appendix 1.

and in the range of products which was produced. In the early 1930s the refinery, in common with the rest of the Company's operations, was subjected to stringent economies and a retrenchment in capital expenditure. However, from 1934 through to the beginning of World War II the installation of new plant to increase throughput and improve the yield of products was virtually continuous. Concurrently, the interaction between theory and practice was encouraged by the appointment of D. G. Smith, the chief chemist at Abadan, to take charge of the research centre at Sunbury in 1935, the same year that a new chemical laboratory was built at Abadan.

During these years, there was continual attention to improving motor spirit quality, evidenced by the commissioning of a new benzine blending plant in 1935, the modification of other plants and the reorganisation of the benzine and kerosene departments, which were combined. Other significant changes included improved treatments for benzine and, most notably, the continued installation of cracking plant. The first thermal cracking units, of American design, had been ordered for the refinery in 1929 and installed by the end of 1931. In the mid-1930s four larger, more advanced thermal crackers were commissioned to produce cracked spirit with higher octane ratings. In 1937, more new plants were added, ranging from washery units, a lime plant and an acid tar burning installation to a benzole pyrolysis plant in

3   The opening of the Kirmanshah refinery, 1935

Table 3.3 *Abadan refinery throughputs, 1930–1938*
(million tons)

	Crude oil throughputs
1930	4.40
1931	4.39
1932	4.98
1933	5.64
1934	6.21
1935	6.80
1936	7.55
1937	9.47
1938	9.66

Sources: See Appendix 1.

which hydrocarbons which had previously been unusable for motor spirit were cracked at very high temperatures to form benzole. In the same year, bitumen manufacture was commenced by adapting some of the smaller and older distillation units for that purpose. The installation of new equipment continued in 1938, in which year the throughput of the refinery was not far short of ten million tons, approximately double the figure for 1932, as shown in table 3.3.

Abadan was not merely a refinery, it was also a large and growing shipping port, where imports of materials were landed and exports of crude and products loaded on an ever-increasing scale. Many port facilities had to be provided, such as repair workshops, a floating dock, jetties, cranes, and accommodation and amenities for those manning the shore installations, the tug and repair services, the ships of the British Tanker Company and other vessels taking bunkers. Manifests had to be checked, regulations observed, customs cleared and tanks dipped. Shipping had to be marshalled in and out of the port and liaison maintained with Iranian officials and the Basra Port Authority, which supervised shipping movements along the Shatt al-Arab waterway. In addition to maritime shipping, there was a fleet of barges, tugs and river steamers which carried goods on the Shatt al-Arab and the river Karun.

By 1935 it was apparent, in view of projected increases in crude production and refining, that marine loading capacity would have to be increased. However, there was insufficient deepwater frontage for

4   The administrative offices of the Company at Abadan in the 1930s

5 View of the waterfront at Abadan showing tankers loading, 1935

additional jetties at Abadan and, besides, the Iranian Ministry of War objected to further expansion there. After investigation, the most suitable alternative site seemed to be at Qasba, some twenty-four miles downstream from Abadan, and after discussions with the Iranian Government authorisation for the acquisition of land was granted in June 1937. However, on closer investigation it was discovered that Qasba would not, after all, be a suitable location because more or less constant dredging would be required for access by large tankers. Attention therefore turned to an alternative site at Khuzalabad, renamed Khusrauabad in 1939, about ten miles upstream from Qasba. Agreement on the use of the site was reached with the Ministry of Finance in November 1937 and installations for the loading of 'black' oils were completed in December 1939.

Meanwhile, concern was being expressed about the limits to future expansion of the Abadan refinery. In May 1937 Pattinson warned Elkington that 'the present limit to expansion in the area is set by the availability of labour' which depended on the provision of accommodation and increased municipal services. The feeling, reflected in plans which were drawn up for the refinery in mid-1937, was that its optimum size was an annual capacity of twelve million tons. Against the background of the concern about the limits to future growth at Abadan, consideration was given to the possibility of refining at a new location served by a marine terminal at Bandar Mashur, five miles from the head of the Khur Musa inlet and ten miles downstream from the port of Bandar Shapur, which was being developed as the terminus of the Trans-Iranian Railway. Bandar Mashur had a deepwater channel capable of taking very large tankers with minimal dredging and it possessed the added attraction of being closer than Abadan to the promising production prospects at Gach Saran and Agha Jari. The case for constructing refining and loading facilities at a new location was explained by Jameson, who wrote, after visiting Iran in 1938, that:

> The question of the ultimate throughput of Abadan is ... not only a matter of plant capacity, but is also a social problem influenced by living conditions ... Any increase in throughput is not a question of simple addition but of multiplication ... Abadan as a manufacturing unit has reached, if it has not exceeded, the economic size from a management point of view.[4]

Commenting on possible alternatives to Abadan, Jameson reported that special attention had been given to Bandar Shapur and the environs of Khur Musa. Contrary to the generally accepted view that it would be impossible to erect large-scale plant in the marshy area of

Bandar Shapur, it had been established, Jameson continued, that there was a suitable refinery site, with good soil conditions and above the highest tides.[5] In 1938–9, further progress was made with the scheme for locating a refinery there. Extensive surveys were carried out and consultations held, land was acquired, plans drawn up and the agreement of the Iranian Government obtained. In the offing, it seemed, was a second centre of operations, similar to that created around Masjid i-Suleiman, Haft Kel and Abadan. Such might have been the outcome had it not been for the intervention of that event which caused widespread upset to the Company's plans: the outbreak of war. The commencement of hostilities scuppered the proposals for a second oilfields-refinery-port complex in southern Iran and the scheme was reduced to simple improvised loading facilities, for which Government permission was granted. Temporary accommodation was made ready, navigational aids installed and provision made for loading up to two million tons a year of fuel oil for the Admiralty using two tanker mooring berths at buoys, which were tested, but never operated.

EMPLOYMENT

Representing a foreign presence which was in economic, social and cultural contrast to Iranian traditions and customs, the Company did not blend inconspicuously into its Iranian surroundings. Abadan, the main centre of employment, stood out as an area of unprecedented industrial concentration in a country which was mainly rural, where the working methods, skills and practices required for Company operations were far removed from the habitual lifestyles and occupations of the local populace. The development of Abadan was the result, not of natural, national growth, but of grafting an alien cutting on to ancient stock.

The rapid growth of the Iranian oil industry posed exceptional administrative problems for both the Company and the local authorities in the province of Khuzistan, where there was negligible industrial infrastructure and little sense of municipal responsibility. A succession of compromises ensued between the Iranian Government and the Company in the fields of health, water, electricity, education and accommodation and later even in subsidising the cost of living. The management of such a complex technical enterprise even in a developed industrial society would have been a formidable undertaking. In a region without relevant technical experience, where life tended to follow seasonal patterns rather than the constant demands of modern industry, the administration of the Company's diverse workforce at a vast distance from the London headquarters posed an immense challenge.

The disturbances of May 1929

By 1928 Abadan had become in some respects a boom town and a centre of opportunism both for officials, who regarded service in Khuzistan as more like exile than promotion, and local entrepreneurs who controlled the bazaars and other commercial activities in the area. An intricate web of local intrigue existed as local personalities promoted their interests and sought to curry favour with the Shah. In such conditions it was not easy to distinguish reality from suspicion and even the Shah reportedly admitted in 1929 that the police and security services were 'always inclined to invent bogies to justify their existence'.[6]

Whatever the immediate causes, the Company's operations were momentarily threatened by an outbreak of unrest on 2 May 1929, when a demonstration was held to make demands and incite disturbances among the Company's Iranian employees.[7] The next day, the Governor of Abadan and the provincial Governor-General decided to take counter measures against the agitators and by 4 May some forty-five people had been arrested, of whom twenty were in the Company's service.[8] The Governor-General told Elkington that he had settled the troubles, which had been organised by the local branch of the Communist Party. His optimism was misplaced, for early on 6 May a body of men began preventing workers from going in to the refinery.[9] Attempts were made to penetrate the refinery fences and in the scuffling one or two of the Company's staff in the grounds became involved, including Andrew S. McQueen, a former Scottish boxing champion. Denied entry, the agitators withdrew and were dispersed by the police. During the incident the refinery kept operating. The failure of the demonstrators to force their way into the refinery and the posting of guards enabled the shifts to change.[10] The crisis had passed, but Elkington admitted to Cadman that 'had the rioters broken into the Refinery chaos would have ensued' and remained concerned at the possibility of continuing communist-inspired activity.[11]

Timurtash, the Shah's Minister of Court, who was informed of the disturbances on 5 May, was congratulated by Cadman on the way in which the Iranian Government was dealing with the difficult situation in Abadan. The British Minister in Tehran, Sir Robert Clive, who saw

Timurtash on 7 May, had taken responsibility for the protective movement from Basra of the sloop HMS *Cyclamen*, which remained out of sight and took no part in the troubles, which were entirely handled by the Iranian civil and military authorities, who succeeded in rounding up many of the instigators of the disturbances and having them transferred outside the area.[12] In London, Cadman requested officials to take a sensible co-operative attitude. The situation in Fields was calm. Nearly all the local Iranian workers had shown no inclination to strike. Most of those arrested came from Isfahan, Shiraz, Bushire and the north.

Although the exact cause of the disturbances was difficult to assess, communist incitement was regarded as a major factor. Clive believed it was 'a carefully prepared Communist plot directed not only against the Anglo-Persian Oil Company but against the Persian Administration generally', whilst Elkington, recognising more local involvement, considered that a 'legitimate nationalist movement had been converted into a Bolshevic intrigue'.[13] The involvement of the Majlis deputy for Abadan, Mirza Husayn Movaqqar, active in the political and commercial circles of Khuzistan, was also alleged.[14] He had German sympathies in World War I and had shown an interest in communist affairs at Ahwaz. He was known to Jameson and Jacks who recognised his ability and his opportunism and were concerned at his arrest for his part in the demonstrations in Abadan in view of the support he had in Tehran.[15] The arrival of the Russian ships, the *Michael Fruntze* on 2 May and the *Communist*, carrying some munitions on 14 May caused some disquiet, adding to the anxiety about opposition to Riza Shah.[16]

In mid-May, Timurtash, who had first ordered that instructions be sent to the Governor 'to deal drastically with the agitators and to insist on Persian employees returning to work', assured F. S. Greenhouse, the acting Company representative in Tehran, that security was adequate and stressed that a 'study of the whole labour question was necessary', the first hearing of an often repeated refrain.[17] Clive told Timurtash on 29 May that 'any idea of settling matters by simply giving another Kran a day to the workmen was childish' and that 'there was absolutely nothing at present to show that wages were the source of the trouble'.[18] Gass arrived in Tehran on 28 May to brief Timurtash, who admitted that the unrest was entirely due to Communist influences and proposed a reorganisation of the secret police.[19] However, the next day he was critical of the role of the local authorities, determined to release those detained and dismissive of the accusations against Movaqqar. He had informed Greenhouse that most of those arrested during the strike had not been guilty of indictable crimes and that it appeared that a reason for discontent had been 'low wages and the lack of specific regulations for their work' which the Company should remedy.[20]

Cadman denied that the disturbances could be attributed to relations between the Company and its employees.[21] In the Iranian press, however, the Company was accused of racial discrimination with complaints that its Indian employees ruled over Iranians, who, 'glorious and noble sons of Darius, who have sacrificed everything, yea, even sons, wives, family, and all in the path of the Anglo-Persian Oil Company, have no better work to do than to carry heavy pipes and material on your shoulders'. Such propaganda was prominent, a ritual prophylactic incantation against malign foreign influence.

Towards the end of June, Elkington felt that the emergency had ended, for the holy month of Muharram had passed peacefully, in spite of some anxiety.[22] Yet, whilst the position in Abadan had stabilised, the situation in the southern provinces with the Qashqa'i tribe was deteriorating. A serious crisis seemed imminent on 24 June when Firuz Mirza Nusrat al-Dowla, Finance Minister, was arrested along with the Governor-General of Fars and others. At the beginning of July most of the Bakhtiari tribe were in rebellion[23] and unrest did not peter out amongst the Qashqa'i until October. Timurtash professed to be satisfied that rumours of British interference in the internal affairs of Iran were unfounded, but he was suspicious, embarrassed by the Abadan affair and his failure to settle a concessionary revision with Cadman earlier in March.[24] Moreover, his relations with Riza Shah were strained.[25] Cadman cautioned Company staff not to meddle in the political affairs of Iran, or even to express any views on political events in the country in case they were misinterpreted as detrimental to the authority of the central Government.

Although the Company's employment terms were at least as good as those of the state railway contractors,[26] Elkington was aware that there had been shortcomings in handling labour relations. At the beginning of June 1929 he proposed better facilities for workers to report their grievances and visualised a co-operative spirit based upon 'firmer foundations of mutual confidence', suggesting that emphasis should be placed upon 'public welfare' in 'vastly improved township conditions'.[27] The attainment of such a goal depended not only on the Company, but also on the municipal authorities who were responsible for the development of the town, the prices in the bazaars, the implementation of sanitary measures and the enforcement of law and order. Nevertheless, Elkington believed that Company efforts to improve labour conditions would help to 'encourage that sense of security which accompanies employment, to stimulate loyalty to the Company and at least go a long way to eradicate the petty grievances which fertilise the soil of discontent'. He reflected that the labour problem was 'irretrievably bound up with the political movement' in Iran and that more thought should be given to Iranian ideals and customs.[28] He did not think that labourers were discontented and they had not generally participated in the turmoil. He believed that wages were adequate provided municipal control prevented shopkeepers from profiteering.[29] He also acknowledged that labour relations were a 'specialised affair' requiring 'the instinctive knowledge which comes from close acquaintance with the people'. He was concerned about the recruitment and role of personnel officers, who he felt should have 'a first class public school background' and preferably the District Officer mentality as he had known it in India, with intensive training and language study in Iran.[30] Colonel Medlicott, in charge of the General Department, was doubtful about 'Varsity' chaps and considered those best suited to be 'sons of clergymen, doctors, land agents, etc., whose parents have sent them to good second-class public schools, and kept them there until they are 18, and then expect them to find their own way in life'.[31]

While the Company's management was generally reluctant to usurp the duties of local government or to become enmeshed in the complexities of local politics, political pressures could not be simply ignored. Towards the end of October 1929 Timurtash suggested that the Company should mark the forthcoming visit of the Shah to Khuzistan by increasing the wage rates of its Iranian employees. Elkington presumed that the Shah was making a bid for popularity and claimed that there was no justification for the request. Cadman thought that he might be satisfied with a general holiday with full pay, but Timurtash wanted more than that.[32] Changing his mind, and with the agreement of Jacks, Elkington recommended a 5 per cent increase to all Iranian workers as a matter of political expediency. Cadman agreed in order to show appreciation of the Shah's 'progressive policy'.[33] The Company then heard that the Shah was not satisfied with the proposed 5 per cent increase, but desired a more substantial advance in the lower rates of pay.[34] He was therefore pleased when the Company marked the occasion of his visit to Khuzistan in January 1930 by announcing not only a general 5 per cent increase but also a minimum wage which provided for a larger rise in the lower wage rates.[35]

While agreeing to the increase in wage rates, Cadman was deeply concerned about the Company's relationship with the Iranian Government. He was in principle 'most keenly anxious' to co-operate with the Government, yet he was opposed to Government interference in the Company's management and troubled by the extent of the Company's responsibilities for services such as security, housing, water supplies, transport, lighting, sanitation, education and medical care.[36] He was at pains to point out to Jacks that the 'Company's raison d'être is purely commercial' and that 'the Company is not a benevolent institution and must be conducted on commercial lines'.[37] However, while economic management was the primary objective of the Company, it was not necessarily so for the Iranian Government. That was, and remained, a constant dilemma as the Company became an increasingly large employer in Iran.

The size and composition of the workforce, 1930–1938

As can be seen from table 3.4, the numbers engaged on Company work, both as direct employees and as contractors, followed the same cyclical pattern as other indicators of the Company's levels of activity in Iran in the 1930s. The reduction in employment during the economic depression of the early 1930s was extremely sharp, the numbers employed being more than halved from 31,312 at the end of 1930 to 15,383 at the end of 1932. Thereafter, recovery was strong, the numbers employed reaching a prewar peak of 48,928 at the end of 1938. At the same time, the composition of the workforce reflected the movement towards 'Iranianisation', meaning increased employment of Iranians relative to foreigners. That was a prominent issue in the Company's relations with the Iranian Government and is described in more detail in the later section on the General Plan. Here, however, it can be seen from table 3.4 that in 1930 about 87 per cent of those engaged on Company work were Iranians, rising to 94 per cent in 1938. The others were foreigners, mostly Indian and British.

The small minority of British staff made up an expatriate community which, drawn together in a foreign climate and culture, inevitably took on some of the characteristics of a social enclave in Iran. Men and women who in Britain would for the most part have blended inconspicuously into their surroundings became, in Iran, a privileged elite, easily caricatured as the stereotype of the British abroad in the age of empire. For those posted to eastern service, the Company issued

Table 3.4 *Company employees and contractors in Iran,*
1930–1938

End of:	(a) Iranian	(b) Indian	(c) British	(d) Other	(e) Total	(a) as % of (e)
1930	27,180	2472	1076	584	31,312	86.8
1931	13,059	1263	775	577	15,674	83.3
1932	13,321	935	704	423	15,383	86.6
1933	15,941	795	749	277	17,762	89.7
1934	22,020	925	901	254	24,100	91.4
1935	25,240	954	1035	119	27,348	92.3
1936	24,948	779	1055	76	26,858	92.9
1937	30,779	786	1185	66	32,816	93.8
1938	45,978	1342	1524	84	48,928	94.0

Sources: See Appendix 1.

advice about living conditions, behaviour and health. For example, in 1928 the newly arrived bachelor at Fields was warned:

> Always wear your topee until the sun is well down in the evening. The habit of running from office to office without a topee is dangerous in the extreme.
>
> Avoid heavy meals and alcoholic liquors during the day time. If you must drink alcohol, do so after sun down and then only in moderation.
>
> Don't sleep under a fan without something extra round your middle to avoid chill.
>
> To keep cool you should perspire slightly.
>
> Flies are the carriers of a large number of Eastern diseases and every effort should be made to keep them from gaining access to food.[38]

Three years later, when the Company was retrenching in the midst of the great depression, Jameson reported, after visiting Iran, that the general management in that country was assuming the proportions of a 'serious overhead' in relation to the operations under its control. As table 3.4 shows, the number of British staff was reduced. 'The Boys', remarked Jameson, 'must be made to realise ... that they have to earn their pay'. He noted disapprovingly that most of the staff in Iran were looking forward to retirement after fifteen years' service, which was

6   Riza Shah visiting Abadan refinery, 1930 (Shah on right; Edward Elkington next right)

the commonly accepted period of overseas employment, and, probably reflecting on the passing of the pioneering spirit which he embodied, he reported that work in Iran was gradually becoming more routine. There was an unmistakably critical tone in his comment that 'everyone in Abadan, quite apart from the salaries they were receiving, was trying to live on the same scale' and he felt that 'we would have to recast all our ideas' on the practice of holding out inducements for continuous service with the Company. C. C. Mylles, the staff manager for Iran, was apprehensive about the effects which a shake-up would have on loyalty, but was concerned that the cost of servants was the largest single item in the domestic budgets of expatriate staff and that social expenses had also increased considerably with the growth in entertainment. Reductions in the number of British staff continued to be made, to the extent that in 1932 A. G. Bell, a staff manager in London, reported on the limited opportunities for promotion, which meant that the staff were in the position of having to wait for 'dead men's shoes'.[39]

With the renewed expansion of activity which took place from 1934, recruitment was resumed and the number of Britons engaged in Iran increased up to the outbreak of World War II. The growing number of staff brought various matters to the fore, such as the provision of married quarters, the payment of allowances, the need for domestic servants and, in general, the social and economic integration of wives into the community of Company employees in Iran. Jameson felt that, rather than relying on domestic servants, wives should involve themselves actively in household duties. He suggested that the Company should therefore provide facilities 'to allow them to do a great deal of cooking', noting that the 'mere fact of us supplying service imposes on the married man a higher standard of living than he would normally be accustomed to in this country'. In similar vein, Elkington was not 'very keen on women messing at the Restaurant as they are a damned nuisance outside their normal surroundings and the Restaurant is absolutely full to capacity'. He was, however, pleased that 'the female typist is doing well ... and, as we are getting the wives and sisters of our own clerks, the question of accommodation does not arise'.[40]

Into this expatriate community were imported not only British mannerisms, but also items of consumption. In 1934, when the average annual salary and cash allowances of British staff in Iran amounted to £712, some monitored examples of staff expenditure in Iran worked out at an average of £390. One lb of marmalade was twice as expensive as a bar of Lifebuoy soap. Tinned fruit was not included, being 'absolutely prohibitive'.[41] Some typical weekly expenditures in 1935 were:[42]

	Rials	Sterling
Groceries and toilet items	139	£2 8s 3d
Tobacco and drinks	86	£1 9s 10d
Bazaar purchases	61	£1 1s 2d
Restaurant	33	11s 2d
Dairy farm produce	35	12s 2d
Domestic staff	79	£1 7s 5d
Laundry	14	4s 10d
Postage	7	2s 5d
Clubs (sports and social)	40	13s 11d

Less evocative than the social imagery of expatriate life, but more germane to the conduct of the Company's operations, were the managerial, administrative, professional and technical jobs performed by Britons. Too numerous to catalogue in detail, they included, apart from general management, a range of occupations as, for example, chemists, geologists, palaeontologists, doctors, dentists, accountants and engineers.[43] In performing those various functions the British staff brought to Iran not only the peculiar social features of an expatriate group, but also a range of experience, skills and qualifications which could not be found in the local populace.

An analysis of the Abadan workforce in 1930 reveals the nature of the occupations in which Iranians and Indians were engaged at the refinery. As table 3.5 shows, the great majority of the labour force of workmen and contractors were Iranians, most of whom were employed in refining and construction work. Indians, who made up only a small minority of the workmen and contractors, outnumbered Iranians in clerical occupations which were carried out mainly in offices and stores. In the foremen category there were more Iranians than Indians, but not by much. More significantly, less than 1 per cent of the Iranian workforce were foremen, compared with about 7 per cent of Indians. The general picture was, therefore, that Iranians made up the bulk of the labour force, mainly as workmen and contractors. Of the much smaller number of Indians a relatively high proportion worked as clerks and foremen.

At the same time only a very small number of Iranians held senior staff positions. In a statement which reflected in microcosm some of the problems of integrating the Company into the social, economic and political life of Iran, it was noted, in relation to the recruitment of Iranians as staff, that there was a type of Iranian who would be:

> probably of good social standing and may even have a British University education, but when in the South [of Iran] he is, both by his natural inclination to seek out his own fellow countrymen and by a peculiar disinclination of the average Britisher to meet on a pleasant social basis with foreigners, driven to seek the society of Junior Customs and other Government officials in Abadan under whose influence he becomes imbued with intensely Nationalistic and anti-foreign prejudices which are common to practically all the juniors in Persian Government billets at the present day.[44]

Despite such reservations, the number of Iranians employed as senior staff increased markedly during the 1930s, rising from 15 at the end of 1932 to 177 at the end of 1938, as shown in table 3.6. During the same period, the general increase in the number of Iranians employed on Company work was accompanied, it would seem, by a rise in quality and changes in attitude fostered, in part, by the Shah's reforms, which included the adoption of western dress and conscription for the army. 'Generally', thought Jameson in 1938, 'labour looks much more workman-like now than they did in tribal clothes and the introduction of conscription has had a beneficial effect on their physique and

Table 3.5 *The Abadan workforce in 1930*

	Workmen		Contractors		Clerks		Foremen	
	Iranians	Indians	Iranians	Indians	Iranians	Indians	Iranians	Indians
Refining	3017	563	203	–	12	19	6	55
Storage and export	795	26	109	–	11	12	67	4
Construction	4976	173	847	–	16	6	27	12
Stores	741	12	166	–	43	38	2	3
Main workshops	407	81	–	–	3	7	–	2
Bungalows and social services	850	191	40	–	5	10	5	8
Guards	356	2	–	–	2	1	–	–
Shipping	721	75	67	–	12	16	8	10
Offices	159	3	–	–	111	199	12	25
Medical and sanitary	280	12	–	–	12	19	1	–
Transport and traffic	863	35	437	–	14	9	12	4
Telephone and telegraph	56	16	–	–	3	13	–	2
Packed oil department	211	9	14	–	1	5	1	1
Total	13,432	1198	1883	–	245	354	141	126

Source: BP 71879.

general discipline'.[45] It was, he noted, not unusual 'to be faced with a complete European meal as well as all the ordinary Iranian courses when being entertained by persons of standing'.[46]

The General Plan

The Company's employment policies in Iran were, like many other aspects of its activities there, closely bound up with its concessionary relationship with the Iranian Government. In employment, the issue between the two sides was at heart straightforward. The Company wished to conduct its business on the criterion of commercial efficiency and not to be fettered by rigid formulae governing employment. The Iranian Government, on the other hand, wished the Company to employ more Iranians and fewer foreigners and sought specific commitments to that end. Though outwardly simple, the matter was complicated by the need for relevant education and training for certain types of job.

The concerns of both sides were reflected, though not resolved, in Article 16 of the 1933 concession which stated that:

(I) Both parties recognise and accept as the principle governing the performance of this Agreement the supreme necessity, in their mutual interest, of maintaining the highest degree of efficiency and of economy in the administration and the operations of the Company in Persia.

(II) It is, however, understood that the Company shall recruit its artisans as well as its technical and commercial staff from among Persian nationals to the extent that it shall find in Persia persons who possess the requisite competence and experience. It is likewise understood that the unskilled staff shall be composed exclusively of Persian nationals.

(III) The parties declare themselves in agreement to study and prepare a general plan of yearly and progressive reduction of the non-Persian employees with a view to replacing them in the shortest possible time and progressively by Persian nationals.

(IV) The Company shall make a yearly grant of £10,000 sterling in order to give in Great Britain, to Persian nationals, the professional education necessary for the oil industry.

The said grant shall be expended by a Committee which shall be constituted as provided in Article XV.

In the light of the potential clash of aims, especially between para-

Table 3.6 Company employment in Iran by category, 1932–1938

End of:	Staff				Labour				Construction	
	Senior		Clerical, technical & supervisory		Artisans & skilled		Unskilled, domestic & contract			
	Foreign	Iranian	Foreign	Iranian	Foreign	Iranian	Foreign	Iranian	Foreign	Iranian
1932	713	15	397	689	539	4992	406	7625	–	–
1933 (May)	739	18	393	697	525	5214	389	6646	–	–
1934	762	29	404	772	578	5647	167	10,485	162	3745
1935	858	45	386	1000	593	7905	19	11,416	251	4513
1936	912	75	265	1223	527	8341	9	11,346	197	3395
1937 (August)	991	82	262	1354	550	9009	4	15,036	189	4806
1938	1283	177	384	1751	762	12,591	–	19,406	439	11,426

Notes: 1. The figures used for this table do not exactly match those in table 3.4 owing to inconsistencies in the sources. The differences are, however, insignificant except for 1933. In that year the data for the month of May, which contain the breakdown needed for this table, vary considerably from the year-end data used in table 3.4.
2. Excludes small numbers of instructors, trainees, specialists and consultants.

Sources: BP 68935; 71191; 68034.

graphs (I) and (III), it is unsurprising that Article 16 became a controversial issue in relations between the Iranian Government and the Company.

The 1933 concession was ratified by the Majlis on 28 May and came into force on receiving the royal assent the following day. Almost immediately, on 31 May, Taqizadeh, the Iranian Finance Minister, requested the Company to appoint a representative to discuss the drawing up of the general plan for the reduction of its non-Iranian employees and their replacement with Iranians.[47] Fraser felt that the subject could not be fruitfully discussed before it had been considered on the spot at Abadan and pointed out that the substitution of Iranian for foreign employees was 'not merely a matter of goodwill', but had to be related to education and training.[48] The divergent views of the Company and the Iranian Government on the preparation of a general plan were apparent in October 1933, when Nasrullah Jahangir, the newly appointed director of the Iranian Petroleum Department, visited Abadan. Jahangir reportedly argued that the Company had 'definitely bound itself not only to study and prepare a general plan to replace in the shortest possible time our non-Persian employees, but to carry out without fail a yearly and progressive reduction of such employees'. He was not apparently interested in the question of training, which he considered an excuse for inaction, and thought that the Government should do its utmost to 'Iranianise' the industry so that it could be operated by his countrymen.[49]

Jahangir's views were not shared by Fraser who was critical of the Iranian insistence on changing personnel 'merely on grounds of nationality without regard to relative efficiency'.[50] 'We cannot', Fraser commented, 'play about with personnel questions as if we were dealing with a lot of lead soldiers'.[51] He had undertaken, he informed Jacks, to study and prepare a *general* plan, whereas the Iranian Government wanted a *definite* plan.[52] He could not, as he told Elkington, commit himself to a programme involving definite figures for the year by year reduction of non-Iranian employees.[53] The difference between the two sides was further emphasised towards the end of November when Ali Asghar Khan Zarrinkafsh, the newly appointed Imperial Delegate, visited Abadan and was reported to claim that doctors were unnecessary in the oil industry and that medical students should not be included in the education allocation.[54] The Company, on the other hand, took the line that doctors were as necessary for the repair and maintenance of the workforce as engineers were necessary for the repair and maintenance of plant and machinery.

The tone of the discussions became more conciliatory after Davar became Minister of Finance, replacing Taqizadeh who was appointed Minister to France. At a meeting with Jacks on 13 December, Davar recognised the importance of efficiency and economy in the Company's operations.[55] Jacks argued that if the Company were to spend more money on training, the Iranian Government ought to accept an obligation to educate Iranians to a higher standard so that they would be better prepared to absorb subsequent training by the Company. Davar agreed and recounted his experiences in reorganising the Ministry of Justice when 'he found the utmost difficulty in providing the personnel necessary'. He had started educational training classes, but found that many applicants were unable to take advantage of the opportunities offered. He assured Jacks that he would not be 'unreasonable or desirous of demanding from the Company the impossible'.

Within a few months a draft plan for Company training of Iranians was prepared in London and submitted to Davar towards the end of March 1934.[56] It was almost two months before Davar discussed the plan with Jacks in mid-May, when Jahangir argued that it contravened Article 16 of the 1933 concession because it did not specify an annual and progressive reduction of non-Iranian personnel and their replacement by trained Iranians.[57] On the subject of education, Jacks maintained that this was wholly the responsibility of the Iranian Government. Davar confirmed that the Company was under no obligation to educate Iranians for employment and that 'on the contrary this was a liability of the Government and one which the Government would discharge'. The importance which the Company attached to that point was reiterated at a further meeting when Jacks told Davar that the success or failure of the plan hinged on the Company's ability to recruit youths educated to the requisite standard. As a concession to Davar, he agreed to produce figures to give the draft plan more quantitative substance, subject to the qualification that in the working of the plan much would depend on factors outside the Company's control.[58] Fraser accepted Jacks' advice and agreed, against his own inclinations, that figures on the operation of the plan over five years should be put forward in a guarded and entirely noncommital manner.[59] Jacks duly submitted figures to Davar, whilst emphasising the importance of the Iranian Government playing its part by providing the necessary educational facilities. Despite the progress which had been made Jacks remained concerned that Article 16 would 'continue a never ending source of trouble'.[60]

Davar might have been inclined to support the draft plan, but he did not seem to want to take responsibility for reaching an agreement with the Company, which might have exposed him to criticism in Iran. Instead, he requested Taqizadeh to open negotiations in London without informing Jacks, who felt that Davar was being 'mentally dishonest and wholly unscrupulous' in trying to bypass him and deal with Cadman and Fraser through Taqizadeh.[61] As it happened, such a course was precluded by Taqizadeh's effective exile after he prudently decided not to go back to Tehran, having been summoned to return from Paris ostensibly for permitting a French newspaper to insult the Shah by publishing a cartoon on a cat (*le chat*). He appealed to Cadman for help and came to London as a lecturer at the School of Oriental and African Studies, where he stayed while Riza Shah remained in power.

Davar again declined to reach a decision on the draft plan on his own on the grounds that he could not commit his successors. Jacks, about to retire, believed it was sheer procrastination. On 26 August Davar appealed to Ala, the Iranian Minister in London, who apologised some ten weeks later for being unable to devote enough time to the problem.[62] When he finally met Company representatives to discuss the draft plan on 23 November, he stated that, having taken legal opinion, he felt that the plan was not in accordance with Article 16 of the 1933 concession because it did not make specific provision for the numerical and progressive reduction of non-Iranian employees.[63] The Company representatives suggested that, as the draft plan was not acceptable to the Iranian Government, it should draw up a plan itself. This suggestion was approved and it was agreed that Zarrinkafsh should prepare proposals for discussion at a further meeting.[64]

At about the same time the Company took legal opinion and was advised that it was not under a concessionary obligation to train Iranians to replace foreign personnel beyond its commitment under Article 16 of the 1933 concession to make an annual grant of £10,000 for the education of Iranians in Britain.[65] The Company thereupon informed the Iranian Government that its offer of training in the draft plan was a purely voluntary one and was not to be regarded as a concessionary obligation. Much argument ensued over this point in the early months of 1935.[66]

In April 1935 Zarrinkafsh, who had not produced a draft plan, reiterated to Neville Gass, who had been recalled to London to take charge of Iranian concessionary affairs for the Company, that the Company's plan was not acceptable for the reason that had already been given: failure to specify the progressive reduction of non-Iranian employees. The Iranian Government therefore wished to have a new plan drawn up.[67] Gass took the matter up and prepared further drafts, on which there were a number of internal Company comments, mainly to the effect that technical developments were increasing the demand for properly trained personnel and that automation and new working methods made it very difficult to foresee future employment requirements.[68] After a series of meetings between Gass and Zarrinkafsh in June, July and August 1935 it looked, by late October, as though a final settlement might be achieved.[69] In November, however, Ala objected that there were still no figures binding the Company to a definite progressive reduction in its non-Iranian employees. Davar appealed to Cadman to intervene, but Cadman declined to do so, pointing out that during an earlier visit to Tehran he had explained to the Shah 'the difficulties of a hard and fast formula and ... that any plan of replacement of non-Iranian employees should be elastic and must be studied in the light of practical experience'.[70] Various amendments were nevertheless made to the draft plan, which was to be discussed with the Iranian Government by Fraser, who was planning to visit Tehran.

Fraser arrived in Tehran on 30 March 1936 having travelled from London via Abadan, Fields, Baghdad and Kirmanshah, where he inspected the Company's new refinery. Much of his time in Tehran was taken up in discussions with Davar about the general plan, the talks beginning on 31 March and continuing for three days. Fraser informed Davar of the Company's programme for additional housing and of its intention to build a technical school for the higher training of Iranian employees. He also repeatedly emphasised that the crux of the problem in attracting and retaining suitable Iranian employees lay in the provision of improved housing and amenities in which the Company 'would be very willing to assist', but which ultimately depended upon the Government who alone possessed the necessary power to execute developments on the large scale which he had in mind.[71] On the subject of the supply and training of suitable Iranians he stressed the inadequacy of educational facilities and explained that until that problem was remedied the Company's training schemes could not come into full operation. Davar, according to Fraser, showed a 'real appreciation' of the problem of amenities and undertook to give his personal attention to an energetic campaign to improve educational facilities.[72]

It was, by Fraser's account, a cordial exchange which resulted in the signing, on 2 April, of two documents: the General Plan and a Procès-Verbal. Under the General Plan the Company undertook to effect a speedy and progressive replacement of its foreign artisans, technical and commercial staff by Iranians insofar as the replacement was compatible with the attainment of the highest degree of efficiency and economy in its operations and to the extent that it could find Iranians possessing the requisite qualifications. It was also laid down that factors which could not be foreseen and which were beyond the control of one or other of the parties might affect the rate of reduction of non-Iranian employees. However, notwithstanding the effect of such factors, the Company stated its express intention to accelerate the reduction of its non-Iranian personnel. Graphs were appended to the Plan, showing the envisaged reduction in foreign employees as a percentage of total Company employees up to 1943. The Company's training programme for Iranians was also set out, it being explicitly stated that the training scheme was voluntary and not obligatory.[73]

Under the Procès-Verbal the Company undertook to extend the new primary school at Abadan by the addition of a secondary section for 120 pupils and also to provide and maintain a technical college in Abadan until 1943. The Company was also to complete its projected housing scheme and to assist with the provision of amenities in the town of Abadan. Davar, in response to Fraser's recommendations for the improvement of primary and secondary education, declared that the Iranian Government took 'the greatest interest in educational questions and you may be assured that it will fulfil its whole duty and do even more than what you have said'.[74]

Although the signing of the General Plan eased the situation regarding the Company's employment of foreigners in Iran, it did not prevent differences arising between the Iranian Government and the Company from time to time. Within a few months, Davar committed suicide, leaving the Plan without its Iranian sponsor. Jahangir left little to trust and in May 1937 asked for a copy of the rules and regulations governing the engagement and promotion of technical, commercial and clerical staff. The Company replied that rigid rules and regulations could not be applied in view of constantly changing operational requirements and the fact that the 'human element' was involved. On the subject of promotion, that, the Company argued, depended on the work and conduct of the individual concerned and the availability of vacancies. The following month Jahangir informed the Company that some of its Iranian employees had complained of discrimination against them in salaries, which the Company denied. Then, in Feb-

Table 3.7 *Foreign personnel as percentage of Company
employees in Iran (excluding unskilled labour):
General Plan versus actual figures, 1936–1943*

End of:	General Plan percentage	Actual percentage achieved
1936	16.63	14.84
1937	16.00	13.63
1938	15.50	14.73
1939	15.00	12.69
1940	14.50	13.41
1941	14.00	11.36
1942	13.75	13.77
1943	13.50	15.12

Source: BP 68184.

ruary 1938 Jahangir asked for minute details of the refining plant and equipment at Abadan and Kirmanshah.[75] Jameson stated that it would take months to make a comprehensive compilation which would be out of date before it was completed, a fair enough point given that, as has been seen, the Company was then making very large capital expenditures on new installations. When asked by Jameson if the Iranian Government wished to see the expansion of the Company retarded, Jahangir was emphatic that it must continue. Jameson informed him that the time required for the construction of two new refinery units would normally be about a year, but in Iran it could not be completed in less than eighteen months because of the dearth of technical and highly skilled personnel.

Though mistrusted by Jahangir, the Company was not neglectful of its undertakings under the General Plan. In 1938 Jameson was emphatic that 'Iranianisation is so important to the Company that everything possible should be done to make the policy fully effective. A senior member of staff should be made responsible for this work'.[76] In January 1939 M. A. C. MacNeil was duly appointed as manager for technical personnel with responsibility for 'ensuring that only such foreign personnel is employed as is necessary for the maintenance of the efficiency and economy of the Company's operations in Iran'. Moreover, as table 3.7 indicates, the percentage of the Company's personnel in Iran which consisted of foreigners was consistently less than that envisaged in the General Plan in all years except 1942 and

1943 when the Company's employment pattern was disrupted by the wartime conditions which are described in more detail in chapter 9.

Training, education and housing

The proposals for Company training of Iranians in the General Plan were not an entirely new initiative, for Iranians had been recruited for training for occupations such as fitters, turners, transport drivers, firemen and pumpmen in the early 1920s. Those early developments were formalised and made more systematic by the opening, in March 1925, of an apprentice training and test shop at Abadan. An electrical training section was opened, cookery classes held and apprentices trained in what was described as a 'centralised scheme for training, testing and grading of all artisans'.[77] In 1933, after the terms of the new concession had been agreed, Elkington instigated a new training scheme whose guiding principle, as he described it, was to produce a 'middle class' of skilled artisans.[78] The scheme was based on divisions to cater for the training of semi-skilled workers, artisans, clerical, technical and supervisory staff and included a mixture of instruction and work experience. At the instruction centre in Abadan men were taught 'the elementary facts and principles common to all skilled labour occupations, that is, simple technical terms in English, the use of a foot rule and dip rod, use of thermometer and hydrometer, mechanism of valves, burners and lubricating devices, the firing of boilers and stills, etc.'.[79] The provision for practical training in refining processes was furthered by the completion, in 1934–5, of process training plants which constituted, in effect, a miniature refinery for the instruction of unskilled labour. Trainees going into jobs at the refinery were thus able to gain prior experience of the distillation of crude oil, chemical treatments and safety measures.[80] For artisans the old three-year apprenticeship was extended to five years, the first two being spent in the training shop learning to handle tools and materials with instruction for two hours a day in English, arithmetic and general mechanical knowledge; then three years spent assisting qualified workmen and attending evening classes.[81] Technical trainees were selected from the best of the artisans undergoing apprenticeships. Training for other occupations such as clerks and telegraphists, requiring literacy and knowledge of English, was also provided. In 1935 the Company began to recruit female Iranian typists and clerks, which was regarded as a successful innovation.[82]

Reference has already been made to the training programme which

Table 3.8 *Iranians in Company training: General Plan
versus actual figures, 1936–1943*

	Artisans		Technical		Commercial	
	General Plan	Actual	General Plan	Actual	General Plan	Actual
1936	300	401	50	61	70	54
1937	300	457	50	71	70	67
1938	350	437	60	70	70	69
1939	350	458	60	83	80	71
1940	–	492	–	67	80	60
1941	–	506	–	66	80	62
1942	–	480	–	65	–	64
1943	–	458	–	58	–	103

Notes: 1. The number of artisans actually in training in 1936–9 includes welders.
2. The number of technical trainees actually in training in 1936–9 includes trainees in Britain.
Source: BP 15918.

was set out in the General Plan of 1936, which included forward estimates of the number of Iranians to be trained as artisans and for technical and commercial occupations.[83] The Company's figures, which were later prepared to compare the Plan estimates with the actual numbers who underwent training, are shown in table 3.8. As can be seen, in the artisan and technical categories the Company consistently exceeded the numbers put forward in the General Plan. The number of commercial trainees, on the other hand, fell short of the Plan estimates. However, in all categories the number of Iranians who completed their training and 'passed out' with qualifications was very much lower than the number who entered training owing to the very high wastage rates as trainees either dropped out or failed to pass their tests.[84]

In the Procès-Verbal which went with the General Plan the Company also undertook to establish a technical school at Abadan. The Technical Institute, as it was called, was duly opened in September 1939, being modelled on the type of technical college to be found in Britain. Under its principal, Dr Riza Fallah, who had himself been sponsored by the Company at Birmingham University, there were

7   Abadan electrical training shop, 1930s

8   Abadan apprentices' hostel – evening study, 1930s

9   Abadan apprentices' hostel – apprentices in their gardens, 1930s

10   Abadan apprentices' training plant, 1930s

seven full-time and seventeen part-time lecturers. The Institute provided training not only in technical subjects such as mechanical and electrical engineering, but also for commercial and clerical work. In addition, a higher level course leading to an intermediate certificate and then a BSc in petroleum technology was introduced. Following the 'sandwich' principle the course consisted of periods of formal study consolidated by practical experience and evening classes.[85] The practical content of the Institute's courses evidently pleased the Shah who visited in March 1940 and was pleased to see young Iranians engaged in workshop activities. According to C. L. Hawker, the superintendent of training, the Shah asked about classes and remarked that these should be the minimum to fit them for their jobs. 'I do not', the Shah reportedly told Hawker, 'want my people to sit down and read books'.[86]

Apart from the various levels and categories of training which the Company offered in Iran, it also sent selected trainees to study in Britain, though as table 3.9 indicates the numbers involved were few compared with the number of trainees in Iran. That was partly because of the much higher *per capita* costs of British training and partly because of the Company's preference for combining training with direct work experience, the basic objective being not to produce a small highly educated elite, but to prepare a larger number of Iranians for employment in practical tasks. The emphasis on practical application as opposed to theoretical abstraction was evident in Elkington's insistence that students sent to British universities should be given the maximum amount of practical work on a 'sandwich' basis. His concern, as he expressed it in 1933, was that 'it is particularly difficult for a Persian graduate on his return to his own country "to get down to it" and get dirty in a practical job and therefore he must have all possible experience of this kind while in the UK'.[87] His concerns were evidently shared by Jacks who told Davar in 1934 that he thought the majority of the men should be 'the practical type rather than the university type'.[88] Mylles, too, felt that students sent to British universities should not 'be spoiled or given false ideas of their own importance'.[89] The same outlook endured into the 1940s, when Pattinson attributed the comparative success of the Technical Institute's training, measured against that of Birmingham University, to the fact that students at the Institute were trained on the job at the same time as receiving academic tuition.[90]

It was not only in education and training that the Company became a large-scale provider of infrastructure facilities and services which in

Table 3.9 *Company-sponsored Iranian students in Britain,*
1928–1938

	University students sent to UK	Non-university students sent to UK	Total
1928	2	–	2
1929	2	–	2
1930	2	–	2
1931	2	–	2
1932	2	–	2
1933	2	–	2
1934	–	–	–
1935	11	31	42
1936	–	15	15
1937	8	6	14
1938	4	3	7

Sources: BP 15918; 68192.

the more developed western economies would not normally have been the responsibility of the firm. As the scale of operations in Iran expanded and the labour force grew, housing became the most difficult welfare problem facing the Company. In the early 1920s Abadan suffered a series of serious epidemics which caused nearly 850 deaths, some 3 per cent of the population. Such health hazards threatened the operation of the refinery. The demolition of much unhygienic accommodation was arranged with Shaykh Khazal and living conditions were improved by re-housing and compensation. However, by 1929 Elkington could claim no more than partial success in dealing with the housing problem because, with control of the town being vested with the local authorities, he found that personal enrichment took precedence over lasting improvements in living conditions and land speculation was rife.[91] He felt that the provision of decent housing was imperative and contemplated the construction of small model townships under local supervision.

In 1934 the preparation of a limited housing plan on the lines of a 'Garden City' in the Bawarda area was entrusted to the Company's architect, J. M. Wilson.[92] Aiming to provide accommodation for about 80 per cent of married supervisors and 25 per cent of married artisans, it reflected the Company's tendency to put the emphasis on quality rather than quantity in its housing provision.[93] Recreational

11   The Braim housing estate at Abadan, 1935

12   The Bawarda swimming pool at Abadan, 1933

13   The Fields hospital at Masjid i-Suleiman, 1932

14   Abadan hospital in the early 1930s

15   The Abadan fire station, built in 1927

16   Abadan new staff hospital, late 1930s

17   Patients in Abadan hospital, late 1930s

facilities were comprehensive and the Company helped the munici-
pality by providing electricity, medical services, running water and
drainage. However, over the next few years there was, as has been
seen, a great increase in the numbers employed on Company work in
Iran as construction work on new capital projects reached new heights.
By the end of 1938 the pressure on facilities in the town of Abadan had
become relentless with newcomers arriving from all over the Persian
Gulf area, sleeping in the open along the roads. Even temporary tented
accommodation failed to make a significant impact on the problem of
providing for the homeless.[94] Although nearly 4,500 staff quarters had
been completed in less than five years up to June 1939 there was, very
obviously, an acute shortage of housing on the eve of World War II.

Leisure activities

The large mixed community which made up the Company's workforce
in Iran covered a wide range of intellects, interests, class distinctions
and professional attainments. Social life and leisure activities therefore
took various forms in a country with its own cultural values, religion
and customs and amongst a people who were naturally polite, but
reserved with their own social distinctions. Above all, until the mid-

1930s Iranian women were excluded from public appearances in mixed company. The reforms of Riza Shah to some extent ended the isolation of women, though relations between Iranians and foreigners were generally discouraged by the Shah. There was no official Company discrimination, but some personal indifference and social conceit could not be completely excluded, however regrettable, for few societies are without some prejudice.

Thus, until the mid-1930s there was little mingling of the two communities outside the circle of Company acquaintances. Gradually, more Iranian staff were employed and the number of local officials increased. In the 1920s there were comparatively few expatriate wives in Khuzistan but a number of them became influential and distinguished figures playing an important role in welfare activities and stabilising the community by their active participation in clubs and societies. Peggy Pattinson, for example, was respected as a sincere and understanding person of discretion and kindness in whose confidence both British and Iranian wives trusted.

There was considerable enthusiasm for sport of various kinds. Local football, rugby, cricket and tennis teams played in the early 1920s and later in the decade the Company undertook to support sporting activities on a regular basis.[95] It provided money for pools, pitches, courts and grounds, so long as there was an active staff association to look after them. Competitive tournaments with winners' cups donated by leading managers were held. The range of activities gradually increased, and hockey, squash, badminton, boating, racing, polo, golf, mountaineering, athletics, gardening and less energetic recreations such as chess, billiards, snooker, photography, bridge and other indoor pursuits appeared. Already in 1927 the first inter-centre competition was held for rugby, followed by cricket and football.

Not only did individual clubs promote their own sports, they also sponsored their own dances and special evenings. The theatrical instinct was strong and Abadan set the example, followed elsewhere, of forming a drama group in which enthusiastic direction and much appreciated performances were given by John Cope of the Shipping Department and his wife, Mabel, the leading thespians of their stage. Entertainment included pantomime, cabaret, burlesque, shows, funfairs, gymkhanas, regattas, galas, children's parties, variety shows, concerts, recitals, *thé dansants*, jazz bands, fancy dress balls and films in the open. There were classes for ladies on health and beauty. Christmas and New Year was a great festive period with all kinds of parties for children and adults. There were dinners for the Greybeards, those who had served fifteen years or more in Iran, the golden oldies of their day, whose motto stopped the clock, 'Talented and charming people have no age' (Oscar Wilde). There were gatherings of the Caledonian clans for Burns' Night and commemorations of service associations on Armistice Day.

As Iranian staff increased they formed their own clubs such as the Bashgah-i Iran in Abadan and the Armenian Sporting Club. Iranians enjoyed playing tennis and riding, were good swimmers and boxers and in the early 1940s made up a rugby team. They enjoyed drama and their children were excellent dancers. The outdoor sporting events, especially the race meetings, were well attended by all sections of the workforce. Armenians were fine musicians and formed several bands much in demand at dances and concerts. There was a great deal of dressing up and formality on the grander occasions, but in the summer informality prevailed and picnics were popular. In Tehran entertainment was augmented by friendship with members of the British Legation, the Imperial Bank of Persia and the foreign residents as well as some of the local society. In Abadan social life was enlivened by the crews from visiting ships, exchanges with Basra and contacts with Iranian officials.

There was, of course, a pecking order of social distinction and professional class, inseparable from expatriate communities, which was generally accepted and does not seem to have been oppressive, though it may at times have been frustrating to the nonconformist. Those living outside Abadan were freer in that respect, partly because of the vastness of the countryside which contrasted with the feeling of claustrophobia around the refinery which dominated Abadan, though even there the river provided an outlet and the later provision of air-conditioning did much to dispel the oppressive heat and humidity. The climate of Fields was more invigorating apart from the wafting traces of gas in the air. There were opportunities for picnicking, shooting and visiting the antiquities. The Church of St Christopher was a centre of pastoral care in which it was joined by the Catholic ministry, other denominations, a Hindu temple and the local mosques. The consulates at Khurramshahr and Ahwaz provided assurances of British official attention, but there was little recourse to them. In general, behaviour was discreet and few *crimes passionnels* rattled the pages of the *Gnatter*, the local news sheet. The variety and interest of the leisure activities and the commitment and enthusiasm of those who organised them contributed to a tolerant society for Iranians and expatriates alike within the social conventions of the time.

= 4 =
Co-operation in the markets,
1928–1939

In the first two decades of the Company's existence, it was Greenway's principal achievement that he succeeded in transforming a vulnerable infant business, founded on the discovery of a single oilfield in Iran, into a vertically integrated oil company which was a significant force in the international oil industry. In performing that feat of entrepreneurship, Greenway faced the considerable difficulty of establishing the Company as a marketer which could treat, on reasonably equal terms, with the longer-established market leaders, Royal Dutch-Shell and Standard Oil (NJ). By pursuing the expansionist marketing policy described in volume 1 of this history Greenway was largely successful in achieving his aims. When he retired as chairman in March 1927, the Company, though still junior to Royal Dutch-Shell and Standard Oil (NJ) in marketing, could at least lay claim to a place at the high table of international oil companies.

The passing of the chairmanship from Greenway to Cadman coincided with a shift in marketing policy as the emphasis on independence, which had characterised Greenway's approach, gave way to a more conciliatory, co-operative relationship with the Company's main rivals. In part, the change was due to personality, Cadman being more inclined to visions of equilibrium between supply and demand and to co-operation with other oil companies than to the cut and thrust of competitive manoeuvres. However, the shift in policy was also associated with changing economic conditions and prevailing trends in industrial organisation. After widespread fears of oil shortages in the early 1920s, new discoveries opened up new sources of production with the capacity to produce far more oil than the markets could absorb. In seeking remedies for the resultant disequilibrium between supply and demand, Cadman, like other British businessmen facing

similar problems in different industries, saw co-operation and combination as viable, indeed preferable, alternatives to a competitive
scramble. In Britain in the late 1920s, such ideas, widely described by
the term 'rationalisation', came to be a vogue which was taken up,
sometimes with near-missionary zeal, by a variety of academics,
economists, management consultants and businessmen.[1] Their ideas
were far removed from those of perfect competition and rationalisation had gained wide currency as a buzz-word in business circles
before the onset of the great economic depression which swept the
world from 1929 through to the mid-1930s. The depression brought
not only severe hardship, but also the collapse of the mechanisms of a
liberal international economy as the abandonment of the gold standard and the erection of increased tariff barriers, coupled with rising
economic nationalism and autarchy, impeded the international movement of goods and capital. In industry, the control of competition by
cartels which regulated prices and/or output became widespread.[2]

The shift towards co-operation in marketing under Cadman first
brought practical results in two agreements which were concluded
early in 1928. The first was the Burmah-Shell Agreement for the Indian
market. In 1927 the Burmah Oil Company and Royal Dutch-Shell
decided to amalgamate their distribution organisations in India into a
joint company, the Burmah-Shell Oil Storage and Distributing
Company of India. Under arrangements which came into force in
1928, Burmah secured for its production in India and Burma the first
call on the new joint company's oil requirements. Rights to supply the
remainder of the Burmah-Shell company's requirements were divided
equally between the Company and Royal Dutch-Shell. In return for its
supply rights, the Company agreed to abstain from entering the Indian
market on its own account and to channel all its supplies to that
country through the new Burmah-Shell company.[3] Under the second
agreement, which was reached early in 1928, the Company and Royal
Dutch-Shell took equal shares in a new joint distribution company, the
Consolidated Petroleum Company, which covered eastern and southern Africa and the other territories shown in map 4.1.[4]

THE ACHNACARRY AGREEMENT AND 'AS IS'

Having settled the terms of the Burmah-Shell and Consolidated agreements, Cadman turned his attention to a more ambitious goal, nothing
less than a wide-ranging international agreement with Royal Dutch-
Shell and Standard Oil (NJ). In February 1928 he suggested to Walter

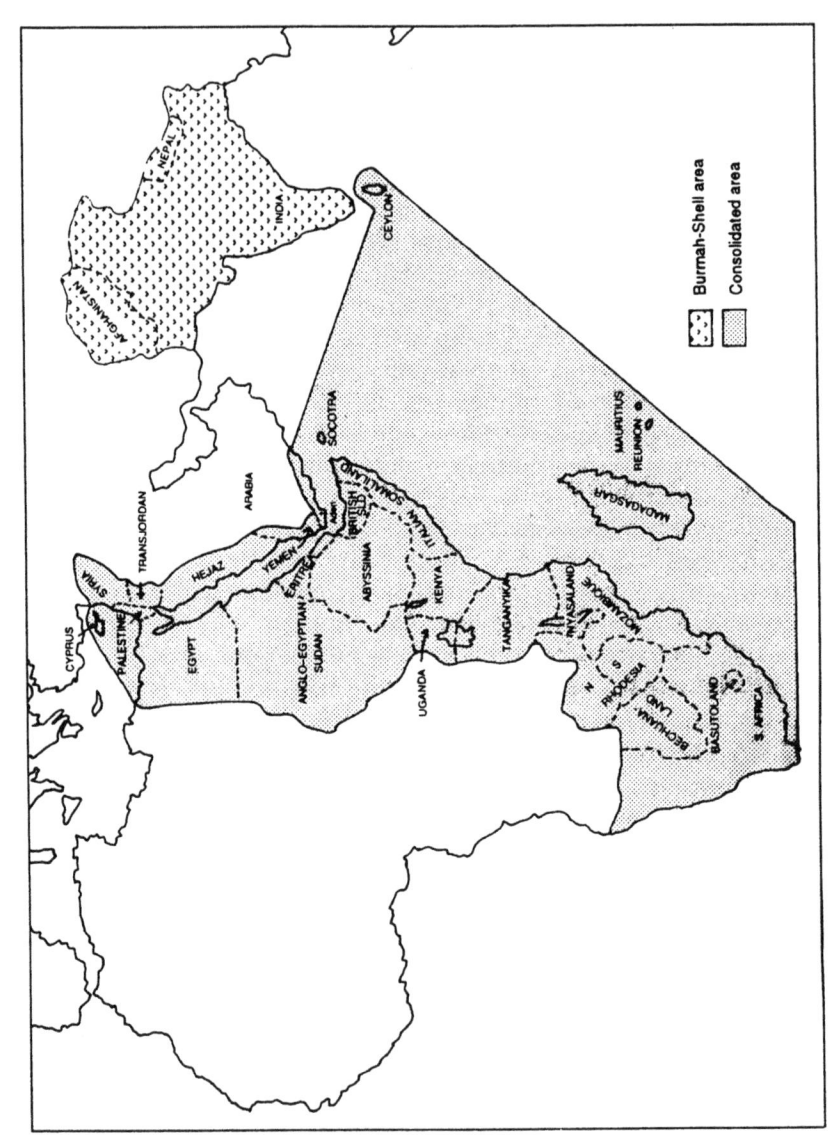

Map 4.1 The Burmah-Shell and Consolidated areas

Teagle, president of Standard Oil (NJ), the establishment of a 'small "clearing house" for matters of the very highest policy' involving the three companies.[5] On the basis of further soundings and exchanges, Fraser drew up a draft agreement, dated 12 August 1928, and showed it to Robert Watson, the managing director of the Burmah Oil Company who was also a director of the Company and friend of Sir Henri Deterding, managing director of Royal Dutch-Shell. Watson approved the draft, though he appears to have been piqued at not participating in the drafting, remarking to Deterding that he always refused 'to be placed in the position of a second editor'. He was reluctant to agree to any kind of definite organisation, 'a clearing house yes', and was convinced that the scope of the proposals should be worldwide.[6]

More meetings were held in London and proposals for agreement were set out in a document, dated 18 August, which is reproduced in Appendix 3. The fundamental objective, as presented in the document, was to bring surplus productive capacity into balance with demand by the control of competition. Using language which had wide currency at the time, it was argued that 'excessive competition' had resulted in 'tremendous overproduction' and that 'money has been poured into manufacturing and marketing facilities so prodigally that those now available are far in excess of those required to handle efficiently the world's consumption'. The efforts of individual companies to increase their sales at the expense of others was described as 'destructive rather than constructive competition' which, the document continued, had resulted in 'non-compensatory returns'. With a high-sounding tone, the preamble concluded:

> The petroleum industry has not of late years earned a return on its investment sufficient to enable it to continue to carry in the future the burden and responsibilities placed upon it in the public's interest, and it would seem impossible that it can do so unless present conditions are changed. Recognizing this, economies must be effected, waste must be eliminated, the expensive duplication of facilities curtailed.

With a view to achieving those goals, seven main principles were put forward, which, in summary, were as follows. First, each company was to accept its existing market share and not seek to increase it. Secondly, participants in the proposed agreement were to make their existing distribution facilities available to other producers on favourable terms, i.e. for a payment below the costs which a producer would incur by creating new facilities for his exclusive use. The purpose of that second principle was bolstered by the third, which stipulated that new facilities were to be created only insofar as they were necessary to meet increases in consumption. By those means, duplication and additions of new capacity were to be held in check. Under the fourth, fifth and sixth principles, each market was to be supplied from the nearest producing area with the object of securing the maximum economies in transportation and, less explicitly, preventing different production areas competing for the same business. Finally, the seventh principle suggested that in the 'best interests of the public as well as the petroleum industry' measures which would increase costs, with consequent reductions in demand, were to be discouraged. The principles set out in the draft agreement were to be applied in all countries except for the US internal market, where they would have been in contravention of anti-trust laws. If they were followed, the result would, it was claimed, be a 'stabilization' of the world market outside the US, which would 'be in the interest of all'.

As for the mechanism by which the principles were to be implemented, it was suggested that an association should be formed to administer the regulations, which would cover all exports (except those to the USA) of crude oil and products apart from lubricating oils, paraffin waxes and speciality products, which were to be the subject of further consideration. The association would allocate to each of its members a quota for each product in each market and would direct oil shipments on the geographically most favourable basis. As regards prices, they would be based on those at the US Gulf of Mexico, generally known as US Gulf prices.

The draft agreement having been drawn up, Cadman, Teagle and Deterding gathered at Achnacarry Castle in Scotland for purposes which outwardly had as much to do with pleasure as business. That, at least, was the impression given by Teagle, who was quoted by a trade journal as saying: 'Sir John Cadman ... and myself were guests of Sir Henri Deterding and Lady Deterding at Achnacarry for the grouse shooting, and while the game was a primary object of the visit, the problem of the world's petroleum industry naturally came in for a great deal of discussion.'[7] The result of their discussions was the Achnacarry Agreement, sometimes known as the 'As Is' Agreement, dated 17 September 1928. So far as can be ascertained, the wording was the very same as the document of 18 August which has been described above.[8]

As the USA was the world's largest oil producer and exporter, it was vital, if the goals of the 'As Is' participants were to be attained, that US exports should be brought under control. Shortly after the Achnacarry Agreement was reached, steps to that end were taken through the formation of two US export associations. The first to come into existence was the Standard Oil Export Corporation which was formed in November 1928 to establish centralised co-ordination and control over the exports of Standard Oil (NJ)'s principal producing and exporting subsidiaries.[9] The second, formed on Teagle's initiative, was the more widely based Export Petroleum Association which was formally incorporated in January 1929.[10] It initially included fifteen US oil companies, to which were added two further companies in 1929–30, when the members of the Association accounted for nearly 45 per cent, by value, of US oil exports.[11] The participants agreed to abide by the decisions of the Association's various committees on prices, quotas and other matters concerning exports and by April 1929 decisions were being taken on export sales of benzine, kerosene and crude oil and export prices for some grades of products.[12]

While those steps were taken to control US exports, the problem of regulating US crude production was studied by a committee of the American Petroleum Institute (API) with the outcome that in March 1929 the committee recommended that production should be limited to the level reached in 1928. The recommendations, though approved by the API, failed to gain the necessary support from the relevant federal authorities in the light of advice from the Attorney-General that they would be in contravention of US anti-trust laws.[13] Early in April, Teagle still hoped that 'the matter can be straightened out soon',[14] but a few days later the federal rejection of the API proposals was confirmed.[15] In the absence of controls, the problem of US surplus productive capacity was greatly exacerbated by the discovery of the huge East Texas field in 1930. Production from the new field rose rapidly and flooded the market with crude oil at the very same time as the great economic depression dampened demand. As was to be expected, US crude prices plummeted, so much so that the average price of US crude, which had already fallen from $1.88 per barrel in 1926 to $1.27 in 1929, dropped to 65 cents in 1931 and was still only 67 cents in 1933.[16]

The failure to bring the underlying problem of surplus productive capacity under control inevitably made it more difficult for members of the Export Petroleum Association to reach agreement on export prices and quotas, especially as decisions by the association required the unanimous support of its members.[17] The competing demands of divergent interests resulted in internal dissension, expressed in acrimonious correspondence between Deterding and some of the other members in 1929–30.[18] Faced with the intractable problem of achieving unity, the Association limped along with declining importance and attention turned away from the worldwide application of 'As Is' towards a more piecemeal approach. Thus, in the spring of 1930 Jackson told Cadman that 'new proposals under which each market would be settled separately offered a reasonably promising basis'[19] and in October the US representatives agreed that little more could be done than 'rationalise in such countries as might be agreed'.[20] The following month, the Export Petroleum Association's export price schedules were cancelled and it became inactive, remaining dormant until its formal dissolution in 1936.[21]

As the prospects of achieving a worldwide application of 'As Is' receded, the Company, Royal Dutch-Shell and Standard Oil (NJ) turned their attention to formulating guidelines for the establishment of local co-operation in individual markets. Their ideas were set out in the Memorandum for European Markets of January 1930 and the Heads of Agreement for Distribution of December 1932. In both documents, the basic principles of the Achnacarry Agreement were preserved, but made more specific and precise in relation to the cardinal issues of quotas and prices as well as related matters.[22] Yet despite the repeated exposition and elaboration of 'As Is' principles, the power of co-operation to stabilise the industry continued to be undermined by the contrary influence of competitive forces emanating not only from the USA, but also from the Soviet Union and Romania.

Before World War I, Russia was a major oil producer, whose production accounted for more than half the world's output for a brief period at the turn of the century. During World War I and the subsequent conflict between the Red and White armies production suffered a severe decline and the condition of the industry deteriorated until an intensive effort of rehabilitation was undertaken in the mid-1920s, linked to a national economic programme in which oil exports were to provide essential foreign exchange. Already in 1924 an importing and distributing organisation, Russian Oil Products Ltd (ROP), was competing in the British market with the local subsidiaries of the three market leaders, namely Standard Oil (NJ), Royal Dutch-Shell and the Company. By aggressive price-cutting, ROP increased the value of its sales, mostly benzine, from £500,000 in 1924 to nearly £4 million in 1929, posing an unwelcome threat to the market dominance of the big three.[23]

By January 1927 discussions had begun between the three principal British marketing companies and ROP with a view to regulating the sales of Soviet oil products in the British market.[24] In many respects their negotiations presaged the 'As Is' principles and procedures. 'The

Combine', as the three major companies were termed in Britain, recognised that it would be impossible to exclude Soviet products. ROP, having made its initial penetration of the market, realised that a satisfactory accommodation with its rivals was in its interests. A settlement was reached in January 1929 when a three-year agreement settled quotas and prices for Soviet supplies of benzine and kerosene. However, the agreement in the British market did not stem the flow of Soviet oil exports and in May 1931 Teagle told Basil Jackson, the Company's representative in New York, that Soviet competition was 'the biggest question facing the industry today ... sooner or later it would be essential to treat' with them.[25] A year later a conference between representatives of the leading international oil companies and Soviet representatives was held in New York in an endeavour to reach agreement.[26] By that time, according to Jackson, the 'free' market for oil in Europe 'had during the past 12 months been taken away entirely from the American companies by the Russians and Rumanians'.[27] That trade had virtually fixed world prices, since 'the free business rate was the price at which any world bootlegger could purchase'. Although the New York negotiations became deadlocked and no agreement was reached,[28] the outcome was less damaging to 'As Is' than might have been expected, for, as it happened, Soviet oil exports reached a peak in 1932 and fell for the remainder of the 1930s when rising internal consumption made increased demands on Soviet production.[29]

At the same time as the negotiations over Soviet exports were taking place, efforts were made to reach agreement on the control of production and exports from Romania, then Europe's largest oil-producing country.[30] In December 1929 it was announced that practically the whole of the Romanian oil industry, which consisted of both international majors and independent producers, had reached agreement to limit production and fix prices in relation to US Gulf export prices.[31] However, the agreement collapsed in less than a year, largely because of the failure of the Export Petroleum Association to control US export prices and the Romanians' fear of competition from US independents.[32] Negotiations were subsequently renewed and in July 1932 an international conference was held in Paris, with the outcome that the Romanians entered into a new agreement on quotas and prices.[33] However, Romanian adherence was conditional on US exports being kept within bounds and prices being maintained, hardly attainable prospects at the very time that the dramatically increased output of the East Texas field was flooding the market and depressing prices. In late 1932 and early 1933 no solution could be found to the problems of reconciling the US and Romanian positions. The refusal of James Moffett, a director of Standard Oil (NJ), to raise his company's US crude prices prompted Fraser to remark, in November, that 'New York does not want to understand the situation . . . Jim [Moffett] seems to have left reason and resorted to threat'.[34] Within a week of the Romanian agreement coming into force at the beginning of 1933 Fraser expressed his continuing concern that its success hinged on 'some forward movement in America'.[35] Only a few days later, the Romanians indicated that unless crude oil prices were increased they would not be able to hold production in check, prompting J. B. Kessler of Royal Dutch-Shell to warn Moffett that if the agreement broke down 'considerable and lasting damage would be done to whole policy of co-operation'.[36] Subsequently, the situation became even more aggravated, when Standard Oil of Indiana lowered crude oil prices further, creating a situation which, in Jackson's words, was 'charged with dynamite'.[37] While attempts to defuse the situation were made at conferences in London and Paris in February–April, Heath Eves and Lloyd reported that all over Europe 'cheap Romanian oil is on offer and is exerting a most depressing influence'.[38] A compromise was reached, but only on the understanding that the Romanian producers were free to terminate the agreement unless US prices were increased and US production kept below two million barrels per day. The continued deluge of oil from East Texas prevented either of those conditions being met and the Romanian producers duly abandoned the agreement for restriction of output in mid-1933. In that year, Romanian oil producers exported 20 per cent more oil than in 1932.[39] The failure to control US production and prices thus undermined the Romanian agreement and remained the Achilles' heel of 'As Is'.

The inauguration of Franklin D. Roosevelt as US President early in 1933 offered the prospect of some relief for the demoralised oil industry.[40] In June, approval was given to the National Recovery Act, intended to create an orderly economic system, and in August a code of fair competition for the US oil industry was approved under the Act. US crude prices rose sharply, the East Texas field price for Grade A crude advancing from 50 cents per barrel in July to $1 per barrel in October.[41] Against the background of continued economic recovery from the economic depression, the average US crude price rose from 67 cents per barrel in 1933 to $1 in 1934 and, after a small fall in 1935, remained above $1 for the remainder of the 1930s.[42]

Meanwhile, in April and May 1934 representatives of major oil companies met in London to revise the earlier Memorandum for European Markets[43] and by the end of June a detailed Draft Memorandum of Principles had been prepared and approved, covering all countries except the USA and others where it was unlawful.[44] The Draft Memorandum was concerned with much the same basic issues as earlier statements of 'As Is' principles, namely the allocation of quotas, agreement on prices and the terms of admission of outsiders. A reduction in 'unnecessary expenditure', including that on advertising, was agreed and a committee was formed 'to cement and coordinate the relations between the participants'.

The Draft Memorandum of Principles was the last comprehensive statement of 'As Is' principles and meetings in the autumn of 1935 were the final occasions on which the original partners, Cadman, Deterding and Teagle, discussed their common problems together.[45] Then and later there were inter-company discussions on matters such as foreign exchange restrictions; the registration and scrapping of tankers; proposals for the acquisition of Soviet distributing organisations; the location of refining capacity; product qualities and exchanges; and quotas.[46] An agreement between Royal Dutch-Shell and Standard-Vacuum was reached in December 1936, the last definite achievement of 'As Is' co-operation.[47] There were more discussions in July 1937 on Caltex (a joint company of Standard Oil of California and the Texas Company), production from Bahrain and marketing in India as well as arrangements between Standard-Vacuum and the Company on exploration in Papua. It was, however, increasingly obvious that formal adherence to 'As Is' principles was weakening. In 1938 Standard Oil (NJ) withdrew from 'As Is'-inspired arrangements and following the outbreak of World War II the administration of petroleum affairs by governments prevented a return to 'As Is' procedures. By 1943 the Company's forward planning for peacetime marketing discounted its revival and by the end of the war it was past resuscitation. It had been marked more by discussions than decisions.

An assessment of the impact of 'As Is' on the Company's fortunes runs into the problem that the data and evidence for a systematic market-by-market analysis is absent and there were, to be sure, local variations which preclude generalisations from isolated cases. In Australia, for instance, the Company could not openly invoke 'As Is' principles because its agreement with the Commonwealth government bound it not to act in concert with other oil companies.[48] In Italy, on the other hand, 'As Is' appears to have influenced the course of marketing development. In the early 1920s the Company had acquired from the Italian government a concession for an ocean installation at Trieste. The Company formed a local subsidiary, took an office in Milan and began to set up a small distributing organisation. New inland depots were added at Bologna and Verona, but the Company's Italian trade remained small and was limited to the north eastern part of the country. So matters continued until 1927, when the Company entered into negotiations to acquire a 30 per cent interest, coupled with a 100 per cent supply contract, in the marketing subsidiary of the state oil company, AGIP. The negotiations had reached an advanced stage by the time of the Achnacarry meeting, at which it was decided that the Company ought not to enter into an agreement with AGIP. An understanding was reached under which the Company undertook to break off the Italian negotiations and Royal Dutch-Shell and Standard Oil (NJ) undertook to purchase 20 per cent of their main product requirements in the Italian market from the Company. The arrangements were not finalised until April 1930, but took effect from the beginning of 1929. It was a part of the terms that the Company was to give up all its direct marketing interests in Italy and sell all stocks and fixed assets to the other parties, with the exception of the Trieste installation and a small number of benzine pumps in Trieste which it was necessary to keep because of the terms of the original concession. It was also stipulated that Royal Dutch-Shell and Standard Oil (NJ) would provide the Company with facilities for delivering bunkers at their Italian ocean installations.[49] It was, therefore, in application of 'As Is' that the Company effectively withdrew from a direct presence in the Italian market and limited itself to making supplies which were to be sold through the local distributing organisations of Royal Dutch-Shell and Standard Oil (NJ).

Turning from isolated examples to more general indicators, it would appear that 'As Is' had a limited overall effect. If it had been successful in bringing supply into equilibrium with demand it would be expected that a stabilisation of prices would have been the result. Local variations in product prices cannot be plotted in the absence of full data, but if US average crude prices are taken as a benchmark it can be seen, both from the preceding text and from figure 4.1, that they were far from stable, going through a deep trough between 1930 and 1934. Given that the Export Petroleum Association proved to be an ineffective instrument for the control of US exports, it would be reasonable to suppose that low US crude prices would have brought down international product prices and so affected the profitability of the Company. The upper line of figure 4.1, which plots the Com-

Figure 4.1   The Company's return on capital employed and US average crude oil prices, 1928–1938
Note and sources: See Appendix 2.

pany's return on capital employed, suggests convincingly that this is what occurred. The figure suggests, not that the Company was making high returns as a member of an effective cartel, but that it was unable to find shelter from the precipitate fall in prices during the great depression.

JOINT MARKETING INITIATIVES

At the same time as efforts at co-operation were made through 'As Is', there were individual attempts between companies to collaborate more effectively, as had earlier been the case with the Burmah-Shell and Consolidated agreements. At the annual meeting of the API in Chicago in 1928 Cadman was approached by W. Rogers of the Texas Company on the subject of co-operation, particularly with regard to lubricating oils.[50] In February 1929 the subject was raised again by Torkild Rieber, European sales manager of the Texas Company, who met Heath Eves to discuss the possibilities of reaching some kind of co-operative association in European markets.[51] Cadman was interested in making arrangements of reciprocal advantage[52] and in London in July a draft agreement was sketched. The tentative proposals never reached fruition, though the Texas Company made a renewed approach in November 1930, prepared to share in any market in the world in which the Company was not already associated with other companies.[53] Those schemes apart, there was some limited co-operation with the Texas Company in Australia resulting from an agreement in early 1932.[54]

While the reaction to the Texas Company's overtures seems to have been somewhat half-hearted, the Company's negotiations with Standard Oil (NJ) were more serious. The idea of some co-operation had already been raised in 1923 and came up again in the first half of April 1930.[55] Cadman agreed to a meeting in New York and broached the subject with Philip Snowden, Chancellor of the Exchequer, who raised no objections, agreeing that the matter was a commercial one which did not concern the Government.[56] In conditions of 'considerable secrecy' Cadman, accompanied by Fraser, Watson and Duncan Anderson (a Company accountant), duly sailed for New York on 27 June, working out a scheme 'which would embrace all European countries and fix up contract for supply of crude'.[57] In recording the subsequent negotiations, Cadman noted that on the first day 'we had outlined the lines of a document for joining APOC [the Company] and Standard of New Jersey'. The whole of the next day was spent on a proposed fifteen-year contract for the Company to supply Standard Oil (NJ) with crude oil, after which 'we finished and signed the papers', leaving the supply contract for further consideration.[58] Cadman, having enjoyed his short visit with 'abounding pleasure', told Teagle he looked forward to a great future, 'for dealing with you and Jimmy [Moffett] is in itself a joy'.[59] On his return voyage, he also had discussions with Watson about Burmah and the Company joining up,[60] noting that 'it seems very difficult', but Watson appeared 'very helpful, offered to lend a hand anywhere ... There seems greater opening for general economy all round'.[61] He optimistically wrote to Jackson that the discussions in the USA would 'in the long run, contribute materially to an enhanced earning capacity of this Company', though he was also concerned that if there was 'a break away in the understanding between the American, Royal Dutch and ourselves ... it would be very detrimental to our interests'.[62] Cadman was to be disappointed. When discussions with Teagle were resumed in London in August, Cadman objected to the restrictive nature of the supply contract and proposals for a monetary payment to Standard Oil (NJ). On 27 August he recorded in his diary that 'I blew up this morning and told Walter [Teagle] I am not prepared to have a rope around us nor was I prepared to pay sum of money for the pleasure of combining with SO'.[63] A little more progress was made at the beginning of September when the prospects of reaching agreement looked more hopeful.[64] Indeed, on 8 September Cadman recorded that 'we almost came to an agreement' and the following day a provisional draft was tentatively agreed,[65] but in spite of further talks nothing definite emerged from the negotiations.

More tangible results came out of the discussions between the Company and Royal Dutch-Shell on joint marketing in the UK, where competitive pressures were mounting in 1931, not only from ROP, but also the Texas Company and the Cleveland Petroleum Products Company who were both seeking to increase their market shares.[66] Faced with the threat of competition from those sources, the Company and Royal Dutch-Shell agreed to combine their UK marketing activities by merging them into Shell-Mex and BP Ltd, which came into being in 1932. Ownership of the new joint marketing company was divided between the two parent organisations in proportion to their market shares, giving the Company a 40 per cent shareholding compared with Royal Dutch-Shell's 60 per cent.[67] During the negotiations on the merger, Standard Oil (NJ) was invited to participate and given an outline of the proposals, but it declined to join in the new company for fear of losing its freedom of action as it would have had only a minority shareholding.[68] In addition, there were, according to Jackson, powerful interests in Standard Oil (NJ) who opposed any association with Royal Dutch-Shell.[69]

At the same time as the Company was engaged in these various co-operative ventures, other oil companies pursued separate schemes of co-operation resulting, most notably, in a merger between Standard Oil of New York (Socony) and the Vacuum Oil Company in August 1931. The next most important agreement was concluded in April 1932 between Standard Oil of Indiana and Standard Oil (NJ), by which the latter acquired large production rights in Venezuela and important foreign distribution outlets through the Pan American Petroleum and Transport Company, including the Cleveland Petroleum Products Company operating in the UK.[70] Thereafter, the incentive for Standard Oil (NJ) to enter into co-operative arrangements with the Company was reduced.

SALES AND MARKETS

The effects of the worldwide economic depression in the early 1930s were felt not only on prices, but also on the volume of the Company's business. As figure 4.2 indicates, the expansion of the Company's sales of its main grades of products during the 1920s came to an end in 1930

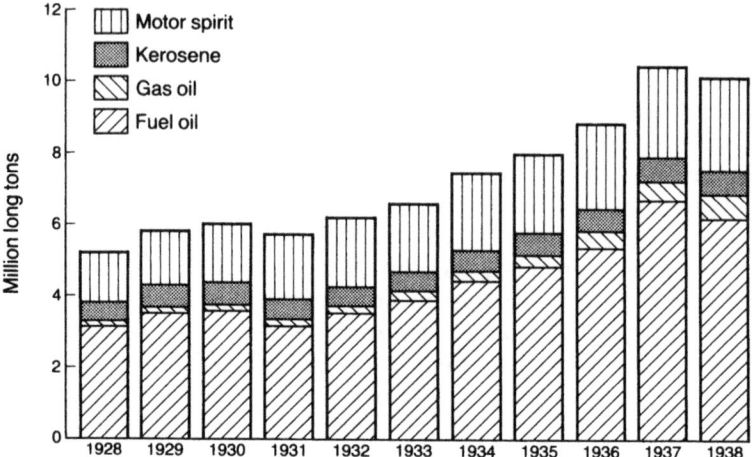

Figure 4.2   The Company's main product sales, 1928–1938
Notes and source: See Appendix 2.

and suffered a reversal in 1931 before recovering strongly up to 1937 and falling back in 1938. The fall in sales from some 6 million tons in 1930 to 5.7 million tons in 1931 was not precipitous. Nor was it long lived. Indeed, by 1932 the Company's sales tonnage already surpassed that achieved in 1930. Nevertheless, the fall in demand at the bottom of the depression must have contributed to the reduction in the Company's profits which has been seen in chapter 1.

Within this aggregate picture, the bulk of the Company's business continued to be in fuel oils, which accounted for nearly 60 per cent of the Company's sales tonnage (including its own tanker bunkers) between 1930 and 1938. The main outlets for this preponderant grade of product were as marine bunkers, both for merchant shipping and for the naval vessels of the British Admiralty, which remained a very large customer of the Company. Between 1930 and 1937 marine bunkers, including sales to the Admiralty, accounted for about two-thirds of the Company's fuel oil business and some 40 per cent of total sales (excluding minor products), shown in table 4.1. Large volumes of fuel oil were also sold for inland consumption in various markets and to other international oil companies, notably Royal Dutch-Shell and Burmah, of which the former was a very large customer indeed. Its purchases of fuel oil were, however, greatly exceeded by those of motor spirit, made under the terms of the Benzine Agreement which was first agreed in 1922, revised in 1927 and modified to accommodate

Table 4.1 *Markets for the Company's main grades of products, 1930–1937 (million long tons)*

	(a) European markets	(b) Extra-European markets	(c) Marine bunkers	(d) Admiralty	(e) Other international oil companies	(f) Sundry	(g) Total
1930	2.16	0.88	1.72	0.62	0.44	0.18	6.00
1931	2.20	0.78	1.64	0.47	0.39	0.23	5.71
1932	2.36	0.76	1.88	0.56	0.41	0.21	6.18
1933	2.46	0.82	2.02	0.60	0.47	0.19	6.56
1934	2.78	0.90	2.24	0.71	0.58	0.23	7.44
1935	2.91	1.09	2.43	0.71	0.66	0.19	7.99
1936	3.15	1.14	2.49	0.98	0.83	0.23	8.82
1937	3.19	1.33	3.03	1.30	1.25	0.31	10.41

Notes: 1. Column (a) includes Company tanker bunkers.
2. Column (f) includes refinery works fuel.

Source: BP 109194.

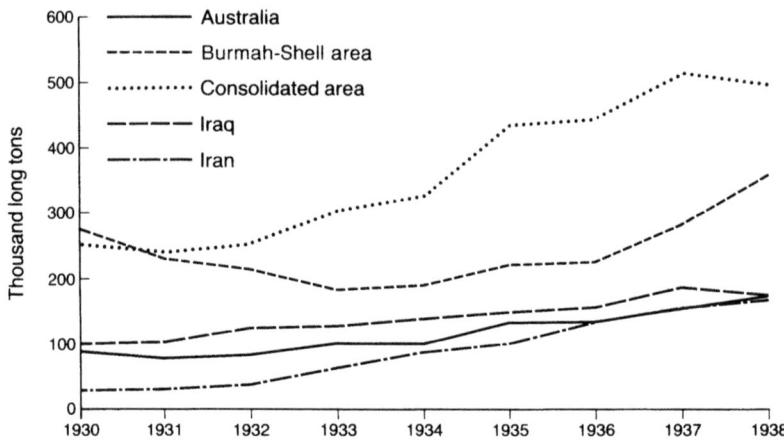

Figure 4.3   The Company's sales in its main extra-European markets, 1930–1938
Note and sources: See Appendix 2.

18   BP service station in Iran, mid-1930s

'As Is' arrangements in 1931. Such was the scale of Royal Dutch-Shell's purchases that between 1930 and 1938 they accounted for 15 to 20 per cent of the Company's total motor spirit sales, the exact proportion varying from year to year.[71]

Geographically, the Company's main extra-European markets were in the Burmah-Shell and Consolidated areas and in Australia, Iraq and Iran, whose respective inland consumption of Company products is shown in figure 4.3. However, the core of the Company's inland marketing activities was in Europe, where the Company had built up a substantial marketing presence during Greenway's time as chairman. Indeed, such was the emphasis on Europe that between 1930 and 1937 European markets absorbed about 71 per cent of the Company's sales of motor spirit, 64 per cent of kerosene and 88 per cent of gas oil. The proportion of fuel oil deliveries which were consumed in the inland markets of Europe was very much less, namely some 12 per cent, the bulk of the Company's fuel oil business being, as has been seen, in marine bunkers. At a lesser level of generality, Europe was not at all a homogeneous market, but rather a mosaic of local markets whose demands for different oil products depended on a variety of factors such as the availability of alternative energy sources, especially coal; climate; national economic policies; standards of living; and, more generally, the extent and nature of economic development in the various countries concerned. Consequently, there were wide variations in the pattern of oil consumption in those European markets in which the Company was active, including the Scandinavian countries of Sweden, Denmark and Norway; the Benelux countries of Belgium and Holland; and others such as Austria, Switzerland, Germany, France and Britain. Of those markets, Britain was easily the largest for the Company, followed by France and Germany in the proportions shown in table 4.2.

In Britain the decline of the old-established export industries, such as iron and steel, coal and cotton, and the accompanying high unemployment cast a shadow of depression over wide swathes of the traditional manufacturing base during the 1930s. Yet the growth of new industries and consumer spending in other sectors of the economy provided expanding opportunities for sales of petroleum products, none more so than sales of motor spirit to fuel the growing use of road transport. Motoring not only caught the imagination of a widening public, but also became more affordable as the new industries created new sources of prosperity and car manufacturers introduced improved production methods and exploited economies of scale to reduce the prices of new cars. At the same time, the price of premier grade motor spirit, excluding tax, fell from 1s 1d per gallon in 1928 to 10d per gallon a decade later, helping to hold down the rise in the retail price caused by increases in the excise tax from 4d to 9d per gallon in the same period.

Table 4.2 *Geographical distribution of the Company's main product sales, 1930–1937*

	% of motor spirit sales	% of kerosene sales	% of gas oil sales	% of fuel oil sales	% of total sales
UK	40.6	41.7	24.2	5.3	18.8
Continental Europe					
Austria	–	0.3	0.9	0.1	0.1
Belgium	2.1	1.0	4.0	0.1	0.9
Denmark	1.9	4.1	1.7	0.2	1.0
France	10.5	2.7	19.4	3.8	6.2
Germany	8.6	5.2	20.6	–	3.7
Holland	–	1.3	2.3	–	0.2
Italy	3.1	3.6	0.6	2.4	2.6
Norway	1.2	1.3	7.8	0.2	0.9
Sweden	1.3	0.7	0.7	0.1	0.5
Switzerland	1.4	0.7	4.7	–	0.7
Other	0.4	1.2	1.2	–	0.3
Sub-total	30.5	22.1	63.9	6.9	17.1
Extra-European markets					
Australia	2.6	1.6	0.2	1.0	1.5
Burmah-Shell area	0.2	4.4	–	4.6	3.1
Consolidated area	5.2	18.6	6.0	2.5	4.7
Iran	1.7	4.4	1.0	0.3	1.1
Iraq	0.8	3.4	0.2	2.2	1.9
Other	1.2	1.3	1.4	0.4	0.8
Sub-total	11.7	33.7	8.8	11.0	13.1
International oil companies					
Royal Dutch-Shell	17.2	2.2	0.4	4.2	7.5
Burmah Oil Co.	–	0.3	2.3	1.4	0.9
Sub-total	17.2	2.5	2.7	5.6	8.4
Marine bunkers	–	–	–	49.3	29.5
Admiralty	–	–	–	16.9	10.1
Sundry	–	–	0.4	5.0	3.0
Total	100.0	100.0	100.0	100.0	100.0

Note: Arabia and Baluchistan included with Iran.
Source: BP 109194.

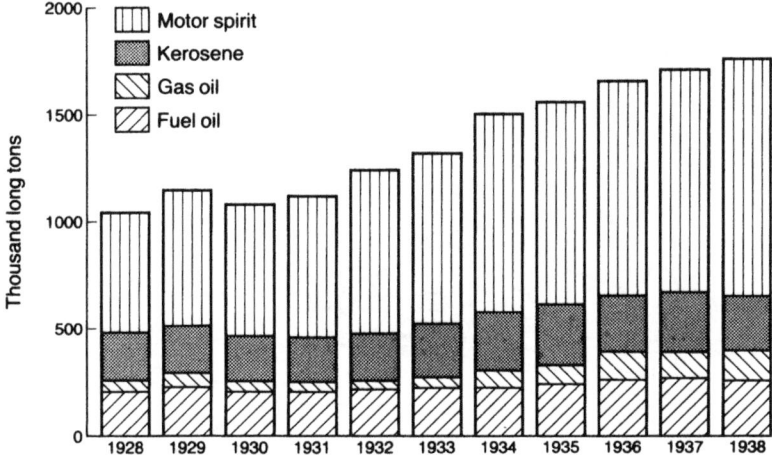

Figure 4.4   The Company's main product sales in the UK, 1928–1938
Note and source: See Appendix 2.

A growing number of people not only experienced the excitement of speed and competition at racing tracks, but also enjoyed the mobility bestowed by car ownership as the number of licensed private cars in the UK more than doubled from about 900,500 in 1928 to some 1,985,000 ten years later. There was also rapid growth in commercial road transport, ranging from local delivery services by laundries, department stores and others to longer-distance haulage which challenged the railways for the transport of goods. Thus, despite the retarding effects of the depression, there was a large increase in commercial traffic and the number of oil-powered licensed goods vehicles in the UK increased from about 300,000 in 1928 to nearly 500,000 in 1938. Within this expanding market the Company's sales of motor spirit constituted a large and growing proportion of its UK sales, as can be seen from figure 4.4.

The development of the motor spirit market was a matter not only of volume, but also of quality. Improved motor spirits were required as advances were made in engine designs to provide greater efficiency and power by, for example, increased compression ratios. The introduction of new qualities was accompanied by imaginative brand-name advertising, with the Shell and BP brands continuing to be separately promoted after Royal Dutch-Shell and the Company merged their UK distributing organisations into Shell-Mex and BP. When tetraethyl lead was introduced as an additive to boost the performance of BP

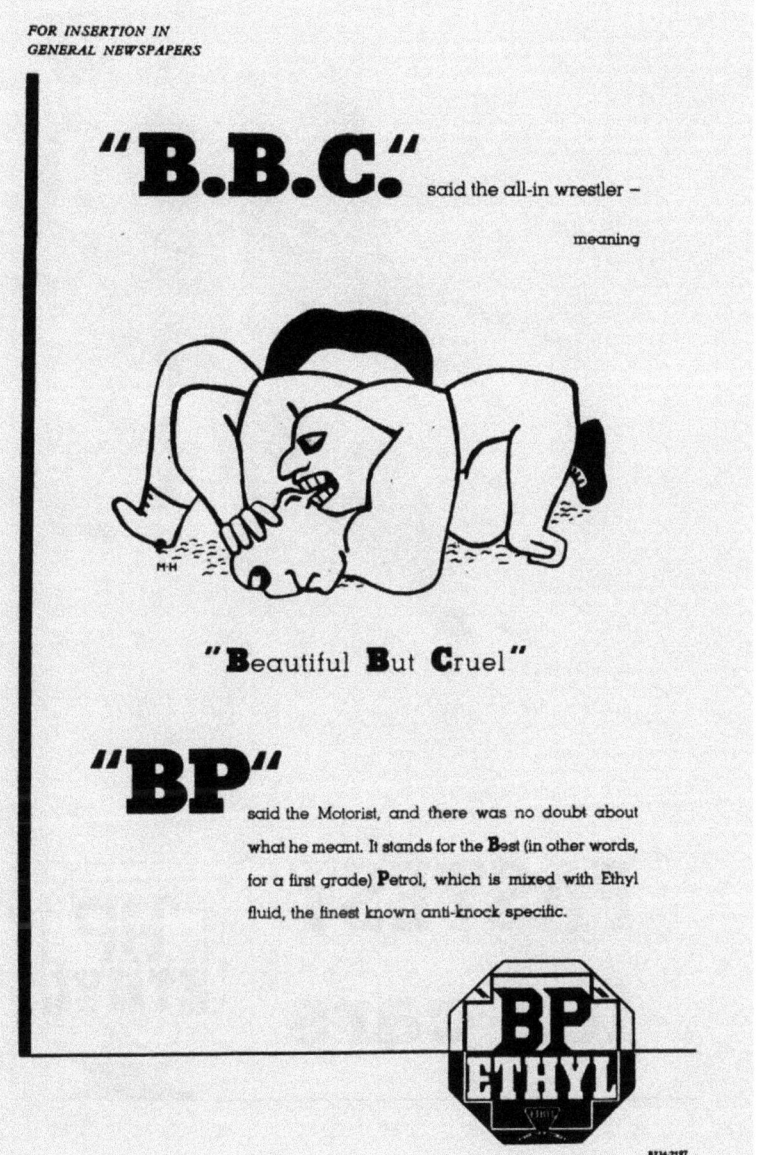

20   Advertisement for 'BP Ethyl', 1934

"*She'll never say 'No' . . . . . on 'BP' Ethyl!*"

FASTEST
for cars

21   Advertisement for 'BP Ethyl', 1938

22   Shell and BP motor spirit brands on sale in Britain, mid-1930s

petrol in 1931, the new grade was given the name BP Plus and advertised with the suggestion that it was 'Plus a little something others haven't got'. In 1934 a new name, BP Ethyl, was introduced and advertised using images by artists such as Edward McKnight Kauffer, Edward Bawden and Frank Dobson coupled with captions, witty in their day, examples of which are shown in the illustrations.

Meanwhile, roadside filling stations and motor spirit pumps became increasingly commonplace, the number of sites and pumps roughly doubling between 1928 and 1938. The allocation of dealer pumps between oil companies reflected the dominance of the market by the local subsidiaries of Royal Dutch-Shell and the Company, operating jointly as Shell-Mex and BP from 1932, and Standard Oil (NJ), which marketed in the UK through the Anglo-American Oil Company and, from 1932, the Cleveland Petroleum Products Company. Between 1934 and 1939 some 45 per cent of the UK's dealer pumps were allocated to Shell-Mex and BP and some 32 per cent to the subsidiaries of Standard Oil (NJ), giving the three parent companies a combined share of more than three-quarters of the available dealer pumps. Their aggregate share of motor spirit sales was only slightly less, hovering between 70 and 76 per cent throughout the 1930s. There was, however, a

Table 4.3 Market shares in motor spirit in the UK, 1929–1938 (percentages)

	BP	Shell-Mex	Shell-Mex and BP	Power	Dominion	Shell-Mex and BP Group	Standard (NJ) Group	National Benzole	Trinidad Leaseholds	ROP	Texas	Others
1929	17.3	29.3	46.6	5.5	1.4	–	27.2	6.0	0.2	6.6	0.8	5.7
1930	14.2	28.1	42.3	4.4	1.2	–	28.8	6.4	0.5	6.3	1.1	9.0
1931	14.8	27.4	42.2	4.5	1.6	–	28.8	5.9	0.6	8.1	2.5	5.8
1932	14.3	27.7	42.0	5.6	1.8	–	30.8	6.5	1.6	6.4	2.2	3.1
1933	12.5	26.5	39.0	5.2	1.7	–	28.9	6.3	3.2	5.3	2.6	7.8
1934	12.7	26.8	39.5	4.9	2.0	46.4	29.7	7.3	4.5	3.5	2.6	6.0
1935	–	–	37.3	4.5	2.4	44.2	30.5	8.4	5.8	3.5	2.5	5.1
1936	–	–	36.8	4.6	2.4	43.8	29.3	9.6	6.4	2.9	2.5	5.5
1937	–	–	35.4	4.7	2.5	42.6	29.2	10.3	6.5	3.2	2.9	5.3
1938	–	–	35.0	4.8	2.5	42.3	28.9	10.8	6.3	3.1	2.9	5.7

Note: Power and Dominion were taken over by Shell-Mex and BP on 1 May 1934.
Sources: BP 7447–53.

tendency for their market shares, especially that of Shell-Mex and BP, to be eroded by their smaller competitors, most notably the National Benzole Company, Trinidad Leaseholds, the Regent Oil Company (a subsidiary of the Texas Company) and ROP, though the latter's market share went into decline after 1931, as can be seen from table 4.3. The smaller companies generally had lower overheads than the big three and tended to pick customers who were cheap to supply, expanding their activities when margins were high and contracting when margins narrowed. In the face of such tactics, Shell-Mex and BP was able to maintain its market share above 40 per cent only by the acquisition of the Power Petroleum Company and the Dominion Motor Company in May 1934.

In France, there was a radical change in national oil policy in 1928, which resulted in the establishment of import and distribution quotas, the development of an independent French refining industry and the creation of French tanker fleets. Prices were fixed and quotas determined annually with encouragement being given to indigenous distribution companies and the disposal of products from local refining companies. With the construction of a refinery at Lavera near Marseilles in 1931 and the establishment of a French tanker company, L'Association Maritime, the Company complied with French laws for its French subsidiary, Société Générale des Huiles de Pétrole (SGHP), whose sales between 1928 and 1938 are shown in figure 4.5. The gradual promotion of French technical and administrative staff meant that by 1938 SGHP was almost entirely and successfully run by its own French staff under the competent direction of Fernand Gilabert, assisted by Joseph Huré. As in Britain, the market leaders were Royal-Dutch Shell, Standard Oil (NJ) and the Company, whose shares of the French market in 1938 were 20, 16 and 11 per cent respectively.

In Germany, the Company had acquired a 40 per cent shareholding in 'Olex' Deutsche Petroleum-Verkaufsgesellschaft in 1926 and in July 1931 it obtained complete control. During the 1930s that control was, however, circumscribed by growing state regulation of the German economy, particularly from 1933 onwards when legislation for the direction of industrial and commercial policy was supplemented by the provisions and objectives of the Second Four Year Plan announced by Hitler in September 1936. The autarchic emphasis of national economic policy was accompanied by subsidies for research into synthetic fuels, the stimulation of exploration for oil and controls over foreign exchange, foreign oil supplies and company dividends.

Co-operation on 'As Is' principles proved difficult and in mid-1933 a

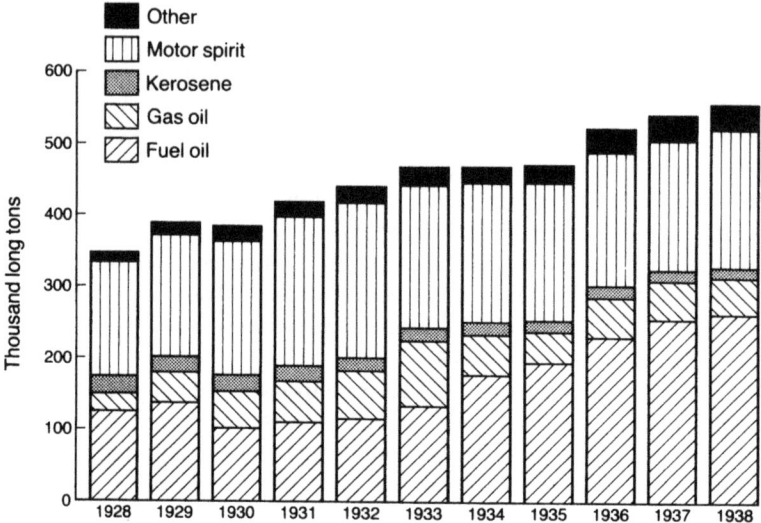

Figure 4.5 Sales of the Company's French subsidiary, Société Générale des Huiles de Pétrole, 1928–1938
Note and source: See Appendix 2.

dispute arose between the 'As Is' participants over state refineries.[72] Although a degree of harmony was restored at a meeting at The Hague in September, when representatives of the three 'As Is' companies agreed that full co-operation should be re-established,[73] there were also differences over Soviet oil imports. Royal Dutch-Shell took a political stance doubting 'whether local agreements with the Russians were of any use', convinced that 'any weakness displayed towards Russia must have its repercussions and must help the Bolshevist wave to spread over the face of Europe'. The other representatives argued that economic interdependence between Germany and the Soviet Union was a fact of life. The main problem, expressed in December 1933 by Dr Krauss, the chairman of Olex, was the preferential treatment given to Soviet oil imports,[74] which meant that 'the future outlook must be regarded with some anxiety'. His concern about future prospects can only have been aggravated by the more certain knowledge that Olex's sales had been in decline for the previous few years, as can be seen from figure 4.6.[75]

    In spite of gloomy forecasts and poor past performance, there was no intention in 1933 of withdrawing from the German market, where the Company's problems were felt to reflect those of the oil industry as

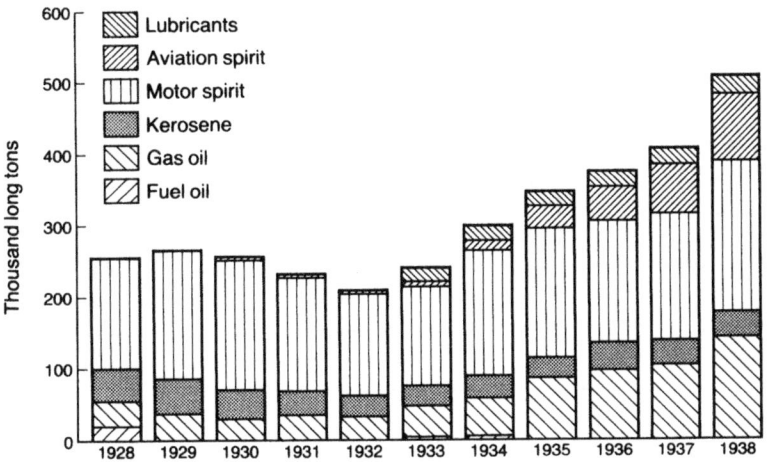

Figure 4.6   Sales of the Company's German subsidiary, 'Olex' Deutsche Petroleum-Verkaufsgesellschaft, 1928–1938
Source: See Appendix 2.

a whole and the longer term potential of the market was held to be considerable. This assessment was vindicated by the subsequent recovery in Olex's sales, despite supply problems arising from foreign currency controls and supplies being switched from Iran to Romania, with whom Germany concluded a trade agreement.[76] The extent of the recovery in sales tonnage is shown in figure 4.6, the most notable features being the growth in sales of gas oil, aviation spirit and lubricants. Market shares followed much the same pattern as elsewhere in Europe, Standard Oil (NJ) and Royal Dutch-Shell being the clear market leaders with 26 and 22 per cent respectively of the German market in 1938. The Company lagged behind them with a market share of about 10 per cent.

In comparison with Britain, France and Germany, the Company's other European sales in Austria, Belgium, Denmark, Holland, Italy, Norway, Sweden and Switzerland were small. In Belgium, the Company's marketing outlets were increased by acquisition when its local subsidiary, L'Alliance, purchased Sinclair Petroleum Société Anonyme in 1937, thereby expanding its share of the Belgian market by 3 per cent.[77] In Italy, on the other hand, the Company was, as has been seen, essentially a supplier to Royal Dutch-Shell and Standard Oil (NJ). There were also varying degrees of 'As Is' co-operation, which was, for example, high in Denmark, but effectively precluded by government

Table 4.4 *Market shares in European markets, 1938*
(percentages)

	The Company	Royal Dutch-Shell	Standard Oil (NJ)
Norway	24.8	27.1	36.2
Belgium	16.6	24.1	22.0
Switzerland	13.5	27.6	33.5
Denmark	13.4	19.0	43.9
France	10.8	20.4	16.2
Germany	9.6	22.1	26.1
Sweden	7.9	34.6	28.5

Source: BP 104332.

regulation of the oil industry in Switzerland. While European markets as a whole thus lacked homogeneity, a consistent feature was that the Company had a considerably smaller market share than Royal Dutch-Shell and Standard Oil (NJ), as is shown in table 4.4. This reflected the Company's comparatively late entry into international marketing, its first overseas European marketing subsidiary, L'Alliance, having been established as late as 1919. In spite of the subsequent expansion of the Company's marketing operations, its position as one of the leading international oil companies continued to be based more on the efficient extraction of oil from its huge reserves in the Middle East than on the cut and thrust of its marketing organisation.

SHIPPING

Notwithstanding the formation of L'Association Maritime in France, most of the Company's international movements of oil continued to be made in the vessels of its main tanker subsidiary, the British Tanker Company (BTC). The sizes of the British and French fleets are shown in table 4.5 from which it can be seen that the whole Company-owned fleet was reduced in size between 1928 and 1930 and showed little sign of renewed expansion until the second half of the 1930s. Indeed, the number of Company-owned tankers did not surpass the 1928 level until 1937, in which year Cadman expressed his pride in the Company's contribution to the revival of prosperity in the British ship-building industry as it came out of the economic depression.[78] Measured by deadweight tonnage, the capacity of the Company's fleet

23   'Olex' BP service station in Germany, mid-1930s

24   BP service station in Sweden, mid-1930s

25   BP service station in Switzerland, mid-1930s. (a) by day ...

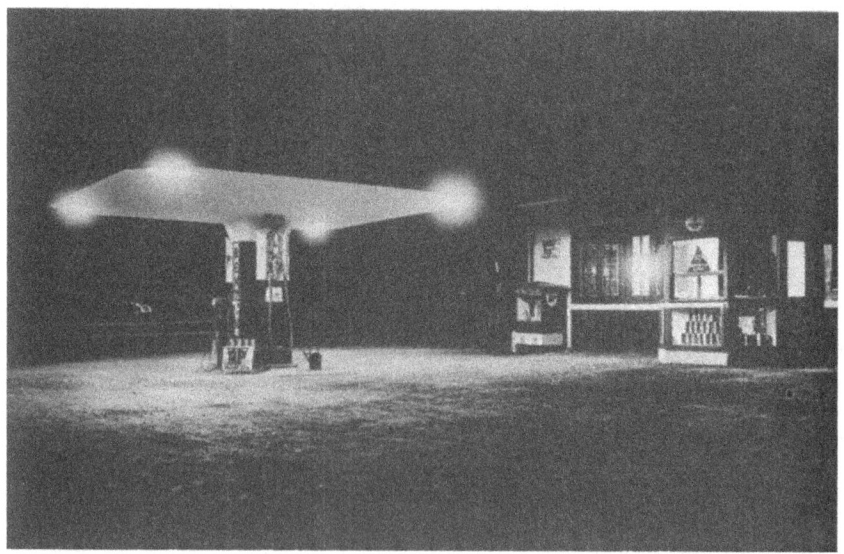

(b) by night

Table 4.5 *The Company-owned tanker fleet, 1928–1939 (d.w.t. = deadweight tons)*

	British Tanker Company		Association Maritime		Total Company fleet		Average size of Company tankers
	Tankers	d.w.t. '000s	Tankers	d.w.t. '000s	Tankers	d.w.t. '000s	d.w.t. '000s
1928	82	790	4	37	86	827	9.62
1929	79	773	4	37	83	810	9.76
1930	76	753	5	48	81	801	9.89
1931	80	794	5	48	85	842	9.91
1932	79	783	5	48	84	831	9.89
1933	79	783	5	48	84	831	9.89
1934	79	783	5	48	84	831	9.89
1935	79	783	5	48	84	831	9.89
1936	81	817	4	40	85	857	10.08
1937	86	885	4	40	90	925	10.28
1938	86	887	4	40	90	927	10.30
1939	92	964	3	34	95	998	10.51

Note: Includes only vessels over 1800 d.w.t.
Source: BP 6B 3099.

Table 4.6 *Company vessel movements by owned and chartered tankers, 1928–1939*

	Company-owned (% of. d.w.t.)	Time charter (% of. d.w.t.)	Consecutive voyage charter (% of. d.w.t.)	Single voyage charter (% of. d.w.t.)
1928	87.0	13.0	–	–
1929	89.1	7.5	–	3.4
1930	90.9	2.1	1.8	5.2
1931	99.5	0.5	–	–
1932	100.0	–	–	–
1933	96.3	–	–	3.7
1934	92.5	–	–	7.5
1935	92.9	–	–	7.1
1936	86.9	1.3	0.5	11.3
1937	80.4	1.8	–	17.8
1938	89.2	0.8	1.9	8.1
1939	91.8	3.7	1.7	2.8

Note: Single voyage charters include some multiple voyage charters.
Sources: BP vessel movement books and chartering divison fixture lists.

recovered its 1928 level somewhat earlier, reflecting the underlying trend towards an increase in the size of tankers. The check on the expansion of the fleet's capacity helped the Tanker Company to weather the depression with most of its ships in service, only six 10,000 deadweight ton vessels being laid up for an average of six weeks. As table 4.6 indicates, little recourse was made to chartering tankers from other owners except in the economic upturn of 1936–7 when there was a significant increase in the Company's single voyage charters, the form of chartering most suitable for meeting marginal requirements.

Easily the most important shipping route in the BTC's operations was that between Abadan and north-west Europe, including Britain, via the Suez Canal, as illustrated in maps 4.2 and 4.3. Along this route were carried crude oil, principally for the Company's refineries at Llandarcy and Grangemouth, and products for European markets. Although the other routes plied by BTC vessels were of lesser individual importance, they added up to an extensive international network, giving an indication of the range and changing pattern of the Company's supply operations during the period. As the maps indicate, the volume of cargoes loaded in the Gulf of Mexico declined consider-

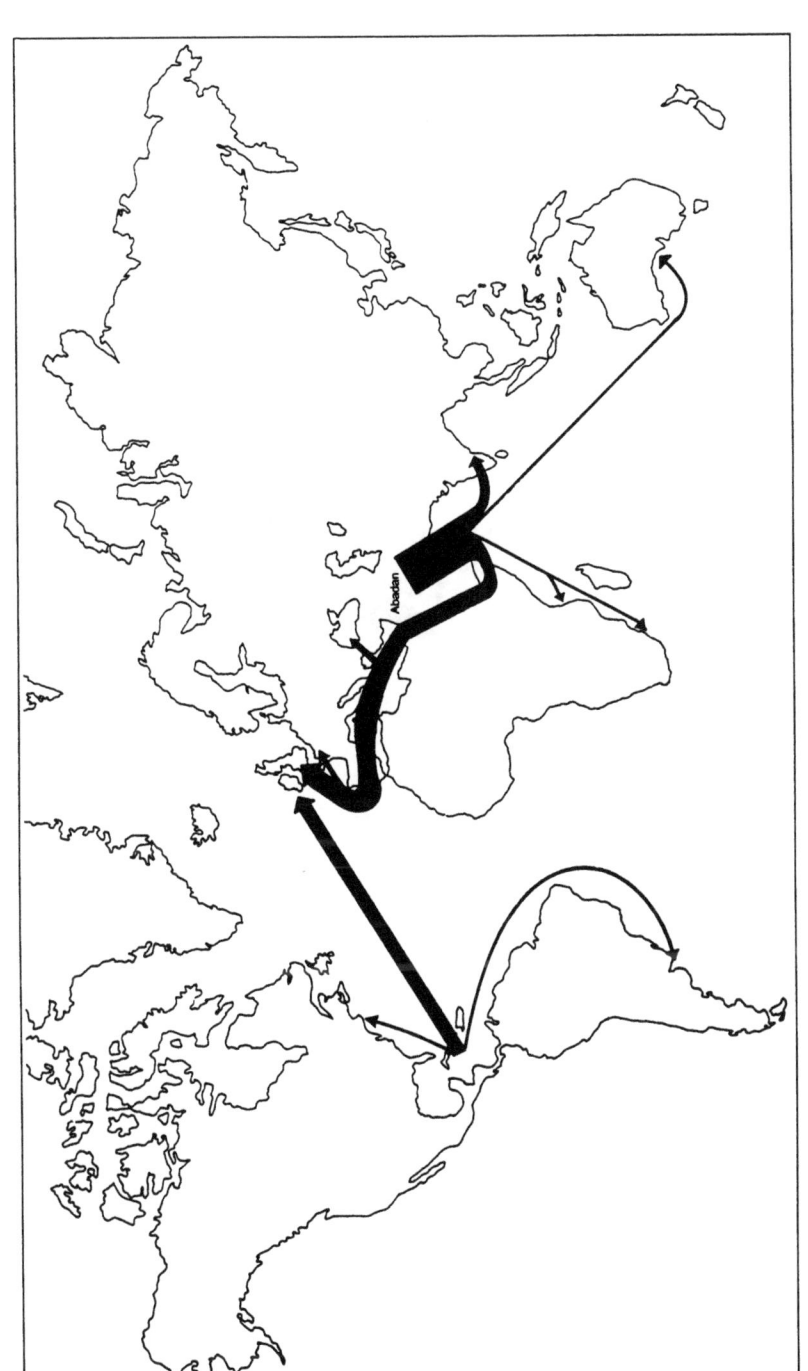

Map 4.2 British Tanker Company cargo routes, 1928

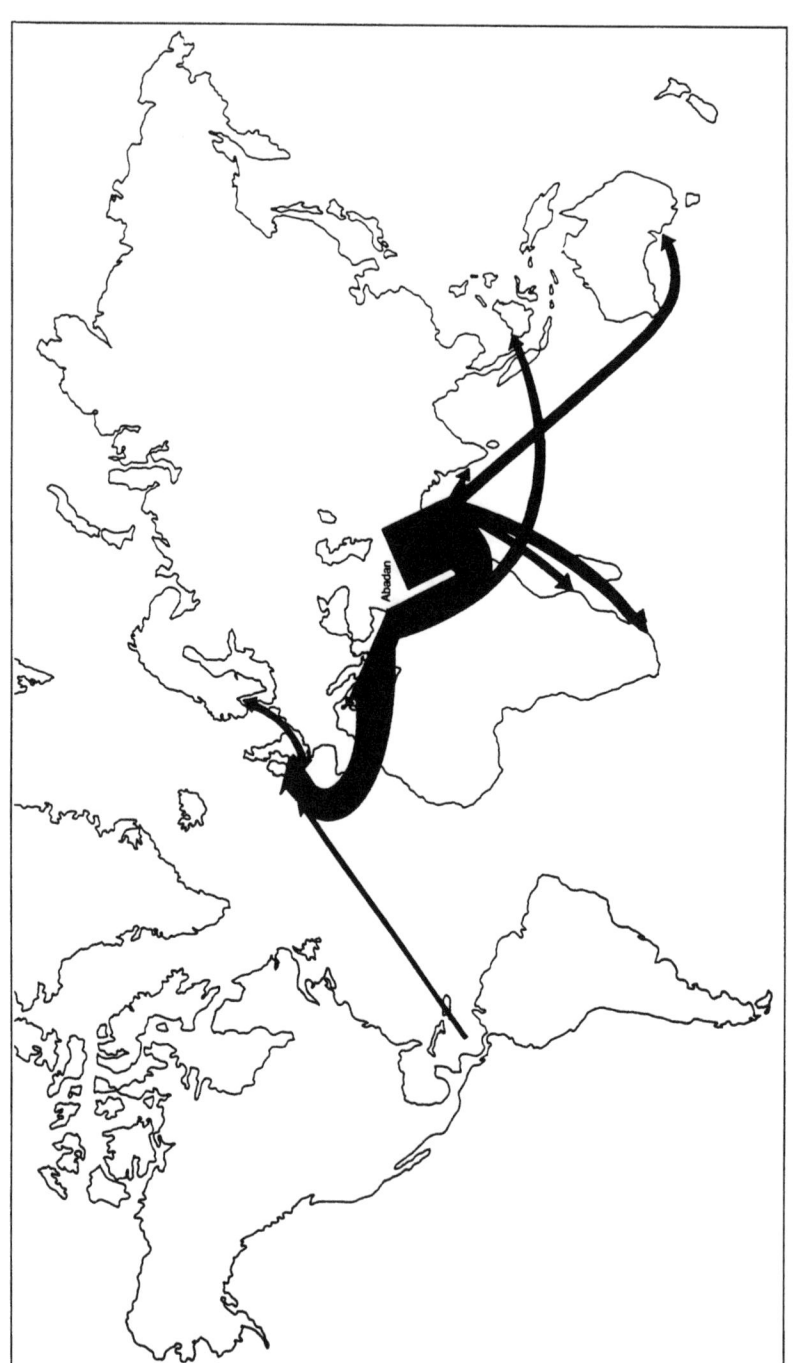

Abadan

Map 4.3　British Tanker Company cargo routes, 1938

ably between 1928 and 1938, while loadings in the eastern Mediterranean started after the pipeline connecting the Kirkuk oilfield in Iraq to terminals on the Mediterranean seaboard was completed (see chapter 5). The main discharge points included bunkering stations and ports, not only in north-west Europe, but also on the shores of the Mediterranean, the Indian subcontinent, Australia and eastern and southern Africa. Thus, while Iran remained the centre of production and refining operations, and Europe the most important inland marketing area, the BTC's cargo routes reflected the wider international dimension of the Company's business. That dimension was evident, not only in the shipping and marketing of oil, but also in the Company's concessionary interests outside Iran, to which it is time to turn.

= 5 =
Concessionary interests outside Iran, 1928–1939

By the late 1920s the Company had already shown interest in concessions in different parts of the world apart from Iran, including the Arabian side of the Persian Gulf and in Iraq in association with the other participants in the Turkish Petroleum Company (TPC), renamed the Iraq Petroleum Company (IPC) in 1929. However, with the exception of Iraq, where oil was discovered at Baba Gurgur in 1927, and a minor and uncertain interest in Argentina, the Company achieved little exploratory success. In the 1930s it failed to obtain any concessions in South America, partly because of the priority which it assigned to Iran, but principally because the British Government's large shareholding effectively disqualified the Company from acquiring concessions in countries where the participation of foreign governments in the exploitation of domestic oil resources was prohibited by law. More success attended efforts in the Middle East, especially in Kuwait where the Company and Gulf Oil Corporation joined hands in obtaining a concession which they held through their jointly owned subsidiary, the Kuwait Oil Company; and in Iraq, Qatar and some of the shaykhdoms on the Trucial coast, where the IPC, in which the Company held a 23.75 per cent shareholding, extended its concessionary interests. Elsewhere, the Company broke little new ground. In Africa, the problem of communications and the unfavourable conditions gave little encouragement to sustained exploration. Minor interest was shown in Europe. There were limited prospects in Australasia and intermittent exploration activities in India and Canada, but there was no real incentive to persevere in these areas. The Company had copious oil reserves in Iran and heeded Riza Shah when in 1938 he warned the Company against neglecting Iran in order to pursue other interests elsewhere.

COLOMBIA

The Company had been interested in the oil prospects of Colombia since January 1918. Exploration was difficult because of the lack of roads, the nature of the terrain and the shortage of labour, but according to a Company survey the country offered 'great possibilities in respect of oil developments', albeit 'at the cost of considerable expenditure and organisation'.[1] Negotiations with the Government commenced in 1926[2] and in September 1927 Hearn, the Company director in charge of concessions, stated approvingly that Colombia was 'excellently situated' as a complementary area to Iran owing to its proximity to western markets in which Iranian oil suffered the competitive disadvantage of high transport costs compared with nearer sources of supply.[3] However, Hearn's enthusiasm on that score was tempered by other considerations of a political nature. Professor Hugo De Böckh, the Company's geological adviser, although impressed with the oil possibilities of Colombia, was less happy about the political situation there. 'Here', he informed Cadman in 1927, 'is everything full with the nationalisation of the oil territories'.[4] Moreover, the Petroleum Law, which stipulated that leases could not be held by companies in which foreign governments were admitted as partners, stood in the way of the Company.[5] In July 1928, Hearn, feeling that negotiations for concessions in Colombia should no longer be contemplated while Colombian legislation operated 'so decidedly to our disadvantage', thought that efforts should be made to get the law amended.[6] It was to that end that Sir Arnold Wilson, the managing director of D'Arcy Exploration Company, visited Colombia in September–October 1928. He felt that José Antonio Montalvo, the Minister for Industry, was sympathetically inclined, but unfortunately the legislation was becoming increasingly enmeshed in the confused political situation prior to a general election. A commission set up to examine proposed oil legislation was dissolved and Wilson had no alternative but to suspend negotiations.[7] Montalvo assured him that when new legislation was enacted, it would include a clause permitting the Company to operate in his country,[8] but enthusiasm on the Company's board was waning. Fraser questioned the wisdom of continuing involvement in Colombia[9] and Cadman was definite 'that we should not consider any proposal until Colombia has done what we asked'.[10]

It was not until 1936 that the Colombian Government passed further petroleum legislation, Law 160, which effectively lifted the restrictions that had stood in the Company's way.[11] The Company was pressed by the Governments of both Britain and Colombia to show renewed interest. The Company geologists, J. V. Harrison and P. T. Cox, successively looked at the possibilities and were in general agreement. Setting Colombian prospects in a worldwide context Cox felt that if 'it should be our policy to prospect for large production outside Iran, I do not know of any country which offers better hope of success than Colombia'.[12] The matter was taken further when Cox visited Colombia in December 1937 and an exploration programme was planned. However, just when the prospects of exploration in Colombia appeared to be brightening the Company was warned by Riza Shah that any further foreign exploration which might prejudice production in Iran was completely unacceptable. Cox was recalled and geological work in Colombia brought to a halt.

VENEZUELA

The Company was interested in the oil possibilities of Venezuela about the same time as Colombia, but in Venezuela much of the most promising oil territory had been acquired by US oil companies and Royal Dutch-Shell. Venezuelan production was increasing rapidly and the Company was anxious not to be excluded from such a prolific oil region which, like Colombia, had the geographical advantage of relative proximity to western markets. However, Venezuela also had problems similar to those of Colombia: vast tracts of difficult terrain, political uncertainty (though President Gomez exercised a virtual dictatorship) and petroleum legislation which forbade the granting of concessions to foreign governments or states, or to corporations which depended upon them.[13]

Since 1920 the Company had received a number of propositions for concessions in Venezuela, but in 1926 the Foreign Office urged caution because 'it would be the height of folly ... to spend a penny in Venezuela if there is the slightest doubt as to [the] legal position'.[14] Legal opinions on the Company's eligibility for Venezuelan concessions were mixed, with Linklaters and Paines, the Company's solicitors, taking a guarded view whereas a prominent Venezuelan lawyer, Dr R. Marciano Rodrigues, was confident that the Company would be taking no legal risk if it operated in Venezuela.[15] On the other hand, Dr Vladimir Idelson, the Company's legal adviser, did not believe it would be easy to show that the Company was entirely independent of the British Government.[16] In April 1927 Cadman suggested that in the circumstances 'it would be expedient to stay our hand in Venezuela, to retreat diplomatically from advances there, and definitely to await full consideration of the fruits of our policy in Colombia'.[17]

However, in 1933 new overtures to the Company were made by Alfred Meyer who, as president of the Caracas Petroleum Corporation, had already tried to interest the Company in a co-operative venture in 1930.[18] Nothing came of that proposal, but Meyer again came to the fore after Orinoco Oilfields Ltd, of which he was a director, was formed in June 1933 and took over the assets of the Caracas Petroleum Corporation. Orinoco needed more capital and broached the idea of the Company joining with it to hold concessions in Venezuela.[19] The idea was pursued in a series of meetings between representatives, including legal advisers, of the Company and Orinoco between November 1933 and April 1934.[20] Meyer wished to reach a rapid conclusion by coming to an understanding with the President of Venezuela himself. The Company, on the other hand, was cautious, looking for a long-term investment and reluctant to base its participation in Venezuela on personal patronage which was not appealing 'either from the point of view of principle or of common sense'.[21]

One factor of the greatest importance was that the geological evidence was favourable. B. K. N. Wyllie, the Company geologist with most experience of South America, was 'satisfied that these concessions hold possibilities of a very high grade and may, in fact, contain oilfields of the first rank. Chances of this order come very seldom in these days and will undoubtedly become rarer still in the future'.[22] There was thus every incentive for the Company to develop an interest in Venezuelan oil and much ingenuity was expended on accommodating the Company's position to Venezuelan legislation or on possible amendments to it. It was Idelson who suggested that the Venezuelan Ambassador in London, a lawyer of considerable standing with whom he was acquainted, should be approached for informal advice on the possibilities of forming a subsidiary company which the Venezuelan Government would give permission to operate in the country. Until the Venezuelan Government had given a positive reaction to such a proposal, which the Ambassador was convinced would be the case, it would be premature for the Company to conclude a definite arrangement with Orinoco.[23]

Meanwhile, towards the end of 1933 another possibility of taking an interest in an existing concession had arisen and Hearn arrived in Venezuela to investigate it in mid-March 1934. On his arrival he learnt from the British Legation that no assurance was dependable unless it was received directly from President Gomez, 'whose health makes accessibility very difficult'. Only one interview at the most could be expected and it would be wasted in general expressions of goodwill unless his support was obtained for a definite project. A visit to Gomez was arranged for 20 May at the presidential palace, Las Delicias, in Maracay. Before the official meeting, scheduled for late afternoon, Hearn came across the President watching a cockfight in mid-morning, 'a very old, tired, grey man, absorbed in the fight and occasionally smiling grimly at some special coup'. Later seated in a very large square reception hall, surrounded by local notables, friends and members of the Government, the President watched dancing and listened to music. Hearn was introduced to Gomez, who was 'smiling affably, said a word or two I couldn't catch – I answered I was much honoured – and the business was over. Absolutely no opportunity for any talk existed ... One would say he was long past business'.[24] The following day Hearn was told by the President's secretary that the President had been favourably impressed by his visit, but was unaware of its purpose. The Minister of Interior said that the Government had changed its mind about welcoming proposals, but gave no indication of the reason. The Company would be able to make suggestions, which would be carefully studied, but there was no time for any modification of the laws. The Minister repeated that the Company would be welcome in Venezuela, 'not only because of its importance but also as British – there being now a preponderance of Americans'. Hearn told the Minister that he 'left with a good many words, but with empty hands'.[25]

Over the next four years the subject of Venezuelan concessions was raised in the Company from time to time, but nothing concrete resulted. The Orinoco option lapsed at the end of 1934 and another proposition, connected with the Ultramar Exploration Company, was rejected by the board of directors in 1938.[26] In commenting upon it, Hearn concluded: 'My colleagues do not see in this plan, or in any possible alternative, sufficient reason or justification for entering Venezuela so long as the Petroleum Laws of Venezuela operate to the disadvantage of this Company'.[27]

KUWAIT

While Venezuelan legislation handicapped the Company's chances of obtaining a concession there, the situation in Kuwait tended to have the opposite effect of working to the Company's advantage in relation to its foreign competitors. In the nineteenth century, when the defence of Britain's position in India was a cardinal feature of British imperial policy, Britain steadily established her authority over the Arab shaykhdoms of the Persian Gulf, which were strategically located on the imperial route to the East. Agreements with the Gulf shaykhs were made at various dates with the objects, broadly speaking, of preventing piracy, slavery and maritime warfare in the Gulf and of keeping the area from falling under the control of another great power. In the case of Kuwait, Shaykh Mubaruk signed an agreement with Britain in 1899 under which Kuwait became to all intents and purposes a British protectorate. In the agreement, the Shaykh undertook not to enter into relations with any other power without British consent and not to grant any concessions or cede his territory to anyone without British approval.[28] The agreement was subsequently supplemented with a specific reference to oil concessions in an exchange of letters between the Shaykh and the British Political Resident, who was the British Government's senior political representative in the Gulf, stationed at Bushire. In the exchange, the Shaykh agreed not to grant an oil concession to anyone other than a person nominated and recommended by the British Government.[29] In other words, the support of the British Government was essential for anyone seeking a concession in Kuwait.

The first recorded expression of interest in the oil prospects of Kuwait was in 1911 when Charles Greenway, who was then the Company's managing director but not yet its chairman, wrote to the Political Resident at Bushire inquiring about the possibility of the Company obtaining an oil concession in Kuwait. Although Greenway's enquiry yielded no immediate concrete result the Company continued to show interest in obtaining a concession in Kuwait in the early- to mid-1920s, encountering stiff competition from the redoubtable Major Frank Holmes who was then very active in seeking oil concessions in Arabia on behalf of the Eastern and General Syndicate (EGS), a British company formed in 1920.[30] After making a disappointing geological reconnaissance in 1926, and unable to reach agreement with the Colonial Office on the terms of a concession, the Company withdrew from negotiations. Holmes, on the other hand, secured the backing of the US oil company, Gulf Oil Corporation, for his negotiations with Shaykh Ahmad, who had become the ruler of Kuwait in 1921 and was to remain so until his death in 1950.[31] Holmes was, however, faced with a difficult obstacle: the insistence of the British Colonial Office that any concession granted in Kuwait should include a British 'nationality clause' which would, in effect, exclude non-

British subjects or companies from holding a concession there.[32] The same restriction was applied to Bahrain which, like Kuwait, had fallen under British sway in the nineteenth century.[33] In the case of Bahrain it took only until January 1930 for the British Government to agree to the US oil company, Standard Oil of California, holding the Bahrain concession through a Canadian subsidiary, the Bahrain Petroleum Company. In relation to Kuwait, the nationality question was less swiftly resolved. In October 1930 the Petroleum Department of the Board of Trade inquired of Cadman whether, in view of EGS/Gulf's desire to obtain the Kuwait concession, the Company was prepared to show renewed interest as 'we do not like to see any area which shows any promise going entirely into American hands'.[34] Cadman was not enthusiastic, replying that the Colonial Office's concessionary terms were too onerous and that 'very little detailed geological work has been done and reports leave little room for optimism'.[35] However, by August 1931 the Company, encouraged by test borings elsewhere in the Gulf area, had undergone a change of heart and was prepared to send a small party of geologists to Kuwait to carry out a survey, to which the Shaykh agreed.

Meanwhile, Holmes was becoming increasingly frustrated by the failure of his efforts to persuade the British Government to drop the nationality clause. On his advice, Gulf Oil took the matter up with the US State Department who in turn put pressure on the British Foreign Office to accord open-door rights to US nationals in Kuwait, as it had already done in the almost identical case of Bahrain. After discussing the matter at a Cabinet meeting on 6 April 1932, the British Government gave way and dropped its insistence on the nationality clause provided that the Shaykh, for his part, was willing to grant a concession without such a clause. Before hearing that the clause had been waived, the Company had decided to abandon its geological reconnaissance in Kuwait and informed the Shaykh that the results did not justify an application for a concession. However, with the nationality clause out of the way, the Company realised the gate was now open for a rival interest to obtain the concession. It therefore reversed its position and informed the Shaykh that it was interested in a prospecting licence.

These events put the Shaykh in a strong bargaining position, for he now had what he had desired all along: competition between two rivals, namely the Company and EGS/Gulf, for his concession. With an obvious interest in raising the stakes, he encouraged the two bidders and fed their eager rivalry, telling the Political Resident that other things being equal he would rather see the concession go to a British concern in view of his relations with Britain, but also informing Holmes that whatever terms the Company offered, 'I have promised the Kuwait concession to you and shall stand by my word'.[36]

If the outlook for the Shaykh was already brightening, it positively shone after oil was struck in Bahrain on 31 May 1932. The discovery, which was the first find of oil in commercial quantities on the Arabian side of the Persian Gulf, confounded earlier pessimism about oil prospects in Arabia and immediately accelerated the tempo of negotiations in Kuwait. In mid-June, Gass, the Company's deputy general manager in Iran, visited the Shaykh and proposed that the Company should negotiate with him for a three-year prospecting licence. This was not, of course, the same as a full concession which, Gass suggested, might await further exploration of the territory. For his part, the Shaykh, having already received the draft of a full concession from Holmes, would accept nothing less from the Company and Gass had no option but to agree that the Company would prepare a full concession document. The Shaykh expressed his satisfaction with the position to the British Political Agent in Kuwait, to whom he confided: 'I have now two bidders and from the point of view of a seller that is all to the good'.[37]

Over the next two months the Company duly prepared a draft concession which was presented to the Shaykh in August by Archibald Chisholm, who had joined the Company in 1927 and worked in Iran until, at the age of thirty-one, he was sent to Kuwait with the Company's concession. From then until the signing of the concession he was to represent the Company in its negotiations with the Shaykh in Kuwait. The Shaykh was not, however, willing to discuss the Company's draft concession in any detail until he had received the Political Resident's approval of it as a basis for discussion. The Political Resident in turn would not agree to the opening of talks on the concessionary terms between the Shaykh and the Company until he had received authorisation to do so from London. The result was a five-month delay while the relevant Government departments in London, of which there were several, examined the Company's proposals as well as those of Holmes. The time-consuming labours of officialdom, though disagreeable to EGS/Gulf, were not unwelcome to the Company which was then preoccupied with its position in Iran, especially after the Shah cancelled its concession in November 1932.

Eventually, in mid-January 1933 the Political Resident authorised the reopening of negotiations between the Shaykh and the two contenders for his concession. For the next five months EGS/Gulf and the Company negotiated separately with the Shaykh, engaging in competitive bidding for his favour. Then, in mid-May the Shaykh suspended negotiations with both applicants, a decision which the Company attributed to a combination of factors, including the strong pressure which was being put on him by the rivals for the concession; his belief that the two competitors might shortly join forces making it unnecessary to decide between them; and his desire to know the terms of the al-Hasa concession which was about to be concluded between King ibn Saud and Standard Oil of California (Socal) in Saudi Arabia.

The Shaykh's belief that the Company and EGS/Gulf might combine to make a joint application for the concession was well-founded, for the two rival camps had been discussing the possibilities of becoming partners in seeking a shared concession in Kuwait since late 1932. On 23 May 1933, little more than a week after the Shaykh's suspension of negotiations, the Company and Gulf agreed to abstain from separate efforts to obtain the concession for the next three weeks. The period of this 'standstill', as it was called, was subsequently prolonged to cover the seven months which it eventually took for the two companies to reach formal agreement to combine their efforts to obtain the Kuwait concession. The formal agreement, signed on 14 December, provided for the formation of a joint company, owned in equal shares by the Company and Gulf, to obtain and operate a concession in Kuwait. The new company, called the Kuwait Oil Company (KOC), was duly incorporated on 2 February 1934 with six directors, half appointed by the Company and half by Gulf. The EGS withdrew from the negotiations and was reimbursed by Gulf for the expenses it had incurred in its efforts to obtain the concession. Holmes, however, remained deeply involved as negotiator for Gulf in Kuwait, his counterpart for the Company being Chisholm.

In mid-February Holmes and Chisholm returned to Kuwait as joint negotiators for the KOC, expecting little difficulty in coming to a rapid settlement on the concession. They were, however, to encounter a number of complications in the labyrinthine negotiations, not to say intrigues, which followed. One of the complications was the agreement, known as the political agreement, between the British Government and the KOC which was signed on 5 March 1934. It stipulated that the KOC was to remain a British company and gave the British Government significant powers, most notably that the concession, if obtained by the KOC, was not to be transferred without the consent of the British Government, which would also have rights of pre-emption over Kuwaiti crude oil and refined products in the event of war. It was, in addition, laid down that should there be any conflict between the terms of the political agreement and the commercial concession agreed between the Shaykh and the KOC, the terms of the political agreement would prevail. When Chisholm and Holmes, acting on instructions from London, proposed to the Shaykh that the concession agreement should include a clause to that effect, the Shaykh raised objections. Not only would the proposed clause mean that anything in his agreement with the KOC could be overruled by the political agreement which had been arrived at without his knowledge, but when the concession agreement was concluded it would become public and it was beneath his dignity to be seen to have negotiated an agreement which was subordinate to a political agreement in which he had played no part. While the Shaykh was adamant that the political agreement should not be mentioned in the concession agreement, there were also differences between him and the KOC on the terms of the concession, especially regarding the level of royalties and the Shaykh's demand that he should have the right to appoint one of the directors on the KOC board in London. Although progress was made on some items, full agreement on the concession agreement had still not been reached by June when the negotiations, which by then had reached deadlock, were suspended for the summer. In mid-June Holmes and Chisholm, whose presence in Kuwait now served no purpose, returned to London. They remained there until late September when they departed for Kuwait, arriving there in mid-October having travelled via Cairo and Basra.

While staying at Basra, Holmes picked up rumours that in his and Chisholm's absence from Kuwait a competitor company had been negotiating with the Shaykh, offering better terms than the KOC and nearly reaching agreement. This was not the first time that mention had been made of a rival to the KOC. As early as April 1934 the Shaykh had told Chisholm and Holmes that he had received a better offer than theirs from a company which was wholly British-owned. The claim was dismissed at the time as a negotiating ploy, but later turned out to be true, the company concerned being the same as that which Holmes got wind of at Basra. He learnt its identity after arriving in Kuwait in mid-October, when he was informed that the KOC's competitor was a company called Traders Ltd, formed in 1932 by a British group which included oil refining, shipping and marketing interests who, inspired by imperial as well as commercial motives, decided to make a bid for the Kuwait concession if it appeared likely that it would be obtained by US or partly US interests. Although Holmes was successful in uncovering that it was Traders Ltd who had made a counter-offer to the Shaykh, he was led seriously astray by Kuwaiti friends and advisers who told him, incorrectly, that Traders was merely a front for the Company, whose hidden hand allegedly lay behind the counter-offer. Holmes, believing what he was told, became convinced that the Company, while ostensibly negotiating with the Shaykh through the KOC, was secretly double-crossing its partner in the KOC by attempting, through Traders, to obtain the concession solely for itself. Without mentioning anything about it to Chisholm, he persuaded the Shaykh that Traders was merely the instrument of the Company and proceeded to hold secret discussions with the Shaykh, excluding Chisholm who was meant to be his co-negotiator for KOC.

It was in this byzantine atmosphere that the final negotiations for the concession were conducted from mid-October. With Traders still actively seeking the concession, the KOC acted swiftly to agree terms and on 23 December the concession agreement was signed by the Shaykh and by Chisholm and Holmes on behalf of the KOC. The earlier draft clauses referring to the political agreement, which had been a contentious issue in the negotiations, were omitted from the concession document and were instead embodied in an exchange of letters between the Shaykh and the British Government. The main terms of the concession were that it covered the whole of Kuwait and was to remain in force for seventy-five years. The KOC was to commence geological exploration within nine months and to perform minimum drilling obligations within specified time limits. An initial payment of 475,000 rupees was to be made by the KOC and the royalty was to be 3 rupees per ton of oil with minimum payments of 95,000 rupees a year until the KOC declared that oil had been found in commercial quantities. Thereafter the minimum was to be 250,000 rupees a year. In the course of the negotiations the Shaykh had given way on his demand that he should have the right to appoint a director to the KOC board in London and in the final agreement he was entitled, instead, to appoint a London representative to represent him in all matters relating to the agreement with the KOC in London, with the right to attend KOC board meetings at which the Shaykh's interests were discussed.[38] These and the other terms of the agreement were, as events turned out, to remain unaltered until 1951 when, as shall be seen in chapter 13, revisions were made.

With the negotiations finally concluded, the remainder of the 1930s was spent putting the agreement into operation. In January 1935 the

26   Shot-hole drill just east of Bahra No. 1 well, Kuwait, 1936

27   The camp at Bahra, Kuwait, where the first well was drilled in 1936

28 Swimming tank at Bahra, 1936

29 Washing cars in Kuwait town after heavy rain, *c.*1937

Shaykh appointed Holmes as his London representative, a post which he retained until his death in February 1947. Meanwhile, in March 1935 senior geologists from the Company and Gulf arrived in Kuwait on behalf of the KOC. Other staff and equipment followed and in May 1936 the first well was 'spudded in' (commenced drilling) at Bahra in the presence of the Shaykh. After reaching a depth of 7,950 feet without producing oil it was abandoned. A geophysical survey in the winter of 1936–7 led to the selection of a second drilling location in the Burgan area where the first well was spudded in during October 1937 and struck oil on the night of 23/24 February 1938. Further drilling confirmed the enormous size of the Burgan field, whose discovery represented the birth of the Kuwait oil industry. After the discovery, progress was made in the equipment and preparation of the field for the production and export of oil, but operations were interrupted by World War II and it was not to be until the postwar years that Kuwait joined the ranks of Middle East oil exporters.[39]

IRAQ

Like Kuwait, Iraq fell squarely within the British sphere of influence in the Middle East in the interwar period. For many years a part of the Ottoman Empire, Iraq was placed under British control in the peace settlement after World War I when the Ottoman Empire was dismembered and its Arab lands assigned to western powers under the mandate system. In the case of Iraq, Britain exercised its control, not through a formal mandate, but through a treaty, concluded in 1922, which, whatever its form, did not really differ from a mandate in its essential provisions. Further treaties later in the 1920s and in 1930 introduced a modest relaxation of British control and in 1932 Iraq attained independence on being admitted to the League of Nations as a full member.

The country's basic characteristics could not, however, be changed simply by the transfer from Ottoman to British control and then to independence. Iraq had been one of the most remote and least-developed parts of the Ottoman Empire. Illiteracy was widespread and educational facilities poor. As for political structure, Britain attempted to endow Iraq with a constitution and the rudiments of a parliamentary system on the lines that had evolved in Britain over the centuries. However, the attempted transplant was rejected by the body politic of Iraq and the external trappings of parliamentary democracy and constitutional monarchy had little real influence on the conduct of

Iraqi politics in which personalities and rival factions were more important than political programmes. In ruling the country, King Faisal I, who ascended the throne in 1921, had to look less to the elected Constituent Assembly than to the real centres of power – the Shia divines, the leading families, the army officers and the tribes. When Faisal died in 1933 he was succeeded by his twenty-one-year-old son, Ghazi, who lacked experience and took more interest in car and motor cycle racing than in affairs of state. The country relapsed into political instability characterised by frequent shifts in the cabinet and outbreaks of tribal and minority unrest. In an atmosphere of suspicion and intrigue, a succession of governments came and went, exalted or deposed by a series of military coups carried out by various army cliques in the second half of the 1930s.[40] Lacking the political, economic and social structures to attract inward investment the country was, however, rich in one resource: oil.

The first commercial oil discovery in Iraq was made by the Company in 1923 at Naftkhana, south of Khanaqin in the border territory transferred from Iran to Turkey (and later by inheritance, Iraq) by the Turco-Persian Frontier Commission of 1913. As concessionary rights over the Transferred Territories had already been granted to the Company as part of the 1901 D'Arcy concession, the Iraqi Government accepted the Company's position as sole concessionaire, which was made explicit in a new concession granted in 1925.[41] The Company duly formed a subsidiary, the Khanaqin Oil Company, to conduct operations in the Transferred Territories and the Naftkhana field was developed for commercial production. Construction of a 4-inch pipeline to a refinery site on the Alwand river outside Khanaqin was commenced in January 1926 and completed in 1927, when the refinery was officially opened by King Faisal. Designed to meet local demand, the Alwand plant supplied northern and central Iraq with oil products while the southern part of the country continued to be supplied from Abadan. In 1932 the Khanaqin Oil Company's functions were reduced to production and transportation after the Company formed a new subsidiary, the Rafidain Oil Company, to undertake distribution and marketing in Iraq.[42]

Although the Khanaqin and Rafidain companies were wholly owned subsidiaries, they were of much less importance to the Company than its minority shareholding in the Turkish Petroleum Company (TPC) which on 14 March 1925 was granted a concession covering the whole of the Mosul and Baghdad *vilayets* (provinces), in other words all of Iraq except for the Transferred Territories and the

Basra *vilayet* in the southern part of the country. At that time, the
Company held 47.5 per cent of the TPC's shares, the remainder being
in the hands of Royal Dutch-Shell (22.5 per cent), Compagnie Française des Pétroles (25 per cent) and Calouste Gulbenkian (5 per cent).
The concession provided, amongst other things, for the selection by
the TPC within 32 months of 24 plots of 8 square miles each for its
own exploitation, with the obligation to let the remaining territory
covered by the concession go to the highest bidders. Other important
provisions of the concession were that the royalty was to be 4s (gold)
per ton; the Iraqi Government was to have the right, under Article 27,
to impose the same taxes on the TPC as were imposed on other
industrial undertakings in Iraq; and the TPC was to build a pipeline
for oil exports as soon as practicable, but in any case within four years
of its plots being fully tested.[43]

At the time the concession was granted, H. E. Nichols, who was
then a director of the Company, was managing director of the TPC as
well as its acting chairman. Optimistic about the oil prospects in Iraq,
he set the administration of the TPC in motion, holding a geological
meeting on 25 June 1925 and a TPC board meeting at the beginning of
July. By mid-March 1926 an office had been opened in Gresham Street
and the nucleus of a staff appointed. Nichols' further involvement was,
however, to be shortlived, for in 1926 Cadman became chairman of the
TPC's board and Sir Adam Ritchie, a chartered accountant who had
been the general manager of Finlay, Fleming and Company, managing
agents for the Burmah Oil Company in Rangoon, was appointed the
TPC's general manager in London. In that capacity, he succeeded
Nichols who remained for a short while on the Company's board
before his death in January 1927.

In Iraq, Professor Hugo de Böckh, geological consultant to the
Company, headed a geological mission in 1926 and on its evidence
drilling sites were selected. Drilling rigs, staff and equipment were duly
despatched to Iraq where the first well was spudded in at Palkhana in
the presence of King Faisal in April 1927. After further drilling at
various sites, a well was spudded in at Baba Gurgur, immediately
north of Kirkuk, in June. On 15 October the drill struck oil which
flowed with such force that its was uncontrollable for several days.
The discovery of the Kirkuk field transformed Iraq from a country of
high oil promise to one of the most valuable concessionary areas in the
world. Although the TPC continued to explore in other parts of its
concessionary area, it concentrated mainly on Kirkuk for the next few
years when further drilling and investigation into the characteristics of

30   Palkhana No. 1 well, Iraq – official opening by King Faisal, 5 April 1927

the oil reservoir revealed that it was a field of immense productive capacity. By the end of 1930 twenty producing wells had been completed, in addition to which others had been drilled for the purposes of observing oil-water and gas-oil levels. At the same time, progress was made with work on ancillary facilities such as roads, water supplies, stores, workshops, offices, laboratories, dwellings, canteens and a hospital.[44]

31 Baba Gurgur No. 1 well, Iraq, where oil was discovered on 15 October 1927

While the physical development of the Kirkuk field went ahead, negotiations were also proceeding on the distribution of the share-holdings in the TPC and on the terms of its concession. As regards the distribution of shareholdings, there was, at the time the concession was granted in 1925, no US participation in the TPC whose shares were, as has been seen, divided between the Company, Royal Dutch-Shell, Compagnie Française des Pétroles (CFP) and Gulbenkian. The exclusion of US oil companies was a controversial issue in which the US State Department became involved in support of US access to Iraqi oil,

32   Baba Gurgur No. 1 well – oil gushing through side of arbor head, 1927

33   A river of oil flowing from Baba Gurgur No. 1 well, 1927

34   Married accommodation in course of construction at Baba Gurgur, late 1920s

invoking the principle of the 'open door', by which was meant equality of commercial opportunity for nationals of all countries in all parts of the world.[45] The matter was eventually settled in an agreement, signed on 31 July 1928, under which the Company, Royal Dutch-Shell, CFP and a group of US oil companies organised in the Near East Development Corporation each received 23.75 per cent of the TPC's shares, leaving the remaining 5 per cent in the hands of Gulbenkian. In other words, the Company's shareholding, which previously had been 47.5 per cent, was split between it and the Near East Development Corporation. By way of compensation for the halving of its interest, the Company was to receive an overriding royalty of 10 per cent on all TPC oil.[46] The agreement of 31 July 1928 dealt not only with the distribution of shareholdings, but also with other matters of which two should be mentioned. First, the TPC, whose functions were limited to crude production and transportation, was to be a non-profit-making enterprise simply producing and supplying crude oil for its shareholders.[47] Refining and marketing by the TPC was to be confined to meeting Iraq's domestic needs.[48] Secondly, the participants in the TPC re-affirmed the self-denying clause which they, or their predecessors, had agreed before World War I. By it, each of the TPC shareholders undertook not to engage, directly or indirectly, in the production or manufacture of oil in the area of the former Ottoman Empire except through the TPC. For purposes of definition the territory of the old Ottoman Empire was outlined in red on a map, as a result of which the terms agreed in 1928 came to be known as the Red Line Agreement.

At the same time as these various matters were being settled between the participants in the TPC, other negotiations were taking place between the TPC and the Iraqi Government on the terms of the concession. As has been mentioned, the 1925 concession gave the TPC 32 months within which to select 24 plots of 8 square miles each for its own exploitation so that areas outside the plots could be offered for open bidding. The selection of plots was, however, impeded by the restrictions on movements in certain parts of Iraq, notably in the northern Mosul *vilayet*, pending the solution of the boundary dispute with Turkey which was under investigation by a commission of the League of Nations and settled by treaty only in July 1926.[49] In the meantime the Foreign Office was very anxious that nothing should compromise a settlement and very sensitive to any unauthorised access to the areas in question. With the time allowed for the selection of plots due to expire in November 1927, a one-year extension was requested and granted in August that year.[50]

Even with the extension, the selection of plots, combined with the evaluation of the Kirkuk field which was discovered in October 1927, remained an arduous task. The negotiation of a further extension of the time allowed for the selection of plots was therefore one of Ritchie's objectives when, in early 1928, he spent three months in Iraq. On 28 March he learnt that an extension was acceptable to the Iraqi Government subject to ratification and modification of the prices for domestic supplies of oil products.[51] Ritchie left Baghdad at the beginning of April and the Iraqi Council of Ministers gave their consent to an extension later that month,[52] but King Faisal refused to approve it at the beginning of May. Why, after several months of discussions, had the King turned down the TPC's request for more time?

The answer, commonplace enough in negotiations of this kind, was that his hand had been greatly strengthened by the appearance of a competitor to the TPC. The rival company which presented itself to King Faisal late in April 1928 was British Oil Developments Ltd (BOD) under the chairmanship of Admiral Lord Wester-Wemyss and with a board generally populated by distinguished industrialists and City notables with a representative in Iraq, Colonel J. H. Stanley. A key figure was de Loys, a former TPC geologist. It was more an investment trust than an oil company, but was not lacking in industrial experience of steel and railway concerns. BOD not only requested that plots should be opened to bidding in accordance with the concession agreement, but also offered to build a railway to the Mediterranean at no cost to Iraq, a project which greatly appealed to the Iraqis.[53] The competitive pressure was intensified by the arrival of Wester-Wemyss and de Loys in Baghdad in May.[54]

With King Faisal apparently receptive to BOD's offers, Ritchie was urged by the Colonial Office to compromise on a railway scheme and to reach an arrangement with BOD.[55] Ritchie, however, informed the Foreign Office that he had no intention of coming to terms with BOD and that he was going to Baghdad, determined to force the issue on an extension of the time allowed for the selection of plots.[56] He duly travelled to Baghdad where, having met King Faisal, he was left under no illusions about the importance which the King attached to the construction of a railway.[57] At a TPC board meeting in London on 12 July divergent views were expressed, but nobody was in favour of railway construction.[58]

What the TPC participants really wanted was to get away from the plot system and to have a larger concessionary area for their own exploitation. They felt that the area open to them under the plot system was too small to be economically attractive in view of the cost of building the pipeline to which they had committed themselves in the concession, let alone a railway. The TPC's desire for a revision of the concession, and the Iraqi Government's wish to improve some of the terms from its point of view, led to the opening of negotiations, with the BOD on the sidelines.[59]

The complicated skein of negotiations between the Iraqi Government and the TPC, which was renamed the Iraq Petroleum Company (IPC) in 1929, continued from 1928 to 1931. During that period each side at one time or another complained about the other in words which illumine, on the one hand, the bitterness and resentment with which the Iraqis tended to regard the oil companies as arrogant foreigners and, on the other hand, the frustration of oilmen bemused by the caprice and volatility of Iraqi politicians. For example, in November 1930 J. Skliros, who had succeeded Ritchie as the IPC's general manager in London, wrote of King Faisal:

> What he, personally, has to complain about, and what the Iraqis resent, is the Company's [IPC] total disregard for Iraqi affairs; we have never attempted to get to know them or to identify ourselves in any way with

the life of the country; it is bad manners to keep aloof like that, and in Iraq it is put down to studied insolence.[60]

A view from the other side was given by Cadman a few months later when, on a visit to Baghdad, he became impatient with the failure to reach agreement on the IPC's liability to pay taxes in Iraq and wrote:

> It is quite impossible to give by telegraph an impression of the position here. The Government is irresponsible from King down and capable of anything. A Cabinet Committee has already expressed the opinion that the Concession is null and void and foolish as it may seem Government quite capable of passing a law to that effect. Cabinet Ministers continually threatening to resign and even talk of suicide.[61]

Somehow, despite such strains in the concessionary relationship, agreement was reached and a new concession signed on 24 March 1931. Under its terms the plot system was abandoned and the IPC obtained exclusive rights for seventy years over an area of about 32,000 square miles in the *vilayets* of Mosul and Baghdad east of the Tigris, but gave up all territory west of the Tigris. On the vexed issue of the IPC's liability to pay taxes, the Iraqi Government agreed to commute the taxes it might have imposed under Article 27 of the 1925 concession into fixed annual payments of £9,000 (gold) up to the commencement of commercial exports of oil, after which £6,000 was to be paid on the first 4 million tons of oil produced and £20,000 on each additional million tons. The IPC agreed that by the end of 1935 it would complete the construction of a pipeline from Kirkuk to the Mediterranean with a throughput capacity of at least 3 million tons a year. There was, however, no undertaking to build a railway. Until the commencement of exports, the IPC was to pay the Iraqi Government £400,000 (gold) of which half was to be 'dead' (i.e. non-recoverable) rent and half recoverable from future royalties. The level of royalty remained at 4s (gold) per ton with a minimum annual payment of £400,000 for twenty years, beginning with the first exports. Finally, the Company's overriding royalty on oil obtained from the concession was reduced from 10 to 7.5 per cent.[62]

After the new concessionary agreement had been signed the IPC moved ahead quickly with the planning and construction of the pipeline. In choosing the route and the terminal locations on the Mediterranean coast, differing views were expressed by the French, who favoured a northern route through Syria and the Lebanon terminating at Tripoli, and the British and Iraqis, who preferred a southern route terminating at Haifa in Palestine. The issue was settled by a compromise which provided for the construction of two pipelines, each with a throughput capacity of two million tons a year, which were to run parallel from Kirkuk to Haditha, where they were to split, with one line taking the northern route to Tripoli and the other running across Transjordan and Palestine to Haifa. The length of the northern line would be 532 miles, that of the southern line 620 miles.[63] There were, of course, other matters which also had to be settled before construction could take place, such as obtaining the agreement of the transit states to the passage of the pipeline, choosing sites for the pumping stations and acquiring land. The construction itself was a major undertaking given the difficult terrain, lack of water supplies, rough desert tracks and river barriers. Yet no serious delay held up the work, which took until late in 1934 when the pipeline came into operation before being officially declared open by King Ghazi at a ceremony at Kirkuk in January 1935.[64]

The opening of the pipeline not only marked the entry of Iraq into the ranks of the major oil exporting nations, but was also an event of some importance in the history of the Company which, with its 23.75 per cent stake in the IPC, now had, for the first time, a source of export crude oil outside Iran (production from the Naftkhana field being entirely for Iraqi domestic consumption). The Company's crude offtake was admittedly much less than its crude production in Iran, as table 5.1 shows. However, the Kirkuk field had ample capacity for increasing production above the four million tons a year which, in approximate terms, was its output in the second half of the 1930s.[65] The limiting factor was the size of the pipelines and, after considering the potential outlets for Iraqi oil, the IPC decided in 1938 that the pipeline capacity should be at least doubled. As it happened, the implementation of the decision was deferred because of the outbreak of World War II.[66]

Meanwhile, work proceeded on the development of the Kirkuk field, with production wells supplying three de-gassing stations and other wells being drilled for later production and for observation purposes. Communications were improved, the power house extended, a new topping plant completed, offices constructed, the hospital enlarged and a crude oil stabilisation plant erected, coming into operation in 1937.[67] At the other end of the pipeline a refinery was constructed at Haifa by Consolidated Refineries Ltd, owned equally by the Company and Royal Dutch-Shell. Despite the outbreak of World War II, the refinery came onstream in December 1939, processing Iraqi oil transported through the pipeline.[68]

35   Drilling for water on the desert route of the IPC pipeline, early 1930s

36   Tribal watchmen in the Iraqi desert, early 1930s

At the same time as production, pipeline and refining facilities were put in place for the commercial exploitation of oil discovered at Kirkuk, the IPC also extended its concessionary area in Iraq. It will be remembered that the 1931 concession included only those parts of the Mosul and Baghdad *vilayets* which lay east of the Tigris, leaving the

37   Pipe-wrapping the IPC pipeline, early 1930s

area to the west vacant. Not long afterwards it was taken up by BOD, which was granted a concession for the area of some 46,000 square miles west of the Tigris and north of the 33-degree line in May 1932. The royalty was the same as in the IPC concession at 4s (gold) per ton and provision was made for minimum payments and 'dead' rents pending the commencement of production. BOD also agreed to

Table 5.1 *The Company's crude oil production and offtake in Iran and Iraq, 1930–1938 (million tons)*

	Iran	Iraq	
		Naftkhana	Company share of IPC production
1930	5.94	0.08	–
1931	5.75	0.08	–
1932	6.45	0.09	–
1933	7.09	0.08	–
1934	7.54	0.09	0.29
1935	7.49	0.10	1.11
1936	8.20	0.10	1.18
1937	10.17	0.12	1.23
1938	10.20	0.13	1.24

Sources: See Appendix 1.

38   Arab women gathering fuel by the IPC pipeline (under construction), early 1930s

39   The IPC pipeline at the Tripoli terminal, eastern Mediterranean, *c.*1934

40   The IPC pipeline at the Haifa terminal, eastern Mediterranean, c.1934

minimum drilling obligations and was given seven years and six months in which to produce and export oil at a rate of at least a million tons a year. For its part, the Iraqi Government was entitled to take up to 20 per cent of the oil produced and to dispose of it either for purposes of local consumption or by reselling it to the concessionary company. For the Iraqis this entitlement represented a notable improvement on the terms of the IPC concession, which included no such provision.[69]

After the grant of the BOD concession in 1932 capital from sundry international sources was raised and a new holding company, Mosul Oilfields Ltd, was formed to acquire the share capital of BOD. Exploratory drilling was carried out at various locations, but the only oil discovered was of relatively low quality. With mounting expenses putting the concessionary enterprise under financial strain, the IPC purchased shares in Mosul Oilfields and by 1937 had acquired effective control. In 1941 the Mosul Petroleum Company, a wholly owned subsidiary of the IPC, received the formal assignment from BOD of its 1932 concession. BOD and Mosul Oilfields were liquidated and the IPC now held the 1931 and 1932 concessions covering the whole of the *vilayets* of Mosul and Baghdad.

The only part of Iraq outside these *vilayets* and the Transferred Territories was the southern *vilayet* of Basra. Here, a concession was granted to the Basra Petroleum Company, a sister company of the IPC, in July 1938. The terms were very similar to those of the 1932 BOD concession, including a royalty of 4s (gold) per ton with a minimum annual payment; 'dead' rents until exports of oil commenced; minimum drilling obligations; an obligation to commence exports of at least a million tons a year within seven years and six months; and an Iraqi Government entitlement to 20 per cent of the oil produced.[70]

To sum up, the result of these concessionary maneouvres was that the shareholders of the IPC came to hold three concessions covering the whole of Iraq outside the Transferred Territories. These were the IPC's concession for the area east of the Tigris granted in 1931; the Mosul Petroleum Company's concession west of the Tigris originally granted to BOD in 1932; and the Basra Petroleum Company's concession granted in 1938. Only in the first of these, where the Kirkuk field was located, had commercial oil production commenced by the outbreak of World War II. In a supplementary agreement which was reached between the Iraqi Government and the IPC group in May 1939, the IPC agreed to make an interest-free loan equivalent to £3 million, recoverable from future royalties, to the Iraqi Government. In return, the Government consented to some modifications of the concession agreements, most notably in the drilling obligations of the 1932 concession and a seven-year extension of the period within which oil from that concession was to be produced on a commercial scale.[71]

CONCESSIONS ELSEWHERE

Although Iraq was easily the most important scene of operations for the IPC, other areas within the Red Line of the 1928 agreement also attracted its attention. Saudi Arabia was one of them, but in concessionary negotiations with King ibn Saud in 1933 the IPC failed to match the terms offered by Socal, which in May that year obtained the concession for the al-Hasa province covering eastern Saudi Arabia. Participation in the concession was broadened by the admission of the Texas Company to a half-share in 1936 and oil was discovered in 1938, the first of a series of discoveries which were to confirm the immense oil reserves of the al-Hasa area.[72]

Although the IPC missed out on the major prize of the al-Hasa concession, it was more successful in Qatar where a concession agreement, reached between the Shaykh and the Company in May 1935, was assigned to the IPC shortly afterwards. Preparations for drilling were put in hand and the first well was spudded in at Dukhan in October 1938. Oil was discovered there in December 1939, but in 1940 operations were suspended for the duration of the war.[73]

To the south-east of Qatar lay the Trucial Coast of Arabia, which included half a dozen small shaykhdoms in which the IPC showed interest in the 1930s, being granted concessions over the lands of Sharja and Dubai in 1937 and Abu Dhabi in 1939, as well as shorter-term exploration permits for the shaykhdoms of Ras al-Khaima and Ajman. The IPC also obtained concessions over parts of the Sultanate of Oman in the south-eastern corner of Saudi Arabia and, in 1936, over the Hejaz province of western Saudi Arabia, with, in addition, an exploration licence, granted in 1938, in the British protectorate of Aden. However, in none of these areas was oil discovered in commercial quantities before World War II.[74]

The IPC was also active in the Levant states of Palestine, Transjordan, Syria and Lebanon which, being parts of the former Ottoman Empire, fell within the Red Line. Exploration licences were obtained in all these territories except for Transjordan and in February 1938 a full concession was agreed with the Syrian Government for later ratification. Varying degrees of progress were made with surveys and drilling in different locations without any commercial oil discovery being made.[75]

Outside the Middle East and South America, the Company was generally less active in pursuit of concessionary opportunities in the 1930s than it had been in the 1920s, when there had been a great deal of Company interest in obtaining concessions across the world.[76] In Europe, proposals for concessions in Romania were virtually an annual occurrence, but aroused little interest. The Company maintained a reduced interest in Albania in the early 1930s without restarting drilling activities. Consideration was given to occasional offers of concessions in Portugal, Greece, Germany, Austria and Yugoslavia, but neither the geological nor the political conditions were encouraging. In Australasia, early hopes in Papua were disappointed, but in 1938 agreement was reached with Royal Dutch-Shell for co-operative exploration in New Zealand. In Africa, intermittent attention was paid to sundry areas, including Abyssinia and Uganda in 1929, Somalia in 1935, Nigeria, Kenya and Tanganyika in 1936. In association with Royal Dutch-Shell, the Company carried out surveys and test drilling in Nigeria between 1936 and 1939. In Canada, New Brunswick was regarded with short-lived favour in the late 1930s.

For the most part, however, the incentive for widespread concession-seeking and exploration was undermined by the availability of
plentiful crude oil supplies from Iran and by the existence of excess
productive capacity in the international oil industry in the 1930s. In
such conditions, the Company and its main international competitors
were naturally concerned less with searching for new sources of oil
than with bringing surplus capacity under control through the application of the 'As Is' principles described in chapter 4. Although confidence in Iran was jolted by the cancellation of the D'Arcy concession
in 1932, the negotiation of a new concession agreement was followed
by the discovery of new Iranian oilfields and by pressure for more
production from the Shah which tended to increase the Company's
emphasis on Iran. In some areas, especially those in South America
which were located close to the main centres of consumption in
western markets, new concessions would have been welcomed.
However, in both Colombia and Venezuela the Company encountered
barriers to entry on account of the British Government's shareholding.
Elsewhere, most notably in Kuwait, the Company had a largely defensive interest in obtaining concessions to prevent them falling into the
hands of competitors who might challenge the supremacy of the
Company and Iran in Middle East production. Such action tended,
however, to create further problems, for it was difficult to obtain
concessions merely to take control over reserves without a commitment to production. With the completion of the IPC pipeline in 1934,
the Company began to receive large quantities of Iraqi crude oil to add
to its Iranian production. Moreover, the new oil discoveries in Kuwait
in 1938 and Qatar in 1939 held out the definite prospect of still greater
production coming onstream, promising to add to the excess of productive capacity over marketing outlets which was to result, as shall be
seen in chapter 11, in a growing emphasis on bulk, high-volume
business in the postwar years.

= 6 =
The British Government and oil, 1928–1939

Before World War I British oil policy had been formulated primarily around the perceived need to obtain secure (effectively synonymous with British-controlled) fuel oil supplies for the Admiralty, which was then engaged in modernising the British navy by replacing coal-burning ships with oil-fuelled vessels. Security of oil supplies was the Government's main motive when, in 1914, it agreed to inject new capital into the Company, which desperately needed a new source of finance as it struggled with the many technical and capital-intensive problems of turning an oil discovery in Iran into a viable commercial business. In providing the new capital the Government gained not only a majority shareholding in the Company and the power to appoint two non-executive directors to its board, but also a contractual commitment from the Company that it would supply fuel oil to the Admiralty. That commitment was not, it should be said, extracted from the Company under sufferance. On the contrary, the Company needed the fuel oil contract as a sales outlet as badly as it needed the new capital to keep it afloat at a time when it might otherwise have been swallowed up by Royal Dutch-Shell.[1]

Although the Government, as majority shareholder with representatives on the board, undertook not to interfere in the normal commercial operations of the Company, its intervention in 1914 was an unusual departure from the British tradition of *laissez-faire* which, if diminishing in influence, still represented the orthodox principle governing relations between state and industry. The very fact that the Government was willing to depart from that orthodoxy in relation to the Company tends only to emphasise the strategic importance which was assigned to oil. Security of supplies was not an aberration from past and future policy, but a consistent concern. The feeling that

Britain should not be dependent on foreign-controlled companies for its oil supplies had already manifested itself in 1904 when the policy was laid down that oil concessions in the British Empire would be granted only to companies under British control. Later on, World War I drove home the importance of oil for warlike purposes and stimulated Government interest in fostering the creation of a new all-British oil company by combining, in various possible permutations, the Company, the Burmah Oil Company, Royal Dutch-Shell and Lord Cowdray's Mexican Eagle Company.[2]

At the same time as the Government was interested in bringing foreign sources of oil under British control, it also pursued the same end of achieving security in oil supplies by investigating and supporting various schemes for the production of oil from indigenous resources. In 1917 a committee was appointed to consider the use of gas as an alternative to petroleum and in 1918 another committee was established to investigate the possibilities of using alcohol as fuel for internal combustion engines. Methods of producing oil from coal were also examined, as were measures to increase the output of shale oil, for which the Company became responsible after it acquired Scottish Oils Ltd in 1919. As for more conventional oil exploration, Cowdray's company began to explore for liquid petroleum in Britain with financial assistance from the Government and found oil in small quantities at Hardstoft in Derbyshire in 1919.[3]

Despite the above efforts, the Government was not markedly successful in adding to the security of oil supplies by increased production in Britain and the Empire. In fact, in the decade following World War I the oil production of the Empire fell from 2.5 per cent of world production to only 1.6 per cent. That poor record could, in some degree, be attributed to two regulatory impediments which stood in the way of a more energetic search for oil. One was the prohibition on foreign-controlled companies exploring for oil in the Empire, which not only resulted in the exclusion of most of the world's leading oil companies, but also invited retaliation from other countries in the form of reciprocal restrictions on British companies. The second was that in Britain ownership of underground oil reserves was vested in the surface landowners, so that the search for oil was hampered by the need to negotiate royalty rights with a multitude of private interests.

In dealing with those and other aspects of oil policy a lamentable multiplicity of Government departments was involved, to the annoyance of Cadman whose strength of feeling on the subject was remarked upon in chapter 2. After the end of World War I the

Petroleum Executive, which had been set up in 1917 under Cadman as part of the Government's wartime machinery for the control of essential supplies and industries, was reconstituted as the Petroleum Department and placed under the Secretary for Overseas Trade. In 1922 the Petroleum Department was absorbed into the Board of Trade and in 1928 it was downgraded when it became a branch of the Mines Department of the Board of Trade.[4] Its lowly importance was illustrated in November that year when the Foreign Office received a request from the Portuguese Embassy for information about the Petroleum Department and replied that it had no statutory functions and acted solely in an advisory capacity.[5] On another occasion, in January 1929, the Foreign Office received a letter from the Finance Department of the Board of Trade indicating that it was not 'in any way connected with the Petroleum Department'. The Foreign Office merely remarked that it would 'certainly have expected' that the two departments of the same Ministry would have collaborated more closely.[6] As a further indication of the Petroleum Department's continuing lack of weight, by 1934 there were only three officials working full-time on oil affairs and three others working part-time.[7]

As the above comments suggest, there was no authoritative body responsible for oil affairs in the machinery of state. Instead, a number of departments had partial concerns with oil. The Admiralty was interested in the supply of fuel oil of the right specification for the navy. The Colonial Office was anxious about the viability of colonial economies, which might be enhanced by oil exports. The Foreign Office was aware of the diplomatic repercussions of oil issues. The Treasury kept itself informed about the Company's activities. Technical developments in exploration, production and refining were noted. Those responsible were not indifferent to oil affairs, but their concern did not manifest itself in any concerted action to establish a positive, comprehensive oil policy.[8]

OIL IN BRITAIN AND THE EMPIRE

Despite the overlapping and sometimes conflicting interests of different Government departments, exploration for oil in Britain and the Empire was facilitated by changes in the regulations in the interwar years. Nationality restrictions and the 'closed door' policy of keeping oil resources in British-administered lands from falling into foreign hands were at loggerheads with the 'open door' policy of the USA, whose State Department successfully put pressure on Britain to yield ground to US oil companies in the 1920s. As a result, US oil companies won admission to the Turkish Petroleum Company in 1928, and in Bahrain the British nationality restriction was relaxed to the point of being effectively lifted in 1930 so that the US oil company, Socal, could hold the concession. In the meantime, the British Government, through its various departments, considered the desirability of allowing foreign companies to explore for oil in the Empire, subject to conditions which would safeguard the Admiralty's supplies of fuel oil and preserve British rights of pre-emption over oil in the event of emergency. General agreement emerged that an open door policy would help to stimulate a more rapid development of the Empire's oil resources and in 1930 the termination of the closed door policy was approved in principle by the Cabinet. However, although licences to prospect for oil were granted in many colonial territories, the results were meagre and production in the colonies remained insignificant up to World War II.[9]

In Britain itself, the small oil discovery at Hardstoft in 1919 did not result in early commercial production and at the end of the 1920s the only indigenous source of oil was the Scottish shale oil industry, whose origins date back to the mid-nineteenth century. After initial expansion, many workings were abandoned because of inferior deposits or uneconomic rates of return. Production reached its peak in 1913 and although the industry had fallen into a 'parlous state' by May 1919,[10] the Mines Department declined to intervene.[11] However, in that year the Company took over Scottish Oils Ltd, composed of the remaining six shale oil companies which had been brought together into one organisation by Fraser, who was then managing director of the Pumpherston Oil Company and chairman of Scottish Oils. The Company acquired sites and selling agencies by which it could distribute its own products in Scotland and helped to modernise the shale oil industry. The Government thus escaped the problem of either subsidising or abandoning the industry. The extraction of oil from the solid clay-like shale, which had first to be dug from the earth, was, however, considerably more expensive than conventional liquid petroleum production and the industry continued to be uncompetitive[12] despite its exemption from the excise tax which was levied on non-indigenous motor spirit from 1928. Although the refining process was gradually altered to produce a better yield of motor spirit, the prospects of the shale oil industry were insufficiently encouraging to place much reliance on it as an additional secure source of oil in an emergency.

Exploration for liquid crude oil reserves held more interesting possibilities. Soon after his appointment as the Company's chief geologist in 1929, G. M. Lees reviewed British oil prospects and expressed the opinion that further exploration was justifiable. During the next few years he, assisted by other geologists, diligently checked the field evidence and relevant literature and began a campaign to persuade his principals to authorise a search for oil in Britain. Lees' later reminiscence suggests that his success in winning approval for domestic exploration was finally achieved when he took Cadman to Lulworth Cove to see the outcropping bituminous sands of the Wealden series. While resting in the course of their climb over the outcrops, Cadman sat on a richly bituminous sandstone. When he came to get up his trousers were well and truly stuck to the softened bitumen and that, according to Lees' later memories, clinched the argument in his favour.[13]

Cadman had already been attracted to the idea of exploration in Britain, reporting to the Board of Trade in August 1929 that on several occasions during the previous few years the Company had been invited by various private interests to carry out exploratory work, but had been deterred by the difficulties of acquiring surface and mineral rights from the numerous surface owners.[14] By 1931 he was convinced that the geological evidence was favourable and within three years, partly at his urging, the Government introduced the Petroleum (Production) Act 1934. The Act vested all property rights over petroleum in the Crown and empowered the Government to issue licences to explore for and produce oil, so that it was no longer necessary to negotiate royalty rights with many private landowners. Under statutory rules and orders issued in furtherance of the Act in May 1935, foreign companies were permitted to apply for licences provided that an operating company was formed and registered in Britain.[15] By the end of 1938 about ninety prospecting licences had been issued, over half of them to the Company and most of the remainder to the Anglo-American Oil Company (a subsidiary of Standard Oil (NJ)) and the Gulf Exploration Company.[16]

Having been granted its first prospecting licences covering 6946 square miles in December 1935, the Company commenced drilling operations in Britain in January 1936. Its British crude oil production started in August 1938 at Hardstoft, where the producing well drilled earlier by Cowdray's company was reconditioned and deepened by the Company, into whose hands it had passed.[17] At Formby in Lancashire drilling was commenced on 2 March 1939 and a small oilfield discovered, from which production commenced in May. In the same year a larger field was discovered near the village of Eakring in Nottinghamshire where a well was started on 26 March and completed as a producing well in July.[18] With the Formby and Eakring fields coming onstream and adding to the very small quantities produced at Hardstoft, the Company thus became a crude oil producer in Britain virtually on the eve of World War II.

OIL FROM COAL

Aside from shale oil and the search for liquid crude oil there was also considerable interest in the production of oil from coal. In Germany, abounding with coal and chemists, much research had been carried out in this field, the most advanced process being the hydrogenation technique invented by Dr Friedrich Bergius. In Britain, too, there was interest in the subject. In 1926 the Board of Trade appointed a committee to advise on the potential for converting different fuels into various forms of energy. In that connection, the Government-funded Fuel Research Station at Greenwich started to investigate the conversion of coal into oil by hydrogenation. The results were sufficiently encouraging for Dr Lander, the Director of the Fuel Research Station, to tell the Imperial Conference held in London in 1930 that there were 'reasonable grounds for supposing that petrol of very good quality could be produced by this means at a price that would not be too expensive'.[19]

It was not only the Government's interest that was aroused. The idea of producing oil from coal was also taken up with great enthusiasm by Imperial Chemical Industries (ICI) after that company was formed by merger in 1926 as the dominant firm in the British chemicals industry, capable, so it was hoped, of holding its own against the might of its much-respected foreign competitors, above all IG Farbenindustrie of Germany. The possibility of manufacturing oil from coal held great appeal for ICI's first chairman, Sir Alfred Mond (elevated to the peerage as Lord Melchett in 1928) whose business acumen was prone to be subordinated to grand imperial designs which, at times, had deleterious effects on the economic robustness of his company's investment decisions.[20] To a businessman of Mond's predisposition there were several reasons to be enthusiastic about venturing into the production of oil from coal: it would strengthen the British economy by saving foreign exchange on oil imports; it would help to relieve the beleaguered British coal industry which was suffering from a loss of export markets; and it would strengthen Britain's defences by promoting self-sufficiency in oil. Set against those benefits was one very considerable handicap: the production of oil from coal was more expensive than conventional liquid petroleum. In other words, it was uncompetitive. However, for ICI that was not a sufficient reason to abandon interest. In the 1920s, with more optimism than foresight, ICI invested huge sums of capital in a large-scale plant at Billingham, near the mouth of the Tees, for the production of nitrogen to make fertilisers for, it was envisaged, the British Empire. After the collapse of the nitrogen market which came with the onset of the great economic depression in 1929, the Billingham plant was greatly under-utilised and any prospect of earning a return on the capital so lavishly expended on it was to be seized upon. As the process of ammonia synthesis by which nitrogen was produced at Billingham had strong technical links with the Bergius process of hydrogenation, an oil-from-coal venture had obvious attractions. Moreover, Mond, an ex-Government minister whose first choice of career had been in politics, would have been more aware than most of the possibilities of appealing to the Government for help with a project which, though uncompetitive, had attractions as a source of employment for coal miners, as a method of import substitution and as a secure source of oil in time of war.

It was not, however, only with the Government that ICI needed to negotiate, for it was not the only company interested in producing oil from coal by hydrogenation. So too, in various ways which it would be excessively complicated and tedious to describe here, were Royal Dutch-Shell, Standard Oil (NJ) and IG Farbenindustrie.[21] With the emerging possibility that the oil giants might invade the chemicals industry and *vice versa* there was every prospect of a clash between the titans of the two industries unless, as was so often the case in such situations, the companies concerned came to a private treaty. Co-operation prevailed over competition and by a series of interrelated agreements signed in 1931 the four companies of Royal Dutch-Shell, Standard Oil (NJ), ICI and IG Farbenindustrie reached the International Hydrogenation Patents (IHP) Agreement, sometimes simplified to the Hydrogenation Cartel.[22] One of its provisions was that ICI was to purchase any oil it needed for hydrogenation and to sell any oil it produced through nominees of IHP, meaning, in effect, Royal Dutch-Shell or Standard Oil (NJ).

In the meantime, ICI, having begun to construct a pilot oil-from-coal plant in 1929, had approached the Government for some form of help. Motor spirit produced in Britain from indigenous coal or shale was already exempt from the duty of 4d per gallon on imported motor spirit which was introduced in the Finance Act of 1928. The value of the exemption was increased when the duty was raised to 6d and then 8d in 1931.[23] After lengthy negotiations between ICI and the Government, it was eventually decided to give oil produced from British coal a guarantee of continued tariff protection against imported oil. That was provided by the British Hydrocarbon Oils Production Act 1934 which gave a statutory guarantee of a preference which was described, somewhat confusingly, as amounting to 36 pence-years. What that meant was that the rate of preference might, for example, be 4d per gallon for nine years (4 multiplied by 9 amounting to 36), 6d for six years (6 multiplied by 6 amounting to 36), or some other combination which came to 36 pence-years. Encouraged by the preference, ICI built a full-scale hydrogenation plant for the production of oil from coal at Billingham. It came onstream in 1936, but was not a commercial success. Coal turned out to be an unsuitable raw material on both economic and technical grounds, and was soon given up in favour of creosote, which gave a more manageable flow through the plant. However, even with creosote the results were disappointing.[24]

The Company, meanwhile, was following closely the development of other processes for the production of oil from coal, particularly the Fischer–Tropsch process. This process had been discovered in Germany, where it was brought into commercial production in 1936. It also excited considerable interest in other countries, including Britain, where the Company and Powell Duffryn Associated Collieries made a joint investigation into the commercial viability of erecting Fischer–Tropsch plant. Much information was collected, but the Company was not persuaded that the process was commercially viable. That view was shared by the Falmouth Committee, which was set up in 1937 by the Committee of Imperial Defence to examine and report on the merits of producing oil from coal in Britain. The Committee concluded that the home production of oil from coal could not be justified on either economic or national security grounds. However, feeling that Britain ought to keep up with technical developments, they favoured the erection of a single Fischer–Tropsch plant.[25] Acting on the Committee's recommendations, the Government provided encouragement for such a plant in the 1938 Finance Act, which increased the amount and duration of the guaranteed preference for motor fuel produced from indigenous raw materials. Despite that measure, the Company remained sceptical about the economic viability of producing oil from coal. Thus after the outbreak of World War II the Company informed the Jowitt Committee – which was conducting a further enquiry into the production of oil from coal – that, having examined the possibilities, the Company had concluded that 'there is no good case for the establishment of such plants in this country'.[26]

AVIATION SPIRIT

Despite ICI's problems with its Billingham plant, the outlook for oil produced from coal by the hydrogenation process was brightened by the growing concern about the availability of aviation spirit which was being shown by those who were engaged in planning for war. As aero-engines of improved performance and higher compression ratios were developed there was increasing demand for aviation spirits of high enough octane numbers to prevent engines 'knocking' (causing not only inferior performance, but also overheating). In the early 1930s aero-engines generally worked satisfactorily on a fuel of about 77-octane, but by 1935 all new engines delivered to the RAF were designed for 87-octane fuel. Spirit of that quality could be produced by blending base spirit with iso-octane, a product of excellent anti-knock value which could be manufactured by the two stages of polymerisation and hydrogenation, described more fully in the next chapter. In 1937 the Air Ministry, wishing to secure adequate supplies, contracted to purchase iso-octane over periods of three to five years from three firms: Trinidad Leaseholds Ltd, Royal Dutch-Shell and Standard Oil (NJ).[27]

However, by 1939 the Air Ministry had to contend with a larger supply problem as the demands for aviation spirit increased in quantity and became more exacting in quality with the introduction of new high performance aircraft which required spirit of 100-octane rating. In autumn 1938 the Government appointed a committee under Sir Harold Hartley, chairman of the Fuel Research Board, to seek a solution to the problem. The membership of the Committee consisted, as would be expected, of those who were considered to have most knowledge of the subject: representatives of the Air Ministry; of Trinidad Leaseholds and Royal Dutch-Shell, who had contracted to supply the Air Ministry with iso-octane; and of the British pioneer of hydrogenation, ICI, whose plant at Billingham produced spirit which, costly though it may have been to manufacture, was of a higher quality than conventionally refined crude oil and therefore of greater suitability for blending into aviation spirit.

The Hartley Committee reported in December 1938, recommending that three new plants should be built to manufacture iso-octane which, blended with a high-octane base fuel manufactured by the hydrogenation process already in use at ICI's Billingham plant, would produce 100-octane spirit. Because of the danger of air attack, the Hartley Committee recommended that only one of the new iso-octane plants should be built in Britain, at Heysham in Lancashire, which was considered to be relatively safe from aerial bombing. The other two plants should, it was suggested, be located in Trinidad, which was British territory and the source of the gas oil which was to be used as the feedstock in the new plants. Capital for the construction of the plants was to be provided by the Government, but to assist it on the technical side the Air Ministry called upon ICI, Royal Dutch-Shell and Trinidad Leaseholds, who formed a joint company called Trimpell Ltd to supervise the construction and operation of the plants. Work on the first plant, that at Heysham, began in the spring of 1939 and was completed in 1941. As for the Trinidad plants, they were never built, it being decided that the number of plants should be reduced from three to two, both of which should be built in Britain. Construction of the second plant duly began at Thornton, but it was abandoned in 1941.[28]

The Company, it will no doubt have been noticed, was not a party to these arrangements, nor did it have a representative on the Hartley Committee. Its absence should not, however, be taken as a sign that it played no part in developing processes for the manufacture of the high-grade aviation spirit which was required for the war effort. On the contrary, as will be seen in the next chapter, its chance discovery of the alkylation process, by which iso-octanes could be produced in one stage rather than the two-stage process of polymerisation and hydrogenation, was an important advance. In fact, as events turned out, the Company's alkylation units were to make a larger contribution to wartime aviation spirit manufacture than the hydrogenation techniques which were the subject of so much attention in the prewar years.

COMBINATION, RATIONALISATION AND PRICES

Apart from the problems of securing supplies of oil products which were essential for war purposes, various other oil-related matters arose, of which two, in particular, raised important questions of public policy: one was the tendency towards increased concentration in the industry or, put another way, the power of a few large firms to control the market; the other, not unrelated to the first, was that issue of inexhaustible consumer interest, petrol prices. On those two issues the

Government's attitude could be broadly summarised as passive approval of increased concentration and non-interference in prices.

The Government's benign blessing of the mergers and associations which increased concentration, and its unwillingness to use the power of the state to hold down prices were consistent with the general leanings of government policy at the time. As was remarked upon in chapter 4, in the late 1920s there was a wave of enthusiasm for industrial rationalisation, a term which, ill-defined as it was, definitely embraced the idea that mergers and combinations were to be encouraged as a means of developing large-scale enterprises which would be able to realise economies of scale and so improve the international competitiveness of British industry. In such a climate of opinion, British industries were commonly criticised, not for being monopolistic, but for the opposite sin of perpetrating the existence of too many competing small firms. Prices, far from being driven downward to unremunerative levels, needed, it was widely believed, to be stabilised.

In that environment there was nothing remarkable in the Government's reception of the agreement reached between the Company and Royal Dutch-Shell in 1928 to establish an equally owned joint marketing organisation, the Consolidated Petroleum Company, to market their products in a huge area stretching from the Middle East to Africa (see chapter 4). In February 1928 the proposed agreement was considered at a meeting of the Committee of Imperial Defence, chaired by the Prime Minister, Stanley Baldwin.[29] When Cadman, accompanied by the Company's two Government directors, Barstow and Packe, joined the meeting, he informed the Committee that with production set to increase in Iran and oil having been discovered in Iraq, it was essential that new markets be found. Cadman presented the proposal for joint marketing as being far more economical than fighting for a footing in the markets of Africa, which were considered to have great potential. Adopting the language of rationalisation, he argued that joint marketing would minimise the duplication of overlapping facilities and result in lower costs of distribution. The Government, feeling that it 'could hardly interfere with the expansion of the Company into new spheres', concluded that it should place no obstacles in the way of the proposed agreement, which was duly put into practice.[30]

When the Company and Royal Dutch-Shell later merged their British marketing subsidiaries into the jointly owned Shell-Mex and BP company, there was very little discussion. The Petroleum Department had been alerted to the outlines of an agreement in November 1931. The Mines Department saw it as a purely commercial arrangement and understood the desirability of reducing distribution costs so that both companies could deal more effectively with recently established competitors selling at lower prices.[31] The Treasury was not concerned, 'a plain case for non-interference on the part of the Government'.[32] The pattern of policy had been set. The Government was not anxious to interfere in the Company's commercial affairs.

On the question of prices, interest focused mainly on motor spirit, the retail price of which was held up more by the Government's imposition of excise taxes than by the oil companies during the late 1920s and 1930s. As has been briefly mentioned earlier, in 1928 Winston Churchill, as Chancellor of the Exchequer, revived a source of revenue which had fallen into abeyance since 1920 by imposing a duty of 4d per gallon on motor spirit. In April 1931 the tax was increased to 6d and in September the same year raised yet again to 8d, before its last prewar rise to 9d from 25 April 1938. As table 6.1 shows, tax revenues from oil used as fuel for road vehicles rose from about £13 million in 1929 to £55 million in 1939.

Although the rise in excise taxes was mostly responsible for pushing up retail prices of motor spirit in 1928–9, concern about the oil companies' control over prices was persistently raised in the House of Commons. 'What', it was asked in the committee stage of the 1928 Finance Bill, 'will happen if the great oil companies get control again of the price of the commodity?'.[33] In the same vein, Lt.-Commander Kenworthy (Conservative), a persistent critic of the Government's attitude to oil, asked Churchill 'what instructions have been given to the Government directors on the Anglo-Persian Oil Company's board with reference to a proposed increase in petrol prices?'.[34] Churchill replied that the Government was under an obligation not to interfere in the commercial management of the Company.[35] More pointedly, a Labour member inquired whether the Government directors made regular reports to the Government about what was happening on the board.[36] Churchill maintained that the Government kept in close touch with its representatives, but did not intervene in the Company's commercial affairs. Two days later he faced more critical questioning on whether the Company's participation in a price fixing ring was a matter of policy affecting the interests of the taxpayer, as distinct from an ordinary commercial question.[37] Churchill answered that the marketing of oil was 'not a matter within the scope of legitimate government interference' and insisted that nothing had taken place which called for Government intervention.[38]

When the price of motor spirit was increased in March 1929, certain

Table 6.1 *Tax revenue from light and heavy oils for road vehicles in the UK, 1929–1939*

Year to 31 March	Tax rate (d per gallon)	Tax revenues			
		Motor spirit (£millions)	Other spirit (£millions)	Heavy oil (£millions)	Total tax revenue (£millions)
1929	4	12.70	0.24	–	12.94
1930	4	14.69	0.27	–	14.96
1931	6d from 28 April	15.58	0.26	–	15.84
	8d from 10 September				
1932	8	28.69	0.47	–	29.16
1933	8	34.66	0.54	–	35.20
1934	8	36.93	0.60	–	37.53
1935	8	38.92	0.65	–	39.57
1936	8	40.69	0.69	0.87	42.25
1937	8	42.56	0.72	1.62	44.90
1938	9d from 27 April	44.06	0.76	2.31	47.13
1939	9	50.71	0.91	3.43	55.05

Notes: 1. Motor spirit includes aviation spirit.

2. Heavy oil was taxed at the motor spirit rate from 8 August 1935. Previously at 1d per gallon.

Source: Annual reports of Commissioners of Customs and Excise.

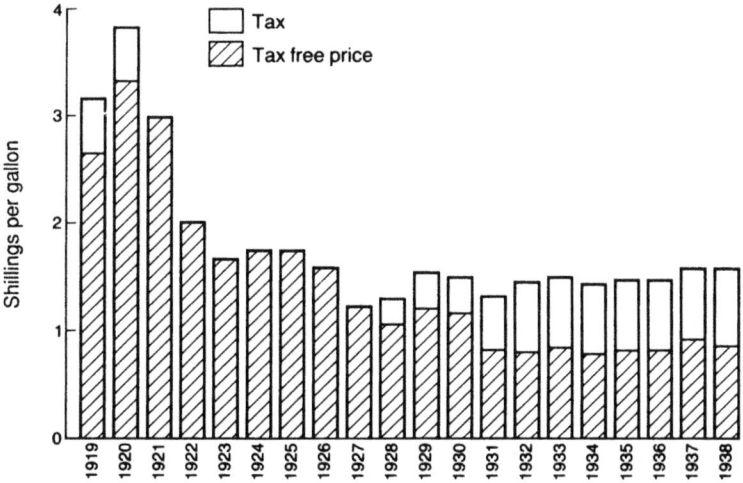

Figure 6.1 UK motor spirit prices, 1919–1938
Source: See Appendix 2.

members of Parliament accused the Government directors of the Company of conniving with the other two main companies in raising prices simultaneously and wondered 'whether any useful purpose can now be served by continuing their services'.[39] Baldwin, though committed to the principle that the public was 'best served by an efficient industry operating freely', admitted that special considerations might arise where a company or group of companies dominated the supply and distribution of an article of common use. The Government would therefore ask the oil companies concerned to explain the reasons for the price rise.[40]

Justifying the increase, the three companies (Royal Dutch-Shell, Standard Oil (NJ) and the Company) claimed that apart from excise duty and a period of exceptionally low prices between April 1927 and February 1929, prices were lower than for a decade.[41] As can be seen from figure 6.1, their claim was fully justified, pre-tax prices for motor spirit having fallen dramatically in the 1920s before later levelling off in the 1930s when tax accounted for a growing proportion of the retail price. The three companies also argued that Soviet oil was sold at cut prices, so there was competition to be met. They explained that prices were based on those prevailing at the Gulf of Mexico which 'constituted the centre of gravity of the world's petroleum trade' and the basis of world prices for oil products. Imports of oil products into Britain represented only 4 per cent of the world's total consumption, quite insufficient to influence supplies or prices. Over two-thirds of British supplies came from the western hemisphere, which dominated the price structure.

The Petroleum Department accepted that posted prices at US Gulf ports constituted 'a proper starting point' and that the Company had 'really no alternative but to work in conjunction with the other two big distributing companies'.[42] The Department also recognised that if the Company became embroiled in a price war there was little doubt that it would be defeated and forced out of distribution, leaving it confined to production and refining. That would give the two foreign-controlled companies of Royal Dutch-Shell and Standard Oil (NJ) a clear field which would be worse than ever for the customer. The Treasury also concluded that there was 'no other alternative to the present arrangement except the ruinous folly of open price-war'.[43] The Government's verdict was that prices were not above the parity of world prices and that recent increases did not involve any discrimination against the British consumer. In those circumstances, the Government considered that no action was called for on its part.

7

Research and technical progress, 1928–1939

In the Company's earliest years the difficulties of making refined products from Iranian crude oil exceeded the scientific knowledge of the day. The chemical composition of the crude determined the product yield in which the major product was fuel oil, quite irrespective of any strategic or political considerations. Indeed, it was the problems of viscosity in fuel oil produced for the Admiralty which prompted the Company's initial research programme and the engagement of its first two research chemists, Dr A. E. Dunstan and Dr F. B. Thole, as described in volume 1.[1]

In 1928 the research centre at Sunbury was limited in scope, with modest facilities and a simple organisation. Jameson was a deputy director and general manager of Production Department, having under him G. H. Coxon in charge of Refineries Branch, Dunstan of Chemicals Branch and S. J. Lister-James of Geological Branch. Dunstan had overall responsibility for research at Sunbury, in Iran and in the refineries in Britain. Under him, Thole, the chief research chemist, was administratively responsible for what happened at Sunbury.[2]

Over the next decade the establishment at Sunbury grew in both facilities and staff. In October 1929 Cadman authorised the expenditure for a new building, which was opened in July 1931.[3] The old laboratory at 'Meadhurst' fell out of use and was demolished in November 1936. Meanwhile, the number of staff at Sunbury was increasing, rising from 76 at the end of 1929 to 99 at the end of 1934 to 197 in mid-1939. The growing staff enjoyed not only the improved facilities afforded by the new building, but also a well-organised programme of social activities.[4] The annual sports day was an event of much conviviality and informality, as were the annual dinner and dance and horticultural show. The atmosphere was generally congenial and friendly. The number of staff was small enough that employees knew each other, but large enough to cover a diversity of interests.

In 1935, however, there was a feeling that there was a need for a new whetstone. Thole, who had done so much to consolidate activities at Sunbury, was succeeded by D. G. Smith, who had been chief chemist at Abadan. Smith, a forceful and practical character who delighted in being regarded as a heretic, stressed a more integrated role for the professional chemist. Not for him a passive presence at the laboratory bench, for he believed the chemist had to be as active a participant in the managerial process as the engineer, an unusual attitude then in British industry, but supported by the Company. Smith was abrasive, energetic, inquisitive and stimulating. Under him, Dr D. A. Howes, who had joined Sunbury in 1932, was appointed deputy chief research chemist. Dunstan remained chief chemist in overall control reporting to Coxon, who sponsored and authorised most of the research activities at Sunbury, with advice from a Research Advisory Committee, consisting of distinguished research advisers.[5]

With a few notable exceptions, the Company's research was pragmatic, the emphasis being on practical application and gradual improvement of processes and plant design, rather than spectacular innovation. Investigations, too numerous to catalogue, were made into technical processes and the properties and uses of the Company's various products from aviation and motor spirits to bitumen and paraffin wax. Research also continued to be undertaken into the technology of crude oil production and into new methods of refining, a field in which US expertise was generally pre-eminent. Jameson and Coxon had close relationships not only with the US contractors whose services they used, but also with senior technical men in major US oil companies. Staff of the Company frequently went on fact-finding visits to the USA, gathering information and making contacts which were of great value. It was primarily by acquisition from the USA rather than its own inventions that the Company kept up with some of the most important new developments in refining.

OILFIELD TECHNOLOGY

As was seen in volume 1, the Company's success in developing the Masjid i-Suleiman oilfield as a complete production unit, rather than as a series of individual wells, resulted in an economic and efficient operation. In applying the principle of unitisation the balancing of production from the various wells was essential and was only made possible by the use of special monitoring equipment such as gauges to measure pressure at the bottom of wells.[6] Apart from the continuing advances which were made in production methods, improved survey-ing techniques were also introduced as geological investigations were supplemented in the late 1920s by geophysical methods.[7] In 1930, Dr J. H. Jones successfully used seismic techniques for a survey of the Haft Kel structure.

There was a constant endeavour to obtain more information about oilfield characteristics and to improve procedures. Investigations were undertaken into, for example, the presence of gas in the oil and the relationships between gas pressures and production potential.[8] Growth of understanding of the conditions in the oil-bearing rocks was, however, a slow process. Within the Company, there was considerable debate as to whether the oil was accumulated in fissures or was retained in the pores of the rock, a subject which was of obvious importance in furthering technical understanding of the processes of production.[9] In February 1929, samples of oil-bearing Asmari lime-stone were analysed for the first time when rock fragments were ejected from a well by an explosion. At that time it was believed that the oil in the reservoir was held in fissures in the limestone, a view which was challenged when the fragments were found to contain oil within their porous structure. This new evidence led Dr C. J. May of Fields Section in Production Department to comment that it appeared probable that the rock itself was a reservoir for large quantities of oil and gas.[10] In March 1930 Jameson initiated further efforts to resolve the issue, writing, after a visit to the Fields area, 'there is no doubt that know-ledge of the reservoir rocks is not as advanced as it might be'.[11] In that year, relevant work was undertaken by G. H. Hubbard, an Abadan chemist, and the geologist, J. McAdam, who examined core samples.[12] Their findings led to the conclusion, expressed at the Company's 1931 production conference, that porosity was a negligible factor in oil accumulation and that further detailed work on the subject was unnecessary.[13] However, a mistake in the calculations was then dis-covered and at the 1932 conference the results were differently expressed that 'large production is principally from cracks and fissures in the rock ... porosity is of secondary importance'.[14]

The matter was not, however, settled. In 1933, after visiting the Asmari mountain, a vast outcrop of limestone, with Company geolo-gists, the Fields manager, Stephen Gibson, who made an outstanding contribution to the development of efficient production techniques (and was later knighted), wrote a comprehensive report on 'The production of oil from the Masjid i-Suleiman Field'.[15] He made the bold statement that although fracturing of the limestone was entirely responsible for the freedom of connection in the reservoir and for the passage of oil to the wells, the fissures contained less than 10 per cent of the oil produced. Porous limestone, stated Gibson, supplied the remainder. His report was not allowed to pass unchallenged and there were criticisms of the calculations by May and his colleague, D. Comins. However, the Sunbury physicist, E. S. L. Beale, was in the Fields area at the beginning of 1934 and devised apparatus and experiments which corroborated Gibson's findings. Gibson's work on the porosity problem and on multi-stage gas separation (see volume 1) were significant contributions to the technical development of production in Iran. Some of his conclusions were amplified in a further report by the geologist, P. T. Cox, in September 1934.[16]

'CRACKING' PROCESSES AND MOTOR SPIRIT QUALITY

In refining, the most important technological development in the prewar years was the growing use of cracking processes by which heavier fractions of oil could be broken down to form lighter products such as motor spirit. The technique, already well established in the USA, was an essential innovation in meeting the growing demand for the higher octane motor spirits which were required as car engine designs improved. In particular, as the power of engines was increased by higher compression ratios there was a tendency for fuel to 'knock' during combustion, causing engines to overheat. Improved resistance to knock, measured by the octane number of the fuel, was, therefore, a central issue in quality competition between motor spirit manufacturers and marketers. In the early days of cracking, the processes which were used to improve both the yield and the quality of motor spirits were thermal in nature, using heat to bring about chemical changes. The more gentle process, known as catalytic cracking, which eventually replaced thermal cracking, was to come later.

Up to the introduction of cracking, the Company had merely applied the pots and pans technology of boiling the crude oil to separate it into fractions of different boiling ranges, which were then purified by further distillation or by simple chemical treatments to produce finished products. Whereas distillation just separated the crude oil, the new thermal cracking processes of the 1920s and 1930s brought about chemical changes by the application of heat and pressure. Such cracking, in those days, was at the frontiers of technology. Sunbury did much work on cracking during the 1920s and a small trial cracker based on a Sunbury process was commissioned at Uphall in 1925. However, when it came to equipping the Llandarcy refinery with cracking plant, the Company took the more cautious course of relying on proven technology, ordering an off-the-shelf unit based on patents by the Cross Brothers and erected by the US firm, M. W. Kellogg and Company, under a contract agreed in 1926. Later that year, another US design, using the Dubbs process, was ordered for Llandarcy from the US firm, Universal Oil Products (UOP).[17] By 1929 further Sunbury research had produced a promising continuous cracking process, and continuous crackers were, in fact, installed during 1929–31 at Llandarcy, Grangemouth and Uphall. However, when four crackers were ordered for Abadan in 1929, the Cross process was again chosen under a package contract offered by Kelloggs.[18] Each unit of 10,000 barrels per day was much larger than the plants previously operated by the Company. The design was also more complex, utilising two reactors operating in series at different pressures. The total cost in March 1930 was £1,978,000. About 50 per cent of the material was purchased in Britain, including the reactor vessels. The foundations were laid in February 1930 and all four plants had been completed by November 1931.

Cracking did not, by itself, yield finished products and further refining of the cracked spirit was needed to remove unstable olefinic compounds, which caused gum deposits in engines.[19] Sunbury had developed a process for treatment of cracked spirit over a zinc chloride catalyst, but trials at Llandarcy in 1929 were disappointing and it was decided not to risk using the process at Abadan. Instead the traditional method of treating the cracked spirit with sulphuric acid was adopted. That was effective, but had the disadvantage that it produced large quantities of acid sludge as waste. After washing with acid, the cracked spirit was neutralised by washing with caustic soda solution and then redistilled. Finally, odorous sulphur compounds were 'sweetened' by treatment with sodium hypochlorite. The acid and soda washing steps were carried out in a new Holley-Mott Washery, commissioned in May 1931. This was an ingenious system, named after the two Company employees who had devised it, for the continuous contacting of oil with chemicals.

By 1931, the possibility of adding small amounts of chemicals to the cracked spirit to enhance its stability was being studied. ICI was consulted and supplied fifty-four possible materials, but none of them was effective.[20] In January 1934 Coxon called a halt to the search, a disappointment to Sunbury after so much work. At Abadan, however, it was demonstrated in May 1932 that a small addition of a light fraction from sulphur dioxide extract (available as a normal refinery component) was effective in suppressing gum formation. That method served Abadan in good stead until 1937, after which a proprietary inhibitor was purchased from the US chemicals company, Du Pont.

The sulphuric acid required for the acid treatment was manufactured at Abadan in a plant, commissioned in 1931, which processed sulphur imported from Texas. Means were soon found, however, of extracting hydrogen sulphide from refinery fuel gas and using it as a local source of sulphur. Cracked spirit, refined by acid treatment, initially caused serious corrosion of the redistillation units, and it was not until 1933 that this problem was overcome.[21] Another difficulty was the disposal of the waste acid sludge, for which various techniques including water washing and burning were tried, but no perfect answer was found.[22] Operation of Abadan's four crackers required a formidable amount of water for steam raising and cooling. The water, around eighteen million gallons a day, was drawn from and returned to the Shatt al-Arab, precautions being taken to prevent river contamination by oil contained in the effluent stream.

By the end of 1931, the four Cross-Kellogg thermal crackers at Abadan had revolutionised the Company's motor spirit production, both in quality and quantities, and six more crackers were built by the end of 1939. Problems such as those outlined above did much to bring operators and researchers closer together in a more complementary relationship during the 1930s, a period of practical consolidation in cracking technology. The basic objective remained the same, namely the processing of such feedstocks as were available, distillate or residual, to make as much gasoline of the highest quality as possible.

It was US research which stimulated the Company to develop its interest in the use of a catalyst rather than temperature in the cracking process. Catalytic cracking had been demonstrated, especially by the Houdry Process Corporation, to be a feasible and viable process. The basic problem was that the catalyst, in time, became coated with carbonaceous deposits or coke which impaired its efficiency and so had to be removed for 'regeneration' by burning off the deposits in a stream of air. Coxon was keenly interested in catalytic cracking and initiated studies covering two aspects of the technology, namely the choice of catalyst and the problems of handling the catalyst within the plant. In November 1937 the Company agreed to join with Kelloggs in a project for the development of catalytic cracking.

Work began at Kelloggs' laboratories in Jersey City and at Sunbury, where experiments using small glass units were commenced in December 1937.[23] In May 1938, Sunbury built larger-scale equipment in which the catalyst (a silica alumina prepared to a formula provided by Kelloggs) was made to flow between a reactor vessel and a separate 'regenerator'. In order to maintain confidentiality the process was described, until 1939, as 'decarbonisation'. In May 1938, seeking to gain operational experience on a larger scale, Kelloggs proposed the conversion of the obsolete Cross cracker at Llandarcy to catalytic operation with a throughput of 500 barrels per day.[24] In August 1939 Kellogg technicians and Llandarcy staff carried out a successful run of ten days on this improvised catalytic cracker, but with the outbreak of war the project at Llandarcy was halted and the team dispersed. Nevertheless, the Company's co-operation with Kelloggs provided an invaluable introduction to a much broader agreement which enabled the Company to share cracking information with other companies such as Standard Oil (NJ), Royal Dutch-Shell, Texas and UOP and to enjoy preferential royalty terms. That was the Catalytic Refining Associates Agreement, legally operative as from 7 August 1942, but effective in practice from 1940.[25]

The installation of crackers during the late 1920s and 1930s was part of the Company's continuing efforts to improve its motor spirit by achieving higher anti-knock qualities. In June 1929, a new blend called 'New BP Spirit' was introduced with much publicity,[26] but early difficulties were encountered with it because of fluctuations in its quality. Although some cracked spirit was available at Llandarcy from 1927, it was not until 1930 onwards, with crackers progressively coming into use at Llandarcy, Grangemouth and Abadan, that cracked spirit became the rule rather than the exception in the Company's motor spirits. The first cargoes of petrol containing cracked spirit were shipped from Abadan in 1931, going to France, Australia and the Consolidated area. Similar shipments were made to Britain from February 1932.

By 1930, the 'New BP Spirit' had been overtaken in quality by competitors' products and the Company, seeking to catch up again, considered the use of a new additive, tetraethyl lead (TEL). This toxic material, extremely effective in raising the octane number of spirit, had been discovered in 1921 by two chemists working for General Motors. In 1924, General Motors and Standard Oil (NJ) formed the Ethyl Gasoline Corporation to exploit the use of TEL.[27] It was introduced to the British market in January 1928 by Standard Oil (NJ)'s local subsidiary, the Anglo-American Oil Company.[28] In September 1930 the president of the Ethyl Gasoline Corporation suggested to the Company that it might use TEL, a suggestion favourably received since the Company's motor spirit sales were down on 1929 figures. After testing at Sunbury and extensive medical consultations over possible health hazards, the new 'BP Plus' was introduced to the market on 15 April 1931.[29] The advantage for the motorist was a marked improvement in anti-knock value, a rise in octane number from 66 to 74. 'BP Plus' for a time advanced the Company's place in the motor spirit rankings,[30] although Sunbury had to deal with various complaints that the new grade caused valve sticking and to convince the Austin Company that such trouble with Austin cars was due to engine design, not the fuel.[31] The use of tetraethyl lead was made explicit in the Company's motor spirit marketing when 'BP Ethyl' replaced 'BP Plus' as the brand name of its premium grade in August 1933. The Company also made use of motor racing competitions at Brooklands, the Isle of Man and elsewhere to advertise its products and enhance its reputation.[32] There was a special Competitions Branch managed by Captain A. B. Rogers who was personally acquainted with many well-known drivers of the day. These included G. E. T. Eyston, H. O. D. Segrave and Malcolm Campbell and their wins, using special BP blends, provided the Company with much reflected glory.

In the search for components of improved anti-knock value Sunbury's work covered not only technical support for tetraethyl lead, but also the use of alcohols and the testing of possible alternative anti-knock 'dopes', all such studies being backed up by thorough testing in engines. Processes for manufacturing gasoline components were studied and treatments with, for example, plumbite or hypochlorite reagents were constantly reviewed in collaboration with Abadan. However, the work with alcohols was disappointing, as was the search for new anti-knock additives.

An important advance was the development of an accurate method of defining and measuring anti-knock quality, using a test procedure devised by the Engine Research Branch at Sunbury under R. Stansfield.[33] Much ingenuity also went into making an instrument with a microphone and amplifier for picking up and measuring sounds emitted by an engine when knocking occurred. The Company had little experience in that area and enlisted the help of R. E. H. Carpenter, a wireless engineer, who devised a 'strobophonometer' for the purpose.[34] Meanwhile, inter-company collaboration on knock testing methods[35] began in 1928, an informal committee being set up which

41   Advertisement: new speed records set using 'BP Plus', 1932

42   Advertising poster: G. E. H. Eyston's new speed record using 'BP Ethyl', 1935

was taken over by the Institute of Petroleum Technologists in December 1930. Close contact was maintained with motor manufacturers and in 1934 Stansfield attended knock testing trials on cars at Uniontown in the USA.

ALKYLATION AND AVIATION SPIRIT MANUFACTURE

While the demand for higher octane spirit for motor cars was expanding, a still greater technical challenge was posed by the increasingly exacting standards to be met in the production of aviation spirit. The growth of aviation provided a strong incentive for the Company to pay increased attention to aviation fuel in the 1930s. It was not, however, a line of business in which, at the beginning of the decade, the Company had much experience as a marketer. Nor, in fact, did it have a product to satisfy the Air Ministry, which in May 1928 complained about the quality of the Company's aviation fuel, which was based on Iranian straight run spirit of low octane number.[36] Sunbury, after much work, recommended in August 1930 a blend based on Romanian light benzine with solvent naphtha (a coal tar product) and benzole. This was of much higher quality and arrangements were in hand by September for its manufacture in the hope of regaining the share formerly held of Air Ministry business. Meanwhile, in February 1930 Fraser chaired a Company meeting at which it was laid down that efforts should be made 'to take a better standing in the Aviation Spirit Trade both in this country and on the continent'.[37] The Company was not starting from scratch, for it already supplied aviation fuel to Imperial Airways for its route to India via Iraq and Iran, met 90 per cent of Lufthansa's requirements in Germany and elsewhere, and supplied the Bristol Aeroplane Company and the Dutch airline, KLM. It was, however, apparent that there was considerable scope for expansion in the fast-growing, but technically demanding aviation fuel business.

In October 1933 the Air Ministry updated its specification for aviation spirit to permit the addition of tetraethyl lead. The Company imported spirit made from Miri (Dutch Borneo) crude oil, added tetraethyl lead, and by February 1934 was making 87-octane aviation spirit, an octane rating which met the Royal Air Force's requirements at the time.[38] The Company was also able to obtain publicity for its aviation products by providing special fuels for record attempts or races, such as the England to Australia Air Race in the autumn of 1934.[39] It was not a high-volume market, but interest in aviation was growing both in the civilian sector and for defence purposes. Moreover, towards the end of 1937 Cadman was appointed chairman of the Committee on Imperial Airways, a position which would have given him close acquaintance with aviation issues.

The incentive to produce high-grade aviation spirits came not only from increasing demand, but also from the availability of surplus gases which had the potential to be turned into valuable liquid components for motor or aviation spirit. The process of cracking generated much gas and one of Sunbury's main tasks in the early 1930s was to seek profitable uses for cracker gases. Of equal importance was the problem of using, rather than burning at flares, the immense quantities of natural gas that were produced along with crude oil in the oilfields. In the USA, with its advanced economy and high standards of living, natural gas was widely distributed and used as a fuel; in Iran, at that time, gas consumption, except as refinery fuel, was minimal and flaring was obviously wasteful.

Before describing the processes by which gases were turned into liquid products of high anti-knock value, some words of simple explanation on the chemistry of petroleum are required. In chemical composition, petroleum is essentially a mixture of hydrocarbons, which are combinations of hydrogen and carbon atoms. Two main classes of hydrocarbons concern us here, namely the 'saturated' hydrocarbons (known as paraffins) and the 'unsaturated' hydrocarbons (known as olefins), so called because they do not contain as many hydrogen atoms as the corresponding paraffins. Olefins (unsaturated) can be transformed into paraffins (saturated) by the chemical addition of hydrogen, i.e. by hydrogenation. Each of these types of hydrocarbon – olefins and paraffins – may be sub-divided into those with the carbon atoms in a single chain (*normal*) and those in which the chain of carbon atoms is branched (*iso-*). So, for example, there is a straight chain 'normal butane' and a branched chain 'isobutane'. Some examples of such hydrocarbons are illustrated in figure 7.1.

There were two main approaches to the utilisation of gaseous hydrocarbons in manufacturing motor and aviation spirit. One was for the gases produced by cracking to be turned into liquids by the polymerisation (combination) of olefins to liquid products. The second was for natural gases to be treated by pyrolysis (high temperature heat treatment) to convert paraffins into liquid products such as benzole.[40] Pyrolysis studies, which included building an experimental benzole plant at Fields, proceeded throughout the late 1920s, but by 1930 still did not show much hope of commercial application. Process requirements were ahead of the metallurgy of the day and furnace tubes which could withstand continuous use at the high temperatures involved (about 900°C) could not be found. Even so, there had been the possibility in 1929 of an interesting collaboration in this field with ICI, who were seeking a process to convert gas from Canadian gas fields into benzole. Discussions on the possibility of using the pyrolysis process for that purpose were held and by mid-1931 it had been proposed that the Company and ICI should pool patents and knowledge on pyrolysis and establish a joint company, the Anglo-Imperial Development Company. In September 1931 a joint committee was set up with Coxon and Dunstan as the Company's representatives and a secrecy agreement was signed on 30 November 1932. The early hopes of the signatories were not, however, realised. ICI, after looking in more detail into the scheme for making benzole from Canadian natural gas, decided that the project was neither practical nor economic.

As it happened, just as the possibility of co-operation with ICI waned there was renewed Company interest in the pyrolysis process after it was found that a two-stage approach could be used which did not require the furnace tubes to withstand such high temperatures. The first stage was to convert Fields gas by pyrolysis at around 700°C into

Paraffins or 'saturated' hydrocarbons	Olefins or 'unsaturated' hydrocarbons
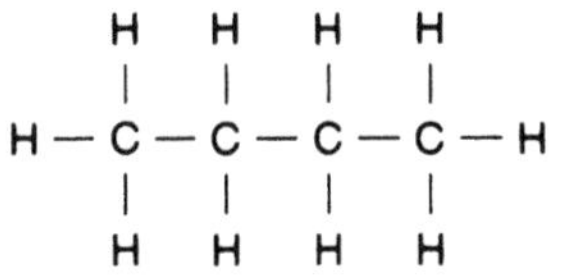 Normal butane (C4 H10) (straight chain)	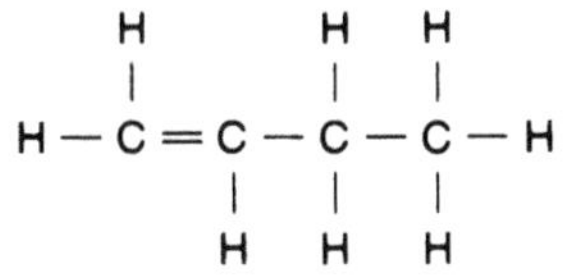 One of the normal butenes (straight chain)
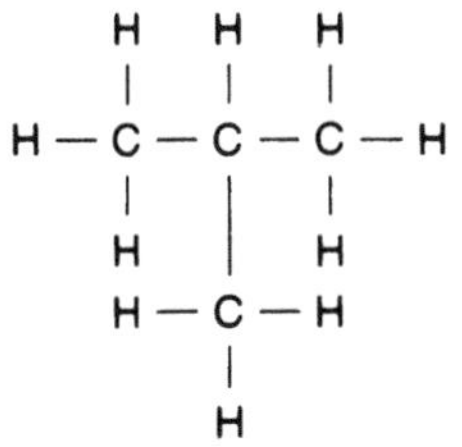 Isobutane (C4 H10) (branched chain)	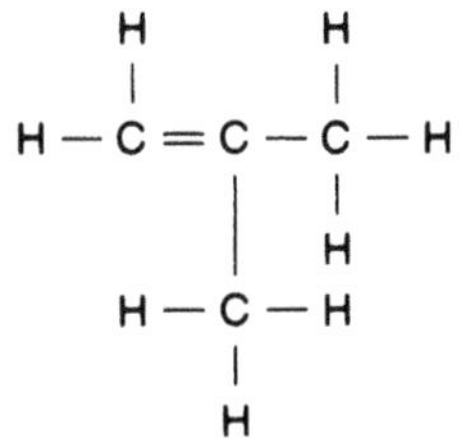 Isobutene (branched chain)
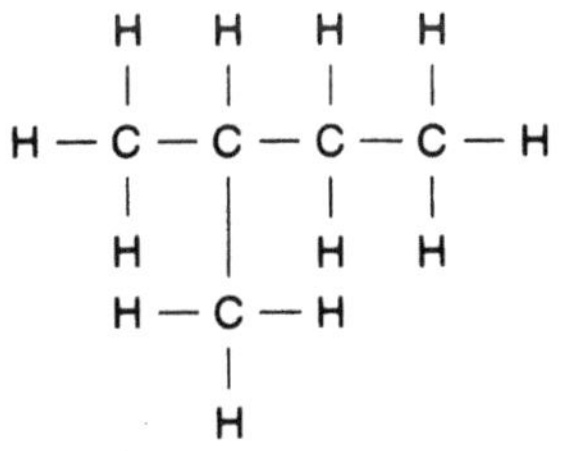 Isopentane (C5 H12) (branched chain)	Note that the olefins shown above contain two less hydrogen atoms per molecule than the corresponding paraffin molecules.

Figure 7.1   Examples of gaseous hydrocarbons

olefins and then, as a second stage, to polymerise the olefins under pressure (at 400–600°C) to produce gasoline. Coxon sanctioned the expenditure of £15,000 for a plant at Grangemouth. Construction started in 1933 and the first runs were made in 1934 under the supervision of Howes, who rapidly became Sunbury's expert on these processes.[41] The thermal polymerisation results were disappointing, but the pyrolysis section of the plant worked well. Moreover, a new alloy which could withstand very high temperatures was found for the furnace tubes and the idea of direct pyrolysis to a benzole product was revived with heavier hydrocarbons, such as butane, being used as the feed. Progress was so satisfactory that at the end of 1935 a contract was agreed with the US firm, Alco Products, for the erection at Abadan of a pyrolysis plant to process two million cubic feet of gas per day. The new plant was brought into operation on 6 January 1938.

Meanwhile Sunbury, using a pilot plant at Grangemouth, had looked into catalytic polymerisation and had devised a cadmium phosphate catalyst for that purpose. The decision was taken to build a catalytic polymerisation unit in Abadan, but the order, placed in June 1936, was for a US process licensed by Universal Oil Products as Sunbury's catalyst was not yet ready for commercial development. The UOP plant, offered on very favourable terms, was commissioned in May 1938. It had already been decided in 1937 to install another polymerisation plant in Abadan for the selective combination of olefins to make a product known as 'codimer', of good anti-knock quality. However, UOP refused to build it on the earlier favourable terms. Coxon therefore invited Kelloggs to design a codimer plant for Abadan based on a Sunbury catalyst. The new plant was commissioned in June 1939.

The Company was also interested in an 'acid polymerisation' process in which two molecules of the olefin, isobutene, combined in the presence of sulphuric acid to form di-isobutene (DIB), a product of around 88-octane rating. Early in 1935 a pilot plant was built at Llandarcy to produce four tons of DIB a day in order to provide design data for a full-scale plant at Abadan. The latter, commissioned in September 1937, had an input capacity of 10,000 gallons an hour of cracker gas, but suffered from leaks and mechanical difficulties which were not cured until December 1938. The DIB produced by polymerisation had quite a good octane number, but if it was treated with hydrogen under pressure, using a catalyst, it could add on two more hydrogen atoms and become the saturated (paraffinic) hydrocarbon known as iso-octane, a product of excellent anti-knock value and an essential component for high performance aviation spirit. A hydrogenation plant, commissioned at Abadan in September 1938, was bought for the purpose of iso-octane manufacture and also to hydrogenate codimer to make another iso-octane concentrate, 'hydrocodimer'.

The manufacture of iso-octane by the two-stage process of polymer-isation and hydrogenation was, however, soon overtaken by the Company's main research success during the period of this volume: the discovery of the alkylation process, enabling 100-octane spirit to be produced from refinery gases by a very simple reaction using sulphuric acid. During 1934 and 1935, the Chemical Research Section at Sunbury, under Dr S. F. Birch, concentrated on polymerisation reac-tions. At Sunbury's annual chemical conference in July 1936 Birch emphasised the need to learn more, not only about reactions between identical molecules as in the formation of di-isobutene, but also about reactions between dissimilar molecules such as isobutene and normal butenes.[42] It was with that purpose that the next week one of Birch's staff, Dr T. Tait, undertook what was intended to be a simple experi-ment. The unexpected result was the discovery of what the relevant report described as a 'fundamental chemical reaction which may prove extremely valuable', the effective birth of the alkylation process.[43]

It was known that by treatment with sulphuric acid of about 60 per cent strength isobutene molecules would combine to give di-isobutene and Tait's hope was that with stronger acid (97 per cent) isobutene might also combine with normal butene. In attempting to achieve that reaction, Tait was faced with the problem that the butenes are gases at ordinary temperatures. Thinking that he might have a better chance of producing the desired reaction with sulphuric acid if he could keep the butenes in liquid form, he decided to dissolve them in a suitable solvent. Pentane, being a paraffin and thus a type of hydrocarbon traditionally regarded as chemically unreactive, seemed ideal for the purpose and was conveniently to hand as a stock laboratory chemical. So Tait put the sulphuric acid and pentane in a two-litre glass bottle, surrounded it with ice, set a stirrer going and slowly added his mixture of iso- and normal butenes. Some two hours later he separated the acid, distilled away the pentane and measured the liquid product which he had left. There was about twice as much as he thought there should have been. Believing he must have made a mistake with his measurements, he repeated the experiment, but again got the same result. Also, the product was paraffinic in nature and not, as Tait had expected, an olefin.

Birch was soon informed of Tait's surprising results and was able to deduce the probable chemistry of the reaction, namely that the pentane, which Tait had assumed to be an inert solvent, had somehow entered into the reaction. On analysis, the pentane used by Tait was found, in fact, to be largely isopentane, in other words a branched chain (iso-) paraffin, which had reacted with the butenes. That was confirmed by further experiments which showed that another branched chain paraffin, isobutane, could also be made to react with butenes. Sunbury reports make no mention of this background and convey the impression that a deliberate (and successful) attempt was made to see if an olefin would react with an isoparaffin. However, interviews with Tait, Howes, Smith and Dr F. A. Fidler in 1976 all confirmed that the result obtained by Tait was not the prime objective of the experiment.[44]

It was an important discovery. It was known that reactions between isoparaffins and olefins had already been achieved in the USA, but only by means of very drastic treatment. The finding that the reaction could readily take place in the cold using a commonly available reagent, sulphuric acid, held out exciting prospects. The reaction of greatest interest was that between isobutane and the butenes to make iso-octanes, the great advantages of the process being that it made iso-octanes in a single step and in better yield than the existing two-stage route of polymerisation followed by hydrogenation. However, before 'acid condensation', as the process was originally described, could be turned into a refinery process, much data had to be obtained.[45]

That was a task for two junior members of the Chemical Research Section, F. B. Pim and Dr Fidler, who worked for months, often in twelve-hour shifts, carrying out reactions with sulphuric acid in a lead-lined autoclave (vessel). By December 1936 enough had been learned to justify the construction at Sunbury of a pilot plant designed for continuous and larger-scale operation than the batch experiments in the autoclave. The pilot plant was the responsibility of E. McNeill of the Development Section, who soon established the importance of recycling unused isobutane back into the reaction vessel and of maintaining a high ratio between the recycled material and the olefins in the feed. His finding greatly aided the development of the process and was duly patented. Action had already been taken to patent the basic reaction, which was done within a fortnight of Tait's discovery, the application being dated 29 July 1936.[46]

A further step forward with the development of alkylation was taken at a head office meeting chaired by Coxon on 31 December, when the decision was taken to instal at Abadan a complete, self-contained pilot plant to operate the process, the objective being to construct a full-scale commercial unit as soon as the process was proved to be economic.[47] The pilot plant, which had a nominal capacity of 100 barrels per day of alkylate, was hurriedly built, largely from salvage material, and began its first run on 8 January 1939.[48] Fundamental studies and pilot plant work were continued throughout 1937–9 and were regularly discussed by the Research Advisory Committee, which imparted a sense of urgency to the work.[49] The term 'acid condensation'[50] remained in use until December 1939, when it gave place to alkylation to fall into line with the terminology used by other companies interested in the reaction. Much of the pilot plant work was aimed at reducing the high acid consumption of the process, an aim partially fulfilled by the end of 1939.

The Company was not alone in 1936 in being interested in iso-paraffin-olefin reactions. Royal Dutch-Shell, the Texas Company, Standard Oil (NJ) and UOP were also active in the same field. However, with its patent confirmed in January 1938, the Company was the first to publish its findings in the technical journals.[51] Howes felt that a mistake was made in 'rushing into print too soon' and that doing so had saved the US oil companies an enormous amount of research effort, leaving the Company vulnerable to others who were clever at exploiting gaps in patents.[52] Certainly, by 1938 a highly confusing patent situation in relation to alkylation reactions had arisen.[53] However, in this as in other fields from marketing to hydrogenation (see chapters 4 and 6), the attractions of co-operation outweighed the appetite for competition. Discussions between the Company and Standard Oil (NJ) crystallised into an agreement on 17 March 1939, by which time a similar agreement had been reached between the Royal Dutch-Shell and Texas companies. It was only a matter of weeks before those agreements were merged into an overall agreement between the four companies to pool their research efforts and work together to expedite the commercial applications of the process. The Alkylation Agreement, which also included UOP, was not finally signed until 11 June 1940, but took effect from 12 April 1939, providing for royalties from licensing the process to be shared between the signatories.[54]

In the tense international atmosphere of 1939 there was, however, good reason for thoughts of future financial gains from alkylation royalties to come second to concern about the increasingly imminent prospect of war. The conflict, when it came, brought huge new demands for high performance aviation spirit, in the manufacture of which the Company was to make arguably its most significant contribution to the British war effort.

8

War,
1939–1945

> Our operations have had to be adjusted, often from day to day, to meet
> the exigencies of the moment. In many respects we have ceased to be free
> agents and commercial considerations have been relegated to the back-
> ground. Government control, British and American, directive and
> restrictive, have influenced our every action. New types of work, often
> quite unconnected with our normal activities have been carried out for
> Government departments.[1]

In those words the Company summed up the improvisation, Government direction and extraordinary demands of the war years, when the
Company, like many others, put its energies and resources behind the
single overriding objective of military victory. For the duration of the
conflict the pursuit of private interest was widely sacrificed to national
duty. Commercial competition was suspended as men and women
from the different spheres of government and business came together
with that sense of common purpose rarely, if ever, achieved in peaceful
times. For this total war, fuelled mainly by oil, the Company was not
entirely unprepared. Its prewar plans were, however, soon overtaken
by events as the struggle spread into a global conflict in which there
was hardly a phase nor a theatre of the war which left the Company
untouched.

ORGANISING FOR WAR

Joining hands

As international tension mounted in the summer of 1938, the leading
oil companies operating in Britain were invited by the Government to
draw up plans for the organisation of the domestic oil industry in the

event of war. The intention was to set up a central body of oilmen through which the policies laid down by Government could be swiftly translated into action; and at the same time to cut out the duplication, considered wasteful in time of war, which resulted from the separate operation of competing firms. Planning the establishment of such a body was greatly facilitated by the very high level of concentration in the oil industry in Britain, where about three-quarters of the market was held by Shell-Mex and BP (the joint marketing organisation of the Company and Royal Dutch-Shell) and the Anglo-American Oil Company, which was the local subsidiary of Standard Oil (NJ). With such a dominant combined position in the market, the three parent companies were well placed to form the core of a central body which would have little difficulty asserting its authority over the smaller firms in the industry.

After a series of meetings, the big three firms joined with Trinidad Leaseholds Ltd, which had been expanding its sales in Britain since the early 1930s, and the National Benzole Company in signing the Petroleum Board Agreement on 8 March 1939. The signatories agreed to pool their distribution networks under the aegis of a Petroleum Board, which would become, in effect, a combined distributing agency for the companies if and when war broke out. The usual paraphernalia of competition, such as separate brand names, would be done away with for the duration of the conflict and the number of grades of oil products would be reduced to a minimum. The member companies' plant and equipment, such as depots and storage tanks, railway wagons and lorries, would also be pooled and each company's share of the Board's trade would be based on its sales of each product in Britain in the last full year of peace. The Agreement, approved by the Government, covered all the major oil products except lubricating oils, which were excluded because of the particularly complicated nature of the lubricating oil trade in which many different qualities were sold for special purposes. The chairman of the Petroleum Board was Sir Andrew Agnew, who had recently retired from active management with Royal Dutch-Shell. Fraser was a member and H. E. Snow, who was then a senior manager in the Company's Distribution Department, was the Board's secretary. In that capacity, he was responsible for much of the preparatory work which went on between March and September 1939, when sundry administrative matters were dealt with, including the selection of members for the various committees associated with the Board.

After Britain declared war on Germany on 3 September 1939 the

43   Pool motor spirit tanker in World War II

planned arrangements for wartime oil administration were immediately put into effect. The Petroleum Board established its headquarters
at Shell-Mex House in London and formally commenced its operations
at midnight on 3/4 September. By the end of the month the Government had introduced rationing to curtail civilian consumption of
motor fuel and by a statutory order dated 13 October the Petroleum
Board was given a monopoly over oil distribution in Britain. A few
months later a further agreement was signed to bring the large number
of small oil distributors who had not signed the Petroleum Board
Agreement of March 1939 into the system of wartime control.

The newly formed Board administered two 'pools' of firms. The
main pool, which dealt with all main products other than lubricants,
took control of all oil distribution facilities on the lines which had
been agreed in March 1939. The second and much smaller pool was
formed to deal with lubricants as a special category. Under the Board's
supervision the number of grades of products was reduced with motor
spirit, for example, being available only in a single grade known as
Pool Motor Spirit, having an octane number of 67/68. Later on, in
1942, it was to be superseded by a leaded grade with an octane number

of 80. Other early moves to suppress competition were the elimination of brand names and abstention from individual advertising. At the same time, various committees were set up to arrange supplies, agree on product specifications and qualities, and consider the myriad other matters involved in the control of the industry for purposes of war. Through those arrangements, the oil firms operating in Britain placed at the disposal of the Government an instrument for the administration of the industry in liaison with relevant Government departments, most notably the Ministries of Transport and Shipping (which merged to form the Ministry of War Transport at the beginning of May 1941) and the Petroleum Department, which was still a branch of the Mines Department of the Board of Trade when war broke out.

At the apex of the wartime administrative machinery for the control of the oil industry stood the Oil Control Board, which was set up in November 1939 with the status of a sub-committee of the War Cabinet. Its functions were essentially to decide on the order of priorities between competing demands for oil and to take whatever action was necessary to conserve and maintain adequate supplies. Under the chairmanship of Geoffrey Lloyd, the Secretary for Mines, it included two other junior ministers from the Admiralty and Ministry of Shipping respectively, various other officials, a former Ambassador, and Cadman, who was appointed expert adviser to the Secretary for Petroleum on the outbreak of the war. It was the sort of role, straddling the worlds of government and industry, in which Cadman felt utterly at ease and he remained a member of the Oil Control Board until his death in mid-1941.

Meanwhile, further changes took place in the administration of the industry. On the Government side, the status of the Petroleum Department was raised in May 1940, when it was taken out of the Mines Department and given a separate identity under Geoffrey Lloyd, who ceased to be Secretary for Mines and became the new Secretary for Petroleum. The Petroleum Department continued, however, to come under the Board of Trade until June 1942, when it became the Petroleum Division of the newly created Ministry of Fuel and Power.

On the industry side, it was felt, even before the commencement of the war, that a more comprehensive body than the Petroleum Board, which was concerned solely with the oil industry in Britain, would be needed for the wartime control of the industry. It was clear, for example, that the availability of oil supplies and tanker tonnage for Britain could not be dealt with in isolation from the requirements of other Empire countries, the Allies and some neutral nations. To meet the need for a more international outlook, a new Trade Control Committee was set up in April 1940 as a parallel organisation to the Petroleum Board with the same chairman, Sir Andrew Agnew, and the same secretary, Snow. Fraser, who sat on the Petroleum Board, was also on the Trade Control Committee, as was Heath Eves. In practice, the Trade Control Committee soon became more important than the Petroleum Board as the peak industry body for wartime control. Under it, numerous other committees such as the Overseas Supply Committee, the Tanker Tonnage Committee and the Technical Advisory Committee dealt with the various matters suggested by their names. They in turn spawned a proliferation of sub-committees as the panoply of wartime administration became ever more comprehensive and elaborate. As the machinery of wartime control expanded, more and more Company employees were selected for membership of one committee or another until the mobilisation for war was so complete that the seam between the Company, other oil firms and the Government seemed scarcely to exist at all.[2]

The evacuation of Head Office

The British declaration of war activated not only the formation of the Petroleum Board, but also other plans which had been made by the Company for the evacuation of staff from its Head Office at Britannic House in Finsbury Circus, London. The plans were characterised by an exacting precision, the civilian counterpart of a military mobilisation. In them, the Company divided its Head Office staff into two categories: essential and reserve. The reserve staff were those employees whose jobs were considered to be inessential in time of war and they were advised to that effect in March 1939. The essential staff were divided into two categories. The first consisted of those who would need to maintain close contact with Government departments and would, therefore, need to be in or near London. That category included the chairman, directors, higher management and their immediate staff. Secondly, there were those who, it was thought, could be located further afield without inconvenience. Each member of essential staff was notified in May 1939 of the centre to which he/she was to report in an emergency, the address where board and lodgings had been reserved, and the exact office at which to work, down to the number of the room. As little as possible was left to chance.

For those who needed to be near London, the research station at Sunbury was chosen as a suitable location on the assumption that

Sunbury's normal research work would cease under the emergency conditions of war. As for the staff who did not need to be near London, it was decided to evacuate them to Llandarcy, where sufficient office space could be provided for them in the main building, augmented by the utilisation as offices of the Llandarcy Institute, the sports pavilion and the bowling green pavilion, all of which were Company property. On the outbreak of hostilities those plans were immediately put into effect, the 913 staff employed at Britannic House being evacuated over the first weekend of September, with the exception of a staff of 15 who remained on duty at Britannic House to look after the premises.

However, for all the exactitude which went into the planning it was found, by the end of 1939, that the dispersal of staff to different locations was inefficient and it was therefore decided to concentrate as many personnel as possible in the vicinity of Sunbury. In order to accommodate the increased numbers, the Company leased property in New Zealand Avenue, Walton-on-Thames, a few miles distant from Sunbury, and arrangements were made to transfer some of the staff previously evacuated to Llandarcy to the Walton and Sunbury area. At about the same time, some staff (in the Share Branch and in Purchase & Supply Branch) returned to Britannic House in October–December 1939. The property leased by the Company at Walton, to which Cadman transferred from Sunbury in April 1940, consisted of two blocks of flats and shops which were converted into offices. The main building was known as Brassey House, the other as Simpson House. However, owing to the recall of reserve staff and the continuing influx of employees, it was necessary to find additional office accommodation at sundry local sites, ranging from 'The Hut', which was a hut erected on ground behind the existing Walton offices, to gas company showrooms on the ground floor of Brassey House.

The improvised arrangements at Sunbury and Walton were further complicated by another upset to the prewar plans which had been made with such precision. Those plans were based, as has been mentioned, on the assumption that technical research and experimental work at Sunbury would cease in wartime. The majority of the research staff at Sunbury would not, it was believed, be required for Company work and would therefore be released for national service or work of national importance. The minority who would continue to be required by the Company would, according to plan, be dispersed from Sunbury. Arrangements along those lines were put into effect after the outbreak of war, when the research station was virtually closed down, the

Table 8.1 *Sunbury technical staff, 1939–1944*

Prewar	86
14 September 1939	26
1 January 1940	47
1 January 1941	99
1 January 1942	96
1 January 1943	106
1 January 1944	156
1 September 1944	190

Source: BP 71483.

number of technical staff being rapidly reduced from eighty-six before the war to only twenty-six by mid-September. However, in October Air Vice Marshal Tedder visited Sunbury to discuss the problem of aviation fuel supplies for the war effort and gave his opinion that aviation spirit would be required not only in very large quantities, but also of 100-octane rating, a very high quality by the standards of the day. It was immediately apparent from his visit that an intensive effort was called for at Sunbury to deal with the technical problems of producing 100-octane spirit from Iranian crude oil. More technical staff were needed and those who had earlier been released were recalled. Some, however, had obtained employment elsewhere, while others had been transferred to Abadan, so the Company had to recruit new technical staff, whose numbers grew as shown in table 8.1. With the increase in research staff and the use of laboratories as offices by Head Office staff evacuated from London, further laboratory accommodation was urgently needed. A new laboratory was erected, but it was the only new accommodation, apart from a new stores building put up in 1942, which was provided until 1943 when a new aero-engine test house was constructed. In the meantime, the Sunbury accommodation was heavily overcrowded.

The situation was eased early in 1943 when the Head Office Refining Branch returned to London, releasing its offices at Sunbury for re-use as laboratories. Moreover, it was decided early in the year to make a general return to Britannic House. Fraser, who had succeeded Cadman as chairman in the midst of the wartime disruption of Head Office, returned there in March and the building in Finsbury Circus was again filled with staff as its prewar functions were resumed. By the end of July 1944, with the worst of the war over, there were 637 Company staff at Britannic House.[3]

The board of directors

Before the outbreak of the war, Cadman was already in declining health, having been seriously ill in spring 1939. After September 1940 he seldom attended board meetings, working intermittently from his home at Shenley Park in Hertfordshire. On 31 May 1941 he died. It had already been decided, at a board meeting some ten days previously, that Fraser should take Cadman's place and on 24 June he was formally appointed chairman. Although the succession came at an exacting time, it caused little upset or discontinuity as Fraser had been the *de facto* chairman for some time. As chairman, he largely perpetuated the policies he inherited and to which he had contributed. He continued, for example, to pursue his predecessor's conservative financial policy, to recognise the crucial importance of Iran and to support a scientific approach to technical affairs. However, though he followed in the path of Cadman, he lacked the diplomatic experience and social panache of his predecessor, whose familiarity with the corridors of power and handling of Government relations were exceptional. Fraser, on the other hand, was not liked in Government circles and his elevation to the chairmanship was, even at the time, frowned upon in Whitehall.[4] From that inauspicious beginning Fraser never succeeded in winning the confidence of ministers and civil servants, who were variously to describe him as secretive, autocratic and, to all intents and purposes, unsuitable for the chairmanship of the Company. The qualities admired by politicians and state officials were not, of course, necessarily those required for success in business. Nor, at an institutional and corporate level, were the Government and the Company always united in their aims during Fraser's term as chairman. Yet, even after allowing for these factors, the period of Fraser's chairmanship was, as will be seen in later chapters, one of unusual strain in the Company's relations with the Government. To a large degree Government criticism of the Company was focused on the exceedingly strong personality of Fraser, whose uncompromising temperament was to raise many hackles not only in Westminster and Whitehall, but also in Washington on the other side of the Atlantic.

Apart from Fraser's elevation to the chairmanship, Cadman's death was not a signal for far-reaching changes to the board. As table 8.2 shows, the only new appointment during the remainder of the war was that of Sir Warren Fisher as an ordinary non-executive director in October 1941. He was a great friend of Cadman and a distinguished civil servant, having been Permanent Secretary to the Treasury in

Table 8.2 *The board of directors, 1941–1945*

The board on Fraser's appointment as chairman in June 1941		Date of appointment as director	Age at end of 1941	Date of resignation	Died in office
Chairman and					
management directors	Sir William Fraser	1923	53	–	–
	H.B. Heath Eves	1924	58	–	–
	Sir John Lloyd	1919	67	–	–
	J.A. Jameson	1939	56	–	–
	N.A. Gass	1939	48	–	–
Government directors	Sir Edward Packe	1919	63	–	–
	Sir George Barstow	1927	67	–	–
Burmah directors	Sir John Cargill	1909	74	1943	–
	R.I. Watson	1918	63	–	–
	G.C. Whigham	1925	64	–	–
Ordinary non-executive	Sir Trevredyn Wynne	1915	88	–	1942
directors	F.W. Lund	1917	67	–	–
	F.C. Tiarks	1917	67	–	–

Subsequent appointments:		Age on appointment			
Ordinary non-executive					
directors	Sir Warren Fisher	1941	62	–	–

1919–39. The year after Fisher's arrival on the board, another of the ordinary non-executive directors, Sir Trevredyn Wynne, died at the age of eighty-nine. He had regularly attended board meetings and although, in Fraser's words, 'he rarely voiced his views', it was recorded that 'his great experience and knowledge of Eastern matters were always readily available in the interests of the Company'.[5] In 1943 the Company lost another of its older 'eastern' directors when Sir John Cargill, then in his mid-seventies, resigned because of ill health, having been a founder-director of the Company in 1909 and chairman of Burmah Oil since 1904. A welcome consequence of the departures of Wynne and Cargill was that the board was thereby reduced to twelve members. In an implied recognition that the board was previously unnecessarily large it was agreed in June 1943 that a board of twelve members was adequate to direct the Company,[6] a view which was formally endorsed at the annual general meeting in September.[7]

If the reduction in the size of the board was intended to make it a more nimble body, the inevitable slowing which comes with the ageing process must to some degree have had an opposite effect. When Fraser became chairman, most of the executive directors who had populated the board as young, relatively new faces when Cadman became chairman in 1927 were still there, but fourteen years had passed in which they had grown older together. During Cadman's chairmanship only one management director, Hearn, had left the board. Two, namely Jameson and Gass, had been added in 1939, but the former was three years older than Fraser and only Gass, who became a director at the age of forty-six, could be said to represent a younger generation of management. He was, in fact, the only director who was younger than Fraser. As for the Government directors, they had remained unchanged since 1927. The net result was that the average age of the Company's directors was sixty-four years at the end of 1941, compared with fifty-six years at the end of 1927. The chairman and management directors tended to be younger than the non-executives, having an average age of fifty-six at the end of 1941 compared with forty-seven at the end of 1927.

Within this experienced but ageing directorate a large burden of responsibilities rested upon the shoulders of Jameson and Gass, who visited Iran for quite long periods during the war.[8] Jameson acted as the technical director, looking after production and technical matters in Iran and elsewhere, including the procurement of materials, equipment and plant in the USA. Gass was concerned with concessionary and administrative affairs in Iran and other areas. Heath Eves dealt with shipping, supply activities and Government liaison and was appointed deputy chairman in October 1943. He was conscientious and conciliatory and had been the general manager for the Burmah Oil Company in India in 1914 before becoming the Company director in charge of distribution and shipping in 1924. Lloyd remained as the director in charge of finance. Fraser, as well as being Honorary Petroleum Adviser to the Government and a member of the Petroleum Board, was chairman of the Petroleum Board Management Committee. At the end of 1943 he decided to revise the responsibilities of the directors in order, amongst other things, 'to effect some reduction in the work which he was undertaking'.[9] It was not a major re-allocation of posts, but it was a recognition of the need to take some of the weight off the chairman's shoulders.

SHIPPING AND THE PATTERN OF TRADE

From their temporary offices at Sunbury and Walton the directors looked on helplessly as the Company's tanker fleet was decimated by enemy action. The scale of the losses was very high, fifty ships manned by the Tanker Company being sunk between September 1939 and January 1945. Of those vessels, forty-four were owned by the Company, representing 46 per cent of the number of ships in the Company's operational fleet at the outbreak of hostilities. As for the crews, 657 sailors died and 260 became prisoners of war. Those grievous losses were not in any way the responsibility of the Company, which had no control over the deployment of its fleet from virtually the beginning of the war when the Government took powers to direct shipping movements by voyage licensing, later replaced by the requisitioning of tankers.[10] Each tanker sunk nevertheless added not only to the loss of life, but also to a problem which seriously affected the Company's wartime operations: the shortage of tanker tonnage.

Before the war the Company's ships, and a small number of tankers which the Company had on charter, were sailing mainly on the arterial supply route from Abadan to western Europe via the Suez Canal and the Mediterranean. Other routes, such as those between Abadan and India, Africa and Australasia, were of lesser importance. Towards the end of August 1939, just before Britain declared war, the Admiralty closed the Mediterranean to British shipping and advised that ships should proceed by the much longer route around Africa by the Cape of Good Hope. The lengthening of routes meant, of course, that voyages took longer to complete and, therefore, that there was insufficient tanker capacity to maintain the previous volume of trade. The problem was exacerbated by other disruptions to normal shipping operations, such as the introduction of convoys, the withdrawal from service of vessels for arming and losses due to enemy action. For the Company, the immediate effect of these various factors was a sharp decline in loadings at Abadan due to the shortage of tanker tonnage. However, on 14 September, Italy not having entered the war, the Admiralty declared the Mediterranean open again. Company tankers which had not passed a certain point in proceeding via the Cape were recalled to the shorter route and, as the organisation of convoys began to improve, the shortage of tanker tonnage became less acute. Loadings at Abadan began to recover.

The crucial shipping route between Abadan and western Europe via the Suez Canal and the Mediterranean remained open for several more months during that period of inaction known as the 'phoney war'. It came to an end in the spring of 1940 when western Europe was devastated by the German attacks on Denmark, Norway, the Low Countries and France. By the end of May the Dutch and Belgian forces had surrendered and in June Britain's outlook was bleakened by the fall of France and Italy's entry into the war on the German side. These events had immediate repercussions on the shipping position. In May, when Italian intervention in the war was imminent, all vessels sailing east of Suez were routed round the Cape instead of through the Mediterranean. In June, when Italy declared war, the closure of the Mediterranean to Allied shipping was confirmed and with the fall of France tankers proceeding via the Cape had to make a long outward sweep into the Atlantic to avoid dangerous proximity to the German-occupied continent of Europe. These route changes increased the sailing distance between Britain and Abadan from roughly 6600 miles to some 11,200 miles.

The lengthening of tanker routes did not initially result in a great tanker shortage because, with much of continental Europe under German occupation, there was much less European demand to be met from Abadan. However, in the second half of 1940, as tanker losses to enemy action mounted, it became more necessary to conserve tanker tonnage by adopting the 'short-haul' principle, by which oil requirements were shipped from the nearest source of supply in order to save tanker tonnage. In the case of Britain, that meant reducing oil liftings from Iran in favour of nearer sources in the western hemisphere, mainly the USA. After a temporary reversal of the short-haul policy in February 1941, it was resumed in March. As a result, oil loadings for

Britain were stopped at Abadan and the last cargo of Iranian oil arrived in Britain in August 1941. Such measures had a considerable impact on the offtake of oil at Abadan, where oil loadings fell dramatically, plummeting to a mere 50 per cent of prewar levels in the last nine months of 1941.

In the meantime, the war spread to the Middle East where, in May 1941, British troops overcame resistance organised by the pro-Axis Government in Iraq and occupied the country, scene of the Iraq Petroleum Company's operations, in which the Company held a stake of 23.75 per cent. In June, the Germans invaded the Soviet Union and in August British and Soviet troops moved into Iran. As will be seen in later sections, these events had considerable effects on the operations of the IPC in Iraq and the Company in Iran.

Then, in December 1941, the struggle spread to the Far East when the Japanese attacked Pearl Harbour, precipitating the USA's declaration of war. As before, the growth of the conflict had repercussions on the Company, caused, on this occasion, by restrictions on shipping in the Dutch East Indies, which resulted in more vessels being routed to Abadan to load with oil for Australia. With the subsequent Japanese invasions of the Dutch East Indies and Burma in the early months of 1942, the Allies lost access to oil from those sources and Iran became the sole remaining large source of oil for the eastern theatre of war operations. Activity at Abadan was raised to cope with the growth in the demands made upon it, which were further increased by the preparations being made for the North African campaign. The invasion of French North Africa in November 1942 not only increased the demand for oil, but also, following the victory at Alamein, held out the prospect that the Mediterranean would soon be reopened to the Allies. In view of the prospective increase in demand, arrangements were made in early 1943 to increase the annual throughput capacity of the Haifa refinery from 2 million tons to about 2.75 million tons by plant modifications. At about the same time, it was also decided to erect new plant to increase the capacity of the refinery to 4 million tons a year, though it was not until 1944 that the new plant came into operation. In the meantime, the allied victory in Tunisia in May 1943 and the invasion of Sicily in July brought about the reopening of the Mediterranean. That had the effect of extending Abadan's short-haul area and so creating considerable additional demand for Iranian oil. The level of offtake from Abadan was further heightened by the growth in demand in India, where preparations were being made for the coming offensive in Burma, and also in Iran and Iraq where large quantities of oil were required by the occupying Allied forces and for the transport of vital war materials to the Soviet Union through Iran.

The effects of these successive phases of the war on the disposal of Iranian oil are shown in table 8.3, from which it can be seen that oil supplies from Iran to northern Europe stopped in 1940, as did those to Britain in 1941, apart from minute quantities delivered in 1942. Supplies to the Mediterranean area were dramatically cut back during the period from 1940 to 1943, that is from the time Italy entered the war to the Allied invasion of Sicily. On the other hand, supplies to eastern markets, including Africa, India, Burma, Australia and the Far East, increased greatly in 1942 after Japanese military successes deprived the Allies of access to the oil resources of the Dutch East Indies and Burma. As for domestic consumption in Iran, it rose rapidly in 1942–3 when the country was occupied by Allied forces and Iran was the main supply route to the Soviet Union. It was all, needless to say, very different from the peacetime pattern of trade in which the Company had been engaged before the war.

WARTIME OPERATIONS

The continuity of the Company's operations and the planned development of new facilities were inevitably disrupted as available resources were channelled into meeting the emergencies of the moment. Improvisation, adjustment and the re-arrangement of normal activities were necessary not only in Iran, described in the next chapter, but also in other Middle East countries and in Britain.

Kuwait and Iraq

In Kuwait, where the Burgan field had been discovered by the Kuwait Oil Company (owned in equal shares by the Company and Gulf Oil) in 1938, further wells were drilled between 1939 and 1942. However, although the great size of the field was confirmed, operations were brought to a halt early in 1942, when the wells were plugged with cement and most of the materials and equipment removed from the Shaykhdom.[11]

Iraq, on the other hand, was the scene of greater activity on the part of the IPC and the Company's wholly owned subsidiaries: the Khanaqin Oil Company, which was responsible for the production of crude from the Naftkhana field and its transportation to Alwand for process-

Table 8.3 *Iranian oil: disposal by destinations, 1938–1943 (thousand tons)*

	1938	1939	1940	1941	1942	1943
United Kingdom	2093	1760	1526	324	1	–
Northern Europe (including France)	982	743	253	–	–	–
Mediterranean (excluding France and Egypt)	472	509	344	61	20	182
Egypt	414	291	299	507	518	797
Africa and Red Sea area	1650	1622	1892	1656	2507	2106
India and Ceylon	919	705	824	969	1861	2431
Burma, Australia and Far East	409	352	402	66	956	707
Local trade (including bunkers)	814	849	865	716	1193	1586
Admiralty	1221	1399	1754	1121	1230	1105
Unknown destinations	145	79	8	–	–	–
Total	9119	8309	8167	5420	8286	8914
Percentage of 1938	100	91.1	89.6	59.4	90.9	97.8

Source: BP 71480.

Table 8.4 *The Company's crude oil offtake in Iraq, 1939–1945*
(million tons)

	Naftkhana	Company share of IPC production
1939	0.14	1.18
1940	0.17	0.77
1941	0.19	0.57
1942	0.30	0.94
1943	0.24	1.31
1944	0.24	1.45
1945	0.28	1.39

Sources: See Appendix 1.

ing at the refinery; and the Rafidain Oil Company, which distributed and marketed the products from the Alwand refinery in northern and central Iraq.

The outbreak of the war caused the IPC to postpone the implementation of its prewar plans to double the capacity of the pipeline from the Kirkuk field to the Mediterranean. Apart from that, the hostilities initially had little effect on the volume of Iraqi crude production, which was absorbed by France and Britain.[12] However, with Italy's entry into the war and the closure of the Mediterranean to Allied shipping in mid-1940, the IPC's production was heavily cut back, the Company's share falling from 1.18 million tons in 1939 to only 0.57 million tons in 1941, as can be seen from table 8.4. At the same time, France, having been overrun by the Germans, was classified as enemy territory and the IPC shares of the French Compagnie Française des Pétroles (CFP) were put in the hands of the Custodian of Enemy Property with the result that the CFP was not allowed to lift its quota of IPC oil until 1945. Calouste Gulbenkian, who had a 5 per cent share in the IPC, was resident in France and was therefore also classified as an enemy until 1943, when he moved to Lisbon.

The slackening of production was reversed by the events of late April and May 1941, when British forces clashed with and overcame the Iraqi forces of the pro-Axis Government of Rashid Ali al-Gaylani.[13] For a brief period in May oil production stopped completely, but after the British had overthrown Rashid Ali the large numbers of Allied troops occupying Iraq increased the local demand for oil. In response, the capacity of the Alwand refinery was increased

44   This group, plus six, shows the entire British and US population of
Kuwait. Taken New Year's day, 1942

to the extent that its throughput doubled between 1939 and 1945, with
crude oil being supplied not only from Naftkhana, but also from
Naft-i-Shah, the Iranian side of the same field.[14] However, even the
enlarged capacity of the Alwand refinery was inadequate to meet
demand. As there was no other refinery in Iraq, the nearest alternative
source of supply was Abadan, but it was difficult to supply the
considerable number of troops stationed in northern Iraq from the
south of Iran over the long and difficult lines of communication.
Arrangements were therefore made to supplement local supplies by
converting stabilisation units at Kirkuk into plant capable of producing 70-octane motor fuel.[15] To cater for the internal distribution of
products new pipelines were laid and tin plants for the manufacture of
petrol containers were erected at Kirkuk and Mosul. Much other work
was carried out by the IPC to help the occupying forces, including any
number of small jobs, such as mounting fuel tanks on military vehicles,
the installation of lorry and rail loading facilities, the erection of
kerbside pumps and tanks, the loan of specialist staff and the provision
of supplies of one kind and another.

   In such conditions, it could not be helped that short-term expedients
took precedence over longer-term development plans, which were

retarded by wartime requirements. Of greatest importance, in 1942 the military command decided that most of the producing wells in the Kirkuk field should be destroyed as a precaution against the possibility that they might fall into enemy hands. All but six of the wells were plugged so effectively by the Royal Engineers as to preclude their being reopened. The war also made it impossible for the IPC to fulfil the obligations to explore for, produce and export oil which it had assumed under its concessions for the Mosul and Basra *vilayets*. After much negotiation an agreement was reached between the IPC and the Iraqi Government providing for a suspension of the IPC's drilling, production and export obligations in the Mosul and Basra concessions until two years after the end of the war. As recompense for the suspension, the IPC made an interest-free loan of £500,000 to the Iraqi Government immediately and promised an additional £1 million at the end of the war, repayable on terms similar to the 1939 loan.[16]

Britain

In Britain, the Company's operations were rearranged to serve the war effort not only by the pooling of its distribution and marketing activities under the Petroleum Board, but also in refining and crude oil production. From early in the war it was clear that the Government did not favour home refining, which was regarded as being wasteful of tanker tonnage on account of the need to import oil to be used as refinery fuel. The Grangemouth refinery, which also had the disadvantage of being on the relatively vulnerable east coast, was one of the first refineries to be shut down, its operations ceasing in March 1940.[17] The Llandarcy refinery had already been brought to an almost complete stop on the outbreak of war so that the staff would be available for starting the can factory which was constructed there to manufacture four-gallon petrol containers for the services. From October 1939 refining operations gradually restarted, processing Iranian and Iraqi crude to turn out a full range of products. During 1940–1 air raids caused considerable dislocation and operations were maintained only with difficulty. However, refining continued until September 1941 when the refinery's crude oil supplies were exhausted and could not be replenished owing to the shortage of tankers. From October 1941 to February 1942 operations were suspended, but in March the plant was brought back onstream for the manufacture of lubricating oils, using feedstocks supplied from the western hemisphere.[18]

As for crude oil exploration and production, the initial reaction to the outbreak of war was to curtail drilling activities in Britain, where operations were confined to drilling production wells on the Formby and Eakring fields which had been discovered just before the war. The reason for the reduction in drilling was the need to conserve steel, which was urgently required for the munitions and shipbuilding industries.[19] However, in December 1940 drilling commenced at Kelham Hills, where oil was discovered. In July 1941, as supplies of Iranian oil were coming to an end on account of the short-haul policy, the Petroleum Department urged the Company to increase its oil production in Britain, pointing out that 'in the present emergency every ton of oil produced in this country is a direct contribution to the national war effort'.[20] The Petroleum Department's target was a production rate of 100,000 tons a year of crude oil, a fourfold increase of the production rate of about 25,000 tons a year which was then being achieved.

On receiving the Petroleum Department's request, the Company directed its efforts mainly to the rapid development of the newly discovered field at Kelham Hills and of a new extension to the Eakring field at Duke's Wood. After achieving the target production rate of 100,000 tons a year in September 1942, a further discovery was made at Caunton, to the east of Eakring, in March 1943, the new field coming onstream in May that year. In addition, a tiny field was discovered at Nocton in Lincolnshire in 1943, production commencing in very small quantities in December.[21]

Aside from conventional oil activities, the Company also undertook all manner of work for the war effort, such as the aforementioned manufacture of petrol cans, the erection of fuel storage depots and various projects for the Petroleum Warfare Department to which a number of Company employees were seconded to work on devices such as flame-traps, flame barrages and flame throwers. The best-known schemes in which the Company was involved went by the acronyms PLUTO and FIDO. The first, standing for Pipeline Under The Ocean, was designed to support the Allied invasion of France in 1944 with petroleum supplies transported across the English Channel through small-bore pipelines laid on the seabed. The second, standing for Fog Investigation Dispersal Operations, was to dissipate fog at airfields so that aircraft could land safely. The story of both has been told elsewhere and need not be repeated here.[22]

Less well publicised, but arguably of greater interest in view of later technological developments in the industry, was the experimentation with horizontal drilling. In March 1940 a Mining Committee was formed under the Scientific Advisory Council of the Ministry of Supply to consider military mining problems. The main challenge was to break the German Siegfried Line, to which end it was decided that the possibilities of horizontal drilling should be investigated to see if it would be a suitable method for placing explosive charges under the enemy-laid mines. C. A. P. Southwell, a Company employee who was a member of the Mining Committee, arranged to carry out experiments using a converted drilling outfit, previously used for geophysical shot-hole drilling, to drill horizontally. A small team of Company drillers carried out secret tests and experiments, a final trial being held in September 1940, attended by senior engineers from the War Office and other Government officials. At the trial, a horizontally drilled hole penetrated a target 5 feet in diameter at a distance of 1000 feet. The boring was enlarged to take explosives and after the demonstration, which was reported to be 'a complete success',[23] it was decided to design new purpose-built boring equipment. Two machines were fabricated, but by then it was apparent that the static line warfare of World War I had been superseded by more mobile manoeuvres. That being so, it was decided not to go ahead with the ordering of further equipment for horizontal drilling.[24]

PROFITS AND FINANCE

For all the anxiety and disruption caused by the war, there was little cause for concern about the state of the Company's finances, which remained comfortingly secure. Admittedly there was a setback to pre-tax profits during the early years of the conflict, when the dislocation of operations caused sales to fall well below the volume of more than 10 million tons which was achieved in 1938. However, from 1941 sales increased rapidly, reaching more than 18 million tons in 1945, the growth in volume being accompanied by the spectacular growth in pre-tax profits which is shown in column (a) of table 8.5. That, though, is before taking account of two factors which had a large impact on the Company's real net profits during the war: the rise in British taxation and price inflation. With regard to the former, the wartime introduction of Excess Profits Tax resulted in a great increase in the Company's tax charge, bringing a corresponding reduction in net profits. As for the effects of inflation on the real value of the Company's profits, the usual reservations have to be made about the accuracy of measurement depending on the weighting of the price deflator which is used. In this case, C. H. Feinstein's index of plant and machinery prices has been employed to calculate the Company's post-tax inflation-adjusted profits. The results are shown as an index in column (d) of table 8.5 and suggest that in three of the war years real post-tax profits were lower than they had been in 1929. In only one year, 1942, were they higher than they had been, on average, in the years 1936–8. In short, the Company made very much less money in terms of real post-tax profits than might be supposed from a cursory glance at its pre-tax profits, unadjusted for inflation.

Nevertheless, the Company remained financially very secure indeed, continuing to practise the financial self-reliance which had become its habit in the prewar decade. As table 8.6 shows, funds generated from operations were more than sufficient to cover expenditure on capital investment and dividends in every year of the war except 1939. The debt repayments of the 1930s had already reduced the Company's borrowings to negligible proportions and there was, therefore, virtually no loan capital left to repay. Instead, the surplus of internally generated funds was applied to increasing liquid resources, which rose from about £17 million at the end of 1938 to some £63 million at the end of 1945. Thus, the Company continued to be self-financed and almost entirely free of debt (excluding debts to current creditors incurred in the normal course of business). Its balance sheet was built on rock.

PLANNING AHEAD

Although the Company's organisation into functional departments remained basically unchanged during the war, a new initiative was taken in 1943, when Fraser, thinking on the same lines as company chairmen in other British firms such as Courtaulds, decided it was time to plan ahead for the return of peace.[25] For that purpose, Fraser relieved George Coxon of his responsibility for refining matters and placed him in charge of a small planning unit, which was independent of the large functional departments, free from operational routine and reported directly to the board.[26] This was a new concept for the Company, the genesis of a long-term planning function which addressed corporate rather than narrowly departmental matters. Coxon was ideally suited to his new task. Rigorous in analysis, technically proficient and intellectually inquisitive, he played a seminal role in the early development of the Company's forward planning, whose usefulness, according to one of his assistants, M. E. Hubbard,

Table 8.5 *The Company's profitability, 1939–1945*

| | Profits before and after taxes and inflation | | | | | Return on capital | |
| | (a) | (b) | (c) | (d) | (e) | (f) | (g) |
	Pre-tax profits at current prices (£millions)	Deduction for tax (£millions)	Post-tax profits at current prices (£millions)	Index of post-tax profits at constant prices (1929 = 100)	Capital employed at current prices (£millions)	Pre-tax return on capital employed: (a) as % of (e) (%)	Post-tax return on capital employed: (c) as % of (e) (%)
Ave 1936–38	8.8	2.3	6.5	155.5	46.4	19.0	14.0
1939	7.4	3.6	3.8	83.3	48.6	15.2	7.8
1940	10.3	5.1	5.2	103.4	46.5	22.2	11.2
1941	11.5	5.5	6.0	106.6	45.5	25.3	13.2
1942	21.0	9.3	11.7	188.5	50.3	41.7	23.3
1943	22.7	15.4	7.3	101.7	54.5	41.7	13.4
1944	27.9	20.6	7.3	96.4	56.8	49.1	12.9
1945	23.4	19.7	3.7	48.9	63.4	36.9	5.8

Sources: BP consolidated balance sheets (see Appendix 1). The price index used in calculating profits at constant prices is that for plant and machinery in C. H. Feinstein, *National Income, Expenditure and Output in the United Kingdom, 1855–1965* (Cambridge, 1972), table 63.

Table 8.6 *The Company's sources and applications of funds, 1939–1945 (£millions)*

| | Internally generated funds less capital expenditure and dividends | | | | Financial movements | | | |
	(a) Funds generated from operations	(b) Capital expenditure	(c) Dividends	(d) Funds generated or (required) after outflows on (b) and (c)	(e) Shares issued	(f) (Increase) or reduction in loan capital	(g) Increase or (reduction) in liquid resources	(h) Total financial movements: (e) + (f) + (g)
1939	8.1	8.8	2.1	(2.8)	–	–	(2.8)	(2.8)
1940	11.4	3.0	2.1	6.3	–	0.5	5.8	6.3
1941	17.0	4.8	2.6	9.6	–	(0.6)	10.2	9.6
1942	24.1	9.0	5.1	10.0	–	0.6	9.4	10.0
1943	21.9	4.2	5.1	12.6	–	–	12.6	12.6
1944	17.4	5.1	5.1	7.2	–	–	7.2	7.2
1945	24.4	16.2	5.1	3.1	–	–	3.1	3.1
Total	124.3	51.1	27.2	46.0	–	0.5	45.5	46.0

Sources: BP consolidated balance sheets (see Appendix 1).

lay 'firstly, in the breadth of vision and the wisdom of Mr Coxon'.[27] He and his small staff paid attention to a wide range of issues, including the development of the Company's marketing organisation, the need for additional housing at Abadan, the extension of the loading facilities at Abadan and Bandar Mashur, a policy for research, the recruitment of new staff, tanker tonnage requirements and the most economic locations for new refining capacity. Above all, what distinguished Coxon's work from anything that had gone before was that he related each individual part of the Company's operations to the long-term development of the Company *as an integrated whole*, from crude production through to the retail marketing of end products.

Yet however novel Coxon's role may have been at the time, the Company's fundamental direction remained unchanged. It was unequivocally set out by the Planning Committee, on which the managing directors were represented, which showed a familiar purpose in agreeing that the cornerstone for future planning should be the steady expansion of exports from Iran at a rate which was at least as fast as the growth of world oil consumption. The primacy of Iran in the Company's forward planning was confirmed.

=9=
Transition in Iran,
1939–1947

After Britain declared war on Germany, Riza Shah adopted the same policy that Iran had followed in World War I in declaring his country's neutrality. That parallel aside, he would not, it may safely be supposed, have wished Iran to undergo a repetition of its experiences in the earlier war, when, despite its neutrality, Iran suffered the humiliations of invasion by foreign troops, a breakdown in the authority of central government and the establishment of a Soviet-backed autonomous republic in one of its northern provinces. Whatever Riza Shah may have wished, much the same pattern of events was to take place in World War II and, caught up in them, he himself was to lose his throne. However, before that happened there was still time for him to go through a last performance of the concessionary routine which had manifested itself earlier in his reign: the threat of action against the Company's concession to extract improved financial terms when his oil revenue was diminished by a downturn in exports.

THE 'MAKE-UP' PAYMENTS AGREEMENT

The reasons for the drop in oil loadings at Abadan when war broke out were enumerated in the last chapter and need not be repeated in detail here. Suffice to say that exports of Iranian oil were reduced because of the disruption to shipping caused by the re-routing of tankers to the long haul round Africa after the Mediterranean was temporarily closed to Allied shipping, the introduction of convoys, the holding of tankers in port while they were armed and the sinking of ships by enemy action. At the same time, the widespread introduction of rationing began to have its desired effect of reducing civilian oil consumption, which meant, of course, less demand. The effects of these factors on Iranian oil exports and, consequently, royalties were explained to the Shah by Cadman[1] and to the Iranian Minister of Finance by Rice, but they were unable to make any impression on the Iranians who insisted that the Company's production ought to be increased.[2] This was made plain by the Shah when he met Jameson in Tehran on 14 February 1940 and asked him to tell Cadman that: 'He must understand that the production of oil from Iran must not be less than in 1937'. Reporting to Cadman on his audience with the Shah and meetings with ministers, Jameson wrote:

> The impression I formed from the audience and also from subsequent interviews with the Minister of Finance . . . and the Prime Minister . . . is that they are not prepared to accept any reasons as reasonable but only as excuses. They all repeat the same lesson which has been dictated to them viz. 'The shipments must not be less than in 1937', and no one dare accept any reason, however justified by conditions, for a lesser amount.[3]

As it happened, Jameson's meetings in Tehran coincided with a recovery in exports from Abadan as the reopening of the Mediterranean and the improved organisation of convoys helped to reduce the tanker shortage during the 'phoney war' period in late 1939 and early 1940. The upturn in exports might have eased the tension between the Company and the Iranian Government had it not been for other matters which caused relations to be soured. In particular, the Iranians were annoyed at the British failure to deliver goods under the Exports Credit Scheme and by the restrictive nature of British controls on the exchange of foreign currencies. As was described in chapter 2, in mid-1939 the Exports Credit Scheme had been held out by Cadman to the Shah as a possible means by which Iran might be able to sustain its expenditures on military and civilian projects at a time of falling oil revenues. The basis of Cadman's suggestion was that if Iran could obtain the necessary export credit licences from the British Exports Credit Guarantee Department, it would be able to purchase British goods for payment out of future rather than present income. That, indeed, was the purpose of the Export Credit Agreement which was concluded between the British and Iranian Governments in February 1940.[4] However, the agreement went sour when it became apparent that Britain was unable to meet Iranian orders for aircraft and railway equipment because, in the deteriorating international situation, it was impossible to supply some of the goods. The Iranian Minister in London expressed his Government's dissatisfaction with the situation.[5] As if to add insult to injury, Iran was prevented from turning to the USA for alternative supplies by British exchange controls under which sterling, the currency in which the Iranian Government received its oil royalties, was not freely convertible into dollars, the currency required for payment to US suppliers. Angered at the state of affairs, the Iranian Government cancelled the Export Credit Agreement in mid-June. About a week later Rice reported that Riza Shah's mood was 'savage' and, ominously, that he would become 'more savage when our drop in export comes to his notice'.[6]

Rice was not scaremongering. A few months earlier, in spring 1940, the 'phoney war' had been transformed into a real one after the Germans invaded Denmark and Norway in April. In the next month they overran Holland, Belgium and Luxembourg. In June, Italy entered the war on the side of Germany, and France accepted German terms for an armistice. In the summer and autumn of 1940, while Hitler launched his great aerial attack on Britain, the course of the war created problems of a different nature in Iran. With Italy's entry into the war the Mediterranean was again closed to Allied shipping with the result that tankers which normally would have plied the Company's main trade route between Abadan and western Europe had to make the much more circuitous and time-consuming voyage around Africa. At the same time, the European market for Iranian oil was greatly reduced because much of Europe, being under German occupation, was closed to the Company. A little later, as the shortage of shipping became more acute, the short-haul policy was rigorously enforced with the result that exports of Iranian oil to markets west of Suez ceased altogether. To the vexed issues of the Export Credit Agreement and exchange controls was added, therefore, the rapid fall in oil exports from Iran which, as Rice was all too well aware, was bound to create friction between the Iranian Government and the Company.

On 2 July Rice was informed by General Amir Khosrovi, the Iranian Minister of Finance, that action against the Company's concession was imminent unless something was done immediately to ease Iran's exchange difficulties.[7] The same day, Sir Reader Bullard, the British Minister in Tehran, cabled Lord Halifax, the Foreign Secretary, with the news that 'Unless Anglo-Iranian Oil Company can somehow satisfy Iranian Government this week, cancellation of their concession will be proposed to Parliament [Majlis], Sunday, 7 July'.[8] The British Government immediately made new proposals regarding the conversion of sterling into dollars, but the crisis did not pass. On 7 July Khosrovi stated in the Majlis that in the Iranian Government's view oil production in Iran could be trebled, instead of which it was falling.[9]

On 16 July he wrote to Rice rejecting the reasons which the Company had put forward for the reductions in exports and stating that the Company should make payments to the Iranian Government as if it had produced 14 million tons of royalty oil in 1940. Khosrovi concluded with the warning that if compensation for reduced royalties could not be arranged the Iranian Government would 'revise [the] oil Concession fundamentally'.[10]

In demanding that the Company should make a payment as if it had produced 14 million tons of royalty oil in 1940 the Iranian Government was, as the Company pointed out, pitching its claims way above the actual tonnage achieved by the Company, whose output of royalty oil was 8,309,707 tons in 1939 and, as it later transpired, 8,167,286 tons in 1940.[11] Nevertheless, it was apparent that something would have to be done to compensate the Iranian Government for the fall in royalties. After discussions with the British Government the Company sent its proposals to Rice on 31 July[12] and he in turn submitted them to Khosrovi on 2 August.[13] In essence, the Company offered to make interest-free loans to cover the difference between its annual payments in the years 1939–41 and £3.5 million per annum. The loans would, it was proposed, be repayable from annual royalties in excess of £3.5 million as and when normal royalties recovered past that amount as a result of an improvement in trade. After Riza Shah had been consulted, Khosrovi and Ali Mansur, the Iranian Prime Minister, informed Rice that the Company's proposals were unacceptable because Iranian demands for a payment based on an output of 14 million tons of royalty oil had been ignored and they would definitely not agree to accept the principle of a loan, insisting that the payment to make up for low royalties should be non-repayable.[14] In London, Halifax thought that 'it was not worth risking some rash action on the part of the Iranian Government for sake of difference between an "advance", which might never be repaid, and an outright gift of money'.[15] For its part, the Company agreed to accept the principle of making an *ex gratia* payment rather than a loan[16] and entered into further negotiations the outcome of which was an exchange of letters, dated 21 August, which constituted the 'make-up' payments agreement, to which, it was later recorded, the Company agreed 'at the express desire of HM's Government'.[17] Under its terms the Iranian Government was to receive £1.5 million to add to the royalty for 1939 and further payments in whatever sums were required to make up a total of £4 million in each of the years 1940 and 1941. By separate agreement, the British Government undertook to contribute half of the payment of

£1.5 million for 1939 and of the subsequent 'make-up' payments for 1940 and 1941.[18]

While the main purpose of the make-up payments agreement of August 1940 was to deal with the immediate threat to the concession, it also provided for the situation to be reviewed after the end of 1941. The matter was duly raised by the Iranian Government in 1942 and again in January 1943, when Taqizadeh, having come out of political exile to be Acting Minister in London, appealed to Fraser to extend the agreement as a wartime measure.[19] As will be seen later in this chapter, the dictatorship of Riza Shah was by then at an end and, with Allied forces occupying Iran, there was no immediate danger to the concession. It was against this backdrop that in January 1943 Fraser discussed the Iranian request for an extension of the make-up payments agreement with Sir Maurice Peterson, Under-Secretary of State at the Foreign Office.[20] Peterson was not in favour of continuing the 'blackmail', as he called it, of 1940 and thought that the agreement should be extended only in exchange for a quid pro quo. Fraser, on the other hand, felt that if Peterson's suggestion was followed it might open the door to prolonged and complicated negotiations on the whole concessionary position, resulting, possibly, in terms less favourable than those which the Company already enjoyed. A simple extension of the make-up payments agreement did not, however, find favour with the Government. Bullard, echoing Peterson, strongly opposed the renewal of the 'blackmail agreement', as he also termed the make-up payments agreement, which he felt 'savoured of extortion'.[21] Eden, who had succeeded Halifax as Foreign Secretary in December 1940 and to whom the matter was referred, was of like view.

Fraser, however, refused to deviate from his preferred course of extending the agreement and, it would appear, pushed the issue close to the point of open rupture with the Government. That, at least, is the conclusion to be drawn from the record of the Company's board meeting on 25 May 1943, when the directors backed Fraser's judgement in favour of extension, but not, significantly, with unanimity. The odd man out was Sir George Barstow, one of the Government directors, who indicated his opposition to the proposed continuation of make-up payments and asked that his view should be recorded in the minutes.[22] Such open opposition was unheard of, there being no previous occasion on which dissent had been expressly recorded in the normally bland minutes of Company board meetings; and no previous chairman who had run so close to having his chosen policy vetoed by the Government. It was not the sort of thing that would have happened

Table 9.1 *Concessional and 'make-up' payments to the
Iranian Government, 1939–1945*

	Concessional payments: royalty and tax (£)	Additional 'make-up' payments (£)	Total (£)
1939	2,770,814	1,500,000	4,270,814
1940	2,786,104	1,213,896	4,000,000
1941	2,025,364	1,974,636	4,000,000
1942	3,427,933	572,067	4,000,000
1943	3,617,917	382,083	4,000,000
1944	4,464,438	–	4,464,438
1945	5,624,308	–	5,624,308

Source: BP 4308.

in Cadman's time, but Fraser, as was noted in chapter 8, was a different kind of chairman: more abrasive, less conciliatory and lacking the polished diplomatic skills of his predecessor. The disagreement over the extension of the make-up payments went no further and the Government accepted the Company's decision to carry on with them until the end of the war, although it declined to continue contributing half of the amounts as it had done for the payments in respect of the years 1939–41. As will be seen in later chapters, it was not, however, to be the last occasion for friction between Fraser and the Government.

It so happened that the extension of the make-up payments agreement until the end of the war did not involve the Company in very large expenditures. As can be seen from table 9.1, in 1939–41 the make-up payments made a very substantial contribution to sustaining the Iranian Government's oil revenue. However, after 1941 the Company's normal concessional payments increased rapidly as oil output in Iran began to recover. As a result, the make-up payments fell to £572,067 for 1942 and £382,083 for 1943 before disappearing altogether as normal payments reached in excess of £4 million.

ENTER THE ALLIES AND A NEW SHAH

In the early years of the war, despite Riza Shah's proclamation of Iran's neutrality, the British Government became increasingly concerned about the extent of German influence in Iran. As was seen in chapter 2, Iran and Germany had developed strong trading links in the 1930s, from which Germany emerged as Iran's main trading partner, accounting for about a quarter of Iran's commercial imports and nearly half of her non-oil exports in the year from March 1939 to March 1940.[23] Germany's economic penetration of Iran was accompanied by propaganda, spread by the ubiquitous German commercial agents and technicians who, according to Bullard, devoted particular attention to the military and merchant classes 'with conspicuous success in predisposing them against the United Kingdom and persuading them of the chances of German victory'.[24]

In late 1940 Britain's anxiety about this state of affairs prompted it to request the expulsion of German personnel from Iran, but to no avail. The Iranian Government responded only with reassurances that it had taken 'every precaution' and that all Germans in the country were kept under close watch.[25] Further British requests were made in the early months of 1941, but they also failed to elicit the desired action from the Iranian Government, which was understandably reluctant to upset Germany, whose resounding military successes in the early stages of the war made it look likely that it would emerge the victor. The British Government's unassuaged anxiety was made more acute by Britain's military contretemps with the pro-German Iraqi Government of Rashid Ali al-Gaylani in April–May 1941. Bullard thought that the Germans might try to set up a pro-German government in Iran as well as in Iraq. Eden agreed and thought that 'immediate and energetic action on our part is necessary with a view to jolting Iranian authorities out of their apparent complacency and inducing them to take what measures they can, before it is too late, to check possible German intrigues'.[26]

Then, on 22 June, the Germans invaded the Soviet Union, raising the prospect that German forces might conquer the Caucasus and threaten Iran from the north. At the end of June Eden tried to impress upon Ali Muqaddam, the Iranian Minister in London before Taqizadeh, that however much Iran might dislike the Soviet Union, if the Germans reached the Caucasus, then Iran would be faced with a 'much more serious menace'.[27] Eden thought that the time had come for Iran to rid herself of the large number of Germans in the country, but Muqaddam claimed that they were essential for Iranian industry.

Ivan Maisky, the Soviet Ambassador in London, saw Eden on 10 July and talked about the number of Germans in Iran, about which they were deeply concerned. It was felt that immediate measures needed to be taken.[28] A ministerial committee discussed the matter the next day and agreed that it was vital to ensure the security of the oilfields and refineries from German sabotage, but undesirable to take drastic action if British objectives could be achieved by persuasion.[29] However, by 19 July Eden, feeling pessimistic about the prospect of winning the Iranians over by power of argument, felt that the possibilities of applying military or economic pressure should be considered.[30] The effectiveness of economic sanctions or a blockade was, as usual, open to doubt. In this case, such measures might well have backfired by endangering oil supplies and/or strengthening Irano-German relations, the very opposite of what was desired.[31] The military option was preferred and its use approved by the War Cabinet on 28 July.[32] The next day Eden set out the lines of British thinking as 'First the concentration of troops, then the presentation of our demands, and in the third place, if the demands are refused, perhaps some military action'.[33] In August this classical plan was duly put into action, the military option being held in reserve while Riza Shah was requested to terminate the unfriendly activities of German nationals in Iran.[34] He made no promise and offered no compromise. Bullard saw him once more on 24 August. There was no change. On the next day two divisions of the British Indian army entered Iran, one with orders to occupy Khurramshahr and Abadan and push on to protect the Fields, the other to cross the Iraqi border to take Kirmanshah and eventually join forces with Soviet troops who invaded Iran from the north.

As the British military operation was aimed at securing the Company's oil installations, there was a strong chance of some of the Company's staff getting caught up in the action, which is what happened. At Abadan the initial landing of Indian troops at dawn on 25 August was resisted by Iranian forces. In the confusion of the engagement there were casualties, some fatal, among Company employees, including three British staff who were tragically killed by the friendly fire of the invading forces.[35] As for the refinery, it was shut down on the afternoon of 25 August because of the difficulty of relieving the shift workers who had been on continuous duty since the previous night. However, it was partly re-commissioned on 28 August, when the Iranian Government ordered the cessation of Iranian resistance to the Allied forces, and normal refining operations were resumed on 30 August. Outside Abadan, some of the Company's pipeline and Fields staff were arrested and held by the Iranian authorities for two or three days, and at Kirmanshah staff were confined to the refinery area.

Once the short-lived military action at Company centres was over, the Allied forces moved towards Tehran. As they approached, on 16 September Riza Shah abdicated and sailed to the Indian Ocean island of Mauritius, where he was to remain until March 1942 when he moved to South Africa. He died there in July 1944. On his abdication, his place as Shah was taken by his son, Muhammad Riza, who was then a few weeks short of his twenty-second birthday.[36]

A few months later, in January 1942, the Allied presence in Iran was formalised in a Tripartite Pact between Britain, the Soviet Union and Iran. Under its terms, the Allies were granted the right to keep armed forces in Iran for the duration of the war and up to six months afterwards. Although it was expressly stipulated that their presence did not constitute a military occupation, this was more a matter of form than substance. The Allies were given rights of passage of troops and supplies through Iran and unrestricted use of Iranian transport and communications facilities, including railways, roads, rivers, ports, aerodromes, pipelines and telephones. For their part, the Allies undertook to respect the territorial integrity, sovereignty and independence of Iran and to defend the country against aggression from other powers.[37]

The accession to the throne of the young Muhammad Riza Shah and the *de facto* occupation of Iran by the Allies introduced a period of heightened economic and political activity in Iran. While the effects on the Company's operations were felt immediately, other events were set in train which were to have an important bearing on the future of its concession.

COMPANY OPERATIONS AND EMPLOYMENT

After the outbreak of war the Company acted quickly to cut back its operations and adjust its output to the reduction in offtake caused by the disruption to shipping and the fall in demand. Exploration activity was suspended, and although the Gach Saran oilfield came into commercial production in 1940 when the pipeline connecting it to Abadan was completed, crude oil production was greatly reduced, falling from the prewar peak of 10,195,371 tons in 1938 to 6,605,320 tons in 1941.[38] At Abadan, the number of grades of products manufactured at the refinery was cut down, particularly in the benzine and fuel oil ranges, as can be seen from table 9.2. Overall, crude oil throughputs at the refinery fell from 9.66 million tons in 1938 to 6.94 million tons in 1941. They would, indeed, have been reduced further but for the fact that

Table 9.2 *Number of products in tankage at Abadan,*
August–October 1939

	August 1939	October 1939
Benzines	14	4
Kerosenes	2	2
Vaporising oil	1	1
Gas oil	3	3
Diesel oil	3	2
Fuel oil	7	4

Source: BP 25553.

aviation spirit was so important for the war effort that refinery throughputs were kept up to produce the maximum possible quantity of that one product. As a result, a surplus of other products was turned out, only to be recycled by return to the reservoirs at Masjid i-Suleiman and, for a time, at Haft Kel.[39]

At the same time, investment and employment were heavily reduced. Capital expenditure in Iran, which had amounted to some £4,429,000 in 1938, was cut to £2,997,000 in 1939, £813,000 in 1940 and a mere £243,000 in 1941. The number of Company employees, together with the employees of contractors engaged on Company work, was reduced from 51,060 at the end of August 1939 to 37,395 by the end of the year and only 26,271 at the end of August 1941. It was, however, not only quantitatively, but also qualitatively that the effects of the war were felt by Company employees in Iran. As the numbers fell, those who remained had to conform with regulations introduced by the Iranian Government in the cause of neutrality, including restrictions on the movement of personnel, the censorship of mail and the suppression of news and propaganda which might, in the opinion of the Iranian Government, have conflicted with Iran's neutral position. As a result, mail was seriously delayed and sometimes destroyed, the showing of newsreels in cinemas was prohibited and the community of British staff in Iran was cut off from much of its normal contact with the outside world.[40]

The entry of Allied troops into Iran in the last week of August 1941 marked a turning point in wartime operations. Local demand for oil was greatly increased by the energy consumption of the British and

Soviet occupying forces, later augmented by the arrival of forces from that most profligate nation of oil consumers, the USA. It was not merely the presence of the Allies in Iran which added to the demand for oil, but their intensive activity in using Iran as a corridor for the supply of materials to the Soviet Union. For that purpose, the route via the Persian Gulf and northwards through Iran, for which the Trans-Iranian Railway provided a most useful connection between the Gulf and the Caspian Sea, was considerably safer than the dangerous northerly route of the Arctic convoys. As a result, Iran became the scene of great wartime activity. Moreover, as the war progressed the demand for Iranian oil was increased further by the Allies' loss of supplies from the Dutch East Indies and Burma after the Japanese invaded those areas in early 1942, and by the preparations for the Allied campaign in North Africa later in the year. Then, in 1943, still more demands were made for Iranian oil after the Mediterranean was reopened to Allied shipping and as preparations were made for the Allied offensive in Burma.

To meet the growing demands there was a great revival and extension of Company activity not only in the usual operations of oil production, refining and distribution, but also in improvised response to the exigencies of war. Defensive precautions at Abadan included a balloon barrage, an improved fire fighting service, the erection of bomb shelters and, at a distance of a few miles, the construction of a dummy refinery consisting of gas flares, lights and other decoy devices.[41] Relations between the Company and the military command were so close that, in the words of the historian Stephen Longrigg, it was 'not possible easily to distinguish between the strictly industrial operations of the Company ... and those designed to help the complex and massive operations of the Military Command'.[42] For example, the Company's barges on the river Karun were used to transport railway engines, tanks and other vehicles destined for the Soviet Union; the capacity of the Abadan tin-making factory for the manufacture of petroleum containers was doubled between 1941 and 1944; and asphalt was supplied in large quantities for military roads and depots.[43]

Expansion also took place in those activities more conventionally associated with oil company operations. For example, in 1943 the Company began pumping gas condensate from Pazanun to Abadan, where it was used to enrich aviation spirit. The following year, a 12-inch pipeline connecting the Agha Jari oilfield to Abadan was completed, and in 1945 the Naft Safid field was linked to the main pipeline system.[44] Output from other oilfields was also increased,

45   Troops at Abadan refinery, *c.*1941

contributing to the spectacular surge in crude oil production, which more than doubled from some 6,605,000 tons in 1941 to 13,274,000 tons in 1944 before rising still further to 16,839,000 tons in 1945 (see table 9.3). Between the same dates, the throughput of the Abadan refinery increased at much the same rate, as can be seen from table 9.4.

Such rapid expansion required, of course, heavy additional inputs of capital and labour. Capital expenditure on Company facilities in Iran, having been cut back to only £243,000 in 1941, leapt dramatically to £3,803,000 in 1942 and then nearly doubled to £7,006,000 in 1944. As table 9.5 indicates, most of the new investment was in refining, which meant, in physical terms, the expansion of facilities at Abadan where easily the most important of the wartime construction projects

Table 9.3 *The Company's crude oil production in Iran,*
1939–1945 (thousand tons)

	Masjid i-Suleiman	Haft Kel	Gach Saran	Agha Jari	Naft-i-Shah	Total
1939	2555	6916	15	6	92	9584
1940	1789	6327	415	12	83	8626
1941	1059	4682	763	9	92	6605
1942	1510	6821	964	3	100	9398
1943	774	7712	956	160	104	9706
1944	1896	9218	1542	519	99	13,274
1945	2671	9810	1905	2371	82	16,839

Sources: See Appendix 1.

Table 9.4. *Abadan refinery throughputs, 1939–1945*
(million tons)

	Crude oil throughputs
1939	9.25
1940	8.77
1941	6.94
1942	10.23
1943	10.49
1944	13.48
1945	16.82

Sources: See Appendix 1.

was the installation of plant for the large-scale manufacture of aviation spirit.

Drawing together the threads of Government policy and Company research on aviation spirit manufacture which were separately traced in chapters 6 and 7, it will be recalled that in July 1936 technical staff at the Company's Sunbury research centre discovered the alkylation process by which iso-octanes with excellent anti-knock value for the manufacture of aviation spirit could be produced. After further work on the process, it was decided at the end of 1937 to erect a pilot plant to operate the alkylation process at Abadan. However, by the time the

Table 9.5 *The Company's capital expenditure in Iran,*
1939–1945 (£thousands)

	Fields	Pipelines	Refineries	Other	Total
1939	345	241	2148	263	2997
1940	5	–	620	188	813
1941	–	15	85	143	243
1942	1	8	3782	12	3803
1943	82	1	3249	26	3358
1944	263	1018	5621	104	7006
1945	1281	687	4633	262	6863

Note: Capital expenditure on refineries from 1942 includes expenditure on aviation spirit plant which was partly recoverable from the British Government.
Sources: See Appendix 1.

pilot plant commenced operating in January 1939 the British Government had come down in favour of the alternative of manufacturing aviation spirit by the process of hydrogenation. The Government was advised on this matter by the Hartley Committee, whose brief was to consider how adequate supplies of 100-octane aviation spirit for the RAF could be assured. The Committee, consisting of representatives of ICI, Royal Dutch-Shell, Trinidad Leaseholds Ltd and the Air Ministry, suggested that three new hydrogenation plants, two in Trinidad and one at Heysham in Lancashire, should be constructed. To supervise the construction and operation of the plants, ICI, Royal Dutch-Shell and Trinidad Leaseholds jointly formed a company called Trimpell Ltd and work on the first of the three plants, the one at Heysham, commenced in the spring of 1939. The Company, which was seeking to produce aviation spirit by the alternative technical route of alkylation, was not involved in the Government-sponsored hydrogenation projects.

However, in mid-1939 Hartley apparently heard of the Company's research into alkylation, whereupon he advised Air Vice Marshal Tedder to get in touch with the Company. Tedder duly did so and, as mentioned in chapter 8, visited Sunbury in October when he emphasised the importance of having large quantities of 100-octane aviation spirit available for the war effort. At that time, Abadan's only alkylation unit was still little more than a pilot plant with a capacity of only about 100 barrels per day. However, in November authorisation was granted for a capital expenditure of £70,000, later increased by £11,000, for the construction of an alkylation unit with a capacity of 900 barrels per day. In erecting the plant, it was possible to use equipment which had been purchased and shipped to Abadan before the war for a project, since abandoned, to improve gas recovery. The alkylation unit was therefore constructed without the need to purchase new equipment, apart from pumps and compressors from the USA, so helping to minimise wartime shipments of heavy materials. The first orders were placed in December and construction work was carried out thereafter.[45] Meanwhile, attention was paid to another bottleneck in aviation spirit production, namely the limited supply of benzole which was produced from the small pyrolysis plant commissioned in 1937 and used as the aromatic component in aviation spirit.[46] A solution was found by modifying solvent extraction plant, which had been installed for the refining of kerosene, to permit extraction of aromatics. This augmented the meagre supply of pyrolysis benzole and allowed aviation spirit manufacture to be increased.

Developments in aviation spirit manufacture at Abadan were given technical support at the Sunbury research centre where process investigations and blending studies were carried out, allied to the testing of aviation spirits in a Bristol Pegasus aero engine which was installed in a laboratory at Sunbury in 1939.[47] The Pegasus, which was replaced in 1942 by a more advanced Bristol engine, the Hercules, proved invaluable in arriving at formulations for manufacture at Abadan. By testing aviation spirit components received from Abadan, the Sunbury research staff were able to establish in great detail the contribution each component could make to a particular blend. With the experience thus obtained, the research centre was able to formulate blends for testing on the Pegasus and submission for official approval. Success in that respect came in June 1940 when the Air Ministry approved the Company's 100-octane aviation spirit.[48] The following month, the first shipment of 100-octane aviation spirit was made from Abadan, where production of the new high-octane spirit amounted to some 23,000 tons in the remainder of 1940.[49] Such an output was soon, however, dwarfed by escalating demand. As the war continued in 1941–2, Abadan became a vital source of aviation spirit supply for the Soviet Union and, indeed, for the whole of the Middle East, East Africa and India after the Japanese overran the Dutch East Indies and Burma, depriving the Allies of aviation spirit from those sources. In order to meet these growing demands, a series of expansion schemes was planned in 1941–3 with the blessing of the British Government, which agreed to bear half of the capital costs. The schemes provided for increases in Abadan's 100-octane aviation spirit output, first to 500,000 tons per annum, then to 750,000 tons and, by a third expansion, to about a million tons.[50]

Such schemes involved not only the erection of alkylation units which were gigantic by the standards of the day, but also the installation of plant to work other processes such as 'isomerisation' and 'superfractionation'. The alkylation process, described in chapter 7, involved a reaction between isobutane and butenes. Isobutane (a branched chain paraffin) was in short supply at Abadan, but there was theoretical scope for its manufacture by rearranging the molecules (isomerisation) of the straight chain normal butane which was available in natural gas derived from crude oil. Research into isomerisation had been commenced at Sunbury in 1938 by just one graduate chemist, using aluminium chloride as the catalyst. Early in 1940, research was resumed and in November that year £2000 was allocated for a pilot plant. The isomerisation team, for it was no longer an individual effort, was directed by E. W. M. Fawcett, formerly a member of ICI's research staff, who had been a co-discoverer of polythene in 1933. By mid-1941 very promising results had been obtained and twenty-three patents filed.[51] The Company, as a participant in the Alkylation Agreement (see chapter 7), suggested its extension to cover isomerisation, but, after encountering problems of one sort and another with that idea, the Company and Standard Oil (NJ) came to an Isomerisation Agreement which was signed by just the two companies in March 1941.

As happened so often in the Company's adoption of new refining technology, it turned to the USA for the design of isomerisation plant. After a visit by Coxon to the USA in September 1941 it was agreed that, using information from Standard Oil (NJ) and Sunbury, the US firm of Kelloggs would design full-scale isomerisation plant to be erected at Abadan. As there were no full-scale isomerisation installations already in operation, Kelloggs had to rely on laboratory data from Sunbury and limited small-scale plant experience on the part of Standard Oil (NJ) to design the full-scale plant. It was a remarkable act of technical faith, especially as the isomerisation plant was to be built, not as a separate unit, but as an integral part of the alkylation plant. Nevertheless, when the combined isomerisation/alkylation plants were commissioned from 1943, they worked well.

Superfractionation, another wartime development, came out of the search for individual hydrocarbons of high octane number, especially isohexanes and isoheptanes, in Iranian crude. In 1937 Professor Merrell Fenske of Pennsylvania State College had carried out a very precise fractionation of Iranian naphtha and in 1940 Dr Birch re-examined Fenske's findings and showed that some 290,000 tons of isohexanes and isoheptanes were potentially available from a crude throughput of about ten million tons. Although those components could only be separated by very precise fractionation, pilot plant superfractionation experiments at Sunbury by Birch and P. Docksey of the Physical Research Section and larger scale trials at Abadan demonstrated that the process was feasible. In 1941 a semi-scale fractionating column (thirty-three feet high) was constructed at Sunbury and used to establish the optimum operating conditions.[52] Docksey then prepared the basic calculations for the Abadan superfractionators which were built by the US firm, Badgers, and brought into use in 1943. The plant, based on a complex series of tall fractionating columns, made an important contribution to aviation spirit manufacture.

By these means, the Company became a very large-scale supplier of 100-octane aviation spirit, the output of which at Abadan rose to 67,000 tons in 1941, 258,000 tons in 1942, 390,000 tons in 1943 and 858,000 tons in 1944. In the year from July 1944 to June 1945, 995,000 tons were produced and by the end of the war the Abadan plant was producing 100-octane aviation spirit at a rate of more than a million tons a year.[53]

While the Company and the British Government expended large sums of capital on new aviation spirit plant, the Company's labour force in Iran grew apace. The number of Company employees in Iran, together with those employed on Company projects by contractors, rose from the low point of 26,271 at the end of August 1941 to about 65,000 at the end of the war, as is shown in table 9.6. Such rapid expansion placed enormous strain on Company facilities and administration in the exceptionally difficult wartime conditions. The supply of housing, for example, was utterly inadequate, as Pattinson, the Company's general manager in Iran, admitted to Rice in 1942.[54] Later in the war he authorised some temporary palliatives, arguing that the expansion of the refinery had reached its limit until the housing problem was resolved and complaining to Jameson that large numbers were still having to camp out in the open.[55] Jameson sympathised, but wrote back that during the war 'no one thinks in terms of actually building permanent housing' and that all requests for building materials were scrutinised by the wartime supply authorities.[56]

Table 9.6 *Company employees and contractors in Iran,*
1939–1945

End of:	Iranian	Indian	British	Other	Total
August 1939	47,678	1709	1616	57	51,060
December 1939	34,938	1154	1234	69	37,395
December 1940	26,484	1158	1056	15	28,713
August 1941	24,292	1036	934	9	26,271
December 1941	28,035	1005	918	9	29,967
December 1942	41,081	1716	1261	234	44,292
December 1943	44,944	2102	1442	279	48,767
December 1944	60,073	2493	1710	476	64,752
December 1945	60,366	2498	2357	240	65,461

Sources: See Appendix 1.

More fundamental even than housing, there were serious shortages
of clothing and food. Within the context of wartime constraints, the
Company took steps to remedy them. For example, when the problem
of clothing became acute for Company employees towards the end of
1943, arrangements were made with the Board of Trade to obtain
ready-made clothing in Britain and ship it to Abadan for sale to
employees and their dependents. Displaying a somewhat quaint and,
some might say, curiously British attachment to the preservation of
social distinctions, 50,000 suits of assorted sizes were sent out for men
and boys of the 'artisan and labourer type' with 'Sports Coats and
Flannel Trousers for sale to members of the Staff'.[57] The Company
also took steps to alleviate the food shortages, importing wheat and
other foodstuffs into Iran, constructing its own bakeries and subsidising the price at which essential goods were sold to employees. With
price inflation utterly rampant in wartime Iran, the Company secured,
for example, stocks of rice, tea, sugar, dates, meat and *ghiva* (local
footwear) and administered their distribution and selling prices. In
addition, with a view to increasing supplies of dairy products for its
employees, the Company acquired the Abadan Dairy Farm from its
private owner at the beginning of May 1943.[58]

Notwithstanding such measures, the Company was encircled by the
abject poverty which was widespread in Iran at the time. In the words
of a Company commentator, there were, in 1942, 'many thousands of
labourers trekking to Abadan, Khorramshahr, Ahwaz and other large

Table 9.7 *Cases of typhus and smallpox amongst Company employees in Iran, 1939–1945*

	Typhus	Smallpox
1939	10	8
1940	10	3
1941	31	3
1942	21	20
1943	1003	37
1944	100	27
1945	20	27

Source: BP 71487.

stations ... in search of work and food. Near-famine conditions had existed in the interior for more than two years. In the wake of the able-bodied came hundreds of starving paupers.'[59] In conditions so conducive to the spread of disease there occurred in the autumn of 1942 an outbreak of smallpox which, as can be seen from table 9.7, affected some of the Company's employees. To prevent its spread, mass vaccinations were carried out by Company medical staff and at the Company's recommendation the military authorities set up a travelling vaccination squad which visited every village on Abadan island.[60] Abadan was also struck by a typhus epidemic which affected more than a thousand Company employees in 1943. Cases were admitted to an isolation hospital where tents were erected to cover the sick after the wards had been filled. Disinfestation centres and vaccinations also helped to bring the epidemic under control. It was, however, recognised that the typhus outbreak was linked to the conditions of partial famine which existed in many parts of the country. As a Company commentator noted: 'Abadan became the Mecca of the starving population of Persia and thus the problem of nutrition became intrinsically allied to the Typhus epidemic'.[61]

Faced with the diminution in quantity and deterioration in quality of food supplies the Company decided in October 1943 to commission an investigation of nutrition standards, for which purpose it secured the services of the nutritionist Dr Arnold Meiklejohn for six months from December 1943 to May 1944. Meiklejohn spent most of that time visiting areas of the Company's operations in Iran and making a study of employees' health. His inquiries, it was reported, 'showed conclu-

46   A street in the Ahmadabad district of Abadan, 1943

sively that malnutrition was very common, in varying degrees, amongst all classes of the community at Abadan, and that diet deficiencies included dangerous shortages of such items as meat, eggs, curds and cheese, fats, fruits and vegetables'.[62]

Another problem, less immediately pressing but still not to be overlooked, was the changing composition of the Company's workforce. Under an Order-in-Council introduced by the British Government in December 1941 the Company's operations in Iran were classified as 'essential undertakings' and British subjects were not allowed to leave their jobs without the permission of Bullard, who was elevated from British Minister in Tehran to Ambassador after the British Legation in Tehran was raised to an Embassy in February 1944.[63] The Order-in-Council did not, however, apply to Iranians, many of whom left the Company to work for the British and American forces, who were large employers of local labour.[64] The loss of manpower, especially skilled workers, posed serious problems for the Company which had to cast its recruiting net widely, taking in a miscellany of recruits, including not only Iranians, but also Czechs, Poles, Palestinians, personnel on loan from the armed forces, Burmah Oil Company employees who had been evacuated from Burma and the wives and daughters of British staff of the Company. There was, as a result, a loss of continuity in employment as the prewar progress in the training and employment of Iranians came to a halt. As Jameson admitted in 1944, the advances which had been made before the war had been 'virtually lost' because of the 'staggering wastage of labour'.[65] Whatever the effects on efficiency, and it is reasonable to presume that they would have been adverse, the wartime disruption of employment left the Company with a workforce which was less settled than before the war and more prone to the labour disturbances which later affected the Company (see chapter 14).

POLITICS, FOREIGN INFLUENCE AND OIL CONCESSIONS

At the same time as the Company's pattern of employment was affected by the war, a chain of events was set in motion which was to have an unsettling effect on the Company's concessionary position. Those events were closely connected with the revival of political currents and personalities which had been suppressed during Riza Shah's reign, and with the foreign domination of Iran by the occupying powers.

After the abdication of Riza Shah and the accession to the throne of his son, Muhammad Riza Shah, politics in Iran entered a new phase. The authoritarian rule of Riza Shah, under whom all forms of political opposition were suppressed, was at an end and the young Muhammad Riza Shah was not able, at that early stage of his reign, to establish autocratic power in the same manner as his father. The power of the Majlis, which had been subdued by Riza Shah, revived. New political parties were formed, such as the Iran Party, composed largely of urban middle class intellectuals; the Tudeh (Masses) Party of the extreme left, formed in 1941 as a successor to the Communist Party which Riza Shah had outlawed; and the right-wing National Will Party led by Sa'id Zia al-Din Tabataba'i, the former Prime Minister who returned from exile to form the party as a means of counteracting the Tudeh. Religious leaders, who had suffered a diminution of power and influence under Riza Shah, also re-asserted their authority. An extreme religious group, Fida'iyan-i Islam (Devotees of Islam), emerged, one of its leading lights being Ayatullah Abul Qasim Kashani, who had a long history of opposition to both the British and the Pahlavi dynasty. He was to play a prominent part in the concessionary crisis of 1951–4, described in later chapters. So too, but to a still greater extent, was Muhammad Musaddiq who, after years of political exile during Riza Shah's reign, came back to the centre of Iranian political life, being elected to the Fourteenth Majlis (1944–6) in which, as will be seen, he

played an important role. Others who came back to political prominence after long years in the wilderness during Riza Shah's reign were
Qavam al-Saltana, who had been Prime Minister in the early 1920s,
and Taqizadeh who, after his years in exile at the School of Oriental
and African Studies in London, became Acting Minister in London
and then Ambassador after the Iranian Legation in London was raised
to an Embassy in 1944.[66]

The re-emergence of political and religious leaders with their
various followings and factions brought a hubbub of political activity
which, stimulating as it may have been after the dictatorship of Riza
Shah, did not produce strong or stable government. As table 9.8
indicates, prime ministers came and went with great alacrity, rarely
holding office for more than a few months at a time. Governments
were not only short-lived, but also had little authority in the outlying
provinces, where separatist movements grew to the point of threatening the integrity of the country. At the same time, Iran's independence
was circumscribed by the presence of military forces from the two
powers which for many years had aroused Iranian fears of foreign
domination: Britain and the Soviet Union. Theoretically, of course,
those two powers were allies of Iran after September 1943, when Iran
declared war on Germany. Moreover, in the Tripartite Pact of January
1942 Britain and the Soviet Union had pledged themselves to uphold
Iran's independence. This principle was re-affirmed at the Tehran
Conference in November 1943, after which Churchill, Roosevelt and
Stalin signed the Tehran Declaration confirming that their Governments were 'at one with the Government of Iran in their desire for the
maintenance of the independence, sovereignty and territorial integrity
of Iran'.[67] Yet the extensive influence and control which the occupying
Allies exercised over Iranian affairs inevitably raised the spectre of
foreign domination which, by offending Iranian pride, was a reliable
formula for releasing the emotive power of nationalism.[68]

In the eyes of Iranian nationalists, the threat of foreign domination
came not only from the presence of Allied forces, but also from the
foreign interest which was shown in obtaining new oil concessions. In
the autumn of 1943 both Royal Dutch-Shell and the Standard Oil
Company of New York (Socony) were known by the British Foreign
Office to be enquiring into the prospects of concessions in Iran.[69] In
November, representatives of Royal Dutch-Shell visited Tehran where
they met the Iranian Prime Minister, Ali Suhaili, who told them that
Iran would 'welcome foreign capital but that everything depended on
the terms as there were other applicants'.[70] Negotiations dragged on

Table 9.8 *Iranian Prime Ministers,*
August 1941–December 1947

Mirza Muhammad Ali Khan Furughi	August 1941–March 1942
Ali Suhaili	March–July 1942
Qavam al-Saltana	August 1942–February 1943
Ali Suhaili	February 1943–March 1944
Muhammad Sa'id	March–November 1944
Murtiza Quli Bayat	November 1944–April 1945
Ibrahim Hakimi	May–June 1945
Sa'id Muhsin Sadr	June–October 1945
Ibrahim Hakimi	October 1945–January 1946
Qavam al-Saltana	January 1946–December 1947

inconclusively until the end of December, when the negotiators returned to London.[71] They returned to Tehran in February 1944.[72] A negotiator for the Standard-Vacuum Oil Company was also in the capital being encouraged by Suhaili. By March both companies had submitted proposals, to which was added an approach from the US oil company, Sinclair, also competing for a concession.[73] The Iranian Government went so far as to retain the services of Herbert Hoover Junior and A. A. Curtice, two US geologists, to study the proposals.[74] In March, Suhaili was replaced as Prime Minister by Muhammad Sa'id and in August rumours began to circulate in the press that Sa'id was secretly, i.e. without reference to the Majlis, offering a southern oil concession to the US and British companies and that Standard-Vacuum would get a northern concession once Soviet troops had withdrawn after the war.[75] The Soviet Union, wishing to press its own claims, despatched an oil mission led by an Assistant Minister for Foreign Affairs, Sergei Kavtaradze, to Tehran in September. In his talks with Sa'id, Kavtaradze demanded an oil concession in Iran's five northern provinces. Sa'id promised to refer the matter to the Cabinet and the Majlis, but Kavtaradze, seeking a more definite commitment from Iran, saw the Shah at the beginning of October and sought permission to explore for oil in northern Iran. He warned that the future of Irano-Soviet relations depended on a favourable reply.[76] It was not a tactic which, in the opinion of Sir Clarmont Skrine, British Consul in Mashad, was likely to further the Soviet cause. As Skrine commented: 'In Persia it is not good manners to say no to a friend; you make vague promises and verbose excuses and do nothing until he gets tired and goes away. Now, however, the Soviet envoy tactlessly insisted on an answer, yes or no, and opposition at once began to build up in the Majlis.'[77] The Iranian Government gave its reply on 8 October when it announced that it would not grant any new oil concessions to foreigners until after the war was over.[78]

Kavtaradze, frustrated that his attempts to force the experienced Sa'id to grant a concession had been unsuccessful, was determined to remove him from power and undermine the Shah. He held a press conference to emphasise the social and economic benefits to be derived from Soviet expertise and exploitation of Iranian oil resources, accompanied both by threats of reprisals against Sa'id for having worsened relations between the two countries and by ill-disguised pressure on prominent Iranian politicians.[79] The Red Army hampered Iranian troop movements and communications and was on the streets of Tehran during anti-Sa'id demonstrations organised by the Tudeh Party in front of the Majlis.[80] Tudeh and pro-Soviet newspapers were also active in the campaign to bring down the Government, and on 10 November 1944 Sa'id resigned.[81] However, in spite of Soviet pressure, his successor, Murtiza Quli Bayat, refused to change Sa'id's oil policy and told the Soviet Ambassador that the matter of the concession would have to await the end of the war and the departure of foreign troops.[82] In the Majlis, the leader of Iranian resistance to Soviet oil demands was the nationalistic Musaddiq who, on 2 December, introduced a Bill to prevent Iranian governments from entering into negotiations or signing oil concessions with foreign interests without the assent of the Majlis.[83] It was passed by 80 votes to 7. Russian reaction was furious[84] and Kavtaradze's mission unceremoniously left Tehran on 9 December.

Soviet pressure for an oil concession did not, however, end with the return of the Kavtaradze mission to Moscow. It soon re-surfaced in connection with negotiations over the evacuation of Soviet troops from Iran and, relatedly, Soviet support for the separatist movement in the northern province of Azerbaijan. After the fall of Bayat's Government in April 1945, the new Prime Minister, Ibrahim Hakimi, began to press for the withdrawal of Soviet troops during his brief period of office in May–June 1945.[85] The matter was discussed further, but inconclusively, during the administration of the next Prime Minister, Sa'id Muhsin Sadr, whose stay in office lasted only the few months from June to October 1945. It was long enough, however, to see World War II come to an end with the surrender of the Japanese on 2 September. It was one of the provisions of the Tripartite Pact of 1942 that within six months of that date, i.e. by 2 March 1946, all Allied forces were to be withdrawn from Iran.

For the Iranian Government, the evacuation of Soviet troops by the due date, if not sooner, was given added importance by the threat to Iran's integrity which was posed by the Soviet-backed separatist movement in Azerbaijan. During Sadr's premiership Azeri claims to autonomy gained momentum under the leadership of Jafar Pishevari and the Azerbaijan Democratic Party, formed in September 1945.[86] After Sadr's Government fell in October, Hakimi formed his second Government which faced mounting problems over Azeri claims to autonomy and the continuing presence of Soviet troops. In November, the People's Congress of Azerbaijan openly rebelled against the central Government of Iran, sending a declaration of autonomy to Tehran.[87] When, in that same month, the central Government sent troops to re-establish its hegemony, they were blocked by Soviet forces.[88] On 19 January 1946, after attempting without success to obtain a Soviet commitment to withdraw its troops, the Iranian Government lodged a complaint with the Security Council of the United Nations that the Soviet Union was interfering in the internal affairs of Iran.[89] The complaint was left on the agenda of the Council pending negotiations between the two sides. Meanwhile, the Hakimi Cabinet fell and on 26 January Qavam al-Saltana, who was believed at the time to be well disposed towards the Soviet Union, was again appointed Prime Minister. In mid-February he set off for Moscow for negotiations which included meetings with Stalin and his Foreign Commissar, V. Molotov. The matters discussed went beyond the withdrawal of Soviet troops from Iran. Molotov suggested that Iran should recognise the autonomy of Azerbaijan and Stalin proposed the formation of an Irano-Soviet oil company for the exploitation of Iranian oil. The Iranians, Stalin suggested, should have a 49 per cent interest in the new company while the Soviet Union would hold the remaining 51 per cent.[90] Despite Soviet denials, it was an inescapable conclusion that Soviet interference in Azerbaijan and the evacuation of troops were being connected with Soviet desires for an oil concession.[91]

While no specific agreements were reached in Moscow, relations between Iran and the Soviet Union became noticeably more cordial as a result of Qavam's visit. It was a sign of their less tense relations that, after Qavam returned home on 10 March, the Soviet Union replaced its hard-line Ambassador in Iran, Mikhail Maximov, with a more conciliatory diplomat, Ivan Sadchikov.[92] On 24 March Sadchikov called on Qavam and formally presented the Soviet decision to withdraw its troops from Iran within six weeks, at the same time making demands for the formation of a joint Irano-Soviet oil company and the recognition of Azerbaijan's autonomy.[93]

Qavam was in a difficult position, for politically he could not afford to jeopardise the prospect of a Soviet troop withdrawal; he could not undermine Iranian sovereignty in Azerbaijan by agreeing to the province's autonomy; and he could not defy the Majlis over oil concessions in the light of the law of 2 December 1944. On 29 March he told the British Chargé d'Affaires, Farquhar, that he felt he had no option but to appear to acquiesce with Soviet demands for the formation of an Irano-Soviet oil company.[94] However, he retained some bargaining leverage by linking an oil agreement to the withdrawal of Soviet troops and by playing on the need for the oil agreement to be approved by the Majlis.

Although Soviet interference in Azerbaijan remained on the agenda of the Security Council, no substantive progress was made on the matter in that procedure-bound forum.[95] Instead, Qavam and Sadchikov reached an agreement which was announced by the two sides on 4 April 1946. The main points were that Soviet forces would be withdrawn from Iran within six weeks of 24 March as previously agreed; the government of Azerbaijan was recognised as an internal Iranian matter; and it was proposed to form an Irano-Soviet oil company for a period of fifty years. For the first twenty-five years the Soviet Union would own 51 per cent of the shares and Iran 49 per cent; for the second twenty-five years each was to own 50 per cent; and profits were to be divided according to the shares held. The Iranian Government was to submit a Bill on the organisation of the company to the Majlis within seven months of 24 March.[96]

In compliance with the Qavam–Sadchikov Agreement of 4 April the Soviet Union completed the evacuation of its troops from Iran by 9 May. As British and US forces had already been withdrawn, Iran was finally free of foreign forces on her soil.[97] A few months later, in December, the movement for autonomy in Azerbaijan was crushed when Iranian troops led by General Ali Razmara entered the province and the separatist regime collapsed.[98] In the meantime, the acute political instability in Iran and the preoccupation with separatist tendencies caused the elections for the Fifteenth Majlis, which should have commenced in March, to be delayed.[99] So too, to the annoyance of the Soviet Union, was the submission to the Majlis of the proposal for an Irano-Soviet oil company.[100]

Eventually the elections, which under the Iranian electoral system were held in various districts over a period of several months, took place in the first half of 1947.[101] The new Majlis, which finally convened in mid-1947, was strongly hostile to the Qavam–Sadchikov Agreement. On 22 October, by an overwhelming majority of 102 votes to 2, the Majlis rejected the proposals for an Irano-Soviet oil company in a single article law which contained five clauses. Clause (A) interpreted the law of 2 December 1944 to mean that Qavam should not have entered into negotiations and drawn up the agreement for the proposed Irano-Soviet oil venture, which was therefore null and void. Clause (B) instructed the Iranian Government to make arrangements for a technical survey of oil prospects in the country so that the Majlis could make arrangements for the commercial exploitation of Iran's national resources by enacting the necessary laws. Clause (C) absolutely forbade the future grant of oil concessions to foreigners. Clause (D) permitted the Iranian Government to negotiate an agreement to sell any oil discovered in commercial quantities in the northern areas of Iran to the Soviet Union, informing the Majlis of the results. Finally, with implicit reference to the Company's concession in the south of Iran, clause (E) ran:

> In all cases where the rights of the Iranian nation in respect of the country's natural resources, whether underground or otherwise, have been impaired, *particularly in regard to the southern oil*, the Government is required to enter into such negotiations and take such measures as are necessary to regain the national rights and inform the Majlis of the result. [Emphasis added][102]

The Soviet Union was furious at the Iranian reversal of the Qavam–Sadchikov Agreement. However, it was the Company's concession as much as the Soviet Union's desire for one which stood to be affected by the single article law. Michael Cresswell, the British Chargé d'Affaires, rightly imagined that the article referring to the Company 'may come to assume major importance'.[103] At a meeting attended by representatives of the Company and officials from the Foreign Office and Ministry of Fuel and Power, Sir John Le Rougetel, who had succeeded Bullard as British Ambassador in Tehran, stated his belief that clause (E), implicitly about the Company, had been included in order to show impartiality between the Britain and the Soviet Union. However, it was also noted, ominously for the Company, that 'the provision may give the Persians the right to nationalise the petroleum industry'.[104]

CONCLUSION

The single article law of 22 October 1947 was the culmination of an eventful period in the internal affairs and external relations of Iran, which rebounded on the Company. The interest shown by US companies and Royal Dutch-Shell in oil concessions in Iran injected an element of competition into the concessionary situation and drew the Soviet Union into the hunt for concessions which, coupled with the presence of Soviet troops in Iran and Soviet support of the separatist movement in Azerbaijan, developed into a crisis in Irano-Soviet relations. The Company was not directly involved, but was dragged into the limelight by clause (E) of the single article law which marked the beginning of a period in which Iranian politics was dominated by the Company's concession.

PART II

POSTWAR EXPANSION, 1946–1951

Management and finance,
1946–1951

From the end of World War II to 1951, the Company's board, organisation and financial policy showed the hallmarks of a somewhat cautious and conservative approach to matters of management and finance. That is not to say that no changes took place. On the contrary, the board of directors, which was becoming rather aged by the end of the war, underwent a very high turnover in the immediate postwar years. Yet the nature of the changes in the membership of the board was not, for the most part, such as to introduce the new blood of a younger generation of executives to that elevated body in the Company. As for organisation, the structure of functional departments in Head Office remained in place, perpetuating a centralised system of management which remained basically unchanged. In finance, the same policy of prudence which had characterised the 1930s and the war years was continued. Unadventurous it may have been, but it left the Company in an extremely secure financial position on the eve of the Iranian crisis which threw the Company's operations into turmoil in 1951.

THE BOARD OF DIRECTORS AND THE COMPANY'S ORGANISATION

If advancement in years bears any correlation to diminishing energy and freshness of ideas, then by the end of the war there was cause for concern about the need for new appointments to the Company's board to stave off the creeping movement towards gerontocracy. As was seen in chapter 8, the average age of the directors had already increased from 56 years at the end of 1927 to 64 years at the end of 1941. By the end of 1945 it was more than 65. Of particular significance, the average

age of the chairman and management directors, having increased from 47 to 56 years between 1927 and 1941, was more than 60 at the end of 1945. Indeed, only two directors, Fraser and Gass, were under 60, as can be seen from table 10.1. It was clear that changes would have to be made, if only because most of the directors were nearing the end of their working lives.

As table 10.1 indicates, there was no shortage of board changes between 1946 and 1954, when there was a very high turnover of directors, with fifteen new board appointments, compared with only four in the much longer period from 1927 to 1945. There was a particularly high shake-out in 1946–8 when, in the space of three years, six of the twelve directors who sat on the board in 1945 either resigned or died in office. By the end of 1954, ten members of the 1945 board had gone and, what was more, so had five of their replacements. The high rate of departures in the latter category was partly the product of the tendency to appoint new directors who were already well advanced in years. Even allowing for prevailing attitudes to the virtues of longevity in the boardroom and for the disruption of the civilian careers of younger managers on account of the war, it was a sign of Fraser's caution (himself having joined the board at the exceptionally young age of thirty-five), that he appointed only one management director, namely Pattinson in 1952, who was under fifty-five years old at the time of becoming a director.

No category of director was exempt from the high board turnover of these years. Of the ordinary non-executive directors, Fisher died in 1948 and within a couple of years Tiarks and Lund, both of whom had served on the board since 1917, retired. They were replaced by more City men, namely William Keswick of Jardine Matheson and the Alliance Insurance Company, who joined the board in May 1949, and Captain Eric Smith, who did likewise in June 1950. Keswick's appointment was inspired by the possibilities of trading in China. Smith died before attending a board meeting, his place being taken by his namesake, Desmond Abel Smith, in December 1950. The latter also had City interests and was chairman of Dalgety and Company, which had extensive Australian connections and had been agents for the Company in the 1920s and 1930s. He was also a director of the National Provincial Bank, of which Fraser became a director in March 1950, and the Equitable Life Assurance Company, amongst others.

Meanwhile, the Burmah Oil Company's representation changed when Gilbert Whigham retired and was replaced in January 1947 by

Sir Kenneth Harper. The other Burmah director, Robert Watson, retired at the end of the year, having served on the board for nearly thirty years. He was not replaced until April 1953, when William Abraham, Burmah's managing director, joined the Company's board.

In the meantime, Government directors also came and went with alacrity. In the previous two decades, Packe and Barstow had ensured a longstanding continuity in Government representation, having sat on the board since 1919 and 1927 respectively. However, in 1946 Packe died and Barstow resigned, their replacements being Sir Percival Robinson and Viscount Alanbrooke. Neither of these two new directors was very effective in fostering close relations between the Company's management and Government officials. Fisher, though not formally a Government director, provided a valuable unofficial link with Government, but this was broken when he died in 1948. The following years, Robinson resigned and was replaced by Sir Thomas Gardiner, who seems to have played a more active intermediary role, but it was not to be longlasting and in 1953–4 both he and Alanbrooke resigned, being replaced by Frederic Harmer and Sir Gordon Munro.

Such frequent changes were not conducive to the maintenance of the informal and relaxed contact between the Company and the Government, which had been a feature of the Cadman years. Instead, the Company's relations with the Government took a turn for the worse at the very time when close harmony was badly needed in view of the Government's close interest in Company affairs in the postwar years, described in later chapters.

Of the management directors, the resignation of Lloyd in 1946 marked a real break with the past.[1] He was the last of the Shaw Wallace connection (see volume 1), having gone to India in 1897 as an accountant after school at Harrow. He had subsequently opened the office of Lloyd Scott and Company, later Strick Scott and Company, at Muhammara in 1909, before being appointed management director for finance under Greenway some ten years later. Lloyd was a powerful influence on the board, guardian of the purse under Greenway, Cadman and Fraser, his word a financial directive. On his departure, his responsibilities were assumed by Fraser. The appointment of Frederick Morris, who had been in charge of distribution in continental Europe, as a director in 1946 gave the board added commercial experience on the marketing side. Heath Eves was already dealing with shipping and insurance and deputising for Fraser, who was under increasing pressure from government business. The appointment of

Table 10.1 *The board of directors, 1945–1954*

The board at the end of 1945		Date of appointment as director	Age at end of 1945	Date of resignation	Died in office
Chairman and *management directors*	Sir William Fraser	1923	57	–	–
	H. B. Heath Eves (knighted in 1946)	1924	62	1953	–
	Sir John Lloyd	1919	71	1946	–
	J. A. Jameson	1939	60	1952	–
	N. A. Gass	1939	52	–	–
Government directors	Sir Edward Packe	1919	67	–	1946
	Sir George Barstow	1927	71	1946	–
Burmah directors	R. I. Watson	1918	67	1947	–
	G. C. Whigham	1925	68	1946	–
Ordinary non-executive *directors*	Sir Warren Fisher	1941	66	–	1948
	F. W. Lund	1917	71	1950	–
	F. C. Tiarks	1917	71	1949	–

Subsequent appointments:

		Age on appointment		
Management directors				
F. G. C. Morris	1946	62	1952	—
E. H. O. Elkington	1948	58	—	—
B. R. Jackson	1948	56	—	—
J. M. Pattinson	1952	53	—	—
H. E. Snow	1952	55	—	—
Government directors				
Sir (F.) Percival Robinson	1946	59	1949	—
Field Marshal Viscount Alanbrooke	1946	63	1954	—
Sir Thomas Gardiner	1950	67	1953	—
F. E. Harmer	1953	47	—	—
Sir (R.) Gordon Munro	1954	59	—	—
Burmah directors				
Sir Kenneth Harper	1947	55	—	—
W. E. V. Abraham	1953	55	—	—
Ordinary non-executive directors				
W. Keswick	1949	46	—	—
Capt. E. C. Smith	1950	56	—	1950
D. Abel Smith	1950	58	—	—

Note: Snow resigned as a director in 1954 when he was appointed general manager of Iranian Oil Participants Ltd (see chapter 19), but was re-appointed to the Company's main board in 1955.

Elkington as director of administration at the beginning of 1948 raised
the number of managing directors (known as management directors
until 1946) to five, with the responsibilities shown in figure 10.1. At the
same time, it was realised that a greater concentration of board effort
was required to direct the various activities of the Company and the
appointment of deputy directors was revived to reduce the pressure on
Jameson, who was in charge of production, and Morris, responsible
for distribution. Pattinson was appointed deputy for the one and Snow
for the other, both of them making regular reports to the board.

Underneath the board, the organisation of the Company's activities
in Iran remained a central managerial issue at a time when, as can be
seen from table 10.2, the great majority of the Company's employees
were located there. However, with Gass in charge of concessionary
affairs in Iran, and Elkington as head of administration, there was no
single direct line of responsibility, as had been exercised by Jameson in
1928–48, between the board and the general manager of operations in
Iran. Instead, the responsibilities for Iran were divided between Production Department, which supervised operational activities, and
Administration, responsible for administrative matters in Iran. There
was, to be sure, a continuing awareness of organisational issues. For
example, Gass, on a visit to Iran at the end of 1947, emphasised the
need for greater management efficiency, better manpower utilisation
and more employment of Iranians, coupled with wider educational
and training facilities. Yet there was a tone of complacency in his
remark that 'the system of management now functioning in Iran ...
seemed to be admirably suited to the present stage of the Company's
development'. Elkington took an interest in management theory and
practice, but although his appointment to the board at the beginning
of 1948 enhanced his authority, he was not to be the architect of a new
approach to the management of operations in Iran. In April 1949, after
visiting Iran, he gave the board a detailed report on the management of
operations in that country, covering matters such as the organisation
of stores and transport, the manpower position, training schemes and
the housing programme.[2] In March 1950, he informed the board that
there was a 'clearer definition of responsibility' in the management in
Iran and that this would make for greater efficiency.[3] The board
approved Elkington's report, satisfied with his handling of the situation, but their very interest in the details of administration was
indicative of an excessively centralised organisation and a lack of
delegation to local management.

Although the numbers employed at Head Office in London were

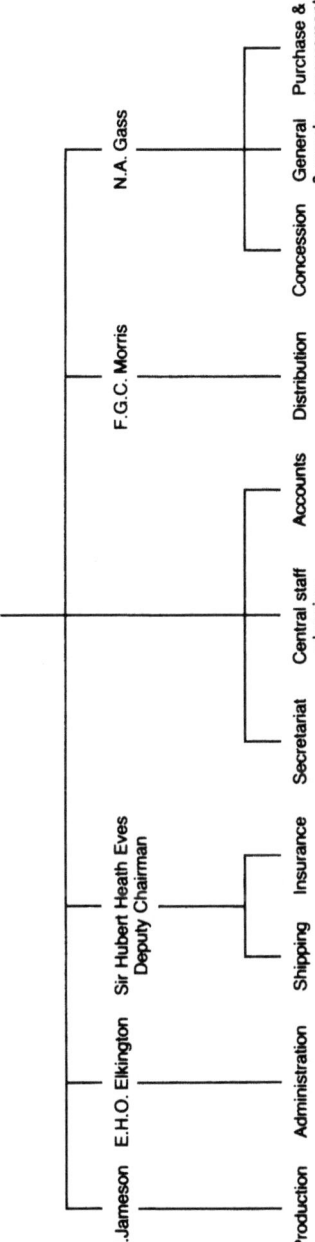

Figure 10.1   Allocation of directors' responsibilities, 1948

Table 10.2 *The Company's employee numbers, 1947*

UK	14,456
Middle East	79,593
Europe	5363
Australasia	527
Africa	2394
Seagoing	4459
Other	477
Total	107,269

Source: BP 67593.

only a fraction of the workforce in Iran, the Company was no exception to that familiar tendency for creeping growth in staff, and hence costs, at the centre of large organisations. This process of self-procreation aroused sufficient concern to provoke a review of what A. G. H. Mayhew, the staff manager, described in January 1948 as a 'very considerable increase in Head Office strength'.[4] In a report submitted to Fraser in March that year, Mayhew surveyed the current staff establishments and organisation charts and made recommendations.[5] Among his principal points, Mayhew referred to the need for an office manager to co-ordinate comprehensively and efficiently the provision of services for Britannic House; the appointment of a welfare officer for staff; the establishment of a section devoted to staff training and education; and the desirability of having a well-informed public relations and corporate advertising unit.

As a result of Mayhew's memorandum, Fraser, concerned about the growth in Head Office numbers, set up an Organisation Committee on 1 July 1948. It consisted of Elkington, Coxon and Sir Humfrey Gale, who, on his retirement from the army, was appointed the Company's adviser on organisation. Robert Norris, who had been appointed office manager in November 1947, was the Committee's secretary.[6] At the same time, in an unusual departure from Company practice, Harold Whitehead and Partners, business consultants, were asked to advise about introducing mechanical aids into office procedures. Their remarks on the conduct of clerical staff in an un-named section of the Company were withering in the extreme. The section, in their opinion, was 'grossly overstaffed' and 'in spite of good wages, absolute security, excellent welfare facilities, and luxurious surroundings, the morale among the lower ranks was unsatisfactory'.[7] The pattern of the day's work was found to be generally as follows:

9.15–10.0	Arriving and titivating.
10.0–10.30	Waiting for the coffee and consuming it. The general feeling seemed to be that it was hardly worthwhile tackling work seriously till coffee was over.
10.30–11.45	The office settled down to get some work done.
11.45–2.15	At 11.45 the first 'lunchers' began making their preparations. This seemed a signal for a general relaxation. Visits were paid from desk to desk. The attitude seemed to be that you could hardly expect people to sit at their desks and work while their fellows were enjoying themselves at lunch.
2.15–3.15	The office settled down once more to work.
3.15	Again an air of relaxation became apparent as tea and cake were expected. After tea there was a general feeling that the back of the day's work was broken and clerks felt themselves at liberty to read, crochet, make out menu cards, draw or gossip until the Mail appeared.
4.15–5.15	Mail dealt with.[8]

Criticisms of the Company's organisation were also made by the Organisation Committee, whose investigation into the Head Office organisation was the most thorough since Cadman's examination of 1925[9] and the first since Fraser's appointment as chairman. In its report, the Committee was critical of the lack of effective cost control and pointed to the need to 'bring back to Management cost consciousness, not only in money, but what is basically more important in manpower control ... [which] has endless repercussions financially'. Arguing that the organisation needed to be streamlined, the Committee suggested that departments should shed their internal administrative and ancillary functions so that administrative tasks could be

Basil Jackson

Frederick Morris   Edward Elkington

47   Newly appointed management directors of the Company, 1945–1954
(also see opposite)

John Pattinson   Harold Snow

separated from technical and operating responsibilities, a move which was felt to be required in the light of the 'immense growth' in the Company's operations and the 'urgent necessity' to establish more direct financial control over administrative, as apart from purely technical, functions. Accordingly, the Committee recommended the establishment of a more structured and authoritative Administration Department, 'a far more business-like organisation', 'a system under which the costs of administration are more closely controlled by policy'. The Committee also suggested that a single organisation dealing with all engineering matters would be more economical and efficient than the diffusion of engineering activities between departments. On the same principle, the Committee felt it was undesirable that operating departments should have their own miniature staff sections, all of which tended to grow and to usurp and cut across the functions of the Staff Department. In addition, the Committee felt that there was insufficient interest in the welfare of staff and noted the absence of a uniform system for appraising staff progress and ability, a deficiency which resulted in an arbitrary approach to promotion and a tendency to tailor the job to the man, 'the negation of sound organisation'. A broadening of the scope of the Staff Department, which was renamed Personnel Department, was therefore recommended. The Distribution Department, responsible for oil supply and distribution, also came in for criticism, having, it was alleged, failed to keep pace with the expansion of the Company's business and requiring a more precise definition of duties to prevent overlapping between the department's twin functions of supply and distribution.

In general, the Committee's examination of the Head Office departments revealed a lack of 'cohesion and symmetry' and evidence that 'responsibilities are confused, over-lapping occurs, and over-centralisation ensues'. The latter point was of particular weight not only because it was a central feature of the Committee's findings, but also because it raised that classic issue in the management of large-scale organisations: centralisation versus decentralisation. In criticising the Company's over-centralisation, the Committee put its finger on one of the typical symptoms: 'too much executive control remains in the hands of a few senior officials – with the result that they are grossly overworked, have no time to devote to mature consideration of the larger issues and future policy, and are cramping the initiative and lowering the morale of competent staff on the less senior levels'. The Committee therefore urged that responsibilities be spread more evenly. Recognising that a highly developed training scheme existed in Iran, the Committee regretted that there was no Head Office branch solely concerned with co-ordinating departmental training activities. On the issue of costs, there had to be a reversal of the tendency 'to get the job done at any price' if rising administrative costs were to be contained.

Although some of the Committee's suggestions were adopted, they had little impact on the overall shape of the Company's managerial structure, where most authority continued to be vested in the centralised functional departments, perpetuating the existence of bureaucratic rigidity in the organisation. A more radical reorganisation was impeded by entrenched attitudes as the weight of tradition and experience inhibited fresh thinking about the structure of the Company by those whose working lives and career advancement in the existing system made them captive to it.

Fraser may have been aware of these managerial constraints and a lack of wider experience among the members of the board when he appointed Basil Jackson as a managing director in October 1948. Jackson was something of a new broom on the board. Unlike many of his colleagues, he was not steeped in the tradition of service in Iran, having spent much of his career as the Company's representative in the USA after opening the New York office in 1929. His American experience had given him ample opportunity to make contacts and build relationships with leaders of the major US oil companies who, together with Royal Dutch-Shell and the Company, dominated the international oil industry. In addition to his cosmopolitan role in New York, Jackson had also been in charge of the D'Arcy Exploration Company in the late 1930s and was closely associated with Cadman. In all likelihood, Fraser already had his eye on Jackson as a probable successor when the latter was appointed to the board in 1948. In any event, the intended line of succession became clear in July 1950, when Jackson was made deputy chairman, with Heath Eves relinquishing that post, though remaining on the board. After that, the only further changes to the board before the end of 1954 were the appointments of Pattinson and Snow as directors in July 1952.

PROFITS AND FINANCE

At the time of Jackson's appointment as deputy chairman the Company was reaching the peak of a period which, in financial terms, was golden. The high demand for oil during the postwar reconstruction of Europe enabled the Company to achieve a rapid expansion in the volume of its sales and in its pre-tax profits. These rose from about £29 million in 1946 to some £41 million in 1949 before more than doubling to approximately £86 million in 1950, in which year the pre-tax return on capital employed was very nearly 73 per cent. Even after allowing, first, for the uplifting effects of the devaluation of sterling in September 1949 on the Company's sterling profits and, secondly, for the price inflation of the period, the years from 1946 to 1950 were still unmistakably ones of unprecedented financial returns, the magnitude of which is shown in table 10.3. The accompanying increase in the funds generated from the Company's operations enabled it to cover dividend payments and the high capital expenditures associated with postwar rehabilitation and expansion, with cash to spare for additions to liquid resources in most years. As table 10.4 indicates, the exceptions were 1946 and 1949, when funds from operations fell short of capital requirements, the gap being filled by drawing on the Company's substantial accumulation of liquid assets, held in cash or marketable securities. The Company was thus able to continue its previous policy of financing expansion from its own resources, without recourse to outside finance from the capital market, apart from relatively minor increases in loan capital which were raised, not by the parent company, but by some of its overseas subsidiaries.

The financial halcyon period of the five postwar years came to an end in 1951 when the Iranian Government's nationalisation of the Company's assets in that country resulted in the cessation of operations there. Deprived of its main source of crude oil and products, the Company faced a major crisis and took emergency action to reorganise its operations which will be described more fully in the next volume of this history. Suffice to say, for present purposes, that one of the effects of the crisis was the fall in profits in 1951 which can be seen in table 10.3.

Table 10.3 *The Company's profitability, 1946–1951*

	Profits before and after taxes and inflation				Return on capital		
	(a)	(b)	(c)	(d)	(e)	(f)	(g)
	Pre-tax profits at current prices (£millions)	Deduction for tax (£millions)	Post-tax profits at current prices (£millions)	Index of post-tax profits at constant prices (1929=100)	Capital employed at current prices (£millions)	Pre-tax return on capital employed: (a) as % of (e) (%)	Post-tax return on capital employed: (c) as % of (e) (%)
1946	28.9	18.6	10.3	136.1	68.5	42.2	15.0
1947	37.3	22.7	14.6	167.2	72.4	51.5	20.2
1948	51.0	28.8	22.2	228.8	82.7	61.7	26.8
1949	41.2	23.5	17.7	179.4	96.5	42.7	18.3
1950	85.7	51.4	34.3	338.5	117.6	72.9	29.2
1951	56.9	28.0	28.9	265.8	147.8	38.5	19.6

Sources: BP consolidated balance sheets (see Appendix 1). The price index used in calculating profits at constant prices is that for plant and machinery in C. H. Feinstein, *National Income, Expenditure and Output in the United Kingdom, 1855–1965* (Cambridge, 1972), table 63.

Table 10.4 *The Company's sources and applications of funds, 1946–1951 (£millions)*

| | Internally generated funds less capital expenditure and dividends | | | | Financial movements | | | |
| | (a) | (b) | (c) | (d) | (e) | (f) | (g) | (h) |
	Funds generated from operations	Capital expenditure	Dividends	Funds generated or (required) after outflows on (b) and (c)	Shares issued	(Increase) or reduction in loan capital	Increase or (reduction) in liquid resources	Total financial movements: (e) + (f) + (g)
1946	15.2	20.2	7.1	(12.1)	–	–	(12.1)	(12.1)
1947	43.9	31.0	7.1	5.8	–	–	5.8	5.8
1948	58.8	39.0	7.1	12.7	–	–	12.7	12.7
1949	43.1	55.7	7.1	(19.7)	–	(1.6)	(18.1)	(19.7)
1950	93.5	42.1	7.1	44.3	–	(0.8)	45.1	44.3
1951	83.2	60.2	7.1	15.9	–	(4.2)	20.1	15.9
Total	337.7	248.2	42.6	46.9	–	(6.6)	53.5	46.9

Sources: BP consolidated balance sheets (see Appendix 1).

The search for market outlets, 1943–1951

It was suggested in chapter 4 that the Company's prewar failure to gain a market share commensurate with its productive capacity was attributable, at least in part, to its comparatively late entry into markets where its main US rivals and Royal Dutch-Shell were already ensconced. The 'As Is' agreements, though imperfect in operation, tended to preserve the status quo, perpetuating the Company's relatively backward marketing position. However, in the changed conditions of the postwar world the Company was presented with substantial opportunities to expand its sales and increase its share of oil markets. Prevailing trends in international oil supply and demand worked in the Company's favour insofar as they helped to create new openings for Middle East oil in fast-growing markets, especially in Europe, whose oil requirements before the war had been met mainly by exports from the USA and the Caribbean. After the war, US consumption grew more rapidly than domestic production with the result that in the second half of the 1940s the USA became a net oil importer, absorbing most of the oil which was available for export from the Caribbean. Meanwhile, Europe turned increasingly to the Middle East for its oil supplies.[1] The Company, with its very large stake in Middle East oil, was well placed in relation to this trend. It also had the advantage of being a producer of sterling oil which was in high demand with customers who found it difficult to purchase dollar oil from US oil companies because of the widespread shortage of dollars. Moreover, postwar coal shortages accelerated the trend to switch from coal to oil, a shift which was given official encouragement by the British Government. The shortages of dollars and coal are described more fully in chapter 12, where it will be seen that the Company had strong misgivings about gaining short-term business as a result of temporary currency and coal problems, especially as these were matters on which the Company came under considerable, and thoroughly unwelcome, pressure from the British Government. The ubiquity of government controls over the British economy was, indeed, a feature of the period and the Company's freedom to pursue its chosen expansion plans was undoubtedly constrained by, for example, steel allocations and foreign exchange regulations. Yet despite the constraints, conditions were fundamentally favourable for the Company to find new outlets for its production.

Market opportunities were certainly badly needed, as the Company's productive capacity continued to grow apace making it ever more difficult to resolve the persistent problem of having more production than markets. Before the war ended, this was explicitly recognised in the forward planning carried out by the Planning Committee which, having been set up by Fraser in 1943, became the focal point for a series of discussions and reports. These amounted, in effect, to integrated planning of the Company's multi-functional operations, including crude production, transportation, refining and marketing. Free from narrowly departmental considerations, stimulated by the non-routine nature of their work and enjoying easy access to Fraser, Coxon and his small band of planning colleagues were both articulate and open in their thinking as they fashioned a stream of memoranda on various aspects of the Company's activities.[2] Their forward planning was, however, subject to a fundamental precept which was laid down by Fraser: that the Company's aim should be to expand Iranian production at a rate which would keep pace with the growth in world oil consumption and that 'in all our actions the development of the Iranian output must be given first place'.[3] Without compromising the primacy of Iranian production, Fraser also acknowledged the need to develop the oil resources of Iraq and Kuwait, which meant, as he pointed out, that the Company would have to find marketing outlets for increased production from those countries as well as Iran.[4] The magnitude of the task was laid bare to the board of directors in November 1943, when Fraser reported his concern that:

> A very major problem before us is the development of our marketing organisation to enable it to find outlets for such a growth [of production] and this is being examined in all its aspects. It is necessary, as a first step towards setting a policy, to discuss with our former marketing partners the possibility of creating additional outlets. Conversations to this end are proceeding.[5]

IDEAS OF ALLIANCE WITH ROYAL DUTCH-SHELL

The conversations to which Fraser referred were no doubt those which the Company was then having with Royal Dutch-Shell, which had nearly a quarter-share of the European market and was represented in all stages of the oil industry across a wide geographical spread. In comparison, the Company was a marketing minnow, but with its vast oil reserves and production and refining facilities in the Middle East, it held a strategic position of growing importance. There was, seemingly, the potential for a classically complementary arrangement which would align the productive capacity of the Company with the marketing power of Royal Dutch-Shell.

The two companies were not by any means strangers to co-operation, having been engaged in joint marketing in the Consolidated area since 1928 and in Britain through Shell-Mex and BP since the early 1930s. Moreover, as was seen in chapter 4, they were both signatories to the Achnacarry Agreement of 1928. Adherence to the principle of co-operation rather than outright competition was re-affirmed in the war years when the Company, Royal Dutch-Shell and the Burmah Oil Company reached an understanding on a 'Joint statement of aims' in September 1942. In the statement, each company agreed that 'it was never more necessary to continue the spirit of co-operation developed prior to the war' and disowned any desire to take advantage of wartime conditions to steal a march on the others. The companies declared that after the war they would, so far as circumstances permitted, help one another to re-establish their prewar trading positions, to which end they would co-operate with other companies wherever possible.[6] Although the statement was no more than an amicable declaration, it was followed by more tangible ideas for co-operation between the Company and Royal Dutch-Shell. For example, in the summer of 1943 the two companies discussed the possibility of appointing a joint public relations officer to represent their combined interests, initially in Britain, but with an eye to including foreign countries at a later date.[7] In relation to the Company's need to find outlets for its production, a more significant overture was made, also in the summer of 1943, when Fraser asked Sir Frederick Godber, chairman of Royal Dutch-Shell, for a broad indication of the supplies which Royal Dutch-Shell would require from the Company pending the rehabilitation of Royal Dutch-Shell's war-damaged production facilities in the Far East.[8] This was followed by discussions between representatives of the two companies, who agreed to exchange statistics on sales tonnages and likely volumes of postwar production.[9]

The possibilities of the Company finding outlets through Royal Dutch-Shell's marketing organisation then took a far bolder turn when, in January 1944, the Company drew up draft proposals for a much closer association, nothing short of a wide-ranging merger of marketing interests covering all products except chemicals and specialities. In more detail, the proposal was to extend joint marketing on a 50:50 basis, as in the 1928 Consolidated Agreement, to all parts of the world except for North, South and Central America and the Caribbean, the Dutch East Indies, Sarawak, British North Borneo, Iran, Iraq and India. Marine bunkering ports and aviation refuelling sites outside the joint marketing area were to be included in the joint arrangements on a worldwide basis. The proposed method of implementation was that each company would purchase a 50 per cent share in the other's assets in the joint marketing areas and thereafter each company would provide 50 per cent of any additional capital required. In all markets where a joint organisation was set up, management was to be in the hands of a separate company, which would operate apart from the parent companies on the same lines as Shell-Mex and BP. The marketing agreement was to last for at least twenty years, continuing thereafter unless terminated jointly by the two companies. Linked with the marketing agreement, there was also to be a crude oil supply agreement lasting for up to twenty years, during which time the Company was to supply Royal Dutch-Shell with crude oil from Iran and/or Kuwait. The quantities to be supplied were to be calculated on a base of two million tons in 1938, indexed to subsequent changes in the volume of trade in the joint markets, subject to a maximum quantity of five million tons in any year. Finally, in relation to refining it was proposed that the companies should investigate the possibilities of making exchange arrangements whereby each company would manufacture products for the other with a view to achieving maximum mutual economies.[10]

By the end of the year, statistics had been compiled indicating the very substantial gain in outlets which would accrue to the Company if the proposals came to fruition and in early 1945 Fraser put the idea of an agreement with Royal Dutch-Shell to the Treasury.[11] In a striking demonstration of the reservations, even hostility, with which Fraser was regarded by the Treasury, his proposals were received without enthusiasm by departmental officials who took the occasion to express, at least in their own circles, extremely strong misgivings about the quality of the Company's management and of Fraser as chairman. Their criticisms are described more fully in chapter 12. On the particular subject of marketing, Treasury officials felt it was a weakness that the Company's sales outlets were underdeveloped in relation to its productive capacity and obviously had a low opinion of the Company's acumen for the aggressive pursuit of market opportunities. In the words of Sir Wilfrid Eady of the Treasury, the Company had 'been content to take its share of what was going but not – so I should judge – fully to make its share'. The Company's marketing policy was, he continued, 'quite reasonable for a profitable jog trot life', but was not suited to the more complicated international situation in which the Company now found itself.[12] The feeling that the Company should pursue a more vigorously independent line in marketing was shared to some degree by Sir John Anderson, the Chancellor of the Exchequer, who thought that the Company would be in 'an unduly weak position' unless it developed at least the nucleus of a 'really effective distribution organisation'. He was, however, willing to agree to Fraser seeking a 50:50 marketing arrangement with Royal Dutch-Shell in certain markets so long as the Company also took steps to strengthen its own marketing outlets.[13] When Fraser was informed of the Chancellor's views, he agreed, according to Sir Edward Bridges, Permanent Secretary at the Treasury, that it would be a 'good thing' for the Company to run its own marketing organisation, but 'more from the point of view of how grand it would be to have a wholly self-contained oil organisation under the British flag than from any commercial or business motive'.[14] Needless to say, such an impression can only have served to reinforce the Treasury's doubts about the Company's commercial vigour. In any event, having received the Chancellor's somewhat qualified blessing, Fraser made his proposition to Royal Dutch-Shell who, Fraser reported in June 1945, had 'gone off to consider it'.[15] They did not, as it happened, consider it for very long and early in August Fraser telephoned the Treasury to say that the negotiations 'had completely broken down' and that Royal Dutch-Shell did 'not want to make a deal'.[16] Fraser's scheme for an instant expansion of the Company's marketing outlets through a joint marketing agreement with Royal Dutch-Shell had thus collapsed and the Company turned to the much more arduous task of building up its own marketing organisation which, as Fraser admitted, was 'a rather long-term job'.[17]

MARKETING AIMS AND ORGANISATION

In seeking outlets for its burgeoning production the Company had to
deal not only with the ongoing expansion of its production, but also
with the legacy of wartime disruptions to its business. During the war
the Company's crude oil production increased greatly, as did its
refining capacity east of Suez, which grew from 10.5 million tons of
annual crude input in 1938 to about 21 million tons in 1945.[18] At the
same time, the previous pattern of sales was severely disrupted by the
wartime loss of European outlets, the enforcement of the short-haul
policy in tanker routes, the demands of the navy and other armed
services for oil, and the destruction of alternative sources of supply in
Burma and the Dutch East Indies. The effects of these various factors
are reflected in the figures shown in table 11.1. As can be seen from
column (j), before the war about 87 per cent of the Company's sales
were made through its own outlets, which provided a relatively secure
and reliable channel for the disposal of production. Supplies to other
oil companies and distributors, which were less secure, constituted
only about 13 per cent of the total sales tonnage. However, by 1945 the
proportion of the Company's sales which was made through its own
outlets had fallen to only about 61 per cent of the total sales tonnage.
Moreover, direct sales included very large supplies to the Admiralty
and other armed services, whose demand for oil was certain to fall
after the end of hostilities. The less secure class of sales, those to other
oil companies and distributors, had increased dramatically, representing about 39 per cent of the total in 1945. Within that category, the
Company's largest customers were Royal Dutch-Shell, Standard-
Vacuum and the Burmah Oil Company, whose demands for Company
supplies were expected to fall as their war-damaged oilfields in the Far
East were rehabilitated.[19] At the end of the war the Company therefore
needed to find assured outlets both for the planned expansion of its
production and to cater for the loss of short-term business which had
come its way during the conflict.

It was no doubt with such thoughts in mind that senior Company
executives discussed marketing policy at a meeting held in Fraser's
office in June 1946.[20] After the meeting Snow drafted a note setting out
an agreed plan for distribution activities, in which it was stated that
the 'primary need' was for the Company to increase its direct outlets
by entering markets where it had not hitherto been represented and by
expanding its market share in areas where it was already active.[21] As a
supplement to sales through its own outlets, the Company would be

Table 11.1 Outlets for the Company's production, 1938 and 1945 (million long tons)

| | Product sales through Company outlets | | | | | Supplies to other oil companies/distributors | | | | | |
	(a) Inland sales by marketing associates and subsidiaries	(b) Marine bunkers	(c) Admiralty and armed forces	(d) Sundry	(e) Sub-total: (a)+(b) +(c)+(d)	(f) Products	(g) Crude oil	(h) Sub-total: (f)+(g)	(i) Total sales	(j) Company outlets as % of total: (e) as % of (i)	(k) Supply sales as % of total: (h) as % of (i)
1938	4.5	3.2	1.2	0.2	9.1	1.3	0.1	1.4	10.5	86.7	13.3
1945	3.8	2.5	4.7	–	11.0	6.8	0.2	7.0	18.0	61.1	38.9

Notes: 1. Column (a) excludes sales in India where the Company had no marketing organisation, although its products were sold there by the Burmah-Shell marketing company. In the source data, sales of Company products in India are classified as the Company's 'Own business', but for the purposes of this table they are included in column (f).
2. Column (b) includes Company tanker bunkers.

Source: BP 106617, Unsigned note on AIOC policy, 18 June 1946.

prepared to supply other local distributors with products provided this would not be detrimental to the expansion of its own marketing activities. With regard to supplies of products to other international oil companies, the Company should, Snow continued, seek to arrange reasonably long-term contracts (of, say, three to five years' duration), with provisions for gradual reductions in volumes so that the Company would not suffer an abrupt curtailment of its business as the likes of Royal Dutch-Shell, the Burmah Oil Company and Standard-Vacuum brought damaged production facilities back into operation. Finally, although the Company's own refining operations were to have first claim on its crude oil production, there was no general objection to selling crude to dispose of surplus production, though individual circumstances would have to be examined to ensure that sales of crude oil would not result in other companies constructing refineries which would be to the competitive disadvantage of the Company.[22]

Having set out the Company's marketing policy, Snow turned his attention to the organisation of the Distribution Department, which he administered under the direction of Morris. The Department was responsible for the twin functions of supply and marketing, the supply side being concerned with matching the availability of crude and products with the Company's marketing requirements. Recognising that the growth in the volume of business since prewar days called for greater devolution of responsibility in the supply function, Snow proposed, in November 1946, that the Department should be structured into four divisions covering: 1) supply, 2) eastern markets, 3) western markets and 4) products which were either of a specialist nature (such as wax, petroleum chemicals, and butane/propane) or traded internationally (such as marine bunkers and aviation fuels).[23] In addition, he suggested that the Department should include a supporting services function which would provide technical support for the sales organisation.[24] This last idea was not a new one, but had been under discussion in the Company for several years, reflecting a growing recognition of the need for marketers to have a channel for technical liaison with customers and manufacturers of oil-fired equipment and for maintaining contacts with Production Department and Sunbury on matters of technical research.[25] The outcome of the discussions was a recommendation, made by an internal inter-departmental committee in April 1946, that a Technical Services Branch should be established and attached to the Distribution Department.[26] Snow's proposals were implemented, with some modifications, in 1947 when the Department was organised into three divisions on the lines set out in figure 11.1.[27]

In 1949 a fourth division was established under W. D. Brown, its main responsibilities being the co-ordination and authorisation of prices, the approval and revision of standard terms of sale and the preparation of special agreements involving departures from standard practice.[28]

In 1950 the effectiveness of the Department's organisation was called into question by the consultancy firm of Harold Whitehead and Partners who, as has been described in chapter 10, were called in on the recommendation of the Organisation Committee to investigate office routines and the possibilities of introducing mechanical aids for clerical work. The identity of the department about which Whiteheads made the damning remarks cited in chapter 10 was kept anonymous and there is no particular reason to suppose that it was the Distribution Department. However, the Department did not escape the consultancy firm's investigations and in 1950 Whiteheads recommended that the Supply Division should be reorganised and the number of its staff substantially reduced.[29] The reaction of A. R. MacWilliam, the manager in charge of supply, was to defend his division with vigour. In a note to Snow he contended that the Whitehead recommendations were based on a 'complete misconception' of the work of the Division and castigated the consultants' report as 'confused and manifestly unsound'.[30] Referring to the growing workload of the division as the Company's business grew in both volume and complexity, he argued that the Supply Division was in fact inadequately staffed, not least because requisitions for new recruits had been suspended for several months while Whitehead and Partners conducted their investigations.[31] The cynical interpretation of MacWilliam's remarks would, no doubt, be to dismiss them as special pleading in defence of a vested interest. However, the work of the supply function was undoubtedly becoming more demanding as new oilfields and refineries came onstream, making it increasingly difficult to match individual crude oils to various refining plants in such a pattern as would meet the marketers' requirements for given quantities and qualities of products in the most economical manner. Snow accepted MacWilliam's arguments and sought the authorisation of the relevant director, Morris, for an increase in staff, adding the rider that he proposed to have a word with a senior partner of Whiteheads regarding 'what is, from our point of view, a thoroughly unsatisfactory report'.[32] Morris, having read the Whitehead report, agreed with Snow and informed him, in January 1951: 'You must make sure that Petroleum Supply Division is adequately staffed ... If new engagements are not made the position may become critical'.[33] Given the pivotal role

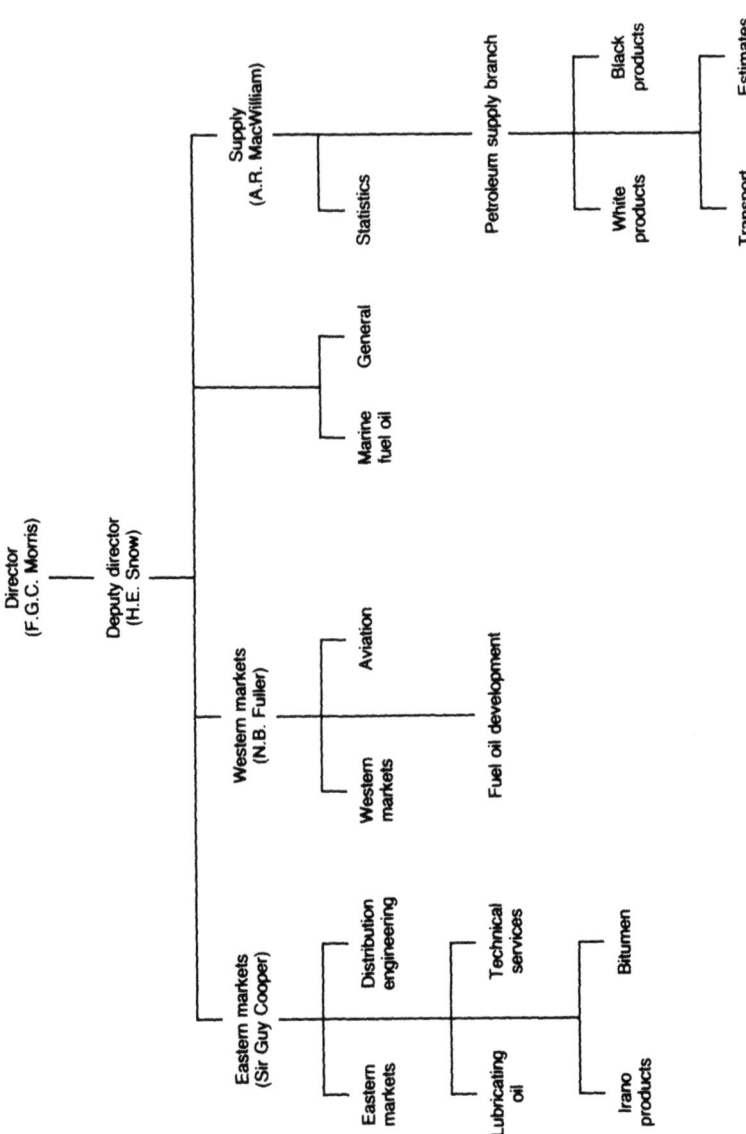

Figure 11.1 Distribution Department, 1947
Note and source: See Appendix 2.

which Supply Division was coming to play as the nerve centre for programming the Company's operations and its growing importance after the suspension of operations in Iran later in 1951, the rejection of the Whitehead recommendations was probably well founded.

MARKET-LOCATED REFINING AND THE EXPANSION OF SHIPPING

While the Company turned to building up its own marketing organisation and outlets after the collapse of the proposals for expanding joint marketing with Royal Dutch-Shell, the adequacy of its refining capacity in relation to future product demand also came under scrutiny. On the basis of projections carried out in 1946 various estimates were made of the additional refining capacity which would be required by 1950. The absolute minimum, which would only meet the Company's 'bare needs' of finished products, was for new plant which would yield an annual output of four million tons of products.[34] As it took four to five years to complete a major refinery project there was little time to be lost.[35] The key question was: where should the new refining capacity be located?

Traditionally, it was the accepted wisdom that the lowest costs were achieved by placing base-load refineries (i.e. those manufacturing standard bulk products) at or near the main sources of crude oil production. Consequently, refineries were not, in general, located in consuming areas except in cases where the plant was designed for the manufacture of specialised products in relatively small quantities or where local refining was fostered by government regulations.[36] In the Company's case, there were several reasons for the lower costs of refining at the source of production, including low labour costs in the Middle East and the saving in royalties and freight costs which would otherwise have been incurred on fuel which was consumed in the refining process.[37] However, above all else it was a matter of economies of scale and transport costs. Before the war, it was uneconomic to locate a main base-load refinery in, say, a European market because local demand was generally too small to absorb the full volume and range of products manufactured in a refinery of sufficient scale to be economic. Large market-located refineries could therefore be ruled out on account of the costs of transporting products which were surplus to local requirements to other markets, in effect a re-transport of oil which had already been shipped as crude from the source of production to the refinery. Re-transport costs could, of course, have been obviated by the construction of small market-located refineries, but the loss of scale economies in refining would then have been very substantial.

The significance of scale economies and transport costs was studied by the Company's Central Planning Department in 1946, with results that appeared to prove the traditional case against erecting new base-load refineries in market locations. The Department found that the capital and operating costs of a base-load refinery dropped progressively as the refinery size increased up to an annual capacity of five million tons, beyond which costs might still fall slightly. However, there were only one or two European ports with sufficient local demand to absorb the output of a refinery with an annual capacity of one million tons, and a further four or five ports with large enough local markets to consume the output of a refinery with an annual capacity of half a million tons. Taking two refineries of similar type and producing similar products, one of half a million tons capacity and the second of five million tons, the capital cost per ton of capacity was found to be more than three times as great in the smaller plant and the operating costs some 35 to 40 per cent higher per ton of throughput. Even if a larger European refinery with an annual capacity of one million tons could dispose of all its output within an economic supply area (i.e. without re-transport costs) the cost per ton of output would still be 5 to 6 shillings more than with a five-million ton refinery in the Middle East because of the lost economies of scale and the freight costs of fuel consumed in the European refinery. The Planning Department therefore concluded that a company which attempted to supply a large proportion of its European market requirements from local refineries would be at a considerable disadvantage when compared with companies refining on a larger scale in the Middle East.[38]

Despite that conclusion, the expansion of the Abadan refinery beyond its existing capacity of about twenty million tons was ruled out for several reasons. One was that the growth in demand for Middle East oil was expected to be mainly in markets west of Suez, from which it followed that new refining capacity would best be located to the west.[39] Secondly, the Company's refining capacity could not be increased at the required rate if construction was concentrated in a single large centre, particularly as plant replacement to cover obsolescence had to be faced. Consequently it was felt 'very strongly' that additional refining installations should be spread in order to limit the amount of growth required in any one location to a reasonable maximum.[40] Thirdly, the costs of labour, with attendant housing and amenities, had 'risen very rapidly' in the Middle East which, coupled with the labour problems which are described in chapter 14, made refining in that area less attractive.[41] In the light of these various factors, the plan which was favoured early in 1947 was one which provided for the construction of a new base-load refinery on the eastern seaboard of the Mediterranean, to which crude oil would be transported through a major new trans-desert pipeline, a project which, as events transpired, never came to fruition. In addition, the refineries at Llandarcy and Grangemouth were to be expanded and a new refinery constructed in the Thames area to supply the London market, its main objective being the manufacture of speciality products.[42]

The internal debate on refining locations was not, however, at an end. At Fraser's request, the Central Planning Department examined the case for expanding refining capacity in Britain and reported, in June 1947, that domestic refining could be financially attractive provided that there was sufficient local demand for products to warrant a plant of reasonable size. As consumption in individual market areas was rising, and product exchanges between oil companies tending to grow, it was possible to envisage large-size local refineries serving 'a very useful purpose'.[43] The balance of the debate was tilted further in favour of local refining in the following year when Coxon assessed the impact of the trend towards larger-sized tankers on the costs of refining at different locations. The advantage of large tankers was that they had lower operating costs per ton of cargo than their smaller brethren. However, whilst they could be used for carrying crude oil to refineries at deepwater ports, they were not generally suitable for carrying finished products as they could not enter many of the normal oil ports. Basing his calculations on the assumption that crude oil would be carried in 28,000 ton tankers, fuel oil in 16,000 ton tankers and other products in 12,000 ton tankers, Coxon found that the overall cost of transporting oil to the British market via a local refinery would be no more than via a refinery in the Middle East, provided that at least half of the British refinery's output was consumed locally. Taking into account the rising costs and labour problems of refining in the Middle East he concluded that it was no longer 'a priori more economical to refine in the Middle East than in Europe'. He felt it was possible to envisage 'fair sized' base-load refineries in countries such as Britain, provided they were accessible to large tankers, assured of a 'good local offtake' and/or situated on a main supply route.[44]

The traditional orthodoxy that base-load refineries should be placed at or near the source of crude production was thus challenged and, to all intents and purposes, overthrown, marking the beginning of the shift towards locating new refineries in the growing markets of Europe rather than in the Middle East. In the Company's case, the most notable early embodiment of the new line of thought was the construction of the Kent refinery which was situated on the Isle of Grain at the convergence of the rivers Thames and Medway with deepwater access and well placed to supply London and south-east England with finished products. Construction of the refinery, which had an initial capacity of about four million tons of crude oil input a year, commenced in mid-1950 and crude oil was first processed there in February 1953.[45] Elsewhere, also, the Company increased its refining capacity in market locations, as will be seen below.

One of the features of the Isle of Grain site was that its deepwater access permitted it to take tankers of 30,000 deadweight tons, drawing 34 feet and having a length of 650 feet. Elsewhere, also, deep marine terminals were brought into operation to handle large crude oil vessels. In Scotland, a pipeline connecting the Grangemouth refinery with the deepwater Finnart terminal on Loch Long was opened in 1951. In Iran, terminal developments at Abadan provided for increased shipments of oil products, whilst further down the Persian Gulf work on a deeper approach channel at Bandar Mashur facilitated loadings of crude oil. In Kuwait, where crude oil production commenced in 1946, the loading jetty of Mina al-Ahmadi which was constructed in 1949 was, at the time, the largest in the world, with virtually unlimited depth for the largest tankers. In Qatar, where crude oil production began in 1949, the Umm Sa'id terminal which was built in 1950 could accommodate tankers of 30,000 deadweight tons.

On the shipping side, the Company decided in September 1948 to order tankers of 28,000 deadweight tons. Vessels of such a size were not, however, representative of the general run of ships in the Company's fleet, which, as a result of wartime losses, was reduced from 95 ships with a total capacity of 998,000 deadweight tons in 1939 to 69 ships with a capacity of 757,000 deadweight tons at the end of 1945. The depleted fleet was clearly inadequate to cope with the growing volume of the Company's business, but the possibilities for expansion and modernisation were limited by the tight capacity of the British shipbuilding industry, shortages of materials such as steel and the rising delivery prices of new tankers. Nevertheless, by various means the size of the Company-owned fleet was substantially increased between 1946 and 1951. In January 1946 the Company ordered 100,000

Table 11.2 *The Company-owned shipping fleet,*
1946–1951 (d.w.t. = deadweight tons)

End of:	(a) Number of ships	(b) Deadweight tonnage (d.w.t. '000s)	(c) Average size of ship: (b) divided by (a) (d.w.t. '000s)
1946	87	964	11.08
1947	103	1202	11.67
1948	112	1311	11.71
1949	124	1468	11.84
1950	139	1655	11.91
1951	152	1850	12.17

Sources: Shipping administration ship history lists. Planning Division's list of tonnage changes.

deadweight tons, consisting mainly of former government ships of 12,000 deadweight tons or less. Between May 1946 and March 1947 fifteen more tankers of 12,250 and 8,000 deadweight tons were ordered, the Distribution Department being reluctant to authorise the purchase of larger tankers for fear that a loss of operational flexibility would result. In addition, in 1947 Robert Gillespie, managing director of the British Tanker Company (the Company's shipping subsidiary), succeeded in purchasing ten American tankers of 16,000 deadweight tons from the American Maritime Commission. Through these and other measures, the Company-owned fleet grew to 152 ships with an aggregate capacity of 1,850,000 deadweight tons by the end of 1951. Judging from the rising average size of Company-owned vessels, shown in table 11.2, it may be presumed that the lower unit freight costs associated with increases in tanker size were realised to some extent. The scope for attaining an economic optimum was, however, substantially reduced by the requirement to ship ever-increasing volumes of oil at a time when the possibilities of replacing old with new vessels were strictly limited. As a result, the Company continued to use shipping which, in other circumstances, would have been regarded as uneconomic. Even so, the Company's own fleet was nothing like large enough to cope with the expansion of business and, as table 11.3 indicates, increasing resort was made to chartering vessels from other owners, so much so that by 1951 the Company-owned fleet accounted

Table 11.3 *Effective deadweight tonnage:*
Company-owned and time-chartered ships, 1946–1951
(d.w.t. = deadweight tons)

	(a) Company-owned ships (d.w.t. '000s)	(b) Time-chartered ships (d.w.t. '000s)	(c) Total effective tonnage (d.w.t. '000s)	(d) (a) as % of (c) (%)
1946	853	182	1035	82
1947	1069	336	1405	76
1948	1229	442	1671	74
1949	1404	657	2061	68
1950	1566	1067	2633	59
1951	1745	1328	3073	57

for only about 57 per cent of effective deadweight tonnage, compared with some 90 per cent in the prewar years.

On organisation, it was partly as a result of Gillespie's belief that there was a limit to the size of fleet which could be efficiently managed, and partly for financial reasons, that around 1950 a separate, but associated shipping company was formed to operate ten time-chartered tankers exclusively for the Company. The new associate, which was called the Lowland Tanker Company, was half owned by the Company, the remainder of its shares being held by Mathesons, the shipping side of Jardine Matheson, and Common Brothers, a Newcastle shipowning firm. The initiative in its formation was taken by William Keswick, the chairman of Mathesons and a non-executive director of the Company. The Lowland Tanker Company was, however, a small-scale affair compared with the much larger operations of the British Tanker Company, which tended to stand apart from the mainstream of the Company's departmental organisation, being largely self-contained with, like much of the British shipping industry, a traditional, conservative air.

PRODUCT MARKETS AND SALES

In quantitative terms, the expansion of the Company's sales was extremely rapid after the war. Having nearly doubled from about 10 million tons to 20.4 million tons between 1938 and 1946, the total sales

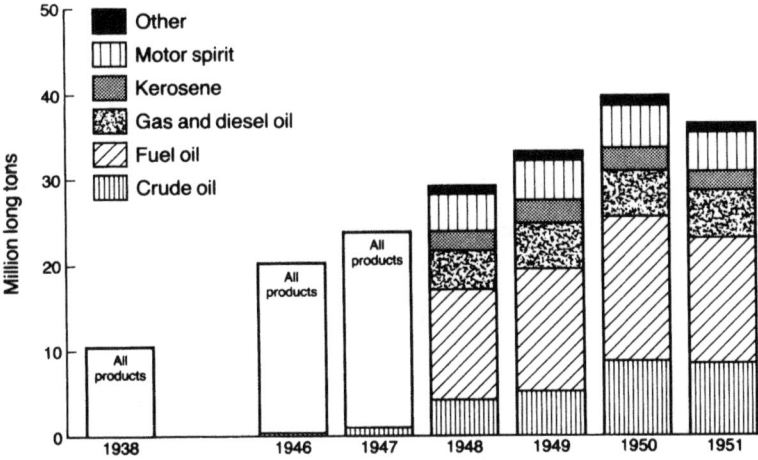

Figure 11.2   The Company's sales of products and crude oil, 1938 and
1946–1951
Notes and sources: See Appendix 2.

tonnage then virtually doubled again by 1950 when sales amounted to
very nearly 40 million tons. The decline in the following year reflected
the upset caused by the suspension of operations in Iran, which threw
the Company's entire operations into crisis. Qualitatively, the Company's sales were less impressive. Looking at figure 11.2, it is apparent
that the growth in total sales tonnage was very largely due to the
spectacular expansion of crude oil sales which rose from a mere
132,000 tons in 1938 to 187,000 tons in 1946 before taking off to reach
8,666,000 tons in 1950, when crude oil accounted for some 22 per cent
of total sales tonnage. At the same time, fuel oil sales also increased
rapidly, reaching 16.8 million tons in 1950. In that year, crude oil and
fuel oil accounted for roughly two-thirds of the Company's overall
sales volume.
  These figures portray an unmistakable characteristic of the Company's growth: that it was based primarily on moving ever-increasing
volumes of low-value output. Driven by the need to safeguard, so far
as was possible, its concessionary position in the Middle East the
Company was increasingly becoming a bulk supplier of raw material
(i.e. crude oil) for processing and marketing by other oil companies.
Low-volume business in more sophisticated products of high added
value was not, it should be said, ignored. In aviation spirit the
Company adopted an aggressive marketing policy aimed at entering

Table 11.4 Outlets for the Company's production, 1946–1951 (million long tons)

| | Product sales through Company outlets | | | | | Supplies to other oil companies/distributors | | | | | |
	(a) Inland sales by marketing associates and sub-sidiaries	(b) Marine bunkers	(c) Admiralty and armed forces	(d) Sundry	(e) Sub-total: (a)+(b)+(c)+(d) Products	(f) Products	(g) Crude oil	(h) Sub-total: (f)+(g)	(i) Total sales	(j) Company outlets as % of total: (e) as % of (i)	(k) Supply sales as % of total: (h) as % of (i)
1946	5.7	3.7	3.3	–	12.7	7.5	0.2	7.7	20.4	62.3	37.7
1947	7.8	4.5	1.8	–	14.1	8.9	1.0	9.9	24.0	58.8	41.3
1948	8.4	6.2	1.7	0.1	16.4	8.6	4.2	12.8	29.2	56.2	43.8
1949	9.5	7.6	2.0	0.2	19.3	8.7	5.2	13.9	33.2	58.1	41.9
1950	11.2	8.5	1.6	0.1	21.4	9.7	8.7	18.4	39.8	53.8	46.2
1951	11.8	8.3	1.2	–	21.3	6.6	8.4	15.0	36.3	58.7	41.3

Notes: 1. Column (a) excludes sales in India and Pakistan where the Company had no marketing organisation, although its products were sold by the Burmah-Shell marketing company. In the source data, sales of Company products in these markets are classified as the Company's 'Own business', but for the purposes of this table they are included in column (f).

2. Column (b) includes Company tanker bunkers.

3. The split of UK motor spirit sales between columns (a) and (f) includes adjustments to the source data for 1946 and 1947 to allow for inconsistencies in classification at source.

Sources: BP 102729; 102738; 102749; 102760; 102771.

markets 'wherever and whenever possible' in full competition with other suppliers.[46] In 1949 and 1950, sales contracts were made with numerous airlines and with air forces in Britain and Italy. A particular marketing success was achieved in February 1950 when the new range of Vickers Viscount aircraft was powered exclusively by the Company's aviation products, similar success being achieved with the de Havilland Comet a year later. The Company also decided to manufacture its own range of lubricating oils and established a sales organisation to sell lubricants through its own retail outlets. Yet, despite these developments, it was mainly the bulk trade in crude oil and heavy products which met the Company's need to increase the volume of its trade.

The depiction of the Company as a bulk producer of crude and products with a relatively undeveloped marketing organisation is confirmed by the figures in table 11.4, which show, in approximate terms, how the Company disposed of its output. As can be seen, the effort to increase sales through the Company's own marketing outlets resulted in considerable expansion both of marine bunker sales, which increased from 3.7 million tons in 1946 to 8.5 million tons in 1950, and in inland sales by marketing associates and subsidiaries, which rose from 5.7 million tons to 11.2 million tons in the same period. These gains were more than sufficient to offset the loss of sales to the armed forces after the war. Yet, notwithstanding the growth of the Company's direct sales to customers, a growing proportion of its output was disposed of to other oil companies and distributors for refining and/or marketing through their facilities. In fact, by 1950 about 46 per cent of the Company's sales took the form of supplies to outside refiners and marketers.

It would, of course, have been expecting much too much for the Company's marketing organisation to catch up with the great increase in production during the war and then to keep pace with the very rapid growth in output afterwards. The proposals for a wide-ranging extension of joint marketing with Royal Dutch-Shell offered a short cut, but after the negotiations had broken down there was no alternative but to accept, as Fraser did, that the expansion of the Company's own outlets would be a long-term undertaking. This was not, as such comments often are, a euphemism for putting off any kind of action. In the shadow of the more spectacular upstream sector, the Company's marketers made substantial progress in expanding the Company's outlets, focusing most of their attention on western Europe.

In Britain, where the Company marketed through Shell-Mex and

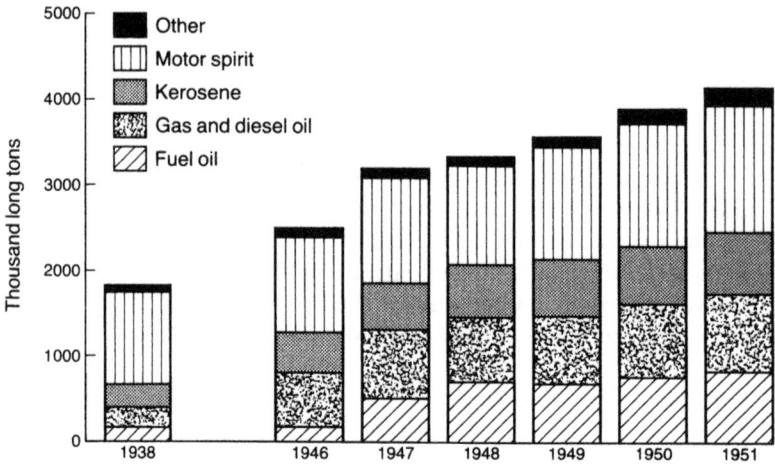

Figure 11.3   The Company's product sales in the UK and Eire, 1938 and 1946–1951
Notes and sources: See Appendix 2.

BP, the oil industry, in common with other sectors of the economy, was affected by the panoply of wartime economic controls which was kept in force by Clement Attlee's postwar Labour Government. The Petroleum Board, through which the oil companies pooled the distribution of their products during the war, was not dissolved until the beginning of July 1948. At the same time it was agreed with the Ministry of Fuel and Power that brand names, the use of which had been dropped during the war, should not be re-introduced for at least six months. Meanwhile, petrol consumption continued to be restricted by rationing, which was not abolished until May 1950. By that time, a decade had elapsed with practically no competition in the British market. Thereafter, the pattern of retailing began to alter with the introduction of service stations which dealt exclusively with the products of particular companies. However, that development, which originated in the USA, came too late to have much bearing on the Company's marketing in Britain up to 1951 when, as figure 11.3 shows, motor spirit sales, constrained by the continuance of rationing, showed weak growth. Sales of heavier products increased much more rapidly, especially those of fuel oil in the years 1946–8 when the chronic shortage of coal encouraged consumers to switch to oil, a shift which, as will be seen in chapter 12, was supported by the Government.

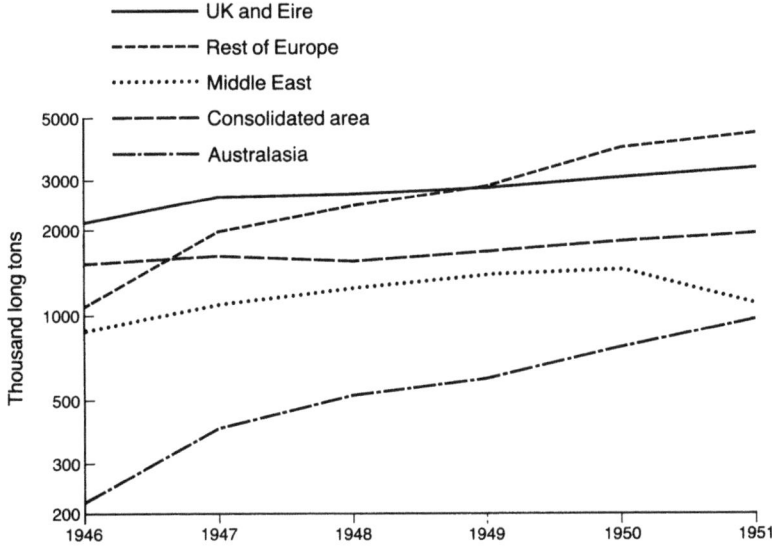

Figure 11.4   Inland sales through marketing associates and subsidiaries, 1946–1951
Notes and sources: See Appendix 2.

Although Britain remained the largest market for the Company's products, it was not by any means the fastest growing. As can be seen from figure 11.4, it was in the continental European countries and, on a lesser scale, in Australasia that the Company achieved a truly dynamic growth in trade through its marketing associates and subsidiaries. In Australasia sales were boosted by the Company's entry into the New Zealand market through a new associate, the British Petroleum Company of New Zealand Ltd, which was 51 per cent owned by the New Zealand Government and 49 per cent by the Company. It commenced marketing in May 1949 and expanded its sales to nearly 220,000 tons in 1951.[47] Meanwhile, in continental Europe the sales of marketing associates and subsidiaries, having declined from 1.4 million tons in 1938 to little more than a million tons in 1946, rose to nearly 4.5 million tons in 1951. Although expansion was to be expected during the postwar economic reconstruction of Europe, it is clear, as will be seen, that the Company was not merely holding its place in continental markets, but was making competitive gains through its pursuit of refining and marketing opportunities.

In France, the Company's refinery at Lavera, near Marseilles, was

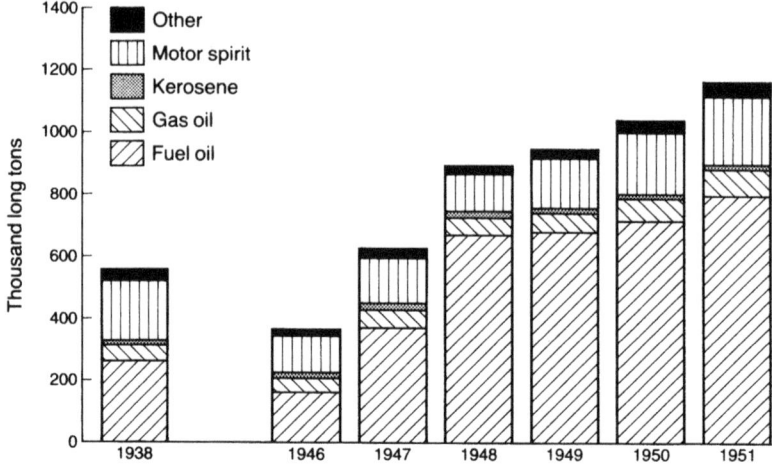

Figure 11.5   Sales of the Company's French subsidiary, Société Générale des
Huiles de Pétrole BP, 1938 and 1946–1951
Source: See Appendix 2.

rehabilitated and a newly constructed refinery at Dunkirk came into
operation in February 1950, replacing the war-damaged plant at
Courchelettes. In conformity with a decision to include the letters BP
in the names of marketing associates, the simple expedient was
adopted in 1950 of appending BP to the French associate's old name,
so that it became known as Société Générale des Huiles de Pétrole
BP.[48] Though in the French case the new name lacked concision and
was somewhat clumsy, the inclusion of BP in associates' names was
indicative of a more modern appreciation of the value of developing a
coherent international marketing identity. As for sales by the French
associate, they increased from little more than half a million tons in
1938 to more than a million tons in 1950, the predominant product
being fuel oil, as shown in figure 11.5.

In Germany, the partition of the country, the division of the western
sector into occupied zones and the acquisition of quotas made for
administrative difficulties.[49] The approval of the British and German
authorities for the purchase of the Eurotank refinery at Hamburg in
January 1949 gave the Company a presence in the local refining
business. In addition, the Company acquired a majority shareholding
in the Oelwerke Julius Schindler refinery, also in Hamburg, in March
1951, after delays because the British Treasury would not sanction the
foreign currency expenditure. In the event, payment was made in oil

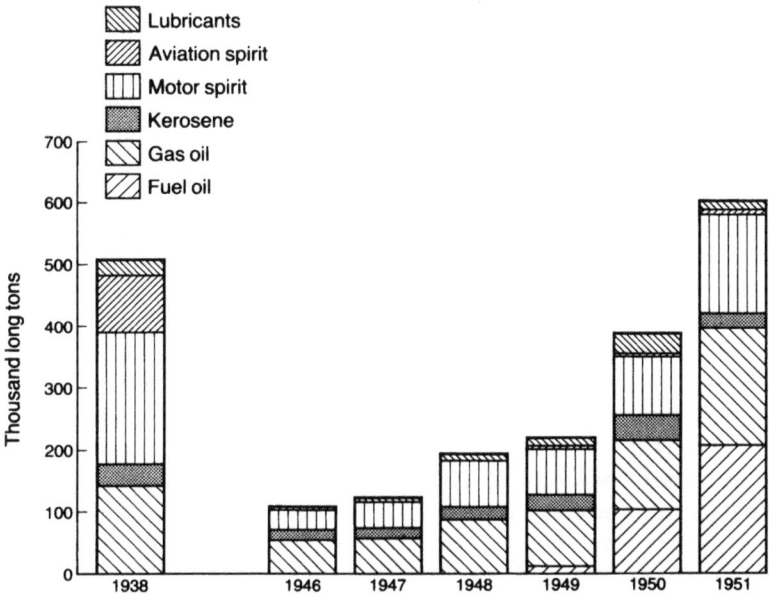

Figure 11.6   Sales of the Company's German subsidiary, BP Benzin- und Petroleum-GmbH, 1938 and 1946–1951
Source: See Appendix 2.

consignments. On the marketing side, in 1949 the Company acquired Runo-Everth, a small Bavarian Austrian distributing organisation.[50] The Bavarian assets were then merged into the Company's old associate, Olex, which in turn combined with Eurotank under the new name of BP Benzin- und Petroleum-GmbH.[51] Sales took much longer to recover their prewar level than in France and, as figure 11.6 indicates, it was not until 1951 that the prewar volume was surpassed. The undoubted architect of the Company's re-emergence in Germany was Hans Ornstein, the local managing director, a man as interesting in personality as he was enterprising in commerce, a combination of businessman, soldier and scholar.

Elsewhere, also, the Company expanded its European refining and marketing interests. In Austria, where distributors were allocated quotas by the government on the basis of prewar trade, the Austrian side of Runo-Everth was combined with the Company's other local assets to form BP Benzin und Petroleum AG.[52] In Belgium, agreement was reached with Petrofina in 1949 to share equally in the construction

of a new refinery at Antwerp.[53] The new plant, to which the Company
had rights to supply crude oil, was commissioned in autumn 1951.[54]
Meanwhile, the name of the Company's original local associate,
L'Alliance, was changed to BP Belgium.[55] Its sales, combined with
those of the Company's second Belgian marketing associate, Anglo-
Belge des Pétroles (formerly Sinclair SA), rose from 105,000 tons in
1938 to 310,000 tons in 1950.[56] In Holland, shares of the market were
effectively fixed by government import quotas based on the position
before the war when the Company's local associate, Benzine en
Petroleum Handel-Maatschappij (BPHM), had only a very small
market share. However, by the postwar acquisition (and absorption
into BPHM) of three independent distributors – Nohaka, Tankopslag
and Fanto – each with an established import right, BPHM achieved a
large expansion in sales, which rose from a mere 19,000 tons in 1938 to
177,000 tons in 1950.[57] In Switzerland, also, the Company expanded
by acquisition, purchasing a distributing company, Noba, in 1949.[58]
With its headquarters at Basle, Noba retained a separate identity to the
Company's other local associate, BP Benzin und Petroleum AG
(BPAG), whose head office was at Zurich.[59] Their combined sales in
1950 were 287,000 tons, a great increase over BPAG's prewar sales
which amounted to 55,000 tons in 1938.[60]

In the Scandinavian countries of Sweden, Denmark and Norway,
subsidiary and associated companies achieved an exceptionally rapid
growth in sales, which rose from less than a quarter of a million tons in
1938 to more than one and a half million tons in 1950. The lion's share
of the new business was in fuel oil sales, which were boosted by the
Company being the first to introduce heavy grades of fuel oil into the
Danish and Norwegian markets.[61] In keeping with the general policy
on names, the title of the wholly owned Danish subsidiary was
changed from Det Forenede Olie-Kompagniet (DFOK) to BP Olie-
Kompagniet (BPOK) in 1950.[62] Following the same principle, the
wholly owned Swedish subsidiary had its name changed to Svenska BP
Olje, but in Norway, where the Company had only a 50 per cent
interest in its marketing associate, the name of Norsk Braendselolje
was not amended to incorporate the BP identity.[63] The expansion of
the Company's Scandinavian business would have been still greater,
but for the fact that negotiations for the purchase of a distributing
organisation in Finland were thwarted by the refusal of the British
Treasury to approve the foreign exchange transaction.[64]

The Company encountered similar exchange problems when it
sought to enter the Greek market by acquiring the distributing organi-

sation of Steaua (Agencies) Ltd, but in this case the Company succeeded in overcoming foreign currency controls by making the acquisition through its French associate.[65] With the purchase of Steaua (Agencies), the number of countries on the continent of Europe in which the Company had marketing associates or subsidiaries was raised to ten, in addition to which it had a small marketing representation in Iceland.[66]

Finally, in connection with downstream operations in Europe, mention should be made of Italy, where the Company lacked a marketing organisation of its own, but entered into an agreement with AGIP, the state oil company, in 1948. Under the agreement, the Company acquired a 49 per cent interest in AGIP's Marghera refinery at Venice with the right to supply the refinery with its crude oil requirements. Products from the refinery were delivered to AGIP, which operated a distributing organisation throughout Italy.[67]

Through these various refining and marketing initiatives the Company succeeded in substantially increasing both the volume of its sales and, more important as a measure of its competitive position, its market share in Europe. As can be seen from columns (a) and (d) of table 11.5, the Company's own outlets (i.e. marketing associates and subsidiaries) increased their share of the European market from about 8 per cent in 1938 to 13 per cent in 1950. Between the same two dates, the share of European markets accounted for by the Company's supplies to other local refiners and distributors increased still faster, from less than 1 per cent in 1938 to nearly 10 per cent in 1950. Altogether, therefore, the Company supplied, either directly or through supplies to other local refiners and distributors, about 9 per cent of Europe's oil consumption in 1938, but 23 per cent in 1950.

CRUDE OIL SALES

Although the Company expanded its existing marketing outlets and opened new ones to increase its sales of products, it had neither the refining capacity nor the marketing organisation to absorb all of its production. Given the unrelenting pressure to increase the offtake from its concessions there was no alternative but to make crude oil sales with the result, already noted, that the Company became, to an increasing extent, a bulk supplier of raw materials for refining and marketing by other oil companies. Contracts were arranged with various companies including other oil majors, the Japanese firm of Mitsubishi and the Argentinian national oil company, Yacimientos

Table 11.5 *The Company's share of European markets, 1938 and 1950 (percentages)*

| | Share of markets in 1938 | | | Share of markets in 1950 | | |
	(a) Sales by associates/ subsidiaries	(b) Supplies to other local distributors	(c) Total	(d) Sales by associates/ subsidiaries	(e) Supplies to other local distributors	(f) Total
Austria	4.6	–	4.6	10.6	6.2	16.8
Belgium	18.1	0.5	18.6	16.3	19.1	35.4
Denmark	13.8	–	13.8	22.6	–	22.6
Finland	–	–	–	–	12.5	12.5
France	11.2	0.5	11.7	12.3	4.8	17.1
Germany	10.2	–	10.2	14.7	3.2	17.9
Greece	–	–	–	9.8	0.9	10.7
Holland	1.9	–	1.9	8.7	8.9	17.6
Iceland	46.0	–	46.0	30.4	–	30.4
Italy	–	3.6	3.6	–	21.0	21.0
Norway	24.8	–	24.8	47.9	–	47.9
Portugal	–	–	–	–	4.3	4.3
Spain	–	–	–	–	6.4	6.4
Sweden	8.0	–	8.0	18.9	20.2	39.1
Switzerland	13.4	0.6	14.0	27.5	6.3	33.8
Yugoslavia	–	–	–	–	50.0	50.0
Overall	8.2	0.8	9.0	13.0	9.9	22.9

Notes: 1. Supplies to other local distributors include sales of products and crude oil to independent distributors and refiners, but exclude sales to other major oil companies and their subsidiaries.
2. Inconsistencies in the source data account for differences between this table and table 4.4.

Source: BP 67664.

Petroliferoso Fiscales Argentinos. None, however, was as significant as the crude oil contracts which were negotiated with the US companies, Standard Oil (NJ) and the Standard Oil Company of New York (Socony).

The possibilities of those two companies taking long-term crude supplies from the Company were raised in 1944–5,[68] but it was not until 1946 that anything concrete resulted. In August that year Orville Harden of Standard Oil (NJ) suggested that, as his company required crude oil mainly for markets west of Suez, the most economical supply route from the Company's centres of production would be via a pipeline running from the Middle East to the Mediterranean. Such a pipeline would, he pointed out, show substantial cost savings over transportation by tanker and would make the Company's crude oil competitive with South American crude in western markets.[69] In October, the idea of a combined crude supply and pipeline agreement between Standard Oil (NJ) and the Company was taken further when Harden, on a visit to London, discussed the subject with Fraser.[70]

Though different in form from the earlier joint marketing initiative with Royal Dutch-Shell, the principles which underlay the scheme for a major crude oil contract with Standard Oil (NJ) had a familiar ring to them: in effect, a strategic alliance which would align the Company's upstream production capacity with the more highly developed downstream refining and marketing outlets of another major oil company. Standard Oil (NJ) was, at the time, short of crude supplies in the eastern hemisphere where its share of less than one-eighth in the Iraq Petroleum Company was nothing like sufficient to meet its downstream requirements.[71] Standard Oil (NJ) was therefore seeking secure crude supplies in the Middle East, to which end it was negotiating for participation in Aramco, which held the Saudi Arabian concession, as well as considering a long-term supply contract with the Company.[72]

The Company stood to gain considerable economic benefits from the construction of a pipeline with sufficient capacity to transport Middle East crude to the Mediterranean not only for the purpose of supplying Standard Oil (NJ), but also for the Company's own downstream facilities in Europe. To give a measure of the transport economies, the estimated cost of moving crude from the Persian Gulf to the Mediterranean through a 24-inch diameter pipeline with an annual capacity of some 15 million tons was 5s 2d per ton at full throughput and 7s 5d per ton at 60 per cent throughput. Shipping costs, on the other hand, were estimated at 16s 10d per ton using the largest tankers available in 1946.[73] If Standard Oil (NJ) entered into a high-volume long-term supply contract and participated in the construction of a new pipeline the initial capital costs of the pipeline would be spread, the US involvement would facilitate the procurement of scarce steel and dollars and the supply contract would provide an assured throughput for the pipeline which, for reasons of economies of scale, would have to be of large capacity if it was to be economically viable.

While the economics of the pipeline were attractive, there was another aspect of the scheme which weighed more heavily in Fraser's mind: the desire to maintain Iran's relative position in Middle East oil production in the interest of concessionary stability. On 8 October 1946 he cabled Basil Jackson (in the USA) informing him that a long-term crude supply contract with Standard Oil (NJ) would help to 'secure desirable regulation' between sources of production.[74] Two days later, in a fuller account of his discussions with Harden, Fraser confirmed his enthusiasm for the scheme and explained to Jackson the reasons for his desire to have the option to supply Standard Oil (NJ) with crude from either Iran or Kuwait at the Company's discretion. In Fraser's words, he regarded a crude supply/pipeline agreement as being,

> in relation to the Middle East situation as a whole and our own position, the most constructive step which could be taken in the interests of stability, and set the greatest store on completing an arrangement ... Moreover, I could not view an arrangement as directed primarily to Kuwait, my fundamental concern being to preserve the position of Iran in the supply picture. Forced development in Kuwait, with the resultant increased quantities in other hands, was capable of defeating the object of stabilisation. My view therefore was that the basis must necessarily be that I had the option to supply from either source, so as to preserve the proper balance.[75]

The expansion of Saudi production posed a still greater threat to the status of Iran as the foremost oil producer in the Middle East, especially as the outlets for Saudi crude would be greatly increased if Standard Oil (NJ) joined Aramco. Indeed, Jackson felt that the 'overriding consideration' in favour of the proposed crude supply contract with Standard Oil (NJ) was that it would provide an 'assured outlet for Iranian crude to largest American marketer thereby affording best insurance against possibility [of] Saudi Arabian production surpassing Iran'.[76] This factor was of such importance in Jackson's mind that he felt it justified the Company agreeing, if necessary, to supply Standard Oil (NJ) on terms which covered costs plus only 'a relatively small margin of profit'.[77] He was relieved to hear from Fraser that Harden seemed to favour a supply agreement with the Company over participation in Aramco.[78] However, by the end of October Standard Oil (NJ)'s board of directors, reportedly divided on the issue, proposed to go ahead with both a supply contract with the Company and participation in the Saudi concession. Their decision was, according to Jackson, 'calculated to contribute to stability in Persian Gulf area'.[79]

Standard Oil (NJ), in conjunction with Socony, duly went ahead and entered into formal agreements for participation in Aramco in March 1947. Meanwhile, outline terms for a combined crude contract/pipeline agreement between the Company and Standard Oil (NJ) were agreed and incorporated in Heads of Agreement, signed on 20 December 1946.[80] In a side letter carrying the same date, Socony also came into the scheme, which thus became a tripartite arrangement between the Company and two of the major US oil companies.[81] The crude contracts were later made legally binding when, in September 1947, the full legal agreements were signed by the three parties.[82]

Under the contracts, the Company undertook to supply the two US companies with, in total, 133 million tons of crude oil from Iran and/or Kuwait (at the Company's option) over a twenty-year period commencing on the date the pipeline became ready for operation.[83] Of the total deliverable quantity, 106.4 million tons was to be for Standard Oil (NJ) and 26.6 million tons for Socony. It requires emphasis that these were very large quantities indeed; as an indication of their scale, the tonnage to be delivered in the first year amounted to about 25 per cent of the Company's total crude production, from all sources, when the contracts were signed in 1947. These large volumes came with a high degree of security, it being a contractual term that the US companies could not terminate the agreements except at the end of the tenth or fifteenth year after giving five years prior notice. As for price, the crude was to be supplied on a cost-plus basis, with the Company receiving a margin of 5s per ton over and above the costs of production, royalty and transportation to the Middle East terminal of the proposed pipeline to the Mediterranean. It was a low margin, but one which could be justified by several considerations including, first, the need to increase offtake from Iran; secondly, the benefits which the Company would get from the US companies' participation in the pipeline project; and thirdly, acceptance that the price of Iranian and Kuwaiti crude, delivered to markets west of Suez, had to be competitive with alternative sources of supply. In agreeing to a price which effectively incorporated a discount to make its crude competitive in western markets, the Company took care to ensure that most of the contract crude would indeed go west of Suez by the contractual condition that at least 95 per cent of the crude was to be delivered at the Mediterranean terminal of the proposed pipeline. Only the small balance of up to 5 per cent could be lifted by Standard Oil (NJ) and Socony in tankers in the Persian Gulf. The contract crude was thus to be virtually excluded from eastern markets which the Persian Gulf was geographically better placed to supply. The importance which Fraser attributed to this aspect of the agreements was emphasised early in 1947 when Socony requested freedom to lift more than 5 per cent of the crude in the Persian Gulf. Fraser refused the request on the grounds that it would be 'contrary to whole principle of deal' if the special price agreed for deliveries to western markets were to be applied in the east.[84]

Under the terms of the agreements signed in September 1947 Socony was, as has been seen, to receive much less crude than Standard Oil (NJ). The disparity in quantities was later reduced after Socony asked the Company, in May 1947, whether it could contract for a further 40 million tons over twenty years, for shipment to the USA.[85] According to Snow, Fraser felt that the terms already agreed for a combined crude supply/pipeline agreement had 'secured his prime objective in relation to our concessional obligations towards Iran'.[86] He was not, therefore, 'an anxious seller for further quantities' and did not contemplate making additional sales on the same low price terms.[87] Jackson, on the other hand, was still concerned that oil production in other Middle East countries would expand rapidly, making it difficult to maintain Iran's relative position.[88] His advice was that, having been offered an additional outlet for 40 million tons, the Company should 'go all out to secure it'.[89] Negotiations ensued, terms for the supply of additional crude to Socony were agreed in principle in August 1947 and the legal agreement was signed in March 1948.[90] Known as the Socony II agreement, it was closely modelled on the earlier contract (Socony I) of September 1947 and provided for the delivery of a further 40 million tons of crude to Socony over the same twenty-year term. The main difference between the two contracts was that under Socony II the Company was to receive a higher margin of between 8s 6d and 15s per ton, the exact amount depending on movements in market prices and costs.

It was an integral part of the crude contracts that the Company, Standard Oil (NJ) and Socony were to participate in the proposed pipeline from the Middle East to the Mediterranean. Accordingly, on 23 March 1948 the three companies entered into the Middle East Pipeline Agreement to share in the finance, construction, ownership and use of a pipeline with an annual throughput capacity of about 26 million tons, starting near the frontiers of Iran and Kuwait and terminating at the eastern shore of the Mediterranean.[91] Shares in the pipeline were to be 60.9 per cent for the Company, 24.7 per cent for Standard Oil (NJ) and 14.4 per cent for Socony.

Thus was put together a series of major agreements which offered the Company two very obvious benefits: a large-capacity pipeline which would reduce transport costs for the westward movement of Middle East oil and, above all, an assured high-volume outlet for Iranian crude. On both counts, the Company was to be frustrated. The first setback came when it became apparent that the pipeline would not, as had been hoped, be completed in 1951 because of delays in obtaining steel pipe and in the negotiation of wayleave rights.[92] In 1948 Fraser and Harden discussed the possibility of activating the crude contracts by lifting the crude in tankers from an agreed date, irrespective of the stage which had been reached with the pipeline project.[93] Jackson questioned whether it was in the Company's interest to supply crude at the low contract price without the benefit of the pipeline, but the overriding factor in Fraser's mind, conveyed by Snow, was the 'vital consideration' of 'securing a long-term large quantity outlet to protect the position of Iran in the development of Middle East production'.[94] In April 1949 the Company therefore agreed to commence deliveries under the Standard Oil (NJ), Socony I and Socony II contracts on 1 January 1952 whether or not any progress had been made with the pipeline by that date.[95] Provided that at least 95 per cent of the crude was shipped west of Suez the contract price was to be unchanged.

A much graver upset than the postponement of the pipeline project (which was later abandoned altogether) was the nationalisation of the Company's concessionary interests in Iran in May 1951 and the suspension of the Company's Iranian operations later that year. It was an ironic reversal that the Company, having committed itself to the long-term contracts in order to gain a secure outlet for its Iranian production, found that it had no Iranian crude with which to meet its contractual obligations. It was, of course, not only the long-term crude contracts which were affected by the Iranian crisis. The loss of Iranian supplies disrupted all of the Company's operations and up-ended the traditional imbalance of more production than markets. Instead, the Company faced the opposite problem of having inadequate supplies to sustain its market outlets. Its response to this aspect of the crisis will be described more fully in the next volume of this history.

= 12 =
The Company and the British Government, 1946–1951

In the aftermath of World War II the British economy and the wider international economic environment in which the Company operated were in a state of disarray. The contrasting fortunes of those countries whose economies had been debilitated by the war and those which had survived unscathed resulted in an economic disequilibrium characterised by very large trade imbalances which prevented the early attainment of the US ideal of a liberal international economy, free of restrictions on the flow of trade and the exchange of currencies. The disparity in economic fortunes was felt, not only between victors and vanquished, but also between the western allies whose economic collaboration during the war, symbolised by the Lend-Lease arrangements between the USA and Britain, was accompanied by major shifts in their respective economic positions. The USA emerged from the war as the world's supreme economic power, with unrivalled industrial productivity and capacity. Britain, on the other hand, was virtually bankrupt as a result of its sales of foreign assets and the accumulation of foreign debts to pay for the war effort, the physical destruction and damage of plant and property, the loss of gold and dollar reserves and the great diminution of its exports as the whole of its economic effort was geared to the prosecution of the war. In common with other war-stricken countries, Britain had no alternative but to look to the USA for the supply of goods which were required for economic relief and reconstruction and for the dollars with which to pay for them. Thus, US hopes for the early creation of a liberal international economic order were overtaken by that now-famous manifestation of the prevailing economic imbalance: the dollar shortage.[1]

US aid to alleviate the problems of postwar reconstruction was made available, most notably through the provision of a large loan to Britain under the Anglo-American Financial Agreement of 1945 and, later, through the more ambitious Marshall Plan for European recovery. Yet the dollar shortage remained an acute problem for Britain, which suffered periodic balance of payments crises, the first of which was in 1947 when the Government honoured its commitment under the Anglo-American Financial Agreement to make sterling freely convertible for current transactions. The result was a run on sterling which forced the Government to suspend convertibility after no more than a few weeks of its introduction. Two years later, in 1949, a fresh balance of payments crisis resulted in the devaluation of sterling.

Clement Attlee's Labour Government, having committed itself to policies of full domestic employment and the nationalisation of key industries when it was elected in 1945, thus found itself bound by the economic imperative of increasing exports to earn much-needed dollars to pay for essential imports. In endeavouring to meet its political priorities and to cope with the country's manifold economic problems, the Government, which remained in office until 1951, became more involved in the economy than any previous peacetime administration in Britain. The nationalisation of institutions such as the Bank of England and industries such as coal, gas, steel, electricity and the railways expanded the frontiers of economic intervention by the state. At the same time, many of the wartime controls over the economy were retained or extended in a panoply of measures too numerous to catalogue, but including controls over foreign exchange and imports, the allocation of materials in scarce supply such as steel, and the rationing of consumer goods including petrol, which was not de-rationed until May 1950.[2] As a result, there were few, if any, corners of the British economy which remained untouched by the hand of Government. The Company, not surprisingly in view of the scale and range of its interests, was affected more than most and found itself subjected to policies and controls which tended to be unwelcome, not only because of their complexity, but also because they hampered some of the Company's plans for expansion; threatened to interfere with the allocation of oil supplies and the Company's competitive relationship with its US rivals; and impinged on concessionary relations with the Iranian Government.

GOVERNMENT CONTROLS AND THE COMPANY'S DOWNSTREAM PLANS

An irksome feature of Government controls over the economy was the sea of red tape which then, as ever, characterised the complex procedures of extensive bureaucracy. The administrative obstacles to the speedy execution of the Company's plans for expansion were too numerous and mundane to justify detailed description, but were a cause of recurring frustration to the Company. For example, applications for the import of capital equipment passed through a process of filtration at various levels of the Government's administrative machinery, during which the projects for which imports were required were assessed for their importance to the national interest and, above all, for their effects on the balance of payments. As a result, delay and inconvenience in procuring equipment was very much the norm and the refusal of approval an ever-present concern for the Company. Matters were further complicated by the advent of the Marshall Plan under which most major projects had to be submitted for approval, via the British Government, to the Economic Co-operation Authority (ECA) which was responsible for the administration of Marshall aid in Washington and which, having assessed individual projects in relation to the agreed European recovery programme, either sanctioned or refused the necessary dollar expenditure.[3] The ECA insisted that all projects had to be fitted into a specifically *European* programme agreed by the Organisation for European Economic Co-operation (OEEC) and this, as might be imagined, resulted in some involved negotiations between OEEC members.[4] As an illustration of the type of difficulties which arose, when the Company sought to obtain the necessary documentation to apply for ECA authority for the Grangemouth petrochemicals plant the Board of Trade was thrown into confusion because 'the general description of the plant is "Chemical Plant" for which there is no ECA classification as such'.[5] In this ocean of administration it was hardly surprising that, in July 1948, Jameson complained to Jackson:

> In the old days, when we were not hampered by such controls, it was relatively easy to take decisions on how, when and where units were going to be ordered ... Under the influence of delays due to Government controls, an enormous amount of work has been created in keeping track of the resulting effects on project development ... Contracts Department's negotiations with contractors have been seriously hampered by the time taken by the various Government Departments who handle our Works Scheme Cases ... Under prevailing conditions the whole development of a project is very much haphazard, and in some instances ... we just give our word to contractors to carry on with design and drafting work without any supporting dollars from the P[etroleum] D[ivision].[6]

The Company's contact with the machinery of control was, in short, frequently a tiresome cause of uncertainty and delay.

Notwithstanding routine administrative frustrations, the Company's plans for expansion of production in the Middle East and refining in Britain seemed, at first sight, to be desirable goals for the Government as well as the Company. Before the war Britain had depended on dollar sources for a large proportion of its oil supplies, which had to be paid for in dollars.[7] However, the Company's production, being British-owned, was classified as sterling oil and the expansion of its large production potential offered opportunities for saving dollars by increasing imports of sterling oil. Furthermore, the establishment of a substantial British refining industry had the balance of payments advantage that imports of crude oil were cheaper than those of products. So, the implementation of the Company's plans would help Britain to economise on dollars and improve the balance of payments, an apparently happy coincidence of private commercial objectives and Government policies. In practice, however, the alignment of Company and Government objectives proved to be less straightforward and relations between the Company and the Government were strained.

One of the causes of difficulty was Britain's shortage of steel-making capacity, which pinched the Company's plans for the expansion of its refineries at Llandarcy and Grangemouth and, by the late 1940s, for the construction of a large new refinery on the Isle of Grain in Kent. The Company also intended to erect a petrochemicals plant at Grangemouth. Although all of these projects were enthusiastically endorsed by the Government,[8] they were held up by steel shortages. In September 1947 the steel allocation for the Petroleum Division's approved programme was cut to 800,000 tons although it was acknowledged that this would mean a two-year delay in the construction of additions to the Llandarcy and Grangemouth refineries.[9] In October, the allocation for the fourth quarter of the year was further reduced from 138,000 to 133,000 tons.[10] In late November, Sir Donald Fergusson of the Ministry of Fuel and Power called an emergency meeting of the heads of the main economic departments to discuss the situation, being concerned, according to Sir Edward Bridges of the Treasury, that 'a very large programme of oil development was in hand, but this has been knocked sideways by the steel shortages'.[11] Fergusson was evidently unable to achieve a material improvement in steel supplies, for in early December a Treasury official remarked on the continued complaints from the Ministry of Fuel and Power about reductions in the steel allocations.[12]

The possibility of alleviating the steel shortage by increasing steel imports ran up against that familiar problem: additional supplies were only available from the USA against payment in dollars. Thus, in the short term additional steel imports would widen the dollar gap which the oil expansion programme was intended to close. That dilemma was expressed by Hugh Gaitskell, as Minister of Fuel and Power, in December 1947: 'Oil seems to me to be the biggest potential dollar earner and saver that we have', yet, 'in the short term the effect of the recent curb in the allocation of steel and dollars ... will be to reduce the production by British oil companies'.[13] In the absence of sufficient dollars to compensate for the shortcomings of British supply industries the oil expansion programme was forced into retardation by late 1947.

The effects of steel shortages continued to be felt by the Company in 1948. Against a total requirement for 580,000 tons of steel, the allocation for the British oil industry was 425,000 tons, of which 120,000 was the Company's allocation. According to the Company's Central Planning Department, that amount represented only the bare minimum of what was required and would do no more than allow the Company's projects to continue on a 'skeleton basis'. As a result, the Company's refineries would, the Department argued, be 'inflexible in operation and inadequately provided with tankage and spare equipment'.[14] As an example of further difficulties, the Company had to choose between up-rating Llandarcy or installing a catalytic cracker at Abadan, the steel position making it impossible to proceed with both projects simultaneously and without delay.[15] To provide further details of the effects of the steel shortages would be tedious and repetitious. Suffice to say that many Company projects suffered delays.

Another issue on which there was friction between Government policies and Company objectives was in the field of international marketing. The Government's attempts to control the balance of payments were exercised not only through import controls, but also through efforts to influence the direction of exports. The basic aim was to steer exports to markets which were deemed to have 'hard' currencies and thus to increase Britain's earnings of the most sought-after foreign exchange, which effectively meant dollars or dollar-convertible units. Exports to countries with 'soft' currencies, which were not convertible into dollars, were very much less desirable and the general objective was therefore to hold them in check. The subject was examined by an official working party on oil exports which sat through 1948, its intention being to single out soft currency markets, exports to which were to be reduced, and hard currency markets, exports to which were to be increased.[16] However, this proved to be a somewhat inconclusive exercise and it is difficult to quantify the extent to which the Government's hunger for hard currencies affected the direction of the Company's sales.

A far more evident pattern of Government intervention can be established with regard to the Company's desire to acquire new overseas marketing outlets. Foreign acquisitions involved international transfers of funds, which were subject to Government controls over foreign expenditure and investment. In a number of cases, the Company's plans for overseas marketing expansion ran into difficulties on that score. In October 1947 Fraser complained to Sir Wilfrid Eady of the Treasury that the Company was having difficulty with the Bank of England over its expansion plans in Belgium, Sweden and Switzerland, where the Company's subsidiaries were felt to be in need of new investment funds.[17] The Bank of England, concerned about the amount of capital being tied up in the foreign marketing subsidiaries of British oil companies, had blocked the application for direct investment, arguing that the Company should seek to raise the necessary capital in the markets concerned.[18] The Company strongly resisted that suggestion on the grounds that a resort to local financing would mean losing exclusive control over subsidiaries. That, in the words of Morris, the Company director responsible for continental marketing, was 'most repugnant to the Company and detrimental to their trade'.[19] A month-long argument ensued and in the end a compromise was agreed whereby the Company was allowed five of the eight million Swiss Francs it had requested, with the rest to be raised locally. In Sweden, the Treasury insisted that a ceiling be placed on the Company's circulating credit, while in Belgium permission for an increase in credits was refused, but the capital of the Company's subsidiary there was allowed to be increased.[20]

When, in 1948, the Company sought to acquire a direct outlet in Greece by purchasing Steaua (Agencies) Ltd, no such compromise with the Government proved possible. In that instance, the Company warned that a failure to purchase Steaua might result in the US oil company, Caltex, snapping up the outlet.[21] However, the British authorities were unmoved, feeling that an increase in oil supplies to Greece was 'not likely to be attractive for some time to come, since the Greek currency seems likely to remain very soft for as long as can be seen ahead'.[22] In the event, the Company succeeded in circumventing the Government's refusal to sanction the purchase by arranging for Steaua to be purchased by a share swapping arrangement with the Company's French subsidiary, SGHP.[23] Although Company sources suggest that this was done with the Treasury's 'tacit approval',[24] a contrary impression is given by the remark of a Treasury official in March 1950 that the Company's acquisition of Steaua Agencies went ahead 'much to our disgust'.[25]

In the meantime, complicated manoeuvres took place on a scheme of much grander design: the possibility of allying the Company, whose upstream production capacity greatly exceeded its marketing outlets, with Royal Dutch-Shell, which had a very extensive downstream marketing network. That idea, which had the magnetic appeal of a classic complementary association, had already come up in 1943–5 (see chapter 11) and was raised again after the end of the war. In December 1946 the Company noted, after a meeting with the Ministry of Fuel and Power, that the Government was anxious to avoid authorising one company to spend dollars on purchasing oil to feed its marketing outlets if another British company was in a position to supply its needs, a clear reference to Royal Dutch-Shell and the Company respectively.[26] A few months later, the matter surfaced more explicitly when, in mid-1947, the Company applied for permission to establish a marketing and distribution company in Finland. At a meeting with senior Ministry of Fuel and Power officials in June that year, Fraser was told that the Ministry found it 'extremely difficult' to support the Finnish proposal, prompting Fraser to ask for guidance on what constituted an acceptable expansion of the Company's marketing business. The Ministry officials replied that the national shortage of currency and resources made it a very awkward time for expansion and that it would be in the national interest if some arrangement could be made between the Company and Royal Dutch-Shell whereby the Company would gain secure outlets for its increasing production and Royal Dutch-Shell would gain the quantities of oil in which it was deficient.[27] Fraser's reaction was that he would be happy to reach an agreement with Royal Dutch-Shell, but previous negotiations had been broken off by that company who, in Fraser's words, were 'very hard bargainers ... [and] were inclined to introduce restrictive clauses on future AIOC activities'.[28] Continuing, Fraser stated that it would be impossible for him to approach Royal Dutch-Shell if the Government was seen to be blocking the Company's expansion, which would give the impression that the Company was desperate for marketing outlets. He therefore requested that the Finnish acquisition be authorised on the grounds that it would strengthen the Company's hand in negotiations with Royal Dutch-Shell.[29] The request was refused, though Fraser was instructed to keep the option to acquire the outlet open for one month.

On the same day, the Ministry officials also met with an executive of Royal Dutch-Shell, who maintained that the Company's expansion of outlets was a 'wasteful duplication', to which the Government officials replied that the Company could not be denied outlets for its production.[30] Their aim was crystal clear, namely to give the Company the impression that it would *not* be allowed to expand whilst giving Royal Dutch-Shell the impression that the Company *would* be allowed to expand. By taking this line it was hoped, in the words of one Government official, to create 'the right atmosphere for constructive talks between the two companies'.[31] On being informed of the scheme, a Treasury official informed two of his colleagues that the 'solution about the Finnish deal is ingenious ... [but] I would press you *not* to agree to its consummation unless we are really satisfied that it has been successfully used as a pawn in the bigger game. We really cannot afford another £300,000 in this way'.[32] As it happened, the Government's angling came to nothing: the agreement between the Company and Royal Dutch-Shell for which officials had plotted did not materialise and the Company was not granted permission to acquire the Finnish outlet. Nevertheless, the idea of bringing the two companies together continued to linger in Government circles, being raised from time to time between 1948 and 1950, without result.[33]

THE COAL/OIL CONVERSION CAMPAIGN

Of the various shortages which afflicted the British economy after the war, the lack of coal was second in importance only to that of dollars. During the war, when miners were called up to the armed forces, Britain's output of deep-mined coal fell year by year. At the same time, there was a fall in labour productivity as the proportion of older workers engaged in the industry increased and maintenance was deferred on account of wartime stringencies. The British coal industry was not, therefore, well placed to meet the country's energy requirements after the war. Faced with the danger that coal supplies might be insufficient to meet demand, the Government launched a campaign for the substitution of oil for coal in British industry.

The campaign commenced in April 1946, when Emanuel Shinwell, the Minister of Fuel and Power, informed the House of Commons that industry would not, for some time to come, be able to rely on adequate supplies of coal.[34] A few days later, Sir Andrew Agnew of the Petroleum Board assured Shinwell's Ministry that the Board was trying to prepare for an increased domestic demand for fuel oil in order to assist industry in view of the tight supply of coal.[35] The matter was of sufficient concern to receive the personal attention of the Prime Minister, Attlee, who sent a minute to the Minister of Supply in June, stating that in view of the 'paramount necessity' of increasing coal production he had asked the Minister of Labour to intensify efforts to increase the intake of new manpower into the coal industry.[36] He also, a few days later, asked Shinwell:

> What progress is being made with substituting the use of fuel oil for coal by big industrial consumers?. . . The effects on our whole economy of a shortage of coal this winter would be so serious, that a determined effort must be made to overcome all obstacles.[37]

In reply, Shinwell explained that the supply and installation of the necessary combustion equipment was a constraint on conversion from coal to oil, as was the cost of fuel oil which was more expensive than coal, being subject to an import duty of £1 per ton.[38] The latter constraint was duly removed in August, when the Chancellor of the Exchequer, Hugh Dalton, announced that he intended to remove the duty in his next budget, but that meanwhile a subsidy of £1 per ton would be paid to users of fuel oil.[39]

The coal/oil conversion campaign did not, however, run smoothly. Although Shinwell announced publicly in August that there would be no fuel crisis,[40] in October he had to admit to the House of Commons that his Ministry's estimates of the coal savings to be effected by switching to fuel oil were way out.[41] His Ministry made an upward revision of its estimates of coal consumption,[42] but was unprepared for the exceptionally cold weather of late January and February 1947, which precipitated a major fuel crisis. As Dalton's biographer has described, at midnight on 28 January Big Ben struck once and then 'symbolically, fell silent'.[43] The next day, the river Thames froze at Windsor and ice-floes formed in Folkestone harbour. By 5 February coal pits were blocked by snow and colliers were unable to leave port. Two days later Shinwell announced to the House of Commons that because of the coal shortage all electricity would be cut off from industry in London, the south-east, the midlands and the north-west, and from all household consumers for five hours a day. Unemployment rose to more than two million and there was a large loss of export production.[44] To private recriminations between ministers were added public criticisms of the Government in the press.[45] It was, in the words of Sir Alec Cairncross, who was Economic Adviser to the Board of Trade at the time, 'a striking example of incompetence in industrial planning by a government dedicated to economic planning'.[46]

As might be expected, one of the effects of the fuel crisis was to put added emphasis on the coal/oil conversion programme. The mood of official thinking in the Ministry of Fuel and Power in early March was conveyed by a memorandum entitled, 'Need for intensifying the scope of the coal oil conversion scheme and accelerating the rate of conversions thereunder'.[47] The 'prime object', according to the memorandum, was 'to secure the maximum practicable substitution of oil for coal both by public utilities and by industry'.[48] By late March the Petroleum Board was reporting that new requests for supplies of oil were 'positively pouring in', to which was added the disquieting comment that in many cases the applications were 'the "most uneconomical types", such as conversions of steam boilers from coal to oil in or adjacent to coal fields and similar uneconomic propositions'.[49]

Yet, despite the added impetus for conversions from coal to oil, there were severe limits to the substitution programme. As has been seen, steel was in short supply, both for the expansion of refining capacity and for the manufacture of oil burning equipment. Furthermore, the shortage of dollars was unrelenting, as was spectacularly demonstrated in July–August 1947 when, as mentioned, the Government lifted restrictions on the convertibility of sterling for current transactions only to be faced with a run on the pound which resulted in the early suspension of convertibility. Thus the attempt to overcome one shortage, that of coal, was beset by other shortages. A manifestation of that type of problem was the acute lack of rail wagons which, by the middle of 1947, was threatening to cripple the coal/oil conversion scheme. In August, Fergusson (Ministry of Fuel and Power) commented on the alarming shortage of wagons and pointed out: 'We dare not sacrifice coal wagons for the sake of putting more tank wagons on oil. Yet without the full programme of tank wagons we cannot distribute the oil and must reduce the oil conversions savings'.[50] Already, the Prime Minister had again become involved, sending a minute to Shinwell on the fuel outlook for the coming winter. With stocks of oil products falling, Attlee was concerned to know whether the requirements of the conversion scheme could be met.[51]

The Government's concern about the low level of Britain's stocks of oil products brought to the surface a highly sensitive issue in Government–Company relations. The crux of the matter was whether the Government could successfully press the Company into reducing oil supplies to other countries in order to maintain supplies to Britain at a time of shortage. This raised a very fundamental question: could the Company, as a *commercial* organisation, be *politically* pressured into accepting responsibility for discriminating between countries in the supply of an essential good?

The Government's desire to see oil products diverted from other European countries to Britain was made clear in a series of meetings starting in the first half of August 1947 between Government officials, Royal Dutch-Shell and the Company, represented by Snow.[52] Although Royal Dutch-Shell was involved, the shortfall in British stocks was attributed mainly to the Company, which was having difficulty maintaining deliveries owing to its acute shortage of tanker tonnage at the time. Royal Dutch-Shell was not suffering the same difficulties and it was understandably the Petroleum Division's view that Royal Dutch-Shell could hardly be expected to reduce its supplies to other countries in order to make good a deficit of British stocks attributable to the Company.[53] It was those factors, rather than the Government's position as the Company's major shareholder, which made the Company the main target of Government pressure in this affair. Indeed, at no stage in the meetings did Government officials so much as mention the Government's shareholding.

By late August Snow was sufficiently concerned about the Government's requests for a diversion of supplies to Britain that he discussed the Company's position with the deputy chairman, Heath Eves. They agreed to tell the Government that the Company could not possibly voluntarily agree to reduce supplies to its European markets to make up the additional quantities required for Britain. Such drastic action would, they thought, be likely to have the most serious repercussions, political as well as commercial, on the Company's position in the countries from which supplies were diverted. Heath Eves and Snow were also aware that if the Company acceded to Government requests for a reduction in supplies to other markets it would undoubtedly find itself involved in legal actions for breach of contracts to overseas customers. If, on the other hand, the Government issued a formal directive which left the Company no choice, then the Company would be able to plead *force majeure* as grounds for breach of contractual obligations. On 28 August, Snow put the Company's agreed line to Victor Butler of the Petroleum Division, but Butler was unrelenting. He continued to press for an increase in Company deliveries to Britain and said he was doubtful whether the Government could issue a directive which would effectively create a condition of *force majeure* in relation to obligations in foreign countries.[54]

With the Company and the Government in deadlock, and the shortage of rail tank wagons a continuing constraint, doubts were inevitably cast on the feasibility of the coal/oil conversion campaign, as was admitted by Butler on 18 September when he informed Snow that he would be telling an inter-departmental meeting that 'it appeared unavoidable to restrict the Fuel Oil conversion programme'.[55] However, the Government's pressure on the Company to increase fuel oil deliveries to Britain did not let up. In the last week of September, Butler told Snow and his Royal Dutch-Shell counterpart that fuel oil stocks at the year-end were estimated at only 300,000 tons, an 'impossible position'. He requested the two companies to arrange for the arrival in Britain of a further 200,000 tons of fuel oil by the end of the year so that estimated stocks would be raised to 500,000 tons. In reply, Snow said that the Company had already taken 'very considerable risks' in reducing shipments to other markets and that no more could be done except at the risk of a 'complete breakdown in other areas' or by stopping shipments of crude oil to the Haifa refinery.[56] Far from being able to meet Butler's request for an extra 200,000 tons of fuel oil, the most that the Company could offer was 50,000 tons, while Royal Dutch-Shell 'found themselves unable to increase their shipments to UK in the absence of the protection of a specific [Government] directive'. When, early in October, Butler expressed his concern at the situation, Snow pointed out the inescapable truth that if the Company tried to do any more by reducing its crude oil shipments to refineries it would merely be 'filling one hole by digging another'. At the same time, the disquieting possibility was raised that the Government might refuse voyage licences for British tanker tonnage scheduled for other countries.[57] That unwelcome prospect for the Company became reality in mid-November when Butler called Snow and a Royal Dutch-Shell representative to a meeting at which they were told that the Government insisted that, in addition to the quantities already programmed, the two companies between them had to bring in ten more cargoes of fuel oil by the end of the year. To secure the additional cargoes, the Government would, if necessary, refuse voyage licences for ships programmed to other destinations or cancel licences for vessels already in voyage and substitute new licences requiring the vessels to come to Britain. In response to that ultimatum, under which the Company and Royal Dutch-Shell were each requested to find five of the cargoes, the company representatives stressed that such emergency measures were 'just borrowing on the future' and suggested that the coal/oil conversion programme should be retarded.[58]

In the face of continued Government pressure the attitude of the two oil companies was visibly hardening, as was apparent from the strong language used in recording a meeting with Government officials in the last week of November, when Snow and his Royal Dutch-Shell counterpart agreed to provide information for a paper on oil supplies which Butler was preparing for Ministers. They made it clear that they were not assenting to any suggestion that cutting out shipments to the other countries concerned was proper action, and that 'there would be a great deal more to be said about this if, in fact, this came to be put forward as the action which the Companies should take (which they certainly could not do on their responsibility)'.[59] After the meeting, Snow wrote to Butler expressing his concern about the effect which a withdrawal of supplies would have on some of the Company's European marketing subsidiaries. In the cases of Belgium, Holland, Denmark and Switzerland, he pointed out that the consequence would be:

> that our companies would either have to go out of business or alternatively to restrict their business so severely as for all practical purposes to have the same effect. That this would be disastrous is too obvious to need elaboration, and we would never ourselves contemplate throwing on to a few areas, to the destruction of local companies which represent years of effort, so disproportionate a share of what is a general burden.[60]

On being asked by Butler how, from a commercial standpoint, they would deal with a supply deficit arising from a tanker shortage, the Company and Royal Dutch-Shell replied that they would aim to 'spread the cut over all areas', paying regard to the various factors affecting each area such as political reactions, the protection of their investments and particular commitments. They were adamant that they 'would not, without Direction, reduce shipments to any country to an extent which put our subsidiary out of business or effectively so'.[61]

On 19 December the matter was discussed at a higher level when Hugh Gaitskell, who had succeeded Shinwell as Minister of Fuel and Power, met oil industry representatives including Fraser and Snow. Gaitskell had requested the meeting in order to discuss the oil position which was causing him 'some considerable anxiety', the immediate problem being a forecast deficit of 1.3 million tons of oil supplies in the first quarter of 1948. The Cabinet had discussed the problem the previous day and had agreed to 'some deferment' of the coal/oil conversion programme, but Gaitskell nevertheless wanted the oil companies to make plans under which oil supplies to Britain and the Commonwealth would have priority over other markets. Fraser was characteristically straightforward in his response. He said that apart from the immediate problem, in the long term it would be very difficult to meet Britain's oil requirements if the coal/oil conversion programme went ahead, to which he added that a number of the conversions were economically unsound. He therefore suggested that the coal/oil conversion scheme 'should be closed immediately'.[62] On the penultimate day of 1947 he reiterated his views in a letter to the Ministry of Fuel and Power.[63]

Fraser's views, in isolation, could not realistically be regarded as a decisive influence, but his hopes for a halt to the coal/oil conversion programme were shortly to be realised. By 1948 it had become inescapably clear that supply inelasticities were such that a shortage of one essential good, in this case coal, could not simply be side-stepped by switching to alternatives. In the first half of 1948 the Government, no doubt with some embarrassment, undid its earlier plans for coal/oil conversion and effectively abandoned them.[64] As a Government official was later to recall: 'We were very badly bitten when at the time of the fuel crisis we sponsored conversion from coal to fuel oil ... It was a nasty blow to planning'.[65]

The end of the coal/oil conversion programme did not, however, signal a return to commercial freedom for the Company. On the contrary, it soon found itself dragged into a new episode in which the Government again sought to overcome a critical shortage in one area by finding supplies from another. In this case, the critical shortage was of dollars, which the Government sought to remedy by substituting oil from sterling sources for oil from dollar sources.

THE CURRENCY CRISIS: STERLING OIL AND DOLLAR OIL

In the first three years after the war oil was in such short supply that dollar oil commanded a high priority for importing countries' scarce hard currency. US oil companies such as Standard Oil (NJ) were therefore able to make shipments to European markets largely unhindered by the near-universal shortage of dollars.[66] Moreover, with the coming of Marshall aid, the USA made dollars available for European purchases of dollar goods, including oil which accounted for more than 10 per cent of Marshall aid expenditure, more than any other single commodity. The aid not only helped to keep Europe supplied with energy, it also assisted US oil companies in retaining markets at a time when their customers would otherwise have been unable to obtain the currency which was needed to pay for dollar imports by US companies.[67]

However, by 1949 the world oil shortage was turning into a surplus and, notwithstanding Marshall aid, the British Government turned to the idea of saving dollars by displacing dollar oil with sterling oil. The policy of substitution threatened to upset the relative market positions of the international oil companies because the distinction between sterling and dollar supplies was made according to the nationality of the supplying company. Thus, oil supplied by US oil companies was classified as dollar oil, whereas supplies from the Company and Royal Dutch-Shell (which was granted the status of a sterling company despite being largely Dutch-owned) were classified as sterling oil. To give an example of the results of that method of definition, the US companies' liftings of the crude oil produced by the Iraq Petroleum Company were dollar oil, while the liftings of the Company and Royal Dutch-Shell from the same source were sterling oil.[68]

British measures to substitute sterling for dollar oil were initiated in the first half of 1949 when bilateral barter arrangements were made with Egypt and Argentina, providing for the supply of sterling oil to those countries in exchange for imports to Britain.[69] It might be expected that such measures to displace dollar oil would have been welcomed by the Company inasmuch as they opened up new sales opportunities for its expanding production. In fact, the Company took the opposite line and sought to distance itself from the Government's policy of substitution, which not only threatened to harm the Company's relations with US oil companies, but also raised, once again, the vexed issue of Government limitation of the Company's commercial freedom.

The Company's views were expressed early in July, when Snow and John Berkin of Royal Dutch-Shell saw Butler at the latter's request. Butler explained that the dollar expenditure of the Sterling Area (essentially, but not exclusively, the UK, the Dominions and the Colonies) was causing the Government the 'gravest concern'. As payments for oil supplied by US oil companies constituted a large item of dollar expenditure there was, Butler explained, to be a drive to displace dollar oil with sterling oil. He recognised that this would involve 'squeezing out the US Companies' and added, no doubt to Snow's disquiet, that it might become necessary – in order to make additional supplies of sterling oil available – to ask the British companies to divert oil from other markets. In view of this situation the

Petroleum Division was not willing to authorise the British companies to take on additional commitments in foreign markets on other than a short-term basis, as it wished to keep the situation as flexible as possible.[70] The immediate reaction of Snow and Berkin was that the Government should explain the situation to the US oil companies since 'otherwise there would be resentment and resistance and possibly even the inference that this situation was ... the result of pressure on HMG by the British companies seeking outlets for their increasing production – which would be very harmful to relations and would be likely to cause a great deal of trouble in the markets'.[71] The following month, Snow and Berkin reiterated their concern that proposals for the displacement of dollar oil should be seen to come from the Government and not from the British oil companies, who might otherwise be accused of trying to steal markets from their US rivals.[72]

However much the Company and Royal Dutch-Shell sought to distance themselves from the policy of substitution, and whatever their reservations about the diversion of oil supplies, the Government went ahead and, after official talks in Washington, informed the US State Department in November 1949 that it intended to begin substituting sterling for dollar oil in the Sterling Area in January 1950.[73] The reaction in the USA was bitter.[74] Amidst protests from the US oil companies, supported by the State Department, that the displacement of dollar oil represented unwarranted discrimination and an attempt to oust US oil companies from oil markets, the British Government delayed the implementation of the substitution policy to mid-February 1950 and offered US oil companies 'incentives' to retain sales which would otherwise be lost through substitution. The incentives took various forms including, for example, allowing US oil companies to sell oil for sterling if they agreed to make purchases for dollars in the Sterling Area.[75] Further discussions in the winter and spring of 1950 failed to produce an overall settlement of the matter, but a series of individual agreements with US oil companies, coupled with an upturn in economic activity and oil consumption later in the year, effectively brought the controversy over substitution to an end.[76] The remnants of substitution nevertheless lingered on into 1951, a continuance which was not welcomed by Fraser. His dislike of the Government's interference in oil markets was lucidly spelt out in a letter to Fergusson at the Ministry of Fuel and Power in November 1950. In Fraser's words, which deserve to be quoted at length:

> The outlet on which the [substitution] policy has involved using our oil is one which we would rather have been without – there is neither

goodwill nor continuity in business directed this way. The arbitrary absorption of our margins [i.e. marginal production] is an impediment to the progressive development of our business on the only sound basis of freely acquired and continuing outlets ... It is very much in the Company's interest that we should be relieved of the potential obligation to provide oil for Substitution, and ... it would be much better for us to be left to find customers ourselves.[77]

Fraser's obvious frustration with the Government's interference in the Company's trade was not without justification. His complaints about the lack of continuity and the arbitrary nature of the business which came the Company's way as a result of Government policies were strongly founded, it being a shared characteristic of the coal/oil conversion campaign and the substitution policy that they were short-term, temporary measures which distorted the Company's development plans and impeded its expansion in more durable directions. Of course, the Government was itself subject to severe constraints, but from a Company perspective its policies were open to the criticism commonly levelled at the economic policies of British governments: they put short-term expediency ahead of long-term economic development.

THE GOVERNMENT'S FISCAL POLICY

While the British Government's economic policies and controls interfered with the Company's commercial freedom, British fiscal measures also impinged on the Company's relations with the Iranian Government. The reverberations of British fiscal policy in that area of the Company's affairs are best described in the context of the concessionary negotiations which took place in the postwar years, as recounted in later chapters. Suffice to say, for present purposes, that there were two aspects of fiscal policy which particularly affected the Company's concessionary discussions with the Iranian Government. One was, quite simply, the increase in corporate taxation levied by the British Government after the introduction of Excess Profits Tax during World War II. Although that particular wartime measure was first reduced and then repealed in 1945–6, it was not long before a new Profits Tax was introduced in the budget of April 1947, the new tax being doubled in the autumn budget that year.[78] As a result of the increased British taxation of corporate profits the British Government's tax revenues from the Company came to exceed Iranian royalties by a substantial margin, as can be seen from table 12.1. Not surprisingly, there was

Table 12.1 *Iranian royalties and taxation versus British
taxation of Company, 1932–1950 (£millions)*

	Iranian royalties and taxation	British taxation
1932	1.53	0.19
1933	1.81	0.46
1934	2.19	0.77
1935	2.22	0.40
1936	2.58	1.17
1937	3.55	2.61
1938	3.31	1.69
1939	4.27	3.32
1940	4.00	4.16
1941	4.00	3.28
1942	4.00	6.60
1943	4.00	12.07
1944	4.46	15.72
1945	5.62	15.63
1946	7.13	15.59
1947	7.10	16.82
1948	9.17	18.03
1949	13.49	16.93
1950	16.03	36.19

Notes: 1. Iranian royalties and taxation in 1930–2 include the retrospective settlement for those years in the 1933 concession agreement; and for 1939–43 they include the make-up payments shown in table 9.1. However, the payment of £1 million made in June 1933 in full settlement of past claims (under Article 23 of the 1933 concession agreement) is excluded.
2. The figures for British taxation given in this table are not comparable with those in tables 1.2, 8.5 and 10.3, which show the tax deduction from the consolidated accounts of the group rather than the actual amounts which were eventually paid by agreement with the British tax authorities.

Sources: BP 4305, Schedules of royalty and taxation payments to the Iranian Government; BP 4308, The case of the AIOC's concession, Memorial (Part 1) submitted by the Government of the United Kingdom of Great Britain and Northern Ireland; BP 9233, UK taxation liabilities on accounts for years 1932 onwards, 20 September 1951.

a strong sense of grievance in Iran at the apparent inequity of the figures.

A second factor in British fiscal policy which was inimical to Iranian interests was the limitation of company dividends. The policy did not, it has to be said, have statutory force, but was urged upon British companies by the Government, anxious to win trade union backing for voluntary wage restraint. Although the Federation of British Industries agreed to recommend dividend restraint to its members, the Labour Government was not satisfied with the results of its voluntarist approach and announced, in the summer of 1951, that it intended to introduce a Bill to make dividend limitation compulsory. In fact, the Bill never materialised, being overtaken by the general election in the autumn, when a new Conservative Government was elected.[79] The announcement that dividend limitation would be made compulsory therefore came to nothing, though Dalton was reported as saying it had 'thrown the Stock Exchange into complete disorder, and that is good fun anyway'.[80] Fraser and the Iranian Government doubtless saw dividend limitation in a less humorous light. Although the restraint was technically voluntary, the Company could not, as one of Britain's most prominent businesses, realistically break ranks with Government policy. Yet its adherence to British policy affected the income of the Iranian Government, whose revenue from the Company was derived in part from a payment representing an agreed percentage of the Company's dividend to shareholders. By pegging that payment and levying taxes on the Company which exceeded Iranian royalties British fiscal policy helped to fuel Iranian grievances about the distribution of the income which was derived from the Company's operations.

FRASER AND THE GOVERNMENT

Relations between the Company and the Government revolved not only around matters of policy, but also personalities. Of particular importance in the postwar years was the dissatisfaction expressed in Government circles about the chairmanship of Fraser. As was mentioned in chapter 8, Fraser's accession to the chairmanship in 1941 was frowned upon in Whitehall, where Sir Horace Wilson, then head of the Civil Service, tried to prevent it.[81] Disapproval of Fraser did not stop there. In February 1945 Eady noted, after a meeting with Fraser and Barstow, that he had reservations about the Company's board and particularly about its chairman. The Company, Eady noted, has always been a dictatorship, but Cadman was a dictator of another order from Fraser. The responsibility of Fraser is too heavy; there is no other oil man on his Board comparable in quality with the best of the Shell men, or the best of the Americans.[82]

Having pointed to what he saw as weaknesses in the Company's management, especially in the sphere of international negotiations, Eady went on:

> Inevitably this ... suggests some doubts about Sir William Fraser personally. The judgment of some other people who have seen more of him than I, is that he is not quite of the quality necessary for this immensely important British institution.[83]

Eady communicated his disquiet to Bridges in July, indicating that both he and the Bank of England were disturbed about 'the poor quality of the Anglo-Iranian Board'. The next month, he suggested that 'we must ... consider as a matter of urgency the reconstruction of the Board of this immensely valuable asset'.[84]

Although the above remarks stemmed from Eady, he was not isolated in his opinions. Not only did his correspondence with Bridges indicate that the Bank of England agreed with his views, but senior civil servants in other departments were also of like mind. For example, in August 1946 an official at the Ministry of Fuel and Power noted that Fergusson, the Ministry's Permanent Secretary, thought that the Company

> Is far less efficiently run than the Dutch and American concerns. The Board of Anglo-Iranian is weak and is left largely in ignorance of what is happening ... Fraser keeps the management of the show and the policy very much in his own hands. He compares very unfavourably with, for example, Godber [of Royal Dutch-Shell]. In Fergusson's view, Fraser ought to be replaced and the Board ought to be strengthened.[85]

Five years of intermittent sniping saw no improvement in relations, for by July 1950 Bridges was noting, after a meeting with Sir Thomas Gardiner, that Fraser 'is a complete totalitarian and does very little in the way of general consultation with his colleagues'.[86] Too many of the other members of the board were, Bridges thought, departmental specialists and he doubted whether there was a single man on the board who was competent to succeed Fraser. Steps ought, he thought, to be taken to appoint younger men as directors.

In the light of this litany of criticisms it is relevant to consider on what grounds the Government disapproved so strongly of Fraser and held a low opinion of the Company's board. To some degree, Fraser was almost bound to suffer by comparison with his predecessor, Cadman, who moved with such ease between the worlds of business and government. Indeed, according to Bridges, Cadman 'used to spend all his time in Warren Fisher's room discussing everything on the face of the earth'.[87] Fraser, on the other hand, was more a straightforward oilman than a businessman cast in the statesman mould. Moreover, he was, according to Gardiner, 'slightly resentful' about the Government's attitude towards him.[88] Bridges, too, alluded to the sense of slight harboured by Fraser, noting in February 1951 that an official of the Foreign Office had treated Fraser a 'little brusquely and had not gone out of his way to ... treat Fraser as the great man whom he feels himself to be'.[89]

Apart from Fraser's personal characteristics, there were also matters of business on which the Government had doubts about the Company. In particular, there were two areas in which the Government felt that the Company's performance needed to be improved. The first of these was the marketing side of the Company's operations in which, as was noted in chapter 11, the Government had strong doubts about the Company's commercial acumen. The second was in the sphere of international relations, on which Gardiner commented in October 1950 that 'the weakest point in the Anglo-Iranian set up was on the side of the Company which dealt with Foreign Relations'.[90] Gass, who was the director responsible for that area, was felt to be 'weak'.[91] Given the growing importance of the Company's relations with the Iranian Government in the postwar years, it could ill afford any shortcomings in that field.

= 13 =
Politics and joint interests in the Middle East, 1946–1951

In the Middle East, as elsewhere, the political and economic environment in which the Company operated was greatly affected by events during World War II and its aftermath. Underlying the temporary dislocations caused by the war, fundamental changes were taking place in the international oil industry, whose centre was shifting increasingly to the Middle East. Changes of great magnitude also occurred in relations between the great powers in the region. The military defeat of Germany and Japan, the devastation of France and Italy and the exhaustion of Britain left the USA and the Soviet Union as the two great world powers, which became locked in the mutual antagonism of the Cold War. The ascendancy of the USA as the main western power was coupled with growing US interest in the Middle East, both on account of its oil and because of the area's strategic importance as a vital link in the attainment of the USA's overriding foreign policy objective: to contain the spread of communism. The diminishing power of Britain, which had previously been the dominant western influence in the region, would, it was feared, leave a vacuum in the Middle East and weaken resistance to communism unless the region could be shored up, politically and economically, as a bulwark against the assumed threat of Soviet expansion. Economically, the oil-producing states of the Middle East were heavily dependent on revenues from oil and the Company, with its great oil interests in the region, was caught up by forces which, though largely outside its control, impinged directly on its business.

THE USA, OIL AND THE MIDDLE EAST

In November 1943 Harold Ickes, the US Secretary of the Interior and Petroleum Administrator for War, sent an exploratory oil mission to the Middle East. The survey team, headed by the well-known geologist, Everett DeGolyer, reported early in 1944 that 'the centre of gravity of world oil production is shifting from the Caribbean area to the Middle East – to the Persian Gulf area'.[1] DeGolyer estimated that the oil reserves of the Middle East were greater than those of the USA, which was traditionally the world's leading oil producer and exporter, and concluded that with much Middle East territory still to be explored 'reserves of great magnitude remain to be discovered'.[2]

DeGolyer's report reflected, and added to, the growing US awareness of the oil potential of the Middle East and the strategic importance of having access to the region's oil reserves. Already, by the time of the DeGolyer mission, the great wartime demands on US oil production had caused Americans to fear that the depletion of irreplaceable domestic reserves would result in the loss of self-sufficiency in oil. Recognising the strategic value of conserving US and other western hemisphere oil, Americans looked increasingly to the Middle East for future supplies. Their attention focused particularly on Saudi Arabia, where the concession covering the al-Hasa area in the eastern part of the country was held by the California Arabian Standard Oil Company (Casoc), a joint subsidiary of two US companies, the Standard Oil Company of California (Socal) and the Texas Company.

Although Casoc discovered oil in Saudi Arabia in 1938, the US Government was not initially inclined to be drawn into granting direct aid to King ibn Saud to help make good his loss of revenues from pilgrimages and customs duties after the outbreak of World War II. However, by 1943 opinion in the USA was changing and in February that year President Roosevelt agreed to extend Lend-Lease to Saudi Arabia. Later that same year, the interest of the US Government in Saudi oil was more bluntly expressed in the formation of the Government-owned Petroleum Reserves Corporation which, under the forceful leadership of Ickes, endeavoured to acquire a direct stake in Saudi oil reserves by purchasing a controlling interest in Casoc. However, terms could not be agreed and the negotiations came to nothing, as did Ickes' subsequent idea of constructing a Government-owned trans-Arabian pipeline to transport Saudi oil to the Mediterranean.[3]

Although Ickes failed in his interventionist schemes for involving the US Government in the ownership of Saudi oil reserves or in the funding of a trans-Arabian pipeline, Middle East oil remained a subject of official US interest. Early in December 1943 the US Secretary of State, Cordell Hull, with whom Ickes was constantly battling for control of US foreign oil policy, invited the British Government to participate in Anglo-American discussions on the future of Middle East oil. The British accepted in early February 1944 and, after Roosevelt and Churchill had exchanged assurances to lay to rest mutual suspicions about each other's intentions in the Middle East, the British delegation, which included Fraser, travelled to Washington to meet their US counterparts in mid-April 1944.[4] Early in May the two sides agreed to a Memorandum of Understanding which propounded a set of principles designed generally to provide for equal, non-discriminatory access to, and 'orderly' development of, international oil reserves outside the USA. More specifically, it was proposed in the Memorandum that adequate oil supplies should be made available to all peaceable countries at fair prices and without discrimination. The producing countries should, went the Memorandum, receive benefits which would encourage their 'sound economic advancement'. In the acquisition of exploration and development rights in areas not already covered by concessions, the principle of equal opportunity was to be upheld. Moreover, subject to considerations of military security, oil operations from exploration to marketing were not to be hampered by restrictions imposed by either of the signatory governments or their nationals. Allied to these principles, which upheld the US goal of establishing a liberal international economy after the war, the Memorandum proposed the establishment of an International Petroleum Commission to prepare long-term estimates of world oil demand and to recommend how that demand might be satisfied by apportioning production among the various producing countries. Although the Memorandum was binational, and the Commission was initially to consist only of US and British members, both sides agreed to work towards an international agreement so that the principles of the Memorandum could be more widely adopted by producing and consuming countries. The objective, in short, was a multilateral agreement which would provide for the regulation of international oil production free from nationally restrictive policies on the one hand and an uncontrolled competitive scramble on the other.

These ideas were taken further in July–August 1944 when a second Anglo-American conference on oil was held in Washington, this time at ministerial level. The British delegation was led by the irascible, energetic and chauvinistic Lord Beaverbrook, the Lord Privy Seal who had already described the proposals contained in the Memorandum of Understanding as a 'monster cartel'.[5] The talks did not go smoothly and indeed nearly collapsed because of British insistence that Britain should have the right to keep out imports of dollar oil because of the precarious foreign exchange position which the country would be in after the war. The Americans, on the other hand, argued that such a move would contravene the all-important principles of equal access and non-discrimination. Eventually, a compromise was reached by noting the position in the official minutes and on 8 August 1944 the Anglo-American Petroleum Agreement was signed by Beaverbrook for Britain and Edward Stettinius, acting Secretary of State, for the USA. Basically, the Agreement incorporated the same principles as had earlier been set out in the Memorandum of Understanding.[6]

While the Agreement was significant as a sign of the US Government's growing interest in Middle East oil, it was never put to a practical test. Fears that it would result in increased federal control over the oil industry fuelled strong opposition from domestic US oil interests and in Congress, so that the Agreement failed to win Senate ratification. Without it, and in the absence of any other inter-governmental accord, it was left to the oil companies to work out their own arrangements in the Middle East. It was not the kind of task to which they were strangers. Between the world wars the major international oil companies had entered into various alliances, agreements and arrangements which had the effect, for better or worse, of concentrating control of internationally traded oil in the interlinked hands of a few very large enterprises. The Company was one of them and, as has been described in previous chapters, participated in joint marketing companies with Royal Dutch-Shell in the Consolidated area and Britain; in the Burmah-Shell Agreement concerning marketing in India; in the 'As Is' Agreement reached at Achnacarry, Scotland, in 1928; and in the Kuwait oil concession, which it shared with Gulf Oil. No less important, the Company was also a signatory of the Red Line Agreement of 1928 under which the members of the Turkish Petroleum Company (TPC) – renamed the Iraq Petroleum Company (IPC) in 1929 – agreed not to enter into concessions except through the TPC in an area corresponding to the old Ottoman Empire and encompassing Turkey, Iraq, Syria, Lebanon, Transjordan, Palestine, Cyprus and all of the Arabian peninsula except Kuwait.

By the end of World War II the old order by which the oil majors had sought to secure the stability of the international oil industry was to some degree obsolete. 'As Is' had come to an end and Saudi Arabia had emerged as a major new source of oil. The exploitation of that oil was in the hands of Casoc, which was renamed the Arabian American Oil Company (Aramco) in 1944. The two parent companies of

Aramco, namely Socal and the Texas Company, lacked the market outlets to increase Saudi production as fast as King ibn Saud wanted. If they had sought to exploit the vast potential of their concession by creating new outlets and forcing Saudi oil into world oil markets the stability of the industry would have been undermined in a competitive battle for markets between Middle East producers. Faced with such an unwelcome prospect the major oil companies reached a series of agreements in 1946–8 which had the effect of balancing increases in production between different sources in the Middle East. Under the agreements the access of Saudi oil to markets was enlarged when the two US companies of Standard Oil (NJ) and Socony bought their way into Aramco, in which they acquired shares of 30 and 10 per cent respectively. Both Standard Oil (NJ) and Socony possessed established marketing organisations whose demands for Middle East oil exceeded their supply. Their entry into Aramco therefore provided ready-made market outlets for Saudi oil. Other agreements reached at about the same time helped to protect other sources of supply from being displaced by Saudi oil. The most notable were the long-term contracts which Standard Oil (NJ) and Socony made with the Company for supplies of Iranian and/or Kuwaiti crude, as described in chapter 11, and a contract which Royal Dutch-Shell made with Gulf Oil for large-scale purchases of Kuwaiti crude. As a result of these arrangements the control of Middle East oil and the balancing of production between sources remained largely in the hands of the international oil majors, the so-called 'Seven Sisters', which comprised the Company, Royal Dutch-Shell, Standard Oil (NJ), Socony, Socal, Gulf Oil and the Texas Company. In short, it was large-scale private enterprise, rather than an inter-governmental agency such as that proposed in the Anglo-American Petroleum Agreement, which was responsible for the regulation of international oil production.

Although the US Government was not directly involved in either the ownership or regulation of Middle East oil production, the Arabian peninsula and the Persian Gulf, which had traditionally been regarded as peripheral to US security, became areas of increasing strategic importance in the Cold War. The Soviet tardiness in withdrawing its troops from northern Iran after the end of World War II increased US fears of Soviet intentions in the region. Those fears were magnified in February 1947 when Britain, suffering severe economic difficulties and overburdened with international commitments, notified the USA that it could not continue providing aid to Greece and Turkey. In March, Harry Truman, who had succeeded Roosevelt as US President, called on Congress for $400 million in aid to the Greek and Turkish Governments, which were opposed by communist rebels. 'The foreign policy and national security of this country are involved', said Truman as he told Congress that the USA must abandon its traditional policy of isolation and intervene throughout the world to oppose communism. In a statement which came to be known as the Truman Doctrine, the President asserted that the USA must 'support free peoples who are resisting attempted subjugation by armed minorities or by outside pressures'.

As British power waned and the USA became more directly interested in the security of the Middle East, Saudi Arabia came to be seen as a crucial link in the defensive arc around the Soviet Union. These conditions gave rise, argues the historian Irvine Anderson, to the formation of a coalition consisting of the US Government, Aramco and King ibn Saud, who shared a common interest in developing Saudi Arabia's oil reserves and securing its economic well-being.[7] A major accomplishment of the coalition was the agreement, signed on 30 December 1950, which modified the 1933 Saudi concession and provided for Aramco's profits to be shared equally between Aramco and the Saudi Government.[8] The 50:50 profit-sharing principle was not without precedent, having already been established in Venezuela, but the Saudi agreement was the first of its kind in the Middle East and rapidly became the benchmark for concessionary revisions in other countries in the region. It had a direct bearing not only on the Company's concessionary negotiations in Iran, as described in later chapters, but also on other Middle East concessions in which the Company was interested through its 23.75 per cent stake in the IPC and its 50 per cent share in the Kuwait Oil Company (KOC).

THE IRAQ PETROLEUM COMPANY

As was seen in chapter 5, the IPC was not a unitary firm, but a consortium in which equal shares of 23.75 per cent were held by the Company, Royal Dutch-Shell, Compagnie Française des Pétroles (CFP) and the Near East Development Corporation (NEDC), which itself had started out as a consortium of several US oil companies, but by the postwar years consisted only of Standard Oil (NJ) and Socony. The remaining 5 per cent of the IPC's shares were held by that remarkable Armenian entrepreneur, Calouste Gulbenkian. With such a composition there was, as might be expected, a tendency for the IPC's affairs to be complicated by the need to reconcile diverse internal interests, while it also conducted external concessionary negotiations which became entangled with Iraqi politics and wider international issues. On account of these factors a great deal of time was expended on negotiations, internal and external, as well as the various operations associated with oil exploration, production and transportation.

Two matters, in particular, caused difficulties among the participants in the IPC in the immediate postwar years. One was CFP's claim for compensation for the loss of its quota of IPC oil during the war, when the fall of France to the Germans resulted in the British sequestration of the CFP's shareholding in the IPC and the loss by the CFP of its share of IPC oil. The CFP's shares in the IPC were restored in 1945, but the French claim for compensation for the wartime loss of its IPC oil quota dragged on for several years before it was finally settled in 1950.[9] A second cause of problems in relations between the participants in the IPC was the desire of Standard Oil (NJ) and Socony to acquire shares in Aramco, whose Saudi Arabian concession lay within the area in which the members of the IPC were pledged, by the Red Line Agreement of 1928, not to take oil interests except through the IPC. In 1946 Standard Oil (NJ) and Socony attempted to get around that restriction by arguing that the Red Line Agreement had been rendered invalid by the technically enemy status of the CFP and Gulbenkian during the war.[10] They were ready to negotiate a new agreement, but not one which included a restrictive provision preventing them from acquiring other oil interests in the Red Line area on their own account, outside the IPC. In September 1946 the matter was discussed at a series of IPC meetings, but no substantive progress was made in reconciling the conflicting views of the NEDC (i.e. Standard Oil (NJ) and Socony) and Gulbenkian, whose legal advisers denied that the 1928 Agreement was invalid and insisted that no agreement which omitted the restrictive covenant would be satisfactory.[11] Discussions and correspondence continued, generating a mass of legal comment without much sign of either side softening.[12] On the contrary, attitudes seemed to be hardening when, in December 1946, CFP, which was short of crude oil and feared that participation by the NEDC in Aramco might retard the development of Iraqi oil, issued writs against the other IPC shareholders apart from Gulbenkian claiming that the 1928 Agreement was valid and binding, re-affirming the Red Line restrictions and demanding that any interest in Aramco acquired by Standard Oil (NJ) and Socony should be shared with the other shareholders in the IPC.[13] An injunction was also requested to prevent Standard Oil (NJ) and Socony acquiring shares in Aramco otherwise than through the IPC.

In the other camp within the IPC, Royal Dutch-Shell and the NEDC entrusted discussions with the CFP to Fraser because it was felt that more progress was likely to be made outside full meetings of all the IPC shareholders. Fraser asked the NEDC whether it would agree to give the other shareholders in the IPC the right to participate proportionately in the NEDC companies' shares in Aramco, but the NEDC rejected this approach.[14] However, as meetings under Fraser's chairmanship continued, a compromise solution began to emerge on lines which would reconcile the NEDC's insistence on freedom of individual action in the Red Line area with French demands for increased IPC production and a more flexible system of allocating IPC oil between shareholders so that they could lift more IPC oil than their shareholding entitled them to under the existing arrangements. Further progress in combining the three elements of increased IPC production, flexible oil allocations and the removal of the Red Line restrictions was made and in May 1947 Heads of Agreement incorporating these features were accepted by the four main IPC shareholders (i.e. the Company, Royal Dutch-Shell, the NEDC and the CFP), but not by Gulbenkian.[15] He had already warned that he would require compensation for the revocation of the 1928 Agreement and he proved to be an obdurate negotiator in the long-drawn-out discussions which followed until, finally, in November 1948 Heads of Agreement were signed between all five shareholders in the IPC.[16] The Heads, which were later incorporated into a full agreement, removed the restrictions preventing IPC shareholders from taking individual oil interests inside the Red Line area, leaving Standard Oil (NJ) and Socony free to enter Aramco. Provision was also made for the expansion of Iraqi production; for IPC shareholders who were short of crude to increase their allocations; and for Gulbenkian to be assured of profitable outlets for his IPC oil.[17]

While the participants in the IPC settled these matters, work continued on the various operations associated with exploration and production in the concessionary areas of the IPC and its affiliates, the Mosul Petroleum Company (MPC) and the Basra Petroleum Company (BPC). The project for increasing pipeline capacity from Kirkuk to the Mediterranean, suspended on the outbreak of World War II, was revived and in 1946 the IPC commenced the construction of a new 16-inch pipeline parallel to the 12-inch pipeline which had been completed in 1934. The northern branch of the new line was, accordingly, to terminate at Tripoli in Lebanon and the southern branch at Haifa in

Palestine. Construction of the southern line commenced first and was nearing completion by early 1948.[18] However, just when it looked as though new pipeline capacity might at last come on stream, the line to Haifa became a political pawn in the turbulent world of Iraqi politics and Arab–Israeli relations on which some brief words of explanation are required.

The armed clash between British and Iraqi forces in spring 1941, described in chapter 8, was followed after the brief life of Jamal al-Midfai's Government by a period of relative continuity in Iraqi politics. The moderate, pro-British Nuri al-Sa'id had an unusually long unbroken spell as Prime Minister from October 1941 to June 1944, when he was succeeded by the similarly minded Hamdi al-Pachachi, who remained in power until the end of January 1946. After Pachachi's resignation the seemingly endemic instability of Iraqi politics was much in evidence as a string of prime ministers and cabinets, which it would be tedious to catalogue, came and went at very short intervals.[19] As rival interests sought to advance their particular causes, Nuri remained the foremost figure in the pro-western camp, but efforts to suppress opposition failed to subdue Arab nationalists, whose views came forcibly to the surface over two main issues. These were the revision of the Anglo-Iraqi treaty of 1930 which, in providing for a British military presence in Iraq, was seen as a symbol of Iraqi subordination to an imperial power; and the partition of the Arab land of Palestine to provide for the new Jewish state of Israel.

In 1947–8 the mixture of these two issues produced a highly volatile atmosphere in Iraq. Negotiations on the revision of the treaty commenced early in 1947 and late in that year the Iraqi Prime Minister, Saleh Jabr, travelled to England, accompanied by Nuri and other senior ministers, for final talks. These resulted in agreement on a new treaty, which was signed in January 1948 in Portsmouth, where the British Foreign Secretary, Ernest Bevin, was on holiday. Meanwhile, the British Government announced its intention to give up the British mandate over Palestine, whose partition between Jews and Arabs was recommended by the United Nations in November 1947. As Arab volunteers were recruited for the defence of Palestine against the Jews, anti-British feelings ran high in Iraq, where Britain was widely seen as responsible for the fate of Palestine. In such a highly charged atmosphere the signing of the draft Portsmouth treaty with Britain aroused violent opposition in Baghdad, where disorder, demonstrations, street marches and riots involving many deaths and injuries broke out. Even after Regent Ilah disclaimed the treaty (which was not yet ratified) the violence continued and when Saleh Jabr returned to Baghdad in late January 1948 he resigned and fled the country, being succeeded as Prime Minister by the nationalist Muhammad al-Sadr, who confirmed the rejection of the Portsmouth treaty.

In Palestine, the situation continued to worsen in early 1948 when armed Arab detachments clashed with Jews and casualties mounted. Then, on 14 May the British mandate over Palestine was officially terminated and the last British forces were withdrawn from the country. On the same day, the Jews in Palestine proclaimed the establishment of the state of Israel to which President Truman extended US recognition only a few hours later. Soon afterwards, Arab armies, including Iraqi forces, entered Palestine only to suffer the humiliation of military defeat in the Arab–Israeli war which followed.[20]

The most immediate impact of these events on the operations of the IPC was the Iraqi decision, which took effect in April 1948, to stop the flow of oil through the pipeline linking Kirkuk with Haifa, located in that part of Palestine which was shortly to become Israel. As a result, the southern branch of the old 12-inch pipeline was closed, the new 16-inch pipeline was abandoned and the Haifa refinery, jointly owned by the Company and Royal Dutch-Shell, was starved of crude oil supplies. Although some cargoes of Venezuelan crude reached the refinery from 1949 they were not enough to allow operation at anything more than a fraction of refining capacity.[21] Meanwhile, production from the Kirkuk field had to be cut back for lack of transportation facilities, output falling from some 4.3 million tons in 1947 to 3.1 million tons in 1948.[22] The fall in production affected not only the IPC, but also the Iraqi Government whose revenues from oil royalties were directly related to the volume of production. Yet, although the Iraqis had a strong financial incentive to reopen the pipeline, no Iraqi Government could afford to run the political risk of making oil supplies to Israel, given the strength of anti-Zionist opinion in the country. For the first time in the Middle East an oil embargo was a feature of an international dispute.

The closure of the pipeline also stood in the way of agreement being reached between the IPC and the Iraqi Government on concessionary matters which were under discussion in the late 1940s. The particular issues which, in various permutations, dominated the negotiations were the level of royalties; the exchange rate between gold and sterling which was to be used in calculating royalty payments; Iraqi participation in ownership of the IPC; the appointment of Iraqis to senior positions in the IPC; and the unification of the three concessions of the IPC, the MPC and the BPC. During the protracted negotiations on these matters the British Government and the IPC persistently sought to bring about the reopening of the Haifa pipeline by making its opening a condition of agreement on other points including, most notably, the IPC's offer of a £3 million loan to the Iraqi Government in 1949.[23] However, even with such a loan on offer the Iraqi Government would not agree to the reopening of the pipeline, which had a political importance transcending purely economic concerns. George McGhee, Assistant Secretary of State in the US State Department, criticised the IPC for insisting that the reopening of the pipeline was a condition of its agreement on other matters, feeling that it was unwise to connect the *economic* question of royalty payments with the essentially *political* problem of oil supplies to Haifa.[24] In mid-1950 the State Department, concerned about Iraqi dissatisfaction with the terms and development of the IPC concession, believed that 'Iraqi complaints are to a considerable extent justified and that Iraq is entitled to improved general IPC performance without strings attached'.[25] It recommended that the US participants in the IPC should 'endeavour to break the current impasse by increasing expenditure and attention to labour conditions, public relations, etc., and by offering increased royalties without demanding Haifa pipeline and unification concessions'.

Eventually, after a visit by Nuri to London in late July, a compromise was reached in August when it was agreed to increase the royalty from 4s (gold) to 6s (gold) per ton, the new rate to be retroactive to 1 January 1950.[26] The revised royalty rate proved, however, to be short-lived. On the last day of 1950, as was noted earlier in this chapter, a new concessionary benchmark was established in the Middle East by the announcement that Aramco had reached agreement with the Government of Saudi Arabia for a 50:50 division of the profits from Aramco's al-Hasa concession. Negotiations on the application of the new principle in Iraq followed in the first half of 1951 with the outcome that on 13 August a new agreement was reached, subject to detailed drafting and approval by the Iraqi parliament. The agreement was signed in its final form on 3 February 1952 and was ratified by the Nuri-dominated Iraqi parliament later that month.[27] Under its terms, which were made retroactive to 1 January 1951, the Iraqi Government was to receive 50 per cent of the IPC's profits calculated by reference to 'posted' prices for Iraqi crude oils, that is the crude prices published by the IPC and its affiliates at export terminals. The Iraqi Government was also entitled to take as royalty 12.5 per cent of the oil produced for export, which it could either sell back to the IPC or dispose of to other buyers. The value of the royalty was, however, to be part of, rather than additional to, the Government's 50 per cent share of profits. Provision was also made for minimum production levels in the IPC, MPC and BPC concessions; minimum receipts for the Iraqi Government; and increased education, training and employment of Iraqis by the IPC and its affiliates.[28]

Although the 50:50 agreement increased Iraq's oil revenues, its negotiation and ratification were opposed by Iraqi nationalist and socialist opposition parties, notably the Independence Party, the National Democratic Party and the United Popular Front, for whom the oil agreement provided a rallying point against Nuri. As the movement towards the nationalisation of the oil industry in neighbouring Iran gathered force, eighteen deputies in the Iraqi parliament demanded on 24 March 1951 that a similar measure should be applied to the IPC and its affiliates in Iraq. Although calls for nationalisation had nothing like the same force as in Iran, the Iraqi Government decided to nationalise the refining and distribution of oil for domestic consumption in which, as was noted in chapter 5, the Company was engaged through its wholly owned subsidiaries, the Khanaqin Oil Company and the Rafidain Oil Company. By an agreement reached in December 1951 the Government purchased the Alwand refinery from the Khanaqin Oil Company and also the assets of the Rafidain Oil Company, which undertook distribution and marketing in Iraq. Opposition to the 50:50 agreement nevertheless continued and when the new concessionary terms were ratified by the Iraqi parliament in February 1952 there were accusations that the Government was subservient to imperialist interests and failing to uphold the national interests of Iraq. However, attempts to undo the agreement by a general strike and street demonstrations failed as Nuri succeeded in steering away from the course of full-blown nationalisation followed by Musaddiq in Iran.[29]

While talks were being held on the concessionary terms, the physical operations of oil exploration, production and associated activities continued. In the Transferred Territories, formerly part of Iran and where the Company had an exclusive concession, lay the oldest Iraqi oilfield at Naftkhana, being the Iraqi half of the single reservoir which was traversed by the Iraq/Iran border. After the war, the Naftkhana side was again operated completely separately from its other half, the Naft-i-Shah field in Iran. Production, which was transported via pipeline to the Alwand refinery, increased from 298,000 tons in 1946 to 437,000 tons in 1951.

In the area of the IPC's concession the Kirkuk field was further developed and delimited by drilling; de-gassing and stabilisation plants were enlarged and new ones added; and many ancillary facilities such as laboratories, workshops, stores and a power house were built.[30] The constraint imposed on production by limited pipeline capacity was eased by the construction of the northern branch of the new 16-inch pipeline, which was completed in 1951, and by a much larger new line running from Kirkuk to a terminal at Banias in Syria. Apart from 90 miles of 26-inch diameter pipe for the most westerly section, the pipeline was half 30-inch and half 32-inch diameter, so designed to minimise shipping costs by nesting the smaller sections within the larger. Construction of the line commenced in November 1950 and it was completed in April 1952.[31]

In the Mosul concession of the MPC the concessionary deadline for the commencement of commercial exports, having been extended in 1939 and 1943, was due to expire in March 1953. Although early exploration results were discouraging, for the oil discovered was heavy and viscous and in relatively small quantities, drilling continued and sufficient oil was discovered at Ain Zala to justify the installation of facilities for oil production and export from the field, which came on stream in 1952, albeit with a very modest output compared to the more prolific Middle East oilfields which had been discovered elsewhere.[32] Meanwhile, in the Basra concession of the BPC exploratory drilling commenced in 1948 after geological and geophysical surveys had been carried out. The drilling resulted in the discovery of oil at Zubair, whereupon a series of wells was completed to establish the dimensions of the field, a de-gassing station was erected and a pipeline constructed to Fao at the mouth of the Shatt al-Arab, where the first loading of oil from the Zubair field took place on 21 December 1951.[33]

As was seen in chapter 5, the IPC also held concessions outside Iraq, including one in Qatar where the Dukhan oilfield was discovered in December 1939. The plugging of the wells during the war delayed the development of the field and it was not until 1947 that drilling was resumed, with oil exports eventually commencing in the last week of 1949.[34] On the concessionary side, a retrospective arrangement for 50:50 profit sharing was reached in August 1952.[35]

KUWAIT

In Kuwait, the concession was held in equal shares by the Company and Gulf Oil Corporation through their joint subsidiary, the Kuwait Oil Company (KOC). The great Burgan field was discovered in 1938,

48   Local notables and British and US staff awaiting the arrival of Shaykh
Ahmad I at the ceremony to mark the commencement of oil exports from
Kuwait on 30 June 1946. The casket which was presented to the Shaykh is seen
on the table.

but, as happened in Qatar, the commencement of commercial production was delayed by the plugging of wells and the suspension of
operations during the war. In the last few months of 1945 staff and
materials were reassembled, buildings reconditioned, the wells (except
the discovery well) cleaned out and the provision of facilities such as
gathering lines, gas separators, storage tanks and pipelines put in
hand. A temporary marine loading terminal was also erected at
Fahahil, from where the first cargo of Kuwaiti crude oil was exported
on 30 June 1946.

Over the next few years extensive drilling was carried out both at
Burgan and elsewhere, including Magwa, where oil was discovered in
1951. Oil from the Burgan field was pumped through large-diameter
pipelines to Ahmadi, which became the KOC's headquarters and was
developed into a new oil town with its tank farm, industrial buildings,
mosque and churches, and residential quarters for European,
American and other staff. From the tank farm at Ahmadi, which lay

49   Preparing a drilling site in Kuwait, 1948

50   The oil and cargo jetty at Mina al-Ahmadi, Kuwait, 1950

athwart a low ridge 14 miles south of Kuwait town, the oil descended by gravity to the coast where a new loading terminal named Mina al-Ahmadi took the place of the temporary facilities which had earlier been constructed in the same area, then known as Fahahil. Also at Mina al-Ahmadi were the main electric power station and the water distillation plant, a vital installation for Kuwait had virtually no indigenous fresh water supplies, having previously relied on cargoes brought by dhow and later by steamer from the Shatt al-Arab. In addition to these installations a new refinery, the construction of which commenced in 1947, came into operation in November 1949. Like the power and water distillation plants it used natural gas from Burgan as fuel. Of its products, the kerosene, gas oil and gasoline were used by the KOC itself and by consumers in Kuwait town, while fuel oil and marine diesel oil were used to supply tankers visiting Mina al-Ahmadi. With regard to the concession, the KOC enjoyed good relations with both Shaykh Ahmad, who died in 1950, and his cousin and successor, Shaykh Abdullah al-Salim. The adoption of the 50:50 profit-sharing principle was accomplished with amity and took effect from 1 December 1951.[36]

CONCLUSION

The emergence of the Middle East as a producing centre of primary importance in the international oil industry came at a time of great political sensitivity and nationalist aspirations in the region. The advent of the Cold War, the Arab–Israeli conflict, the exercise of national sovereignty and the interplay of commercial rivalry and collaboration combined to produce an environment in which the conduct of the Company's business was unavoidably affected by forces of great power of a geopolitical as well as economic nature. During the war, when Anglo-American co-operation was at its apogee, the US and British Governments attempted to formulate a joint approach to secure the 'orderly' development of the international oil industry, but without success. In the event, much of the responsibility for postwar international oil arrangements fell upon the major oil companies in their relations both with each other and with concession-granting governments.

Many of the issues faced by the oil companies in the Middle East were similar, including, for example, the desire of host governments for increased revenues, guaranteed payments and/or production levels, greater training and employment of the local populace, more employee

Table 13.1 *The Company's crude oil offtake in Iraq, Kuwait and Qatar, 1946–1950 (million tons)*

	Iraq		Kuwait	Qatar
	Naftkhana	Company share of IPC production	Company share of KOC production	Company share of IPC production
1946	0.30	1.27	0.40	–
1947	0.34	1.25	1.11	–
1948	0.37	0.86	3.15	–
1949	0.37	0.98	6.09	–
1950	0.38	1.68	7.37	0.38

Sources: See Appendix 1.

welfare, more open accounting procedures and greater appreciation of local conditions. Yet national governments also had their own particular characteristics and concerns which affected their relationships with the oil companies and sometimes made it difficult to reconcile political and commercial considerations in the changing circumstances of the postwar world.

In such conditions the Company's directors must have been aware of the very strong case for spreading political risk by developing diverse sources of oil supplies and so reducing the Company's traditional dependence on Iran for nearly all its supplies of crude oil and products. Some progress was, indeed, made in that direction. In refining, the trend, described in chapter 11, was for new capacity to be located not at the source of the crude oil, but in the Company's main markets. In crude production, the Company's share of output from Iraq, Qatar and Kuwait was making a growing contribution to its crude availability in the postwar years. As can be seen from table 13.1, Kuwait was easily the largest non-Iranian source of crude, followed by Iraq, where production was set back in the late 1940s by the closure of the Haifa pipeline, and Qatar. Yet, despite the growing production in these joint concessionary interests in the Middle East, the Company remained very heavily dependent on supplies from Iran and was utterly committed, as was seen in chapter 11, to giving the expansion of Iranian output first place in its forward planning. In seeking to explain the continuing emphasis on Iran, it may be that simple conservatism played its part, the more so, perhaps, as the board of directors was far from being a youthful body, as was noted in chapter 10. There can, however, be no doubt that the Company faced a real quandary in that the more interest it showed in developing its non-Iranian production, the greater the criticism it could expect from Iranians for failing to maintain Iran's relative position in Middle East oil production and the greater the risk of precipitating a concessionary crisis. It was not, therefore, without reason that the Company put Iran ahead of its other concessionary interests in the Middle East.

=== I4 ===

Operations and employment in Iran, 1946–1951

In seeking to satisfy Iran by keeping it in the forefront of the Middle East oil industry, the Company set itself a goal which left little room for complacency in the postwar years. The Abadan refinery was the largest in the world and Iran was the leading oil producer in the Middle East, but the explosive growth in production from nearby Arab states, led by Saudi Arabia, threatened Iran's traditional oil supremacy in the region. The preservation of Iran's position could be achieved only by a rapid expansion of the Company's operations, for which added inputs of both labour and capital in very large amounts were required. The growth in employment is shown in table 14.1, from which it can be seen that the numbers employed, either directly by the Company or by contractors working on Company projects, rose from about 65,000 in 1946 to a peak of nearly 83,000 in 1949, before falling

Table 14.1 *Company employees and contractors in Iran, 1946–1950*

End of:	(a) Iranian	(b) Indian	(c) British	(d) Other	(e) Total	(a) as % of (e)
1946	60,807	2557	1869	169	65,402	93.0
1947	67,190	2461	2134	89	71,874	93.5
1948	76,994	2126	2472	81	81,673	94.3
1949	78,162	1977	2698	82	82,919	94.3
1950	72,681	1744	2725	34	77,184	94.2

Sources: See Appendix 1.

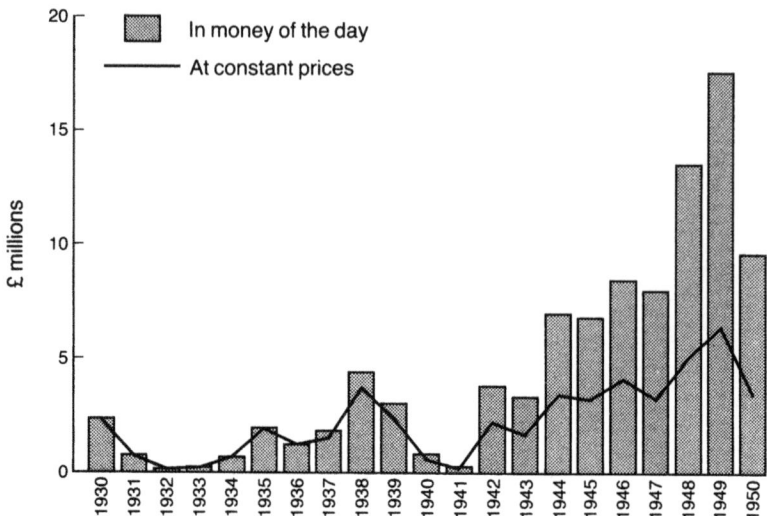

Figure 14.1   The Company's capital expenditure in Iran, 1930–1950
Notes and sources: See Appendix 2.

back to 77,000 in 1950. As for investment, the Company poured capital into Iran in amounts which would have been unimaginably large before the war. To some extent the increase in capital expenditure was more apparent than real owing to the diminution in the value of money which came with the appearance and persistence of price inflation during and after the war. As an approximation, figure 14.1 shows capital expenditure in Iran both in money of the day and after making an approximate adjustment for inflation. The graph suggests that there was a substantial real increase in the Company's capital expenditure in the second half of the 1940s, evidence of the Company's undiminished commitment to expanding its operations in Iran up to 1950 when investment continued, albeit at a reduced level during the mounting concessionary uncertainty of that year.

OPERATIONS

After the suspension of exploration activity in Iran on the outbreak of World War II, the Company's geologists there were redeployed to other duties until, early in 1944, Pattinson and Lees decided to re-activate the Geological Branch. Over the next five to six years the Company not only undertook development drilling in the fields dis-covered before the war, but also made extensive geological and geo-

51   Pipelines in the hills at Agha Jari, 1949

physical surveys, using refraction and reflection techniques to investigate possible drilling sites in the foothills and plains of the central part
of Iran. However, although much information was obtained, its interpretation was often problematical, a puzzle rather than a solution.
Consequently, the surveys were not conspicuously successful in terms
of confirmed results. Exploratory drilling was confined to areas accessible to the pipeline network, most of the effort being concentrated on
the Ahwaz anticline. In addition, new exploration was undertaken at
Batwand near Masjid i-Suleiman, where drilling began in July 1947
and ended as a dry hole in March 1948, and at Khurramshahr, where
work was abandoned in 1949. Further drilling at Marmatain and
Chiah Surkh was unsuccessful, a disappointing exploration record in
the last years of the Company's operations in Iran under the 1933
concession.

Meanwhile, the field which had earlier been discovered at Lali was
connected to the pipeline network by a 10-inch line in 1948 and came
into commercial production. Elsewhere, the capacity of the Company's
pipelines was increased by additions to the Gach Saran line and, most

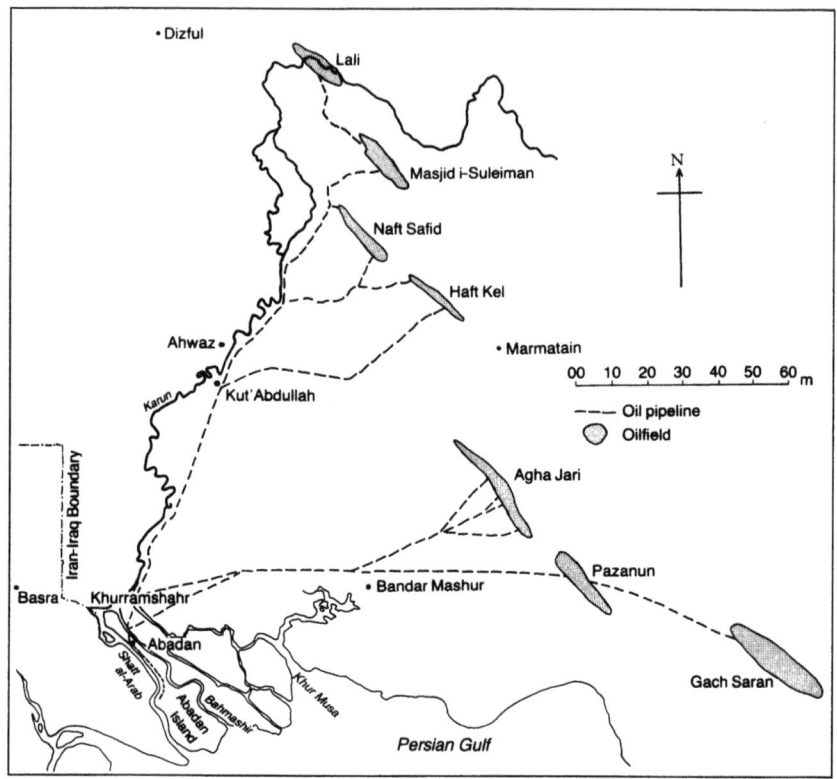

Map 14.1   The oilfields and pipelines of southern Iran, 1951

notably, by the laying of new lines for the transport of crude from the prolific Agha Jari field. Not only was the original pipeline from Agha Jari to Abadan tripled, but the field was also connected to Bandar Mashur by first a 12-inch line and then a 22-inch line, the largest in Iran at the time. As a result of these various pipeline projects, the Company's main pipeline system, shown in map 14.1, included 1990 miles of pipe and had a capacity of more than 30 million tons a year by 1950.[1]

While the pipeline system expanded, the number of productive oilfields also increased to the point where, after the opening of the Lali field, the Company had six producing oilfields in southern Iran and the gas field at Pazanun. With new oilfields also coming on stream in Kuwait and Iraq, the Company had available a growing number of crude oils, each with unique characteristics. With a view to co-

52   Agha Jari, 1949

ordinating the technical evaluation of the various crudes a Crude Oil
Advisory Group under the chairmanship of P. Docksey was set up at
Sunbury in July 1948. In 1949, at Docksey's instigation, laboratories
were set up at Masjid i-Suleiman for on-site analysis of crude oils by
J. A. E. Moy and for fields gas analysis by J. H. D. Hooper.[2] At the
oilfields, production arrangements were not uniform, but had to take
account of individual conditions such as the nature of the oil-bearing
rock, its porosity and permeability, the physical dimensions of the
reservoirs, their temperatures and pressures, gas saturation levels, the
position of water tables and gas domes, all of which influenced the
spacing and depth of wells, rates of extraction and other aspects of
production.[3] At Naft Safid, which incidentally was the scene of a
dramatic fire in May 1951, production was limited by the lack of
adequate natural fissures for the free passage of oil through the reservoir.[4] The Lali field, which was divided into two by the Karun river,
also had poor productivity, in this case because of the low porosity of
the oil-bearing rock. Agha Jari, on the other hand, was a very large
field with great reserves and production potential. Its output, shown in

Table 14.2 *The Company's crude oil production in Iran,*
1946–1950 (thousand tons)

	Masjid i-Suleiman	Haft Kel	Gach Saran	Agha Jari	Naft-i-Shah	Naft Safid	Lali	Total
1946	3675	8751	1879	4093	93	699	–	19,190
1947	2692	8908	1766	6043	131	653	2	20,195
1948	3313	9790	1981	9093	141	165	388	24,871
1949	2695	9403	1959	11,169	145	1039	396	26,806
1950	2869	9160	2065	15,621	159	1188	689	31,751

Sources: See Appendix 1.

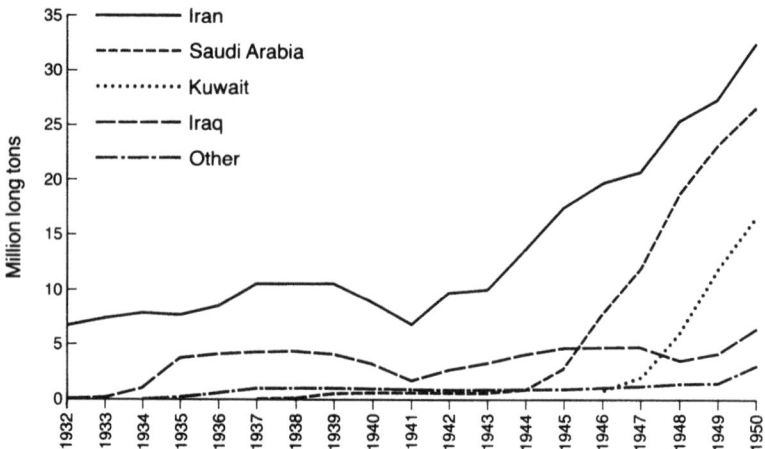

Figure 14.2   Crude oil production in the Middle East, 1932–1950
Note and source: See Appendix 2.

table 14.2, increased spectacularly from 4,093,000 tons in 1946 to 15,621,000 tons in 1950, by which time it accounted for nearly half of the Company's total crude production in Iran, which rose from 19,190,000 tons in 1945 to 31,751,000 tons in 1950. At that level of output, Iran remained, as figure 14.2 indicates, the foremost oil producer in the Middle East despite the rapid growth in crude production in other countries, most notably Saudi Arabia, Kuwait and Iraq.

The individual characteristics of the various fields made for a range of factors which affected costs of production, but the single most decisive influence on unit costs was as evident in production as in other

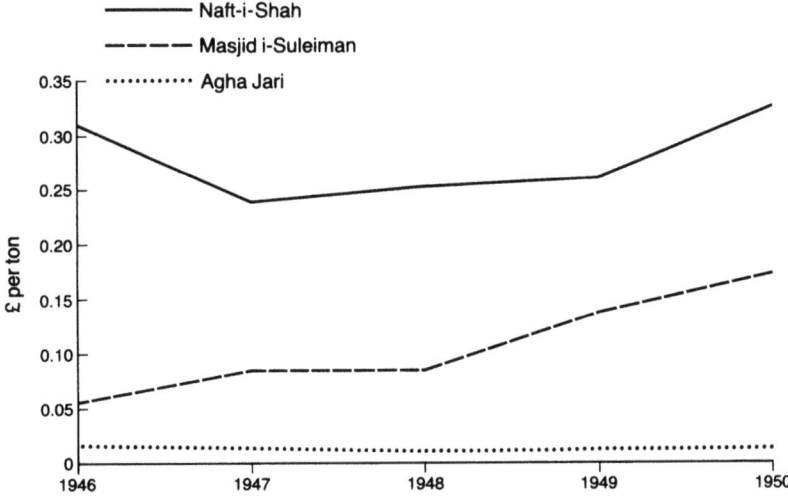

Figure 14.3   Unit production and process costs of selected Iranian oilfields, 1946–1950
Note and sources: See Appendix 2.

activities such as refining and transportation: economies of scale. To illustrate the point, figure 14.3 plots the unit production and process costs, excluding drilling and so-called 'establishment' charges, at three fields between 1946 and 1950. As the graph shows, the small, low-output Naft-i-Shah field had easily the highest unit costs. At Masjid i-Suleiman, a larger mature field with declining production, unit costs steadily increased and were substantially higher than those at the prolific new field of Agha Jari, which had by far the lowest unit costs of the three fields. Taking all the fields together, the productivity of the Company's oilwells in Iran was undoubtedly extremely high. In 1947, for example, Iranian production, from only a few dozen wells, averaged 280,000 tons per year per well compared with an average for the whole Middle East of 180,000 tons per well. The comparable figure for the USA, with about 500,000 wells, was 610 tons and for the Caribbean area 7850 tons.[5]

Meanwhile, at Abadan, as the *ad hoc* demands of the war came to an end it was time to take stock of the state of the refinery and to adjust its operations to peacetime requirements. The yield of products, which had reflected the heavy demand for aviation and motor spirits during the war, was altered to meet the postwar demand for heavier products as industrial activity recovered and the shortage of coal accelerated the

Table 14.3 *The Abadan refinery: throughputs and
yields, 1939–1950*

| | Throughputs | | Approximate yields | | |
	Crude oil (million tons)	Spirits (%)	Kerosene (%)	Gas, diesel and fuel oils (%)	Other (%)	Total (%)
1939	9.25	18	9	66	7	100
1940	8.77	18	13	58	11	100
1941	6.94	23	4	58	15	100
1942	10.23	22	7	56	15	100
1943	10.49	27	7	51	15	100
1944	13.48	27	6	51	16	100
1945	16.82	18	12	65	5	100
1946	17.67	18	10	65	7	100
1947	18.57	17	9	63	11	100
1948	20.94	18	9	63	10	100
1949	23.25	18	9	63	10	100
1950	24.05	18	10	64	8	100

Note: Refinery throughputs shown here for 1940–4 exceed the crude oil production shown in table 9.3 because some products were recycled from the refinery to fields and not counted as part of crude oil production.
Sources: See Appendix 1.

long-term trend for consumers to switch to oil. The adaptation of refinery yields to meet the changing pattern of demand can be seen in table 14.3. At the same time, plants were overhauled, maintenance, having been deferred during the war, was carried out, and ancillary services such as electricity supplies were checked and augmented. More accommodation, welfare and municipal amenities were arranged, subsidies were continued in the face of rising costs of living, and recruitment and training were put in hand. The main developments in the improvement and updating of plant were the ordering of a large catalytic cracking plant in 1946 and, in the following year, a lubricating oil plant to manufacture automotive and industrial lubricants. The installation of both plants was delayed because of the shortage of dollars, a constant constraint in the postwar years. Ironically, the new units were on the point of being commissioned in

53 View of the Shatt al-Arab waterway and Abadan refinery from the Bawarda housing estate, c.1949

mid-1951, when the flow of Iranian oil came to a stop following the Iranian Government's nationalisation of the Company's assets in Iran, as described in later chapters. In 1950, the last year before nationalisation, the throughput of the Abadan refinery was more than twenty-four million tons.

Feeling that further expansion could not easily be accommodated at Abadan, the Company kept the possibilities of constructing large-scale refineries at other locations under review. The scheme for a refinery at Bandar Mashur, suspended on the outbreak of war, was re-examined in 1943. Although it was decided to develop Bandar Mashur as a loading terminal, new refining installations were, it was thought, best postponed until after the war. By 1944 permanent loading facilities with the capacity to handle two million tons a year had been put in place and in 1946 work began on proper accommodation and amenities and modifications were completed on the loading facilities. However, while the construction of the crude oil pipelines from Agha Jari to Bandar Mashur in the late 1940s confirmed Bandar Mashur's importance as a loading terminal, it never became the site of a refinery. Instead, in seeking alternatives to Abadan, the Company turned to the idea of locating refineries in its main markets outside Iran. As was mentioned in chapter 11, one of the factors which was taken into account in assessing the economics of alternative refinery locations was the rising cost of employing Iranian labour.

EMPLOYMENT

Labour costs

Doubts about the cheapness of Iranian labour had been expressed very much earlier by Jameson, who observed in 1929 that:

> cheap labour is a mixed blessing at times, and although it may have been in the past the cheapest most economical method of undertaking certain work, with the changing conditions throughout the world, the limitation of hours, the demand for special accommodation and considerations of all kinds, our so-called cheap labour is a bit of a myth.[6]

The matter was not, as the Company was well aware, just one of wage levels and earnings; it was also to do with the high costs of providing employment infrastructure which would not have been its responsibility in other, more developed economies. As Elkington foresaw in 1935:

> Production costs will in time certainly rise in this country. Whilst even today their cheapness is largely offset and will be further offset by the enormous burden which the Company has to carry in connection with its housing, hospitals, schools: in fact, the maintenance and performance of all those functions normally furnished by state or private enterprise.[7]

A few years later, the outbreak of World War II and the subsequent Allied military occupation of Iran unsettled the previous pattern of employment. As skilled workers were lost and the continuity of service disrupted, wages and salaries, which had been reasonably stable in the five years before the war, increased sharply, as table 14.4 indicates. In the absence of consistent indices of price inflation in Iran for the period covered by the table, the movement in real wages/salaries (i.e. after adjustment for changes in the cost of living) cannot be ascertained. However, it is apparent that the wages and salaries paid by the Company in Iran continued to rise very rapidly after the war. More importantly, from a comparative point of view, was that the average weekly wage of the Company's Iranian labour was rising very much faster in the 1940s than average weekly earnings in other countries

Table 14.4 Wages and salaries of Company employees in Iran, 1934–1949 (selected years)

	Iranian labour: average wage		Non-graded Iranian staff: average salary and cash allowances		Graded Iranian staff: average salary and cash allowances		British staff: average salary and cash allowance (Abadan)
	Rials per week	£ per annum	Rials per month	£ per annum	Rials per month	£ per annum	£ per annum
1934	57	37	626	94	2570	386	712
1940	62	50	857	158	1971	365	788
1945	186	76	3095	290	6438	604	980
1949 (October)	539	314	6246	838	14,233	1910	2140

Source: BP 109667.

Table 14.5 *Index of weekly wages of the Company's*
Iranian labour compared with average UK weekly wage
earnings, 1934–1949 (selected years)

	Average weekly wage (sterling) of Iranian labour	UK average weekly wage earnings
	Index: 1934 = 100	Index: 1934 = 100
1934	100	100
1940	135	145
1945	205	198
1949	849	252

Sources: BP 109667; C. H. Feinstein, *Statistical Tables of National Income,*
Expenditure and Output of the UK, 1855–1965 (Cambridge, 1972), table 65.

such as Britain, which was the Company's main centre of employment
outside Iran. In the light of such figures as those shown in table 14.5, it
was no wonder that when the Company was debating where to locate
new refining capacity after the war the rising costs of labour and
attendant amenities in Iran influenced the decision to expand refining
capacity elsewhere.

The General Plan

The war years brought not only rising prices and wages, but also a
haemorrhage of Iranian skilled labour from the Company's employment with the result that the substitution of Iranian for foreign personnel, agreed under the General Plan of 1936, was put into reverse. As
table 14.6 shows, the percentage of Company employees (excluding
unskilled labour) who were foreigners in Iran, having fallen from 14.84
per cent in 1936 to 12.69 per cent in 1939, rose to 15.56 per cent in
1944. However, after the war the downward trend in the proportion of
Company employees who were foreigners resumed, the percentage
represented by foreigners falling to 10.45 per cent in 1950.

Meanwhile, as the period covered by the Plan drew to its close
towards the end of 1943, the question of renewal arose. On 1
December, Elkington handed the Imperial Delegate a note indicating

Table 14.6 *Foreign personnel as percentage of Company*
employees in Iran (excluding unskilled labour),
1936–1950

End of:	Percentage
1936	14.84
1937	13.63
1938	14.73
1939	12.69
1940	13.41
1941	11.36
1942	13.77
1943	15.12
1944	15.56
1945	14.57
1946	11.84
1947	11.85
1948	11.61
1949	10.92
1950	10.45

Source: BP 68184.

that the Company was willing to prolong the training schemes, continue paying for the Abadan Technical Institute, resume the British educational programme and prepare plans for a postwar situation. There were talks but no decisions.[8]

At the end of May 1947 the matter came up again when the Company received an official request from the Ministry of Finance to participate in discussions.[9] In October, the Company agreed to extend the General Plan on training schemes, education and amenities.[10] At the end of November, Gass attended meetings in Tehran, but they were generally inconclusive because of the political situation of Qavam's Government.[11] Gass regretted the Iranians' 'outrageous indictment' of the Company's conduct in relation to the employment of foreigners, feeling that it was 'entirely unjust and full of inaccuracies and irrelevancies'. He also rejected the Iranian contention that the General Plan, signed by Davar as Minister of Finance, was illegal because it had not been approved by the Council of Ministers. The Iranian Government's case was clearly influenced by Jahangir, then a director of the Bank Melli, and Dr Husayn Pirnia in the Concessions

Department. However, legal advice confirmed that the General Plan was not inconsistent with the provisions of the 1933 concession and that Davar had effectively bound his Government by signing it. The Company was irritated by the growing Iranian tendency to hold it responsible for every improvement in Khuzistan which required funds and by allegations that it was discriminating against Iranians in its employment practices.[12]

In an attempt to counter criticism Mustafa Fateh was appointed assistant general manager in charge of labour affairs at Abadan in January 1948 and Riza Fallah, who had been the first Principal of the Abadan Technical Institute, was made manager of the staff department in Tehran.[13] The Company was convinced that its 'programme of voluntary measures undertaken to accelerate the progress of Iranian-isation would stand the test of any impartial examination' and that its terms and conditions of employment were comparable to any in the industry.[14] G. N. Gobey, who was a senior Company manager in Iran, met Pirnia six times between 2 and 15 August 1948, but the talks foundered on Iranian demands that the reduction of foreign employees should be calculated on a strict arithmetical decrease and that the Company should take responsibility for educational facilities in Khuzi-stan.[15] More exchanges took place during Gass's visit to Iran in October 1948, but it had become clear that agreement on the General Plan was linked to the general concessionary situation, which is described in later chapters.[16]

At the heart of the matter was a divergence of outlooks which, though unique in particular aspects, was not unrepresentative of the sort of tensions to which relations between direct foreign investors and host governments are sometimes prone. The Iranian Government sought to impose conditions on the Company with a view to ensuring that Iranians would not be treated merely as unskilled labour, but would have opportunities to develop their own managerial and tech-nical skills. For the Company, on the other hand, the efficiency and economy of its operations was a fundamental concern which it was reluctant to compromise by submitting to rigid regulations on employ-ment. Between these polarities, there was room for the sort of personal appreciation of the other side's view which was expressed in 1947 by C. C. Mylles, the Company's staff manager in Iran: 'There is more joy in Iran over the appointment of one Iranian Chemist/Engineer/ Accountant/Doctor/Labour Officer than there is over the appointment of 100 Iranian artisans or 1000 Iranian cooks'.[17] Understanding at individual working level was not, however, a sufficient condition for

Table 14.7 *Company-sponsored Iranian students in
the UK, 1944–1950*

	University students sent to UK	Non-university students sent to UK	Total
1944	12	–	12
1945	15	–	15
1946	6	18	24
1947	6	12	18
1948	5	14	19
1949	9	14	23
1950 (June)	9	14	23

Sources: BP 15918; 68192.

bringing the two sides together on an issue which continued to gener-
ate friction in the postwar years.

Training and education

As was seen in chapter 3, the General Plan of 1936 provided not only
for an increase in the Company's employment of Iranians relative to
foreign personnel, but also for the training of Iranians as artisans and
for technical and commercial jobs. After the war, the Company continued with its training schemes in Iran and with sponsoring Iranian
students in Britain in the numbers shown in table 14.7. In the 1940s the
Technical Institute at Abadan also began to provide courses for graduates of Tehran University, the first graduate intake consisting of
nineteen engineers in the autumn of 1945. At the Institute, the graduates went through a one-year, later two-year, course based on British
models. Apart from practical experience in their respective fields as
mechanical, mining and chemical engineers, graduates attended
English classes, their previous foreign language study having generally
been in French.[18] In 1947 the Company furthered its links with Tehran
University when it endowed the Engineering Faculty with £150,000 for
the purchase of laboratory equipment.[19] The Company also seconded
three English lecturers to the University as well as supervisors from its
training department, who were to train Iranian lecturers in the use of
the new equipment so that they, in turn, could teach the students.[20]

54   The Abadan Technical Institute, late 1940s

The success of the Company's training efforts depended not only on Company funding and facilities, but also on the availability of applicants who, in attitude and prior education, possessed the attributes required for training for employment in an increasingly technical industry. However, the Company's emphasis on practical skills and discipline did not mesh easily with Iranian traditions or with the Iranian educational system. The number of schools, colleges and universities in Iran did, it is true, increase substantially in the 1930s and 1940s, as did the number of pupils and students.[21] Yet the achievement of the deep-seated changes which would have been required to align Company objectives with Iranian customs, culture and education was another matter. In 1948 Jameson remarked that it was an Iranian tradition to despise the man who worked with his hands.[22] His views were broadly corroborated in the survey of education in Iran which was made by the US firm, Overseas Consultants Inc., in their contemporaneous report on the Seven Year Plan for the economic development of Iran.[23] The US consultants felt that the Iranian school system had accomplished with relative success the aims of its founders three-quarters of a century earlier: 'that of producing a distinguished intellectual elite'. The Iranian educational philosophy was reported to be characterised by, amongst other things, an emphasis on theoretical rather than practical studies and offered 'very little of a technical nature which is worthy of retention'. The Iranian people were found to have at least five outstanding educational assets: a high tradition of culture, a developed craftsmanship, a great appreciation of beauty, a high natural intelligence and an intellectual inquisitiveness. Those possessing such qualities would, it hardly needs saying, have found little relish in getting, as Elkington put it, 'dirty in a practical job'. Moreover, Overseas Consultants felt that weaknesses existed in, for example, Iranian standards of public honesty, tendencies to seek excuses and scapegoats, preoccupations with short-term plans and exorbitant profits, deprecation of physical labour, lack of personal initiative and extreme social stratification. According to their report, the Iranian school system appeared superficially to offer all necessary branches of learning, but on closer examination the consultants found that schools were not filled to capacity, the standards of instruction were very poor and the secondary level technical schools did not fit students for employment in any local industries. Moreover, teachers below university level were underpaid, poorly prepared and frequently distracted from their educational tasks by the need to earn a part of their living from other employment.[24]

It is unsurprising, in view of the foregoing, that the Company found it necessary to take a hand in the provision of educational facilities. The first involvement came in 1923 when the Company opened a primary school at Abadan and took over another at Ahwaz.[25] Further schools followed and by the early 1930s there were five schools at Company centres: the primary school at Abadan, a primary and a middle school at Ahwaz and two primary schools at Fields. During the remainder of the 1930s and continuing in the 1940s, concern about the Iranian educational system was repeatedly expressed in and by the Company on matters such as the provision and funding of educational facilities, the inadequate and irregular remuneration of teachers, and schools' curricula.[26] Increasingly, the Company was drawn into educational provision with the result that by 1951 it had built, or housed in Company buildings, twelve schools at Abadan, one at Bandar Mashur and eighteen at Fields. Another five schools were at the planning stage.[27]

Although the Company's education and training schemes were extensive, the Company and the Iranian Government disagreed about the speed with which trained Iranians could replace foreign employees.

55   The Bahmashir school, Abadan, late 1940s

56   Machine shop in the Abadan apprentice training centre, *c*.1947

57 Physical training for apprentices at Masjid i-Suleiman, *c.*1947

In essence, the Iranian Government would have liked more instant results than the Company felt able to deliver, given the long lead-times which were involved in the training and promotion of Iranians to senior positions. As was pointed out to Jahangir in 1933, apprentices completing five years' training still had to do two years of military service, so that a minimum of seven years was required between the commencement of training and continuous employment.[28] Even then, as Elkington noted in 1935:

> There is a vast gap between the most finished of our training establishments and a man really competent to take over from our British and Indian workmen. This gap can only be bridged by the improver who, as he gains the experience which only time can bring, gradually reaches the necessary standard of efficiency. It must therefore be remembered always, that whilst we may attain an output of 10, 12 or even 50 artisans a year this does not imply that we can dispense with our equivalent number of foreigners.[29]

Differences of opinion about the speed with which Iranians could be trained and employed nevertheless persisted, as was apparent at a series of meetings held in 1948, when Pirnia reportedly said that what was needed was 'an emergency, short-term intensive training programme, irrespective of facilities that existed or did not exist'. In reply, Gobey pointed out that adequate replacements for foreign employees could not be trained on a short programme and that the Company was 'primarily an Oil Company, not a training organisation'.[30]

Another problem was that the training effort was continually diluted by the high wastage rate which occurred not only amongst trainees and skilled labour, especially during the war, but also amongst the students who were sent to Britain for training. The climatic and social conditions in Khuzistan were less inviting than in other parts of the country and it was not easy for returning students to settle down to the disciplines of industrial employment when life in, say, Tehran looked easier and more socially interesting. As Pattinson remarked in 1941, in a vein which many British industrialists might have felt applied with equal force to British graduates: 'The dilettante attitude of the educated Iranian towards life seems to me to render him particularly unsuited to a successful career in industry under the arduous conditions imposed by the Company's business and by the climate in which we have to work'.[31] Moreover, the Company expected those who gained qualifications to demonstrate their worth on the job and gradually to work their way up to more senior positions, whereas a predominant characteristic of those with degrees was reportedly 'an impatience for promotion before the acquisition of adequate experience'.[32] A Company report noted in 1948 that while the Iranian students at Birmingham University had recently done very well:

> The trouble is that when they get their degree they think they are professionally capable of being considered for the most senior appointments in the Company's service and owing to the dearth of this type of educated man in Iran, the wastage when they arrive back in their own country is very heavy, as they generally filter back to the more congenial climate and life of the capital.[33]

Finally, and again it may be noted that it was not without parallel in Britain, the Iranian student was often attracted towards Government service in which, as Pattinson remarked, 'his reward will include a certain amount of prestige and where pre-eminence confers a sense of power over his fellow men and the ability to ease his own private circumstances by taking advantage of his position'.[34]

Housing and medical care

After the wartime shortages the difficulty of providing sufficient Company housing for employees was as acute as ever. The problem, as Sir Reader Bullard, the British Ambassador, noted in 1945, was not so much the standard of the accommodation, but its insufficiency. 'So far as it goes', he observed, 'the housing of the subordinate Persian employees is very good'.[35] In October–December 1946 both Sir Frederick Leggett, who visited Iran as a consultant to the Company after retiring as Permanent Under-Secretary at the Ministry of Labour, and K. J. Hird, the newly appointed Labour Attaché in Tehran, expressed appreciation of the quality of the Company's housing. However, they suggested that the Company should rent out reasonably priced accommodation and wondered whether more building at a lower standard might not be more appropriate than the emphasis on maintaining high standards in fewer houses.[36] The Company's general management took the different view that a 'permanently lower standard' was not acceptable.[37] Leggett, on his visit to Iran in October 1946, recognised that 'a large proportion of the ordinary work people are living at high cost in bad conditions' and suggested the formation of a co-operative housing authority for Company employees. That, however, raised the question of whether it was the Company's responsibility to act as 'universal provider at Abadan'.[38] The Company was prepared to make low interest loans on a 50:50 basis for municipal schemes, but did not accept the idea of a co-operative housing venture because, if the houses were sold, it would gradually lose control to private individuals.[39]

Meanwhile, the Company continued with its own housing programme around Abadan. With the labour force increasing rapidly as operations expanded after the war, the housing shortage affected newly engaged British staff who had to share accommodation if single, or wait for it if married.[40] Although an emergency accommodation scheme was introduced, rents in the private sector continued to rise as the Company was unable to satisfy housing demand. The problem of accommodation persisted, with the Company, as Leggett noted in 1950, insisting on 'quality rather than quantity' and wanting the municipal authorities to take more responsibility for housing.[41] Although the Company was unable to meet demand it had by that time become a large supplier of housing and amenities. Between 1936 and 1950 it constructed about 21,000 houses, of which just under half were built between 1946 and 1950.[42] The Company also provided shopping, restaurant and leisure facilities, including nineteen cinemas, twenty swimming pools and various sports grounds.[43] These added up to a major aspect of the Company's activities in Iran, on which the International Labour Office (ILO) commented in 1950:

58   A two-storey house in the South Bawarda estate, *c*.1949

59   Staff bachelor flats in the Braim district of Abadan, *c*.1949

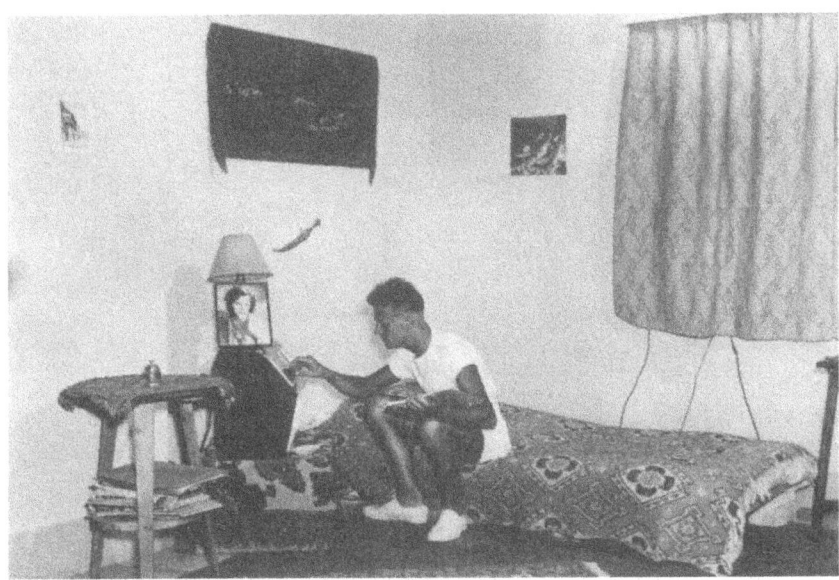

60   The type of room provided for junior bachelors on the Company's staff
at Abadan, 1950

61   Senior bachelor staff accommodation at Abadan, 1950

62   Artisans' houses in the Bahmashir district of Abadan, *c.*1949

63   The Company's housing estate for artisans at Bahmashir, Abadan, *c.*1949

64   The 'Taj' cinema, Abadan, late 1940s

65   Swimming pool, Masjid i-Suleiman, 1949

66 Display of flowers, Gymkhana Club, Abadan, 1945

67　Abadan symphony orchestra, 1946

68   Iranian apprentices practising basketball, 1950

Looking objectively and soberly at the manner in which this problem
has been tackled, the observer cannot fail to be impressed by the vast
number of modern houses and amenities which the Company has been
able to provide in a comparatively short time in spite of exceptionally
unfavourable circumstances.[44]

69   No Ruz (Iranian New Year) sports day, Masjid i-Suleiman, 1949

No less demanding than the housing problem was the provision of medical care and attention on which the ILO also remarked: 'No one who visits the Company's areas can fail to recognise the effort which the Company has made in organising its health and medical services' for employees and their dependents.[45] The assistant secretary of the British Medical Association was also impressed, reporting in March 1951 that the Company's clinics and health centres were the 'fulfilment of a general practitioner's dream'.[46] Statistics may be the bare bones of analysis, but they do indicate the extent of the Company's medical services, even if they convey nothing of the spirit of patient care. The medical staff in 1951 consisted of 101 doctors of whom 10 were specialists, 7 dentists and 130 nursing staff. There were 4 hospitals with 853 beds and 35 clinics. In 1950 there were 12,162 hospital admissions and 1,530,815 dispensary attendances, 3111 major and 15,080 minor operations performed. Trachoma treatment and a preventive medicine programme were practised on a wide scale to eradicate malaria, cholera, dysentery, typhus and smallpox. The annual cost was approximately £2 million.[47]

Industrial relations

Early in 1944 some British staff were prevented from leaving Iran because of the provisions of the wartime Order-in-Council regulating employment.[48] They suffered from a shortage of housing, were temporarily obliged to work longer hours on a shift system, became discontented and threatened to strike.[49] Miss (later Dame) Irene Ward, MP, happening to be on a visit to Iran at the time, reported that while the staff were being unreasonable, the management had 'little knowledge of modern industrial organisation'.[50] Pattinson admitted that the problem would probably not have arisen if there had been machinery for the joint representation of staff and management on matters of mutual concern. Leggett, in consultation with the Company, persuaded Bevin to authorise the visit to Iran of two officials of the Ministry of Labour, whose report of April 1944 discounted many of the staff complaints, but confirmed that much of the friction which had developed arose from the absence of suitable means for joint consultation between management and staff.[51]

The first meeting of a joint consultative committee, designed to represent the interests of staff to management on welfare matters, took place on 22 April 1944.[52] Initially there were differences of opinion over the subjects to be discussed and the respective roles of the members.[53] Bevin, with the approval of Fraser, sent out two more British officials from the Birmingham Manpower Board and the Ministry of Labour to examine practical ways of meeting labour difficulties and to foster the maintenance of good industrial relations. They submitted a long report on their visit in spring 1945, commending the activities of the joint consultative committee, but also suggesting that efforts should be made to reach junior as well as senior staff and to provide consultative machinery for Iranian employees.[54] At a meeting in mid-September the report was generally approved and the Company was advised to encourage the emergence of representative figures who might act as 'a bulwark against the political and less responsible kind of "trade unionism", which was only too liable to be used for purposes of political agitation'.[55]

There was no cause for complacency as Fraser realised during his visit to Iran in autumn 1945.[56] S. K. Kazeruni, Deputy Labour Superintendent, wrote a memorandum in late 1945 suggesting the formation of departmental labour councils to make it possible for employees to make representations to management, and for the management to inform employees more fully regarding Company plans, policies and problems.[57] He warned presciently that 'intended improvements should be carried out before any communal protests force the Company to effect improvements'. Political unrest was surfacing as the cost of living rose.[58] Elkington realised that wages and conditions of employment had to be examined.[59] On 18 February 1946 it was agreed that the creation of 'Welfare Councils or Committees' would be the most effective way of providing for employee representations. A fortnight later with a growing number of aggravating stoppages and all kinds of minor grievances, there were signs of unrest as Ivor Jones, who succeeded Pattinson as the Company's general manager in Iran in 1945, warned that large scale demonstrations or strikes in Abadan in the near future should not be excluded after the withdrawal of British troops which was to take place on 2 March. There were no specific complaints over wages, although there was criticism of the Company by officials in Tehran.[60] In April, Elkington pressed for the organisation of welfare committees in association with the Iranian Government, but it was too late.[61] The outbreak of a series of strikes at Agha Jari at the beginning of May deprived the Company of the initiative. The Tudeh Party had caught both the Iranian Government and the Company off balance. Although its political control was not absolute and discipline was fragmented, the Tudeh had consolidated its position to the extent that there was, according to Elkington, 'a real danger that they could at any given moment bring our entire operations to a standstill'.[62] A general strike was declared on 14 July 1946, associated with demands for the dismissal of the Governor-General of Khuzistan, Misbah Fatimi, disarmament of the Iranian tribes, stoppage of alleged Company interference in Iranian internal affairs and payment for the rest day. The episode involved a complicated skein of issues, events and personalities, opportunism and indecision.[63] The strike was settled in a whirlwind visit to Abadan by Muzaffar Firuz, Minister of Labour, who conceded all demands and had all the detainees released. It was the opinion of the British Labour Attaché in Cairo, after discussions in Abadan and Tehran, that the attitude of the local Tudeh Party leaders was 'based upon political motives and not industrial Trade Union policy as we understand it'. If they had wanted to negotiate on the lines of the agreed joint machinery many of the labour problems would, the Labour Attaché thought, have been solved, but they 'were determined to go their own way', wanting to secure a political success before the Iranian Government's Commission on the Minimum Wage made its decision.[64] By October, in meetings between Government officials, representatives of trade associations, trade union officers, British labour advisers and members of the Company's industrial relations management, labour terms and conditions and the minimum wage were eventually settled.[65] Trade unionism, which had been banned by Riza Shah, was now tolerated, although it was regarded with suspicion by the rulers of Iran.[66]

In 1946 the Company appointed Clifford Tucker, who took a sympathetic view of trade union rights, as senior industrial relations officer in Abadan and the following year strengthened its industrial relations management by forming a separate department to deal with labour issues. However, while the Company accepted trade unionism in its workforce, it disliked being caught in the crossfire of rival unions led by A. Q. Muhammadi and Yusuf Tftikhari respectively. It took time before Company relations with local labour officials settled into a pattern based on the joint consultative machinery, which was weakened by the general Iranian tendency to place greater emphasis on personal responsibility than on the collective bargaining process. The general impression of the Company's management was that 'the vast majority of the men have little trust in their own leaders on matters which affect their every day life and have little interest in or understanding of the major issues'.

Nevertheless, the joint consultative process became the main medium of industrial relations and worked reasonably satisfactorily until a crisis occurred in March 1951 over the reduction of outstation allowances. These were paid by the Company to employees working in areas of new development which lacked the same amenities as could be found in more established places of employment. The procedures for phasing out the allowances had been discussed in the consultative programme and they were scheduled to end at the beginning of the Iranian calendar year on 21 March 1951. Unfortunately, at that time the political situation in Iran was highly charged. On 7 March, General Ali Razmara, who was then Prime Minister, had been assassinated and there were mounting calls for the nationalisation of the Company's operations (see chapter 16). Despite that, the Company decided to proceed with its decision to phase out the allowances.

Initially, the reductions were generally accepted, except by the Iranians employed at Bandar Mashur. Over the New Year holidays dissatisfaction spread to Agha Jari, where workers went on strike on 24 March not only for the restoration of the allowances, but also for an increase in wages and other demands. Then the students at the Abadan Technical Institute demanded a lower pass mark in examination results, which had nothing to do with the Company, but the Ministry of Education. The original issue of the outstation allowances was lost sight of as protests and intimidation spread. On 26 March the Iranian Government imposed martial law and on 7 April a three-man Commission was sent to Khuzistan to report on the situation. By that time, work had been resumed in the Fields apart from Agha Jari. However, on 12 April a riot in Abadan caused bloodshed, including the loss of three British lives, and increased military measures were taken by General Shahbakhti, the military commander. The Commission suggested that strike pay should be granted, but the Company's management in Iran was reluctant to authorise it. However, influenced by the gradual return to work which took place from 23 April, improvements in the security position and advice from London on the wider issues involved, the general management agreed to the Commission's suggestion. General Shahbakhti issued a proclamation for all the workmen to return to work on 24 April and arrested some of those engaged in intimidation. By 26 April the situation was practically back to normal and meetings were arranged between all those concerned, leading to a settlement on 27 April, when the Company agreed to grant strike pay and postpone action on the outstation allowances pending full resumption of work.[67]

The settlement of the strike did not, however, inaugurate a period of harmony in relations between the Company and the Iranian Government. On the contrary, at the end of April the Company was on the verge of being plunged into a crisis of unprecedented magnitude by an event which shook its business to its foundations: the nationalisation of its assets in Iran. The nationalisation did not come out of the blue, but was the outcome of earlier developments in Irano-Company relations which are described in the next chapter.

PART III
CRISIS IN IRAN, 1947–1954

The Supplemental Agreement, 1947–1950

Amid rising Iranian nationalism, the single article law of 22 October 1947 committed the Iranian Government to look afresh at the Company's concession, which became a dominant issue not only in the Company's boardroom, but also in political life in Iran for the next few years. Of the Iranian politicians most closely associated with the nationalistic energy which was directed with great vigour against the Company one man, in particular, stands out: Muhammad Musaddiq. Born into the Iranian elite in 1882, Musaddiq came from a privileged background. His landowning family was related to the Qajar dynasty which then ruled Iran and his father was a senior government minister in the late nineteenth century. Musaddiq as a young man was a supporter of the Iranian constitutional revolution which sought to limit the despotic power of the Shah in the 1900s. After periods in higher education in France and Switzerland, he held various political posts in the early 1920s as provincial Governor of Fars and later of Azerbaijan, and as Minister of Finance and Minister of Foreign Affairs. Elected to the Majlis in 1924, he opposed Riza Shah's dictatorial consolidation of power, for which he was exiled from active politics and imprisoned for some of the time during Riza Shah's rule. After the abdication of Riza Shah, Musaddiq re-emerged as a central politician figure in Tehran, being elected to the Fourteenth Majlis where, as was described in chapter 9, he played a prominent part in the passage of the important law of 2 December 1944, which made it illegal for the Iranian Government to negotiate or enter into concessions with foreigners without the assent of the Majlis. In the rigged elections for the Fifteenth Majlis, Musaddiq was denied a seat and was therefore absent from that body when the single article law of 22 October 1947 was passed. However, as will be seen, he was soon to win re-election to the Majlis at the head of a new political group, the National Front, which rallied opposition to the Company and called for its nationalisation.[1] In the chapters which follow, much more will be heard of Musaddiq, the champion of Iranian nationalism and *bête noire* of the Company in the early 1950s.

In the meantime, Iranian political life was also the scene of a clash between democratic (Majlis) and authoritarian (Shah) tendencies. The Shah, lacking his father's authoritarian grip, was exasperated by his comparative lack of power. His opponents, however, resented his monarchical disregard for constitutional democracy. At the same time, traditional elements in Iranian society, such as the religious hierarchy, were opposed to the plans for the economic modernisation of Iran which were drawn up after the war with advice from US consultants and embodied in the Seven Year Plan for the economic development of the country.[2] Thus, nationalistic feeling against the Company grew to the accompaniment of other stresses and strains in Iranian political and economic life as constitutionalists tangled with the Shah and traditionalists with modernisers.

The resultant instability of Iran, which occupied a strategic position bordering the Soviet Union, was a cause for concern in the USA, whose foreign policy, as briefly described in chapter 13, became utterly dominated by the containment of communism as the Cold War superseded the wartime alliance with the Soviet Union. In relation to Iran, the USA was by no means in complete harmony with its other wartime ally, Britain. US anti-communism was of a much more virulent strain than that to be found in the postwar British Labour Government headed by Clement Attlee. That factor, coupled with the USA's traditional distrust of old world imperialism, predisposed the USA generally to take a more sympathetic line than the British towards Musaddiq and the Iranian nationalists, who seemed to be an alternative to communism. Moreover, the USA and Britain were in utterly dissimilar economic positions. Britain, drained by the war, was in a state of virtual bankruptcy and could ill-afford to lose the Company's assets in Iran, which were Britain's largest single overseas investment. The USA, on the other hand, had no direct investment at stake in Iran, while its virile economy generated massive financial surpluses which gave the country a seemingly bottomless purse compared with its impoverished ally across the Atlantic. It followed from those differences in foreign policy and economic situation that the USA was more inclined than Britain to look upon the Company as expendable in the greater geopolitical struggle. The Company, not surprisingly, had no desire to indulge in self-sacrifice as it entered into negotiations with the Iranian Government to update the 1933 concession by reaching a supplemental agreement.

PRELIMINARY NEGOTIATIONS

In June 1947, some months before the Majlis passed the single article law of 22 October, Qavam warned Sir John Le Rougetel, the British Ambassador in Tehran, that the Iranian Government might before long feel obliged to 'attack' the Company so that it would appear even-handed in opposing Soviet concessionary aims in northern Iran. Opposition to the Company was presented as a tactical move which would not, according to Qavam, weaken the Iranian Government's desire to maintain the closest possible relations with the Company and with the British Government.[3] Le Rougetel was assured that Qavam had no intention whatever of suggesting any change in the concession. Clause (E) of the single article law of 22 October nevertheless raised Company misgivings about Iranian intentions. As Norman Seddon, one of the Company's staff in Tehran, warned on 29 October, the Iranian Government would want to show that it was protecting the national interest and 'we shall, at least for a time, find it more difficult to deal with them than we have done in recent years'.[4] As if to confirm that opinion, in November Qavam suggested to Gass, who was on a visit to Tehran, that the Company should make a contribution to the welfare of Iran over and above the strict terms of the concession.[5]

Politically, the Company's concession was a difficult issue for Qavam, who had his back to the wall as he struggled to retain his position in power. Appealing to nationalist sentiment, he broadcast a speech on 1 December, championing Iranian claims to British-protected Bahrain and asserting that when he informed the Soviet Government of the Majlis' rejection of the proposed Irano-Soviet oil company he had brought up the question of the Company's concession and would insist on satisfaction for the Iranian people.[6] His public statements contradicted the private assurances he had given to Le Rougetel who, while realising that Qavam was in an awkward position, deplored the lack of consistency between his private and public utterances.[7] However, Qavam's nationalistic rhetoric failed to save his political career and he resigned after a lukewarm vote of confidence on 10 December, leaving for Europe almost immediately 'for medical reasons'.[8] Gass, meanwhile, reported that Iranian Ministers were, for the sake of their own political futures, using the single article law as 'a stick with which to belabour us'.[9] Leaving Tehran in December, Gass planned to return for further discussions. As rising discontent and economic stagnation played into the hands of extremists of various persuasions Le Rougetel, at least, regretted the fall of Qavam in whose absence 'there seem[ed] to be no personality in any way capable of stopping the rot'.[10]

Ibrahim Hakimi, who succeeded Qavam on 22 December, was a respected but ineffective figure, who gained Majlis approval for platitudes rather than a programme.[11] There was little enthusiasm or confidence in the new Government, which was attacked by two Tudeh deputies because it failed to promise to revise the concession or to establish Iranian sovereignty over Bahrain.[12] In an atmosphere of growing nationalism, the coupling of references to the Company and Bahrain helped to promote the impression that the Company was an instrument of British imperialism. The political situation became even more unstable in 1948 as relations between Iran and the Soviet Union deteriorated. In February, the Majlis passed a Bill sanctioning the purchase of $10 million worth of arms from the USA. The Soviets objected most strongly, fearing that the USA might use Iran as a base for attacking the Soviet Union. Trade between Iran and the Soviet Union suffered, the Soviet army began to hold manoeuvres near the border and an anti-Iran radio campaign fuelled ill feeling.[13]

On 8 June the Hakimi government tottered to its demise, to be replaced by the short-lived government of Abdul Husayn Hazhir, voted in by a majority of only six votes. Hazhir's domestic programme seemed, in British eyes, to be reassuring and worthy of support, as it included the intentions to expedite legislation for the Seven Year Plan and to raise the standard of living, while also negotiating with the Company.[14] In Tehran, however, Hazhir was denounced from the start by Ayatullah Kashani, who had emerged as an influential religious leader with close ties to the Tehran bazaar.[15] Kashani looked upon the Seven Year Plan as a 'godless enterprise',[16] his dislike of Hazhir's modernisation programme being matched only by his animosity towards the British. At the opposite end of the Iranian political spectrum, the communist Tudeh Party was also growing in strength[17] and the political atmosphere was tense.

Like other Iranian politicians of his time, Hazhir looked upon the Company as a potential source of increased revenue for the financing of Iran's economic development. On 23 June he told Norman Roberts, the British Embassy's commercial counsellor, that a loan of some £10 million from the Company would help to finance the Seven Year

Plan and to raise the standard of living.[18] He referred in the Majlis and in the press to the revision of the oil concession, but regarded the Company and the Government as partners and did not dispute the concession's validity.[19] He claimed that Iran's oil revenue was proportionately less than that of the Venezuelan and Iraqi Governments and that the official gold exchange rate against sterling no longer corresponded with the free market price of gold, to Iran's disadvantage.[20] If the concession were modified then it would provide the revenue he required.

Iranian dissatisfaction with the level of oil revenues was greatly aggravated by the growing annoyance that the British Government was extracting more income from the Company through British taxation than the Iranian Government was obtaining from the exploitation of Iran's national resources. The grievance was not without foundation for, as was seen in chapter 12 (especially table 12.1), British taxation of the Company rose very rapidly during and after the war with the result that Iranian royalties and taxation, which in prewar years had exceeded British taxation of the Company, were left far behind in the postwar years. On top of that, the postwar British policy of dividend limitation had an adverse effect on Iran's annual oil revenues, which consisted not only of a fixed royalty per ton, but also of a sum equal to 20 per cent of the dividend paid to the ordinary shareholders in excess of £671,250. Consequently, the lower the dividend, the lower the annual payment made to the Iranian Government. Under the terms of the 1933 concession the Iranian Government was also to receive 20 per cent of the sums which were allocated to the Company's General Reserve between the end of 1932 and the end of the concession. However, the prospect of that payment can have been of little consolation to the Iranians in 1947, as it was not to be made until the expiry of the concession in 1993, unless the Company surrendered the concession earlier, in which event payment would fall due at the date of surrender.

Fraser, aware of the strength of Iranian feeling on the matter of dividend limitation, raised it with the Treasury in May 1948. However, Sir Stafford Cripps, Chancellor of the Exchequer, was unmoved and wrote to Fraser that he expected the Company to conform with the dividend limitation policy, which was an essential element in the postwar Labour Government's economic policy. If it operated to the disadvantage of Iran, Cripps presumed that Fraser would discuss the matter with the Iranian Government 'with a view to avoiding hardship to that Government'.[21] When the Company announced its dividend on 1 June, Fraser wrote to the Imperial Delegate, Imami, expressing his view that the Iranian Government's position was 'well secured' by the financial provisions of the 1933 concession, but offering to discuss any hardship caused by the British policy of dividend limitation.[22]

The Foreign Office was more sympathetic than the Treasury to Iranian grievances about British taxation and dividend limitation. Eric Berthoud, the Assistant Under-Secretary who supervised the Economic Relations Department of the Foreign Office, accepted that the Iranian Government had been 'mulcted' of revenue as a result of British fiscal policy and argued in favour of a 'reasonable increase' in royalties 'having regard to the repercussions over the whole Middle Eastern area'.[23] Ernest Bevin, the Foreign Secretary, was also concerned about British fiscal policy in relation to royalty payments. He instructed Berthoud to draft a letter to Cripps requesting Treasury co-operation on the grounds that the Iranian Government might have a legitimate grievance on account of British taxation and dividend limitation.[24] Gass, accompanied by Rice, met Treasury officials on 27 August 1948.[25] They agreed that the Company should offer some £5 million short-term accommodation to the Iranian Government rather than the £10 million requested. Cripps did not feel that the Company should be pressed to do more than that.[26]

Another vexatious matter in the concessionary relationship was the 'Iranianisation' of the Company, by which was meant the increased employment of Iranians and the reduction of foreign, mainly British and Indian, personnel. Moves in that direction had been provided for in the General Plan of 1936 (see chapter 3), but progress had been disrupted by the war. The Concessions Department within the Iranian Ministry of Finance was under the influential direction of Dr Husayn Pirnia, who was adamant in calling for a fixed programme under which non-Iranian Company staff would be replaced with Iranians.[27] Talks on the General Plan took place in July and August 1948. Pirnia was the representative of the Iranian Government and Nigel Gobey, the general manager in Iran, represented the Company. Although a considerable measure of agreement was reached, two issues could not be resolved: the amount of aid for education that the Company was expected to give in Khuzistan outside its areas of operation; and the question of whether, as the Iranians wanted, the Company would make an *absolute* reduction in the number of its non-Iranian employees or, as the Company wanted, the reduction would be in the *percentage* of its employees who were foreigners. The distinction was more than a splitting of hairs, for if, as was likely, the Company's overall employee numbers continued to increase, it was, of course, conceivable that the percentage of foreign employees could fall while the absolute numbers rose. Pirnia disputed the validity of the agreements reached between Davar and Fraser over the General Plan in 1936 and though the negotiations were 'most friendly throughout', Pirnia informed Fu'ad Ruhani, the Company's legal adviser, that both Hazhir and he were gravely disappointed at the Company's attitude over the General Plan and its lack of sympathy for the interests of Iran.[28] The issue depressed the atmosphere for the talks on the concession which were shortly to be held between Gass and representatives of the Iranian Government.

Gass arrived in Tehran at the end of August and, after nearly a month spent in trying to agree on procedures, formal discussions opened on 28 September with the presentation of a 25-point memorandum compiled by Pirnia and his associates. It linked the talks directly to clause (E) of the single article law of 22 October 1947 and set out Iranian dissatisfaction on a number of scores. The most important was the claim that Iranian royalties compared unfavourably with those in Iraq, Kuwait and Venezuela, a country where the principle of 50:50 profit sharing between the Government and the oil companies had already been introduced. Other complaints of significance were concerned with British taxation and dividend limitation; the sterling/gold exchange rate used in calculating royalty payments; the prices charged by the Company for its oil products in Iran; the Iranianisation of the Company; and the Iranian demand that the Company ought not to export crude oil, but should refine it into products in Iran.[29] In discussions on the memorandum in the first half of October, Gass tried to convince the Iranian Government that the Venezuelan 50:50 profit-sharing agreement was not a suitable one for comparison as the oil revenues of the Venezuelan Government were derived solely from oil operations conducted in Venezuela, whereas the Company's profits were earned both inside and outside Iran. The Iranianisation of the Company was also discussed at length, but no specific agreements were reached and after the release of a non-committal press statement and an audience with the Shah, Gass left Tehran for London on 18 October.[30]

Shortly after Gass's return to England the political situation in Tehran was complicated by the resignation of Hazhir on 6 November after he had failed to get the budget, the Seven Year Plan or labour legislation passed. There was little conviction that the Government of

Muhammad Sa'id, appointed Prime Minister on 10 November, would be more effective than its predecessor.[31] Moreover, by the end of 1948 the Fifteenth Majlis, which had splintered into fractions, had only another six months to run before its dissolution, which cast a shadow over legislative activities. Political instability was accompanied by rising resentment of the Company. In December, Hazhir, who had become Minister of Court, warned Le Rougetel that the 'feeling was growing that Persia was being swindled' and had 'only to take a really strong line to ensure a much better deal' with the Company.[32] Hazhir's remarks can only have added to the concern which Le Rougetel had already expressed to Bevin about the mounting feeling of resentment in Iran that a much larger share of the Company's profits was appropriated by the British Exchequer than was paid to the Iranian Government.[33] Alerted to the growing tension in Iran, the Foreign Office held to the view that the Company should be given a free hand to negotiate with the Iranian Government.[34] In the words of Michael Wright, the Assistant Under-Secretary supervising Middle Eastern affairs at the Foreign Office, there was no intention 'to abandon the position that the Company's operations and policy are not in any way controlled by HMG as this attitude would, we hope, relieve us of any direct responsibility for the breakdown of any negotiations'.[35]

It was on that basis that Gass planned a further visit to Tehran after Ernest Northcroft, who had succeeded Rice as the Company's chief representative in Tehran, informed Abbas Quli Gulshayan, the Iranian Finance Minister, on 11 January 1949 that the Company was ready to resume negotiations.[36] Gulshayan agreed to renew the talks and announced that he would be the spokesman for the Iranian Government, with full powers to negotiate.[37] Before leaving for Tehran Gass attended a meeting at the Foreign Office, where he expressed the Company's desire to reach a definitive settlement.[38] In keeping with the earlier Foreign Office line Bevin did not wish to intervene and hoped for a fair outcome.[39] Afterwards, in a renewed expression of his sympathy for the Iranians' financial grievances, he wrote to Cripps about the effects of British taxation and dividend limitation and the need for the Iranians to be satisfied that they were receiving an 'equitable' return from the exploitation of their oil resources.[40]

In Tehran, Sa'id renewed criticisms of the Company and the British-owned Imperial Bank of Iran, whose joint presence, he said, was incompatible with Iran's sovereignty and natural rights. He was looking for a larger share of the profits of the Company, whose wealth was not, he thought, 'in accord with the poverty of the people'.[41] On

23 January 1949, Abbas Iskandari, a very persuasive orator and vehemently anti-British, censured the Government for failing to take action against the Company and bitterly attacked Taqizadeh for his complicity in the 1933 concessionary negotiations, as a result of which 'oil, the life blood of Persia, was being stolen by the British'.[42] Iskandari spoke for three whole sessions of the Majlis on the theme of the foreign exploitation of Iran and provoked Taqizadeh into disowning his part in the 1933 negotiations, claiming 'I had no part whatsoever in this business, except that my signature appears on the paper'.[43]

An even larger shadow was cast over the approaching talks by the attempted assassination of the Shah at the University of Tehran on 4 February 1949. He retaliated by taking reprisals against the Tudeh Party (the would-be assassin appeared to be connected to the party), curbing the powers of the Majlis, creating a Senate, half of whose members were appointed by the Shah, and increasing his own authority by constitutional changes he had long contemplated.[44] Le Rougetel informed Gass, who arrived in Tehran on 9 February, that the attempted assassination was not an isolated fanatical action, that future prospects inspired little confidence and that a left-wing government might cancel the concession.[45] Gass determined to renew discussions without delay.[46]

In the negotiations, which opened on 13 February,[47] the Iranian negotiators, led by Gulshayan, objected to the sterling/gold exchange rate used in calculating royalties, the adverse effect of the British Government's dividend limitation policy and rising British taxation in comparison with Iranian royalties. Gass took note and wrote to Fraser expressing his view that the British Government should recognise that the Company's position was a special case,[48] but Fraser reminded him that the subject could only be dealt with at governmental level and that the Company could not seek exemption from British law. 'I know', wrote Fraser, 'that we should not get very far if we decide to approach the Government on this matter. I am sure that the right course is for the Iranian Government to raise the question of taxation direct with HMG if it wishes to pursue the matter'.[49]

Gulshayan in turn stressed the obligation of the Iranian Government to conform to the single article law of 22 October 1947, around which Iranian oil policy now revolved.[50] He expected increased revenues and hoped that the Company would answer the matters raised in the 25-point memorandum. He requested a fundamental alteration in the methods of paying royalties and protested that they compared unfavourably with those in other oil-producing countries. He wanted to have specified articles of the concession revised every fifteen years, which Gass found unacceptable. However, Gulshayan's key proposal, based on the Venezuelan precedent, was that the Iranian Government should receive 50 per cent of the Company's profits. Gass argued that the Venezuelan model was inapplicable because it applied only to producing operations carried out in Venezuela, and not to the international operations of an integrated organisation such as the Company, whose profits were made outside as well as inside Iran. Gulshayan, however, responded with the argument that the Company's establishments outside Iran had been created from the profits earned from Iranian oil. Gass countered that the capital of the Company had been raised for operations outside Iran as well as for the Iranian operations and that profits earned outside the country had been ploughed back into the business. He spent over five meetings expounding his arguments to the Iranian delegates, noting in his diary that Pirnia and Gulshayan 'possessed only a very elementary idea of accounts'.[51] More suspiciously, he came to believe that the Iranians' apparent lack of understanding was a show, a deliberate bargaining tactic. Gass feared that a delay in putting forward the Company proposals might create the impression that it was holding up the talks. On 21 February he therefore put forward proposals for an increase in the Company's royalty per ton and tax payments.[52] Gulshayan rejected the proposals, continuing to insist that Iran should have a 50 per cent share of all the Company's profits.[53] Gass, however, would not agree to the Company giving up to one country 50 per cent of the profits which it made in all the countries in which it operated. The impasse, in short, persisted.[54]

On 9 March, Sa'id demanded immediate acceptance of the 50:50 profit sharing principle, claiming that most Iranians were 'in a state of acute distress, while great wealth was flowing out of the country with little benefit to them'.[55] Gass wrote to Fraser, warning that unless something could be done to meet the 50:50 demand the talks might end in deadlock.[56] Fraser replied by telegram, suggesting that a solution might be found by segregating the Company's interests in Iran from those elsewhere through the formation of a new subsidiary, whose activities would be limited to Iran and be subject to a 50:50 division of profits.[57] Gass, having agreed the wording of a press release announcing the adjournment of the talks, left Tehran on 20 March.[58]

Coinciding with Gass's return to England, the Iranian Ambassador in London, Muhsin Ra'is, approached Bevin towards the end of March, describing the Company's offer as an insignificant increase and repeating the complaint that the British Government's tax revenues from the Company greatly exceeded the payments received by Iran. He hoped that the British Government would advise the Company to come to an agreement and reduce the impact of British taxation. He assured Bevin that while the Iranian Government did not wish to cancel the concession or nationalise the Company, it was essential to raise the standard of living to resist pressure from the Soviet Union. Funds for economic development could only come from Iran's share of her oil revenues. Bevin was on the point of leaving London for Washington to sign the North Atlantic Treaty, but promised that on his return he would examine the effect of British fiscal policy upon royalty payments.[59] He did not think that the Company's offer was unfavourable and did not intend to 'give the Persians all they ask for', but he was uncomfortable with the situation, feeling that 'whether we like it or not, the company is generally considered to be under government control, and we cannot go on treating it like a purely private concern'.[60]

While Bevin was beginning to hint at a more interventionist approach and was open to the idea of at least examining the problems arising from British fiscal policy, the Treasury took a much less flexible financial view. Its adherence to the established tax procedures was confirmed when, in the last week of March and the first week of April, Gass attended meetings at the Foreign Office and Treasury, by whom he was informed on 6 April that there was no way in which the Inland Revenue could make a direct contribution to the solution of the Company's problems.[61] When Fraser and Gass visited the Treasury on 8 April there was no change. The Company's proposed offer was considered to be substantial and the Treasury thought that 'they should not be pushed any higher'.[62]

Before Gass left for Tehran on 14 April to be followed by Fraser, at the Shah's request, on the 29th, he visited the Foreign Office to explain the Company's proposals for increased payments to Iran, which he considered superior to any Venezuelan precedent.[63] The Foreign Office realised that the Company's offer was 'a fair one' by existing standards. However, it was below Iranian expectations[64] and an increase should, it was thought, be kept up the Company's sleeve to clinch an agreement.[65]

THE SUPPLEMENTAL AGREEMENT

Gass reached Tehran on 15 April 1949 and held his first meeting with Gulshayan on the 18th, followed by five further meetings between the

19th and 26th of the month.[66] The discussions covered a number of topics, including Iranian requests for an enlargement of the Imperial Delegate's powers; the Iranian proposal that the concession should be revised every fifteen years; the prices which the Company charged for its products in Iran; and that recurring bone of contention: the Iranianisation of the Company. In that connection, the Company was asked to produce a new General Plan within three months, a reduction in non-Iranian employees was requested in every category and there was insistence on a financial commitment for the technical and professional training of Iranians. Gass pointed out that the Company had not defaulted on the provisions of the concession and that the existing Company training schemes were on a greater scale than those of any other oil company in the world. Nevertheless, Gulshayan accused the Company of failing to find enough employment for Iranian graduates. Although Gass was relieved that the proposal for 50:50 sharing of profits had not been revived, he protested that the proposed revisions struck at the root of the concession and called into question whether the talks should be continued further. By 26 April all aspects had been examined, though disagreement persisted on most issues.[67] At a friendly meeting on 28 April, Sa'id hoped that Fraser's forthcoming visit would bring about an agreement and promised that every endeavour would be made to complete the talks in one week.[68]

After Fraser arrived in Tehran on 29 April the negotiations were resumed on 1 May. Fraser made it clear that he could accept changes in the concession only if they were consistent and reasonable in relation to changes which had occurred in world economic conditions. After an inconclusive meeting on 2 May, he decided to submit the text of the Supplemental Agreement, as discussed with the Treasury and the Foreign Office, to Sa'id.[69] Unlike the Company's earlier proposals, the new terms provided for payments to the Iranian Government from the Company's General Reserve to be made annually, instead of accumulating for later payment when the concession expired. The amount of the payment for 1948 on account of dividend participation and General Reserve would be £5.9 million. The Supplemental Agreement, General Plan and all ancillary documents were given to the Iranians on 3 May. Four days later Sa'id, who seemed genuinely to want a settlement, declared that he was prepared to accept the Company's proposals if they were modified from 1949 onwards by increasing the tonnage royalty, or by guaranteeing that the amount payable from dividend participation and the General Reserve would not be less than the £5.9 million offered for 1948, in effect a guaranteed payment.[70] Sa'id was satisfied with the proposals for 1947 and 1948, but he feared that payments might drop in 1949 and cause the Government embarrassment in the Majlis and elsewhere. As in 1933, the Iranian Government wanted to be insulated from the vagaries of the market. Fraser, however, pointed out that profitability 'was not what the Majlis, or the public or the Government expected, but what was permitted by the position of oil in the world's markets'. Nevertheless, the following day he offered to guarantee that the sum payable for 1949 on account of dividends and General Reserve would be not less than £2.5 million.

The Company's offer was presented to the Iranian Cabinet on 9 May, after which Sa'id confirmed the Government's acceptance of the offer for 1947 and 1948, but renewed his appeal that the payments for future years should not be less than for 1948.[71] If that could be settled, then agreement could also be reached on the other outstanding matters of Iranianisation and product prices in Iran. Fraser was cautious. 'If the Company', he asked, 'could not gain the same profits in future years how could it make the same payment to Iran?' It was a familiar impasse and Fraser left the negotiating table without a definite result. Nevertheless, there was now relatively little separating the two sides and the Company negotiators were in confident mood on the evening of 10 May, when the Iranian Cabinet met to consider the proposals.[72] Fraser, however, was not called by Sa'id to learn the result of the Iranian Government's deliberations, as had been expected. The Iranian Cabinet met again, but inconclusively, on the morning of 11 May.[73] Fraser had a long audience with the Shah in the early evening before attending a social function, in the middle of which he was asked to call on Sa'id, who prevaricated and, excusing himself with professions of goodwill, stated that the Cabinet needed 'to study the matter more closely'.[74] Fraser, disappointed, left Tehran the next day.

Shortly after his arrival in London, Fraser reported to the Company's board, who endorsed his proposals and agreed that the guarantee to make annual dividend and General Reserve payments of at least £2.5 million should apply to the remainder of the concession.[75] On 18 May, the Iranian Ambassador informed Bevin that the Iranian Government would accept the Company's proposals if the overall level of payments agreed for 1948 could be maintained in subsequent years.[76] When Fraser and Gass visited the Foreign Office on 19 May, they learnt the views of Le Rougetel and Hazhir that the whole matter could be settled if the annual guarantee of £2.5 million could be raised to

£3.5 million. Bevin regarded the Company's offer as a generous one and he was not prepared to press them to raise it.[77]

In a new development on 18 June the Iranian Government told Le Rougetel that it now requested a minimum guarantee of £4 million annually from dividend participation and the General Reserve, or payments from the Company on a sliding scale related to the world oil price.[78] It was decided that Gass should return to Tehran to complete the negotiations on the basis of the Iranian proposal.[79] Bevin wrote to Le Rougetel that 'a major British interest is at stake, and that failure to reach a settlement on the basis of this offer would have serious consequences both for ourselves and in Persia'.[80]

Gass arrived back in Tehran on 1 July and had a long discussion with Le Rougetel, who thought that 'the majority of responsible authorities would welcome an enhanced settlement'. The alternatives of the sliding scale and the increased guarantees were the proposals of Gulshayan and the Shah respectively. The next day Gass met Gulshayan who, after some discussion, was persuaded to discard the sliding scale formula in favour of a guaranteed annual payment of £4 million.[81] Gass said he would accept the guarantee provided that agreement was reached on every other point. Gulshayan promised to put the agreed terms to the Majlis to be ratified or rejected as a whole without any modifications. He was 'confident of ratification during the life of the present Majlis'. That was the most definite commitment made by Gulshayan during the negotiations, in which he often appeared to be a reluctant participant, more prepared to spin out the proceedings than to precipitate a decision. Sa'id implied two days later that an agreement was in sight.[82]

However, it was only a couple of days before the optimism began to evaporate as Gulshayan began to doubt whether the ratifying legislation could be passed before the Majlis rose.[83] On 6 July Gass learnt that some of the Cabinet were still studying documents which they should have read already, raising new issues and waiting for more consultations.[84] On 8 July Gulshayan, who had postponed three meetings and was still delaying, met Gass and after discussion accepted the guarantee proposal, but raised further objections to the General Plan, refused to accept the agreed settlement on rates of exchange and suggested that the Company's exports of crude oil should be limited to 25 per cent of its production.[85] Gulshayan no longer seemed eager for an early settlement and Gass had 'the haunting thought' that he was actually opposing one.[86] Gulshayan insisted on redrafting certain articles. Gass was ready to put off his departure, but declared that if the Iranian Government changed its mind on the General Plan there could be no Supplemental Agreement.[87] Gulshayan did not appear to be dismayed.

There was much speculation on Gulshayan's change of mind.[88] Some attributed it to a desire to stall so that he might strengthen his position in the Majlis. Others suggested that Pirnia was influencing him against the proposed settlement. There were rumours that the Shah was intending to change the Government, which made the Ministers very disgruntled and disinclined to put through such an important matter with only a short time left before the Majlis dissolved. Le Rougetel was sure that opponents were seeking to prevent the agreement being signed until it was too late to obtain ratification by the Fifteenth Majlis.[89]

Gulshayan complained that Gass had prevented discussion on the General Plan, but Gass assured Sa'id that there was no truth in the complaint and that Gulshayan had agreed that there were no more outstanding issues.[90] Sa'id was troubled and admitted to Gass that, while he was very much in favour of concluding the agreement during the last few days of the Fifteenth Majlis, he was in the hands of his colleagues. The next day, 12 July, Gass repeated to Sa'id that he was ready to sign an agreement on the basis which had been discussed.[91]

On Sa'id's initiative a meeting was held, attended by himself and other Iranian ministers, Gass (accompanied by Derek Hobson, assistant Company representative in Tehran) and Le Rougetel. The Iranian Ministers contended that the Company's proposals on the General Plan infringed the provisions of the concession and would have to be submitted to the Majlis for approval, for which little time remained.[92] Gass protested that the Company's legal advisers disputed this entirely. However, the Iranian Government continued to insist on an absolute reduction of foreign personnel as opposed to the percentage basis proposed by the Company. Eventually it was a suggestion made at a luncheon meeting between Le Rougetel, Sa'id and his Foreign Minister, Ali Asghar Hikmat, which tipped the balance in favour of a settlement.

Under the compromise of Le Rougetel, Hikmat and Sa'id the Iranian Government undertook to study within a period of two months any outstanding points of detail in the General Plan for approval by the Council of Ministers.[93] Gass informed Fraser that the compromise fell short of Company expectations, but, knowing that there was a lack of unanimity in the Iranian Cabinet, he thought that the new formula might help those in favour of the agreement to get it approved.[94] It seemed the only possibility of resolving the problem before the end of the Majlis' session. Gass realised that Fraser's instructions were for a simultaneous agreement on the Supplemental Agreement and the General Plan and was not absolutely confident that a settlement would be reached on the General Plan within two months. However, he thought the compromise was reasonable and agreed to it on the basis that the Company's commitment to the General Plan would lapse if the Supplemental Agreement was rejected by the Majlis.[95]

On that basis, the Supplemental Agreement was signed on 17 July 1949 by Gulshayan and Gass. Under its terms the royalty per ton was to be raised from 4 to 6s, retroactive to 1948. The dividend limitation problem was to be circumvented by the provision that the Company would make an annual payment to the Iranian Government equivalent to 20 per cent of the amount placed to General Reserve each year, retroactive to 1948. Moreover, the annual payment was to be grossed up by the standard rate of British income tax. In addition, of the £14 million which constituted the General Reserve at the end of 1947, the Company was to pay £5,090,909 to the Iranian Government within thirty days of the Supplemental Agreement coming into force. The new payments in respect of General Reserve were to supersede the provision of the 1933 concession under which the Company would eventually (at the expiry of the concession) have paid the Iranian government a lump sum equal to 20 per cent of the amounts allocated to the General Reserve after the end of 1932.

With regard to minimum payments, the Company guaranteed that the Iranian Government would receive at least £4 million a year from its participation in dividends and the sum placed to General Reserve. In return for continued exemption from Iranian taxation, the Company agreed to increase its commutation payments on production of more than six million tons a year from 9d to 1s per ton, retroactive to 1948. Finally, it was agreed that the Company would increase its discounts on prices of products sold for consumption in Iran. Apart from the one-off payment of £5,090,909 to be made from the General Reserve as it stood at the end of 1947, it was calculated that the royalty and tax payments under the Supplemental Agreement would be £18,667,786 for 1948 and £22,890,261 for 1949. For comparison, the sums for those two years under the 1933 concession would be £9,172,245 and £13,489,271 respectively.[96]

Having signed the Agreement containing these terms, Gass left Tehran on 18 July and arrived back in London on the 20th. The

Agreement still, however, had to be ratified by the Majlis before it could become effective.

THE FATE OF THE SUPPLEMENTAL AGREEMENT

Although Le Rougetel had been told that there would be no serious opposition to the Agreement, he was uneasy about the attitude of the Majlis, whose session was due to end in only a few days' time.[97] The Supplemental Agreement was presented to the deputies on 19 July and passed through the committee stage before being debated in public session on the 23rd. The debate was notable for the obstructive tactics employed by Muzaffar Baqa'i, Amir Timur and Abdul Husayn Haerizadeh, seasoned opponents of the Shah, as well as for the filibustering address of Husayn Makki, who did his utmost to drag out the proceedings. The President of the Majlis announced a request from Sa'id for evening sessions to settle the outstanding business, but opposing deputies walked out to prevent a quorum. The next day, Sa'id unsuccessfully attempted to limit speeches on the Bill. With time running out opponents had only to keep talking. For the majority there was an end of term weariness. Makki held the floor with interminable speeches and readings. Sa'id announced that the elections for the Sixteenth Majlis would commence on 6 August and the Fifteenth Majlis ended on 28 July without a vote being called on the Supplemental Agreement.[98]

In August, the financial effects of the non-ratification began to be felt by the Iranian Government, which claimed a payment of £6 million from the Company.[99] This was the sum from General Reserve (rounded up) which was incorporated in the Supplemental Agreement. Hikmat, the Foreign Minister, argued that the Company should pay, as it would now be only a short time before the Agreement was brought before the Sixteenth Majlis and without doubt quickly ratified.[100] The Company agreed to the Iranian request and the offer was made of a £6 million interest-free advance against future royalties.[101] However, on 15 September Gulshayan rejected the offer, as the funds were required for current expenditure and if they took the form of an advance they would be appropriated by the Seven Year Plan organisation under Iranian accounting procedures.[102]

On 18 September sterling was devalued and a few days later the Iranian Ministry of Finance asked the Company if it would make good the resultant reduction in the value of Iran's receipts from dividend participation and General Reserve under the Supplemental Agreement.[103] After consultation with the Foreign Office and the Treasury, the Company refused.

Sa'id arrived in England on a private medical visit in October, during which time he met Bevin and accepted an invitation to lunch at Britannic House, the Company's head office. He reiterated his claims for financial assistance and left London having assured both Fraser[104] and Bevin that he intended to submit the Supplemental Agreement to the next Majlis and would defend it strongly but, as he told Bevin, he would need help.[105] He had as yet received no economic aid from the USA. Oil revenue was vital to Iran and if the Iranian economy collapsed, political stability would break down and recovery would be difficult. Bevin, though sympathetic, was firm that the British Government had done its best to ensure that the Supplemental Agreement was fair and reasonable,[106] and Sa'id left empty-handed. The Shah was just as unsuccessful in obtaining financial aid during his visit to the USA in November. President Truman promised sympathetic consideration to reasonable requests for assistance, but the Shah returned to Tehran without receiving any immediate economic help.[107]

Meanwhile, in October, during the elections for the Sixteenth Majlis and just before the Shah's visit to Washington, Musaddiq and others participated in a *bast* (freely translatable as 'sit-in') in the grounds of the imperial palace in Tehran to protest at the rigging of the elections.[108] The group, consisting mainly of politicians, university students and bazaar traders, elected a committee of twenty, which formed the hub of the new National Front, headed by Musaddiq, with a programme in favour of honest elections, a free press and improved economic conditions. The election campaign was further disrupted when a member of an extremist religious group, the Fida'iyan-i Islam (Devotees of Islam) assassinated Hazhir while he was visiting a mosque. When the Sixteenth Majlis eventually convened in February 1950 there were eight National Front deputies including Musaddiq in a parliament that had a royalist majority, despite a growing feeling in some quarters that the Shah was gaining too much political power. Failing to win a vote of confidence from the new assembly, Sa'id was forced to resign on 19 March in favour of royalist Ali Mansur, who had been Prime Minister under Riza Shah.[109]

Mansur approached the Company on 6 April to renegotiate the advance of £6 million. He complained that the terms of the Supplemental Agreement were inadequate and warned that the Agreement would not go through unless a gesture were made by the Company.[110]

The Company agreed, provided that the £6 million was repaid by deductions from future concessional payments, but made no other conditions as it wished to avoid its action being misconstrued as an attempt to exert pressure on the Majlis.[111] This time the advance went through.[112]

On his part, however, Mansur procrastinated and it was not until mid-June that he submitted the Supplemental Agreement to the Majlis. The proposals were attacked by the National Front deputies, who made accusations that the Company had paid inadequate royalties, avoided local taxes, refused to train Iranian personnel, obtained the 1933 concession by force, interfered in national politics and deprived the country of its full sovereignty. They demanded nationalisation of the industry. Instead of putting the Agreement to a vote, Mansur appointed a special committee to examine it and report to the Majlis.[113] The committee was headed by Musaddiq, with Makki as its secretary. Of its eighteen members, five were National Front supporters.[114]

Mansur resigned as the oil committee convened and was replaced by General Ali Razmara, Chief of the General Staff, who had played a prominent role in the restoration of central control over Azerbaijan in December 1946. His appointment almost coincided with the arrival of a new US Ambassador, Henry Grady, who was thought by the US State Department to be 'the ideal man' to solve the 'perennial problem of aid to Iran'.[115] The State Department hoped Razmara and Grady would be able to stabilise the political and economic situation in Iran and remove the incipient communist threat by a combination of strong government and financial assistance. Grady's appointment indicated a much more active phase in Irano-American relations, and in October 1950 Iran received £500,000 under the US Point Four aid programme. Negotiations also took place for a loan of $25 million from the US Export–Import Bank, though as events turned out the loan did not materialise.[116]

In the meantime, Razmara wanted to establish his authority before putting the Supplemental Agreement to the Majlis.[117] Grady approved of Razmara's plan to leave the Agreement until he had improved his prestige and popularity and was in favour of the Company making a further advance against a guarantee from Razmara that he would put the Supplemental Agreement through within a given time. Grady thought that if Razmara failed to bring stability to Iran, the country would be a lost cause. 'This was the last effective chance for Iran', Grady told Northcroft, and 'the risk to the Company in his view was small'. The Company, Grady believed, could save Iran from communism or allow communism to triumph by its attitude towards cooperation with Razmara.[118]

A pious appeal was accompanied by the rattling of the collection box. On 25 July Razmara approached Northcroft for a further advance of £25 million against which he offered to guarantee that the Supplemental Agreement would be passed in six months' time provided the Company agreed to certain amendments. Northcroft replied that the Company could not agree to any changes in the terms of the Supplemental Agreement, which had been entered into by the duly constituted Government of Iran, and would not contemplate any advance against the Supplemental Agreement until it was a proper legal document, i.e. had been ratified.

In London, Fraser took the same line in a meeting at the Foreign Office on 2 August. It was chaired by Michael Wright and attended by other officials from the Ministry of Fuel and Power, Treasury and Foreign Office. The Foreign Office recognised Iran's need for economic assistance, as did the Company, but Fraser emphasised that it was answerable to its shareholders, who might protest if it were to subsidise Iran's economic development beyond the amounts provided by royalties. In Fraser's view it was for governments, not oil companies, to lend money to governments. He felt the Iranians had no grounds for refusing to sign the Supplemental Agreement.[119] The Americans took a different view, believing that the Company ought to make further concessions and criticising it for treating the problems of Iran from a purely commercial point of view, without regard to political and strategic considerations. Grady was annoyed that the British Government and the Company were not taking the same line as the USA.[120] Bevin, however, stuck to the view that the Company should not yield to Razmara's demands for unspecified further concessions. 'It was', he wrote, 'the bazaar method of negotiation . . . we had made a good offer to Persia and I felt we should stick by it'. Any further offer would, he thought, only be taken as a sign of weakness and lead to further demands.[121]

With Britain and the USA divided, the Iranian Ambassador left London for Tehran with a promise from Fraser that if, once the Supplemental Agreement was ratified, the Iranians wanted to consider a different method of payment, such as 50:50 profit-sharing on Iranian operations, then that possibility would be explored.[122] However, as the events of autumn 1950 unfolded there was no indication that

Razmara made use of Fraser's offer and the Company did not know whether it had been passed on to him.

Negotiations for the £25 million loan continued in Tehran, nevertheless, and by the beginning of September Northcroft and the Iranian Finance Minister, Taqi Nasr, had compromised on a lower figure of £8 million, payable in stages.[123] The advance would be used to meet immediate requirements and to keep the Government's head above water.[124] Northcroft justified the loan on the grounds that the Iranian Government was financially embarrassed, which might weaken its position when the Majlis re-assembled, and that there was little hope of the Supplemental Agreement being ratified by an alternative Government; so it was prudent to help Razmara remain in office. The proposal was approved and the advance paid, despite Fraser's earlier opposition to making a further advance before the Supplemental Agreement was ratified.[125]

On 1 October Razmara told Sir Francis Shepherd, who had succeeded Le Rougetel as British Ambassador in Tehran, that he intended to recommend the Supplemental Agreement for the approval of the Majlis. The oil committee was due to report in a very short time and would, according to Razmara, make suggestions more or less on his advice. However, before recommending the Agreement to the Majlis, Razmara wanted the Company to agree to some further concessions which he specified. They included Company undertakings that Iranian royalties would never be less than those of Iraq; that the Company would make free supplies of oil and gas in Iran; and further commitments to the Iranianisation of the Company.[126] Razmara was obviously engaged in tactical manoeuvres, preparing for opposition in the Majlis by getting the Company to agree to further concessions in advance. Northcroft was conciliatory, but unsure whether Razmara could successfully steer the oil committee in his chosen direction in view of the consistent opposition of the National Front.[127] On the other hand, he realised that an uncompromising rejection of Razmara's requests carried the risk that Razmara would put the Agreement to the Majlis without Government support, in which case it would very probably be rejected. He asked Razmara for an indication of his support for the Agreement, to which Razmara replied that he would publicly support it when replying to a censure motion against him, probably on 19 October.[128] However, to make certain that the Majlis would accept the Agreement, Razmara said he needed some face-saving concession which need not be incorporated in the Agreement itself, but could take the form of an exchange of letters and be kept secret before its disclosure to the Majlis. Northcroft was encouraged that Razmara had promised to give his personal backing to the Supplemental Agreement and consulted with London regarding the further concessions which Razmara was seeking.[129]

Razmara appeared before the oil committee at its fifteenth meeting in mid-October. He advised them that he was engaged in negotiations with the Company, but would not reveal details. He indicated that he supported the continuance of the concessionary relationship by saying that he doubted whether the single article law of 22 October 1947 was intended to be a call for the cancellation of the 1933 concession. Musaddiq expressed an opposing view, casting doubt on the validity of the 1933 concession by stating that the consent of both sides was required for an agreement to be valid, whereas the concession had, according to Taqizadeh's own public statements, been signed under duress.[130]

The Company was coming under increasing pressure to make further concessions not only in Tehran, but also in London. By the middle of October, Bevin, while holding to the view that the Supplemental Agreement was 'fair and even generous', felt it was now imperative to make further concessions 'in the interests of the Company themselves'.[131] On 20 October he told Fraser that he hoped the Company would be able to find something extra to offer 'even if it were somewhat painful'.[132] Fraser, apparently to the surprise of the Foreign Office, agreed with Bevin's assessment. Accompanied by Gass, he visited the Foreign Office on 24 October and agreed to make further concessions to the Iranian Government on matters such as the provision of gas supplies, reduced prices for products sold for consumption in Iran and a donation of some £1 million to the University of Tehran. On top of those items, he was also prepared to consider 'some further offer of a non-recurrent character'.[133]

Although the Company was prepared to make further concessions, Razmara's ability to deliver the ratification of the Supplemental Agreement was not as great as he had led the Company to believe. In early October he had been confident that the oil committee would make suggestions more or less on his advice. On 3 November he told Northcroft of his optimism that he could persuade the committee to frame its report to the Majlis around a number of general recommendations.[134] But as the committee's meetings continued it became clear that the rejection of the Supplemental Agreement was the only proposal on which all members agreed.[135] The Company's reading of the situation was that the ratification of the Supplemental Agreement was in the hands of Razmara, who viewed the oil committee as a safety valve for the Majlis, but not as part of the decision-making process.[136] The members of the committee, on the other hand, behaved as if they did have decision-making powers and, after discussing a motion put by the National Front in favour of oil nationalisation, on 25 November they unanimously passed a resolution that 'the Supplemental Agreement does not secure the rights of Iran sufficiently'.[137]

The resolution was passed to the Majlis for debate, but Razmara's efforts to influence deputies were unavailing. The National Front and powerful members of the clergy, like Kashani, who exhorted the Majlis to nationalise the oil industry on 13 December, were not to be cowed. The intervention of the *mullahs* was significant in turning the issue from a partisan political dispute into an emotional, religious and national issue. Kashani proclaimed that all the sorrows of Iran were created by the British and that nationalisation of the oil industry was the sole remedy for their ills.[138] After the Majlis debates of 19 and 20 December, with their almost uninterrupted criticisms of the Supplemental Agreement and the British by members of the National Front, the political atmosphere was very charged. Unable to win the approval of a majority of deputies for the Agreement, Razmara withdrew it from the Majlis on 26 December, with the consent of the Shah.

THE SAUDI ARABIAN 50:50 AGREEMENT

The protracted negotiations which led to the signing of the Supplemental Agreement were not the only concessionary discussions affecting the Middle East in the postwar period. Rumours of negotiations in Saudi Arabia were rife during the controversy over the Supplemental Agreement, which was withdrawn from the Majlis only four days before a 50:50 profit-sharing agreement was concluded between the Saudi Government and Aramco, the jointly owned subsidiary of four of the major US oil companies: Socal, the Texas Company, Standard Oil (NJ) and Socony.[139] Despite the rumours, the Company was unable to obtain hard information about the terms being negotiated in Saudi Arabia until after Razmara had withdrawn the Supplemental Agreement from the Majlis.[140] Under the provisions of the Saudi 50:50 agreement, which was signed on 30 December 1950 and went into effect on 1 January 1951, the Saudi Government was to receive 50 per cent of Aramco's profits by introducing a new income tax for that purpose. The same terms could not, however, have been

70   Ernest Northcroft, the Company's chief representative in Tehran from
1945 to 1951

71 An Iranian cartoon (November 1950) showing Northcroft playing the Iranian children's game, 'The wolf and the cattle'.

The caption starts:

Northcroft: 'I am the Lord and carry away the oil.'
Iranian shepherd: 'I am the shepherd; I will not allow it.'
Northcroft: 'My pounds are more valuable.'
Iranian shepherd: 'My oil is more delicious.'

applied to the Company's concession in Iran without complications. For one thing, under the US tax system the parent companies of Aramco were able to deduct the tax they paid to the Saudi Government from their tax liabilities in the USA, so that the new agreement cost them nothing. The British tax authorities offered no such tax offset. Secondly, Aramco's operations were confined to Saudi Arabia and the Saudi tax was levied only on profits earned in that country. As has been seen, the Iranians accepted no such limitation and when they raised the 50:50 principle they demanded a share of the Company's profits not only in Iran, but throughout the world.[141] Fraser, as has been mentioned, was open to, indeed suggested, the possibility that the Company might form a new subsidiary whose activities would be confined to Iran and be subject to a 50:50 division of profits. However, he remained adamant that he would not agree to the 50:50 principle unless the profits to be shared were only those arising from the Company's operations in Iran.[142] Otherwise, the Company would have been in the position of paying the Iranian Government 50 per cent of the profits which it earned from, say, the oil resources of Kuwait and Iraq, a proposition which, it may be presumed, would have been as patently unacceptable to Kuwait and Iraq as it was to the Company.

CONCLUSION

The episode of the Supplemental Agreement began with the single article law of 22 October 1947 and an agreed acceptance that the Company's concession required updating. It ended with Razmara's humiliating withdrawal of the Agreement from the Majlis in December 1950. What, to sum up, had happened in the meantime to bring about that result?

The failure of the Agreement was not, to be sure, caused by a breakdown in negotiations between the Company and the appointed representatives of the Iranian Government. The negotiations admittedly took a long time, but that was not unusual in such matters, and after terms were agreed and signed by Gass and Gulshayan in July 1949, it was left to the Iranian Government to secure the passage of the Agreement through the Majlis, to which end the Company was prepared to make further concessions. However, by the time the Agreement was presented to the Fifteenth Majlis only a few days of the Majlis' session remained, making it relatively easy for nationalist deputies who opposed the Agreement to delay its consideration by filibustering tactics. The action of the opposing deputies on that occasion was not, however, decisive in defeating the Agreement which remained, in effect, on the agenda for the Sixteenth Majlis after that body had been elected.

In the course of the elections Musaddiq emerged at the head of a recognisable group, the National Front, which was united in opposition to the Company and secured a small, but vociferous, representation in the new Majlis. The National Front succeeded in gaining a disproportionately large representation on the oil committee which was set up in June 1950, during Mansur's premiership. That proved to be a decisive event in passing the initiative from Mansur and his successor as Prime Minister, Razmara, to the National Front. Musaddiq and his supporters succeeded in turning the oil committee into a platform from which to attack Razmara and the Supplemental Agreement which, as time passed, looked more and more like a case of too little too late. In punctual confirmation of that assessment, the Saudi 50:50 agreement was concluded within a few days of Razmara's withdrawal of the Supplemental Agreement from the Majlis. The financial benefits offered in the Agreement suddenly looked less attractive and the judgement of those who opposed it appeared to be vindicated. It had simply been left behind by political developments in Iran and the terms which the US oil companies, enjoying tax offsets not available to the British, were prepared to offer elsewhere in the Middle East.

= 16 =

Nationalisation,
January–June 1951

At the time the Supplemental Agreement was withdrawn from the Majlis in December 1950, there was no knowing where events would lead. Quite obviously, the Company had worrisome concessionary problems, but that was not a new phenomenon and matters had not reached the point of breakdown which occurred in the earlier concessionary crisis of 1932–3. On that occasion, the concession had been cancelled, the dispute was referred to the League of Nations and negotiations seemed to have reached a dead end when Cadman and the Shah came together at the Shah's palace in Tehran and succeeded in reaching agreement (see chapter 2). What, it is relevant to ask, had changed since 1932–3 to stop the pattern repeating itself?

The Company had grown a great deal, but in some respects it did not seem to have changed all that much. Arguably, it had changed too little. As was seen in chapter 10, there was not much sign of a rising generation of young executives being elevated to the board, which was an ageing body, criticised for its 'out-moded attitude' by the Eastern Department of the Foreign Office in February 1951.[1] An official of that Department noted the stronger language used by Sir Frederick Leggett, the Company's labour adviser, who reportedly condemned the directors as 'helpless, niggling, without an idea between them, confused, hide-bound, small-minded, blind' and 'generally ineffective'.[2] In April 1951 doubts were voiced in the Foreign Office about suspicions that the Company was 'still following a 19th century line', while in the USA there was, according to Sir Oliver Franks, the British Ambassador in Washington, an impression that the Company had not passed 'the state of Victorian paternalism'.[3] If such impressions were accurate, it would seem that the Company's board had become somewhat old-fashioned, tending to react to events as they occurred instead of looking forward, anticipating trends and acting upon them.

While attitudes in the boardroom had not perhaps changed very much, the world outside it had changed a great deal. Apart from the general rise of nationalism in the non-western world, there were three factors, in particular, which differentiated the Company's situation in Iran in 1951 from that in 1932–3. First, in 1932–3 political power in Iran was concentrated overwhelmingly in the dictatorial hands of Riza Shah, who had the unchallenged authority to reach an agreement with the Company. In 1951, on the other hand, Iran was politically fragmented, the fractures being far more complicated than a simple split between the communist Tudeh Party, the National Front and the royalists of the Shah's court. In Iran, political allegiances tended to follow a shifting pattern of personal loyalties and factions rather than the more institutional, durable and organised parties to be found in, say, Britain. Most importantly, Musaddiq's National Front was not a political party in the western sense of the term, but a coalition consisting of disparate interests including the traditional, right-wing religious hierarchy represented by Kashani; the National Party led by Dariush Foruhar, representing the secular side of the right-wing National Front supporters; the more centrist Iran Party, led by Allahyar Saleh; and on the left, the Toilers Party, led by the French-educated Muzaffar Baqa'i.[4] The one thing that united this collection of right-wing clergy, secular nationalists and leftists was vehement anti-British sentiment, which found an outlet in the issue which dominated Iranian politics: the oil question. Given the nature of his power base, Musaddiq, having risen on the rallying cry of oil nationalisation, was left with very little room for manoeuvre, unable to accept any form of compromise with Britain, let alone to take any positive initiative to reach a settlement, without being charged with betrayal by his own supporters. In short, Musaddiq's power was of an essentially *negative* nature, as has been acknowledged by various commentators and historians. For example, the US historian James Bill, in taking a generally sympathetic view of Musaddiq, described him as a 'magnificent negativist in that he had the courage to challenge but lacked the capacity to construct'.[5] Similarly, W. R. Louis, another US historian, has remarked that 'anti-British sentiment was one of Musaddiq's sources of strength and a sustaining negative inspiration'.[6] Political fragmentation and the nature of Musaddiq's power base made it much more difficult to arrive at a settlement in the early 1950s than twenty years previously.

A second factor which differentiated the position in 1951 from that in 1932–3 was at the international level. The earlier dispute was essentially a binational affair involving only Britain and Iran. In 1951 the USA, having vastly extended the definition of its vital interests to embrace almost any cause which helped to contain the spread of communism, was worried that Iran might fall prey to its powerful northern neighbour, the Soviet Union, or to the Tudeh Party. The involvement of the USA gave the dispute a triangular international dimension which made it much more complicated than the earlier concessionary crisis of 1932–3. Within that triangular relationship, Britain and Iran were diametrically opposed. Britain, a traditional imperial power beleaguered by economic problems in the postwar world, looked upon the Company's interests in Iran as a vital British asset. Moreover, with major British foreign investments elsewhere, the British Government felt that if it failed to take a strong stand against the nationalisation of the Company's assets in Iran it might open the door for other countries to follow Iran's example. That concern was repeatedly expressed in British Government circles during the Iranian crisis.

While the British Government, and still more the Company, opposed Iranian ownership and control of the oil industry, Musaddiq and his supporters would not accept anything less. The resultant confrontation between Britain and Iran placed the USA in a dilemma. The USA could not ignore the interests of Britain, its closest western ally. Moreover, the USA had its own oil interests in the Middle East, which it wished to preserve. The US Government therefore wished to avoid a settlement being reached with Iran that would upset the stability of the 50:50 profit sharing agreements under which US oil companies operated in other Middle East oil producing countries. On the other hand, the US Government looked upon Iranian nationalism as a bulwark against communism and feared that Iran might fall under communist rule if the nationalists failed to achieve their objectives and fell from power. The only settlement that would meet these sometimes conflicting elements in US policy was one which accepted the Iranian nationalisation of the oil industry, but upheld the 50:50 division of profits as the basis for future operations. In seeking to bring about a settlement on these lines, the USA repeatedly took on the role of mediator between Britain and Iran. Each of these countries, realising that US support was vital to its cause, constantly tried to win the US Government over to its side. The British, who took the view that the US fear of communism was exaggerated, sought to gain US backing by emphasising the sanctity of international contracts and the unreasonableness of Musaddiq's nationalist Government. Musaddiq, on the other hand, tried to win over the USA by playing on the threat of communism.

Caught in this web of international diplomacy, the Company faced the very real danger that its interests might be sacrificed for the sake of international strategic concerns, particularly as the US Government tended to look upon the Company as an expendable pawn which could be given up if that would facilitate a settlement which, by meeting Iranian nationalist aspirations, would strengthen the country against communism.

Thirdly, and finally for present purposes, however inexorably the tides of change may have influenced matters, the contribution of personalities to the course of events, the speed and manner of their occurrence and, possibly, their outcome cannot be overlooked. Fraser, whatever his competence as a practical oilman, was not by nature a conciliatory diplomat. His relations with Whitehall were, as has been noted in chapter 12, at best tense, at worst overtly hostile. He had, in the words of Louis, a 'fire-eating contempt for civil servants'[7] and there can be no doubt but that the antagonism was reciprocated. Musaddiq was equally uncompromising. A controversial personality, his advanced years, physical frailty, tendency to faint in public and habit of negotiating from his bed wearing pyjamas, made him an easy subject of ridicule and lampoon. He was distrusted and seemingly regarded with contempt by Sir Francis Shepherd, the British Ambassador in Tehran, who wrote of Musaddiq in May 1951:

> We shall have to watch him carefully because he is both cunning and slippery and completely unscrupulous ... He is rather tall but has short and bandy legs so that he shambles like a bear, a trait which is generally associated with considerable physical strength. He looks rather like a cab horse and is slightly deaf so that he listens with a strained but otherwise expressionless look on his face. He conducts the conversation at a distance of about six inches at which range he diffuses a slight reek of opium. His remarks tend to prolixity and he gives the impression of being impervious to argument.[8]

Yet whilst the British often looked upon him as a demagogic agitator and lunatic extremist, Musaddiq was regarded by many Iranians as a national hero and has been described by various historians as a liberal, constitutional nationalist, a valiant champion of Iran's independence from foreign domination and upholder of democracy against the royal dictatorships of the Pahlavi Shahs.[9] On one point, however, there is consensus: Musaddiq was an anti-British nationalist. As this brief depiction of personalities makes clear, it was inconceivable that Fraser and Musaddiq could meet and reach agreement where others had failed, as Cadman and Riza Shah had done in 1933.

COUNTDOWN TO NATIONALISATION

On 11 January 1951 the Supplemental Agreement was finally buried by the Majlis, which passed a motion rejecting the Bill that contained the Agreement and instructed the oil committee to make recommendations about the oil question within two months.[10] The oil committee was thus vested with renewed authority, while the National Front continued to gain ground, assisted by demonstrations and propaganda for nationalisation in both Tehran and the provinces. For example, National Front ribbons bearing the slogan 'Nationalise the oil industry' were being worn, even by government officials.[11] Some of the support for the National Front was due to the endorsement of nationalisation by the clergy, most notably Ayatullah Kashani, who was able to mobilise mass rallies and, in general, to raise the temperature of opinion against the Company.[12]

After visiting London for discussions with Company colleagues and British Government officials,[13] Northcroft returned to Tehran at the end of January and informed the Shah and Razmara that the Company was prepared to make an advance of £25 million to help the Iranian Government in its financial difficulties. The offer, which took the form of £5 million for immediate payment followed by ten monthly payments of £2 million in the rest of 1951, was accepted. In the event, however, only the initial and the first monthly payments were made.[14] Northcroft also told Razmara that the Company was prepared to discuss a new agreement on the basis of a 50:50 division of the profits arising from its operations in Iran, or for that matter any other reasonable proposals.[15] Razmara urged that the Company's offer to discuss 50:50 terms be kept secret, giving the reason that if the Majlis deputies knew of the offer they would hold out for more. Both Shepherd and Northcroft tried to persuade him to make the offer public, but he was adamant in refusing.[16]

The Company's willingness to discuss a 50:50 arrangement was therefore undisclosed when, on 19 February, Musaddiq proposed the nationalisation of the Company to the oil committee.[17] Arguing that nationalisation was the only proposal worth debating, he claimed that the Company was the source of all the misfortunes of 'this tortured nation'.[18] However, Razmara, speaking to the oil committee with the deliberate exclusion of the National Front members, argued that nationalisation might prove disastrous. Feeling that it would be politically dangerous to oppose it outright, he suggested a compromise formula, which did not rule out nationalisation as a long-term solution, but which requested the Government to consider talks with the Company for a 50:50 agreement to cover an interim period.[19] He also submitted the question of nationalisation to a panel of four Supreme Court judges and experts in various ministries.[20]

In London, on 21 February, Ernest Davies, Parliamentary Under-Secretary at the Foreign Office, in answering a parliamentary question, expressed again the Foreign Office opinion that the Supplemental Agreement was fair and reasonable, and hoped for a satisfactory settlement.[21] In view of the firm stance taken by the British, and after requests for guidance by members of the oil committee, Razmara now tried to persuade the non-National Front members of the committee to take a strong line against nationalisation.[22] However, Geoffrey Furlonge, of the Eastern Department at the Foreign Office, returning from a visit to Iran, reported to his colleagues that the 'parrot cry of nationalisation had attained a considerable hold' and that Razmara showed no signs of being at all strong.[23] The accuracy of that assessment was substantiated on 3 March, when Razmara again refused to disclose the Company's offer to discuss a 50:50 arrangement or to acknowledge the Company's financial assistance in case it was imagined that he was being bribed.[24] Instead, he appeared before the oil committee and read reports by the panel of Iranian experts from different ministries, who were generally against nationalisation as impractical from technical, political, financial and legal points of view.[25] The next day, his opponents alleged that the experts had been bribed by British officials and Musaddiq claimed that nationalisation was not only possible, but would considerably increase the country's revenues.[26] On 5 March, Razmara told a press conference that precipitate action on the oil question which ignored the advice of the experts would be treasonous. On the 6th, Musaddiq issued a manifesto attacking Razmara.[27] The conflict between Musaddiq, in favour of nationalisation, and Razmara, against it, had become uncompromising. It was brought to a violent end when, on 7 March, Razmara was assassinated by a member of the Fida'iyan-i Islam (Devotees of Islam), which had links with Ayatullah Kashani[28] and which openly supported nationalisation.[29]

The deliberations of the oil committee were swamped by the

passionate appeals of Musaddiq and Makki to nationalise at once.[30] With Razmara gone, on 8 March the oil committee passed a resolution adopting the principle of nationalisation and asking the Majlis to give the committee two months to study ways and means of implementing it.[31] The resolution was followed by a large demonstration organised by Kashani in Baharistan Square outside the Majlis, and other meetings were held by supporters of the National Front and Kashani's religious organisation.[32] In such a charged political atmosphere the Shah not surprisingly found it difficult to find a suitable candidate who was willing to replace Razmara as Prime Minister. However, the veteran statesman, Husayn Ala, was prevailed upon to form a caretaker government on 11 March. Three days later he was given a note from the British Government claiming that the Company's activities could not be legally terminated by nationalisation and pointing out that the terms of the 1933 concession provided for arbitration in the case of a dispute. The Company could not, in the British Government's view, reasonably be expected to negotiate a 50:50 agreement under the threat of nationalisation.[33] Before a reply was received, the British Government and the Company learned that on 15 March the Majlis had passed a Bill adopting the oil committee's resolution of 8 March espousing the principle of nationalisation.[34]

On 19 March Dr Zanganeh, Minister of Education under Razmara, was shot and fatally wounded by a theological student with a grudge against him. The assassin claimed that he had shot Zanganeh for political reasons after the Minister had enforced a law banning Tudeh propaganda in universities and schools. His death heightened the feeling of terror and uncertainty in Tehran where, on 20 March, the Senate followed the example of the Majlis and adopted the principle of nationalisation by passing a single article Bill approving the oil committee's resolution of 8 March. The single article law of 20 March was a milestone on the path to nationalisation. It declared that the oil industry throughout Iran was nationalised and that all operations of exploration, extraction and exploitation would be carried out by the Government.[35]

While the movement towards nationalisation gathered pace and momentum in the excited political atmosphere of Tehran in the first three weeks of March, other important developments were taking place in London. At the Foreign Office, Herbert Morrison, having succeeded Bevin as Foreign Secretary on 9 March, considered possible courses of action, not excluding military and naval movements.[36] On 19 March, Sir William Strang, Permanent Under-Secretary at the

Foreign Office, agreed with Fraser that the concession could not be legally terminated or varied unilaterally by the Iranian Government. According to Strang, Fraser was prepared to consider any proposal which would leave the management of operations in the Company's hands, be it a 50:50 arrangement or any other possibility the Iranian Government wished to discuss.[37]

On 20 March, at an important interdepartmental meeting held at the Foreign Office under Strang's chairmanship, it was agreed that the British Government would, in effect, assume responsibility for negotiations with the Iranian Government.[38] The meeting agreed that new proposals on the Company's future in Iran should be drawn up; there should be constant consultation between the Company and the Government; the Company should not put forward proposals to the Iranian government without the knowledge of the British Government; a Working Party consisting of senior Government officials should be established to keep developments in Iran under constant review; and, finally, a joint Anglo-American initiative was needed. In effect, the meeting recognised that events had moved past the point where the Company's position in Iran could be left to simple bilateral negotiations between the Company and the Iranian Government and that the dispute over the Company's activities in Iran could no longer be regarded as a commercial matter: it was an international diplomatic issue.

The actions agreed at the interdepartmental meeting of 20 March were duly set in motion. In London, Government officials and Company representatives joined in the Working Party and considered various schemes involving acceptance of 50:50 profit sharing between the Company and the Iranian Government, but not the principle of nationalisation. For that reason the lines on which the British were thinking did not, in the US view, go far enough.

The difference between the British and US views was apparent when British officials, but not Company representatives, visited Washington in April for Anglo-American talks on the oil problem in Iran. Franks outlined the principal British objectives, which were that the period of the Company's concession should remain unchanged and that any new arrangements should safeguard British control over the Company's assets.[39] George McGhee, the US Assistant Secretary of State, saw the problem from a different angle. For him, the top priority was to prevent Iran from falling to communism, to which end he believed it was necessary to make concessions to the Iranian Government by paying lip service to nationalisation. Franks mentioned the possibility of transferring the Company's concession, assets and operations in Iran to a new British company whose profits would be shared with the Iranian Government, in effect a 50:50 agreement. However, US officials felt that a proposal on those lines would fail to satisfy the Iranian desire for nationalisation. McGhee, favouring an approach which would give the façade of nationalisation whilst retaining control of operations, suggested the idea of a form of management agreement under which the Iranian Government would own the assets and contract with an oil company to control operations and receive 50 per cent of the profits.[40] Geoffrey Furlonge, head of the Eastern Department of the Foreign Office, reported that the main US concern was to prevent Iran from falling under Soviet or communist domination, for which the Americans were prepared to sacrifice almost anything, including the Company.[41] As Strang put it: 'The main difference between us and the Americans in this affair seems to be that to the Americans, in the fight against Communism in Persia, the Anglo-Iranian Oil Company is expendable. It is not possible for us to start from this premise.'[42]

At odds with the USA over Iran, the British Government approved proposals which were presented to the Iranian Prime Minister, Ala, by Shepherd on 26 April. In essence, the proposals provided for a 50:50 division of profits, but not nationalisation, by the formation of a new company whose profits would be divided equally with the Iranian Government and whose board of directors would include representatives of the Iranian Government. It was also suggested that the distribution of oil in Iran be transferred to an Iranian national oil company and that an agreement be reached to provide for the replacement of non-Iranian employees in the Iranian oil industry by qualified Iranians.[43]

Musaddiq was a step ahead. On the same day that Shepherd presented the British proposals to Ala, the oil committee under Musaddiq passed a nine-point resolution on nationalisation for submission to the Majlis. The next day, Ala resigned and on 28 April Musaddiq was appointed Prime Minister. The oil committee's resolution, which came to be known as the nine-point law, passed the Majlis on the 28th, the Senate on the 29th and received royal assent on 1 May.[44] Whereas the single article law of 20 March had adopted the *principle* of nationalisation, the nine-point law provided for the *implementation* of nationalisation. It was another milestone on the path which now seemed to be leading to the takeover of the Company's assets in Iran.

The nine-point law provided for the appointment of a mixed committee consisting of Majlis deputies, Senators and Government representatives, who were charged with the tasks of carrying out the nation-alisation of the Company, forming a new National Iranian Oil Company (NIOC) to take its place and making preparations for the gradual replacement of foreign employees with Iranians. The law also declared that the Company's revenues since 20 March, when the principle of nationalisation was sanctioned by the Senate, belonged to the Iranian Government. However, recognising that compensation for nationalisation would be payable to the Company, the law provided for 25 per cent of the net revenues from oil operations to be deposited in a bank acceptable to both sides for the settlement of the Company's claims. The law also declared that purchasers of products from the Company would be able to continue buying the same quantities from the nationalised concern at a 'just price'. As for timing, the mixed committee was given three months to complete its work.

APPEAL FOR ARBITRATION

To add to the problems which it faced in Tehran, the Company had for some weeks been faced with unrest in the province of Khuzistan, the scene of its operations in southern Iran. There, as was described in chapter 14, special outstation allowances had been paid to staff and labour to compensate them for the relatively poor facilities at their places of employment which, being areas of new development, lacked the amenities available at more established centres. By March 1951 the facilities had been improved and the Company's management in the south of Iran, with an imperfect understanding of the situation in Tehran, decided to cut back the allowances. The first of the reduced pay packets was distributed on 20 March, the last day before the Iranian New Year holidays. When work was due to resume, on 24 March, nearly all labour in Bandar Mashur and Agha Jari went on strike, followed by general unrest and strikes throughout the area.[45] The unrest was spread by agitators, whether National Front or Tudeh supporters, 'young men and boys who bicycled from place to place threatening those who wished to work with later gang violence either to themselves or to their families'.[46] Martial law was declared and the Iranian Cabinet sent a three-man commission, who negotiated with the Company to restore the outstation allowances and to look on the strike period as part of the strikers' annual leave entitlement. However, as the situation improved in the outstations, in Abadan it deteriorated. In early April, several people, including three Britons, were killed in riots and work at the refinery was brought almost to a standstill. On

23 April, realising that intimidation was the cause of much of the disturbance there, Elkington wrote to Eric Drake, the general manager, that instead of making reference to 'strike pay' the Company should use the term 'absence from work' to describe the strike period, on which basis the Company agreed to pay all workers in Khuzistan full wages for the days they were absent.[47]

The situation in the south appeared to calm down as attention was concentrated on events in Tehran after the passage of the nine-point law on 1 May. That day, Morrison, speaking in the House of Commons, recognised the right of the Iranian people to take a greater share in their major industry, but did not accept that the Company's great investment in Iran could be unilaterally nationalised.[48] His tone was deliberately temperate, but the Government regarded the situation with concern. On 2 May, James Callaghan, then Parliamentary Secretary to the Admiralty, in a written answer to a parliamentary question on the stationing of vessels in the Persian Gulf, announced that the deployment of British ships was such 'as to facilitate the taking of any action that the situation might demand'.[49] Morrison spoke to the Iranian Ambassador the same day and informed him that the British Government did not accept the nationalisation of oil as a *fait accompli*.[50] Musaddiq's reply to this was a letter, dated 8 May, stressing the sovereign right of every nation to nationalise its industries and enclosing details for the practical implementation of the measure.[51]

Also on 8 May, the Company formally notified the Iranian Government that it requested arbitration in accordance with the terms of the 1933 concession and that it appointed Lord Radcliffe as its arbitrator. The Company awaited notification of the appointment of the Iranian Government's arbitrator.[52] The Iranian Government continued to implement nationalisation regardless and on 9 May the Senate elected its representatives on the mixed committee which was to carry out the nationalisation of the Company. On 13 May the Majlis followed suit.

The US reaction to the nine-point law was slower to emerge. In mid-May representatives of the major US oil companies made it plain to McGhee that they opposed the Iranian nationalisation and 'concession jumping'.[53] Following their representations, on 18 May the State Department issued an official statement opposing the unilateral cancellation of contractual relationships. The statement doubted whether it was in Iran's best interests to expel the Company and declared that US personnel would not be available or willing to replace the British technicians in Iran. The USA wanted the two parties to sit down at the conference table and work the matter out: in other words, to negotiate.[54] The Iranian response ranged from disappointment to indignation and the Iranian Foreign Ministry handed Grady an *aide-mémoire* stating that the US declaration constituted an interference in the internal affairs of Iran. Makki reportedly called it a 'stab in the back'.[55]

In Britain, tension mounted. Morrison replied to Musaddiq's letter of 8 May with a strongly worded message from Westminster eleven days later. He stated that while he did not wish to interfere with the sovereign rights of Iran, the nationalisation of the Company was not a legitimate exercise of sovereign rights. The 1933 concession, he pointed out, was a contract between the Iranian Government and the Company, concluded under the auspices of the League of Nations and ratified by the Majlis, so becoming Iranian law. The concession agreement provided for arbitration in cases of dispute between the parties and grievances should be settled by that procedure. If the Iranian Government refused to accept arbitration then, Morrison warned, the British Government would bring a complaint before the International Court of Justice at The Hague.[56]

The message had no impact on the Iranian Government, which not only failed to appoint an arbitrator, but invited the Company immediately to nominate representatives to attend a meeting of the mixed committee formed to put the nine-point law into effect.[57] Sir Eric Beckett, Foreign Office legal adviser, held a meeting on 22 May to examine the options open to the Company: 1) refusal to attend the meeting of the mixed committee; 2) attending as observers and 3) protesting and reserving its position, but attending the meeting and co-operating. None of those courses was thought to affect the rights of the British Government to declare a dispute between itself and the Iranian Government and to take the matter to the International Court of Justice.

The Company's failure to make an immediate response to the Iranian invitation drew forth a curt note from Ali Varasteh, the Finance Minister. On 24 May he wrote to Norman Seddon, who had succeeded Northcroft as the Company's chief representative in Tehran: 'I am waiting every day in the Finance Ministry for your representatives'. If they were not nominated by 30 May the Iranian Government would, Varasteh continued, proceed unilaterally with nationalisation.[58] Fraser replied that Seddon would attend the meeting as an observer only and announced that the Company was making application to the International Court of Justice for the appointment of an arbitrator in accordance with the provisions of the 1933 concession.[59] That was immediately followed by the British Government's institution of proceedings against the Iranian Government before the International Court. The British Government asked the Court to rule that Iran must agree to arbitration or be found guilty of violating international law. Shepherd notified Varasteh of the institution of proceedings in a lengthy note which concluded by emphasising that the British Government would rather settle the matter by negotiation and that if negotiations were successful the proceedings at the International Court could be halted before a judgement was given.[60]

Musaddiq, however, contended that the matter was not one for arbitration, as nationalisation was an exercise of Iran's sovereign rights. On 29 May he met Shepherd and Grady for discussions which Shepherd described as 'most unsatisfactory'. Musaddiq made it quite clear that he would accept terms of reference for negotiations only if they specified that the nationalisation law 'could not be questioned in any way, nor could any amendment of it be envisaged'. He seemed to have no idea that a marketing organisation was required to dispose of Iranian oil, thinking that the Iranians 'would simply make large contracts at market rates with those countries to which their oil had previously gone'. When Grady remarked that it would be difficult to replace British technicians, Musaddiq enquired 'whether the British were gods that others could not do what they could do'. After Grady referred to the unemployment and distress that would be caused if oil operations came to a stop, Musaddiq replied 'so much the worse for us'.[61]

THE JACKSON MISSION

So long as Iran insisted on nationalisation and Britain refused to concede it there was little prospect of the two sides coming to the negotiating table, let alone reaching a settlement of the dispute. However, in the last days of May and the early days of June the chasm between the two sides closed just far enough for them to agree to come together in negotiations. In mid-May, Attlee had already expressed his view that Britain would have to accept the principle of nationalisation.[62] Moreover, the US Government was, Furlonge reported on 28 May, 'pressing us to declare categorically that we are willing to accept nationalisation in principle'.[63] The next day, Morrison announced in the Commons that the British Government was prepared 'to consider a settlement which would involve some form of nationalisation, provided – a qualification to which they attach importance – it were satisfactory in other respects'.[64]

72 Anti-American banner depicting Uncle Sam being thrown out of Iran, carried by Iranian demonstrators in Tehran, May 1951

73 Muhammad Musaddiq (right) and Henry Grady (left), June 1951

On 30 May Varasteh wrote to Seddon reiterating the Iranian Government's intention to maintain oil supplies to the Company's customers and its willingness to provide for compensation to the Company by depositing 25 per cent of net oil revenues in a mutually acceptable bank. In other words, as Varasteh made clear, the Iranian Government did not propose to expropriate the Company without compensation, nor did it propose to hinder sales to buyers of Iranian oil. Varasteh went on to outline the arrangements by which the Iranian Government intended to implement nationalisation, which were: the mixed committee was to have full powers to administer all of the Company's operations; under the supervision of the mixed committee, a provisional board of directors of the NIOC was to be nominated; pending approval of the constitution of the NIOC the rules and procedures of the Company would remain in force, except where they conflicted with nationalisation; all Company employees, foreign and Iranian, would continue in their jobs, being regarded as employees of the NIOC. Finally, Varasteh invited the Company to make proposals, which the Iranian Government would consider so long as they did not conflict with the principle of nationalisation. The proposals were to be submitted within five days.[65] On the following day, 31 May, Truman sent messages to Attlee and Musaddiq advising them to negotiate. The USA, he wrote to Musaddiq, was a friend of both countries and was anxious that a solution be found by negotiations.[66]

Seddon, replying to Varasteh on 3 June, said that the Company and the British Government were ready to try to solve the dispute by negotiation. As the Company felt it could not submit proposals on such a complicated matter in five days, and as it believed that face to face discussions were preferable to written communications, Seddon proposed that Company representatives should visit Tehran as soon as possible for discussions with the Iranian Government.[67] After consultations with the Foreign Office, a Company delegation including Basil Jackson (deputy chairman), Sir Thomas Gardiner (one of the Government directors), Gass and Elkington made preparations in early June for a mission to Tehran with the full approval of the British and US Governments.

Before the Jackson mission arrived in Tehran the focus of attention swung back to the centre of the Company's operations in Khuzistan, where the Iranian Government lost no time in taking action to nationalise the Company. Amir Ala'i was appointed Governor-General and Special Representative of the Government in Khuzistan and took up his new post in Ahwaz on 3 June. On the same day Hassibi (a French-trained engineer), Murtiza Bayat (a planning agriculturist), and Dr Abdul Husayn Aliabadi (a lawyer) were appointed members of the three-man provisional board of the NIOC, under the supervision of the mixed committee. In fact Bayat and Aliabadi were members of both. Hassibi was replaced almost immediately by Mihdi Bazargan, another engineer, who years later was to become the first Prime Minister of the Islamic Republic.[68]

When Drake paid a courtesy call on Ala'i on 6 June, he learnt that some twenty-five officials, including members of the mixed committee as well as the three-man provisional board, were on their way from Tehran to Ahwaz.[69] They arrived in Khurramshahr on 10 June and were accommodated by Drake as guests of the Company.[70] There were speeches, the ritual slaughter of a sheep and the raising of the Iranian flag over the Company's main building by Ala'i. The office was designated 'Office of the provisional board of directors come to nationalise oil' and Ala'i informed a crowd of about a thousand that oil was now nationalised. A naval band played the Iranian national anthem.[71] Drake had a stormy interview with the provisional board after he was met at the door of his office by a soldier with a fixed bayonet. The board contended that Drake, like the rest of the Company's staff and workforce, was now an employee of the Iranian Government and that any refusal to co-operate with the Government representatives was illegal. They demanded that he hand over the organisation charts of the industry, a statement of oil export sales proceeds since 20 March and 75 per cent of all money received in Iran or London on account of oil sales after that date. Drake explained that he was responsible to the Company's board of directors and welcomed his visitors as officials of the Imperial Iranian Government, but not as the provisional board of the NIOC. He protested vigorously about the raising of the Iranian flag, the police posted at the doors and the speeches that had been made. Refusing to comply with the provisional board's instructions, he asked for them to be put in writing so that he could refer them to the Company's Head Office in London. Matin-Daftari, who was one of the members of the mixed committee and Musaddiq's son-in-law, stated that if the Company refused to obey the law, the committee would be obliged to notify the Government, which would apply sanctions and prevent the Company from exporting oil.[72]

While Drake was coping with the mixed committee and the provisional board in Khuzistan, the Jackson mission arrived in Tehran to try to negotiate an agreement which would meet the practical requirements of maintaining efficient oil operations, while also recognising

74   The Iranian flag being raised over the Company's offices at Abadan, June 1951

the principle of nationalisation. At the first meeting on 14 June, Varasteh declared that any settlement would have to be within the terms of the nine-point law. There was, he said, no possibility of reversing the Iranian Government's oil policy and as a condition of continuing the talks he demanded that the Company should recognise the authority of the provisional board in all matters concerning oil operations in Iran; give the provisional board a statement showing its transactions; and hand over to the Iranian Government its net revenues from the sale of Iranian oil since 20 March on the basis that the Iranian Government would deposit 25 per cent of the proceeds with a mutually agreed bank to provide for compensation to the Company. Jackson, while stating that he was ready to reach a settlement consistent with the principle of nationalisation, asked how the Iranians planned to market the oil internationally when Iran did not own any tankers. Varasteh contended that those were only details to be examined by experts in sub-committees once the main principles had been conceded. When Jackson suggested that this was putting the cart before the horse, Varasteh said in that case there could be no basis for further

75   Husayn Makki addressing a crowd at Abadan, June 1951

discussions. Jackson still wished to reach a settlement and they agreed to meet three days later on 17 June.[73] The Foreign Office and the Company were in general agreement that the Iranian demands were 'wholly unacceptable'. The State Department also regarded them as 'completely unreasonable ... designed to remove all hope of negotiations except on terms of complete capitulation'.[74]

When the talks resumed on 17 June, Elkington tried to persuade Hassibi of the immensity and complexity of the Company's operations, but Hassibi was unimpressed, believing that the Iranian oil industry was a 'child which the Western world would fondly nurture in any circumstances' and could not afford to lose. Hassibi was certain that Iran would have no problem because after nationalisation Iranian oil revenue would, he thought, greatly increase. Musaddiq summoned Mustafa Fateh, the Company's most senior Iranian employee, to his house on 18 June and informed him that his patience was running out, that he was not interested in the international aspects of the oil industry and that he would dispossess the Company from 20 June if it did not accept the Iranian terms.[75]

At the third meeting on 19 June the British delegation, having consulted with London, presented counter-proposals. They offered, on behalf of the Company, to advance the Iranian Government £10 million on the understanding that the Government would not interfere with Company operations while discussions were proceeding and, in addition, to make monthly advances of £3 million while the dispute lasted. As for arrangements which would maintain efficient operations while being consistent with the principle of nationalisation, the British delegation presented the following proposal: the Iranian assets of the Company would be vested in an Iranian national oil company which would grant the use of the assets to a new subsidiary to be established by the Company. The new subsidiary would have Iranian directors on its board and would, in effect, operate the Iranian oil industry using the assets owned by the national oil company. The Company's distribution business in Iran would be transferred to an entirely Iranian-owned and operated company.[76] The Iranian delegation rejected these proposals as 'completely at variance with the laws enacted in Iran for the nationalisation of oil'.[77] Grady had appealed to the Iranian Cabinet to support the Company's proposals, but Musaddiq reportedly answered that he would rather see the wells run dry.[78]

The situation in Khuzistan was becoming critical as Drake endeavoured to keep the oil flowing following threats from Makki, who was one of the Majlis deputies on the mixed committee, to shut off the valves. On 19 June, Drake used the imaginative negotiating tactic of recounting a made-up 'dream' to a meeting of the mixed committee and the provisional board. In Drake's 'dream', representatives of the Company's staff and the Iranian Government, together with an *Ittila'at* photographer, were flying through very bad weather in a plane that Drake was piloting. During the flight Makki told him that a law had been passed making the plane the property of the Iranian people and that he, Makki, would now take over as pilot even though he had never flown a plane before. Drake refused as he was responsible for the very valuable plane and the safety of the passengers. Makki returned to the cabin and made a long speech explaining that workers must make sacrifices for Iran to achieve its aims and that he must exercise his right to pilot the plane. Although some of the workers were impressed, certain members of the committee were uneasy at the bumping of the plane in the bad weather and advised that nothing should be done until the plane reached the ground. Makki returned to the pilot's cabin and asked that, if he could not pilot the plane, he be allowed to press one particular black knob on the instrument panel to satisfy himself

76 The Jackson mission returning from Tehran, June 1951. From right to left: Basil Jackson, Sir Thomas Gardiner, Edward Elkington (front) and Neville Gass (behind)

that he was sharing some of the control of the machine. Drake pointed out that although he sympathised with Makki, in fact pressing that particular knob would cut off the supply of petrol to both engines. At this Makki looked thoughtful and asked what he should do. Drake suggested that when the plane was safely on the ground the *Ittila'at* photographer should take a photograph of Makki at the controls in such a way that the plane should appear to be flying at a great height. It was at this point, Drake concluded, that he woke up.[79] The Government representatives were greatly amused by Drake's account, but it appeared to have little effect on Makki, whose enthusiasm for shutting down the valves was curbed later the same day by Daftari.[80]

TAKEOVER

The takeover of the Company's business in Iran came formally into effect the following day, 20 June, with the passing of decrees to the effect that no operating instructions issued by the Company's management would be valid unless countersigned by the provisional board; that Iranian officials would take over the Company's installations at Kirmanshah and Naft-i-Shah in western Iran; that Iranian officials would assume the supervision of the Company's Tehran office and its sales organisation in Iran; that the Company's information departments in Iran would be closed; that the Company's name would be replaced by that of the NIOC on all Company name boards in Iran; and that the Company's revenues from internal sales in Iran would be deposited in Government accounts.[81]

At Kirmanshah, the refinery was indeed shut down and the manager confined to his house. At the Abadan refinery, Iranian flags were raised and Musaddiq made a broadcast to the nation announcing the end of fifty years of the Company and imperialism in Iran. He blamed the Jackson delegation for the failure of the talks because they had refused to be guided by the nationalisation law.[82] On the same day the British Government, in consultation with the Company, recalled the Jackson mission from Tehran and began making arrangements for evacuation from Abadan, should it be necessary. In the Commons, Morrison warned that the British Government would not stand idly by if the lives of British citizens were placed in jeopardy.[83]

The State Department, while recognising that Musaddiq was only prepared to negotiate on the basis of complete capitulation and liquidation of the Company, advised not closing the door on negotiations completely or using force or sanctions, because the security of Iran

77   British staff dependents returning to England from Iran, June 1951

took precedence over everything else. Julian Holmes, the US Minister in London, met Strang on 20 June and expressed concern about the possible use of force. Strang replied that it could not be excluded.[84] When Holmes saw Strang again three days later, he told him that the US Government would do anything it could to help, but had not been able to think of any useful course to keep Iranian oil flowing.[85]

78   Iranians removing the illuminated sign at the Company's information
office in Tehran, June 1951

Activities against the Company increased in Tehran. On 21 June the
police surrounded the chief representative's office and refused to allow
documents to be taken out. A large crowd forced their way into the
Company's main office and destroyed the electric sign bearing the
Company's name. Police closed the information office and stopped all
mail, and Company signs on offices and road tankers were demolished
or obliterated. The occupants of the information office, P. A. Stockil
and B. G. Stiles, moved temporarily to the chief representative's town
house until it was invaded on 30 June.[86]

There was similar activity in the south as the brass nameplate was
taken off the Khurramshahr office at 8 am on the 21st. In the course of
that day Drake received no less than nine letters of instructions from
the provisional board. All staff leave was cancelled by the board and
he was warned that movements of British staff between Company
areas must cease from 22 June. Drake was instructed to hand over the
proceeds of all sales of oil in Iran to the local government office
representing the Iranian Finance Ministry.[87] In addition, the pro-

visional board posted a proclamation stating that Company staff and workers were now employees of the Iranian Government and their salaries, allowances and position would remain unchanged.[88] Drake was ordered to state in writing whether or not he was willing to serve the NIOC and carry out its instructions. Jackson had advised him to hold his post as long as possible, writing 'We are confident that you will know how to achieve this without surrendering anything essential'.[89] Management at Khurramshahr decided that it should be business as usual. They would not dissolve the Information Department, nor pay the proceeds of sales to the Iranian Government. However, as a precaution, exit visas were to be obtained for all foreign employees, and staff were advised to pack their kit ready for despatch if need be.[90]

On the evening of 21 June the Company's printing press was commandeered for the purpose of printing receipts for tanker captains to sign, stating that their oil cargoes were received from the NIOC, not the Company.[91] Drake tried to have a disclaimer added so that the receipts would not concede that the NIOC owned the oil. He informed Fraser that he thought he would be dismissed 'since I can see no means of complying with the majority of the temporary board's instructions consistent with my duty to AIOC'. Under exceptional personal pressure, he wrote that he would carry on to the last possible moment.[92]

Throughout 23 June constant efforts were made to reach a compromise over the matter of the receipts to be signed by tanker captains, but Makki and the Chief of Customs would not accept anything other than receipts which acknowledged the NIOC's title to the oil. A further serious development was the introduction of a Sabotage Bill into the Majlis on 21 June. Under its terms any person found guilty of malicious designs and treacherous plots resulting in an interruption to operations would receive a punishment ranging from imprisonment with hard labour to the death penalty. Two days later the provisional board wrote to Drake accusing him of ill intention and sabotage in failing to produce valid oil receipts.[93] The warning could not be ignored and Drake doubted if it was fair to ask the British staff to continue working if the Bill became law. Capper, the Consul-General, asked Shepherd for urgent guidance, in view of the fact that Drake was accused of a capital crime. Capper thought Drake should go to Basra and Shepherd agreed. Shepherd promised to inform the Iranian Foreign Minister that he would be unable to advise British subjects to work in Iran if the Bill became law.

79   Crowd in Tehran carrying one of the signs torn down from Company premises, June 1951

At a meeting on 25 June the mixed committee refused to withdraw the letter to Drake accusing him of sabotage and would not even speak to Drake in the presence of Capper, who was obliged to withdraw.[94] On the advice of Shepherd and Capper, Drake left immediately for Basra. He informed Fraser of his action, stating that he was determined to direct affairs from Basra 'unless I hear to the contrary from you'.[95] Drake had been asked repeatedly to confirm his acceptance that he would henceforth be an employee of the NIOC, not the Company. Drake's response was always that he had passed the matter on to the Company's Head Office in London and was awaiting a reply: The Iranians' patience having finally run out, Drake was given three days, until 28 June, to comply with all their demands or be dismissed. The notice was received in Drake's absence by Alick Mason, his deputy, who took over as the Company's representative for the purpose of contact with the provisional board.[96]

On 26 June, Morrison reviewed the situation in the House of Commons. He made it clear that if the Iranians persisted in their measures, operations at Abadan would come to a stop in a few days. There were naval vessels already stationed nearby, two frigates off

Bahrain and a cruiser in the Red Sea. The tank-landing ship, *Messina*, arrived off Basra on 26 June and in addition troop movements towards the Mediterranean were attracting the notice of both the US and Iranian Governments.[97] Moreover, the cruiser, HMS *Mauritius*, had been sent to the vicinity of Abadan to protect British subjects in Iran should the Iranian Government 'prove incapable of discharging that task'.[98] The Shah was greatly displeased and told Shepherd that if the British Government attempted to violate Iranian sovereignty he personally would lead Iranian forces to 'resist the aggressors'.[99]

On 27 June the British staff unanimously rejected the Iranian Government's offer of employment with the NIOC. The USA had already stated that US technicians would be unwilling to help the Iranian Government and their lead was followed by refusals of assistance from Sweden, Belgium, the Netherlands, Pakistan and Germany. Italy was the only country to send an adviser.[100] Jackson had made it clear in the British press that the passage of the Sabotage Bill would mean the withdrawal of the Company's staff from Iran and the Iranian Government announced on 28 June that the Bill would be withdrawn if British technicians would reconsider the offer of employment by the NIOC. In the event, the Bill was withdrawn on 1 July without any change of mind on the part of the British staff.

As Drake was deemed to have resigned on 28 June, the mixed committee and provisional board arrived at his office at 9.20 that morning and demanded that Mason, who was occupying the office on Drake's instructions, should vacate it. After protests, Mason and his staff were forced to leave their offices in the hands of the Iranian authorities.[101] In consultation with the Company, British ministers reassessed the situation and decided that all tankers should be withdrawn from Abadan until the Iranian attitude on receipts was modified. The withdrawal of tankers meant that oil exports would cease, storage capacity at Abadan would soon be filled and the refinery and oilfields would have to be closed down. Personnel in the Fields would be withdrawn to Abadan when their presence in the oilfields was no longer required.[102]

On 29 June, Morrison broadcast a message to the British staff still in Iran, warning them that 'the refinery and field operations may have to be closed down at any rate for the present'. In an endeavour to boost their spirits, he praised them for their fortitude, called for dignity, assured them that all practical measures would be taken for their protection and wished them good luck. Drake reported to Fraser that morale was high and staff would remain at their posts until there were serious signs of insecurity. However, after the failure of the Jackson mission, the departure of Drake, the seizure of the Company's main offices and the stoppage of oil exports, the evacuation of the Company's staff looked ever more likely.

CONCLUSION

In the events which followed the withdrawal of the Supplemental Agreement from the Majlis in December 1950, the Company's position became increasingly enmeshed in the domestic political struggles which were going on inside Iran and in the international relations between states who felt their interests were affected by a dispute which went far beyond the bounds of commerce and was, without exaggeration, of geopolitical concern.

Inside Iran the sides for and against nationalisation were personified by Musaddiq and Razmara respectively. The struggle between them ended with the death of Razmara and a series of political triumphs for Musaddiq in March–April 1951, marked by the single article law of 20 March adopting the principle of nationalisation, Musaddiq's appointment as Prime Minister in the last days of April and the adoption of the nine-point law of 1 May on the implementation of nationalisation.

In a manner characteristic of periods of revolution, the speed of Musaddiq's rise on a wave of nationalistic fervour left little time for calculated response. Britain seemed consistently to be a step behind Musaddiq, as Iranian demands outstripped anything that was offered by the Company and the British Government. After failing with the Supplemental Agreement, which was made obsolete by the Saudi 50:50 agreement, the Company offered to discuss a 50:50 division of the profits arising from its operations in Iran. The British Government gave its approval to the 50:50 terms embodied in the proposals which were submitted by Shepherd to Ala in the last week of April, but it was too late for an agreement on those lines. By that time Musaddiq and the oil committee which he chaired were on the verge of winning Majlis and Senate approval for nationalisation.

In June, the Jackson mission made proposals which accepted the principle of nationalisation by suggesting that ownership of the Company's assets in Iran should pass to an Iranian national oil company. However, by proposing that the use of the assets should be granted to a new subsidiary to be established for the purpose by the Company, the Jackson mission's offer, if accepted, would have meant that the control of operations remained, in effect, with the Company. That was not acceptable to the Iranians.

Increasingly, it was becoming apparent that the British and Iranians were, for the most part, on different wavelengths. The British Government and the Company tended to lay emphasis on the *technical* and *economic* problems of running the Iranian oil industry, whose efficient operation depended, in their view, on managerial, commercial and technical ability and experience. From that standpoint it seemed reasonable, as Shepherd expressed it in October 1951, that the need for Iran was 'not to run the oil industry for herself (which she cannot do) but to profit from the technical ability of the West'.[103] For Musaddiq and his supporters, on the other hand, nationalisation had a *political* importance which transcended purely economic considerations. They therefore tended to be dismissive of operational matters and to act as if technical and commercial problems were minor irritations in the greater cause of anti-imperialism and the assertion of Iranian control over Iran's national resources. The gap between the two sides yawned wide and at the end of June 1951 there was no obvious mechanism to bring them together.

= 17 =
Eviction,
June–October 1951

THE INTERNATIONAL COURT'S INTERIM RULING
AND THE HARRIMAN MISSION

Although the British Government had referred the Anglo-Iranian dispute to the International Court of Justice on 26 May 1951, the case had still not been heard by the time the unsuccessful Jackson mission departed from Iran on 22 June. That day, the British Government applied to the International Court for an interim ruling to protect the Company's position, which might otherwise have suffered irreparable damage before the case was heard.[1] The Iranians, however, argued that nationalisation was an exercise of rights of national sovereignty which lay outside the Court's jurisdiction and that the British Government was not a party to the dispute, which was between the Iranian Government and the Company. They therefore refused to be represented at the Court's hearings, which opened on 30 June.[2]

Without prejudging the question of whether the dispute fell within its jurisdiction, the Court issued an interim order on 5 July. It provided, in essence, for a return to the state of affairs before the nine-point law was passed on 1 May 1951. In particular, the interim order stated that: 'no measure of any kind should be taken designed to hinder the carrying on of the industrial and commercial operations of the ... Company ... as they were carried on prior to 1st May, 1951'.[3] On the matter of managerial control, the Court's ruling was explicit that the Company's operations should continue under the direction of its management as it was constituted before 1 May, subject only to such modifications as might be agreed by a Board of Supervision, which was to be formed with two members appointed by the British and Iranian Governments respectively and with a fifth member from a third country. The British Government accepted the Court's interim order, but the Iranian Government refused to recognise the injunction and withdrew its earlier acceptance of the Court's compulsory jurisdiction over Iran.[4]

In the meantime, on 4 July Sir Oliver Franks, the British Ambassador to Washington, met Dean Acheson, the US Secretary of State, and other members of the State Department in Washington. The contrasting opinions expressed by Franks and Acheson reflected a fundamental difference in outlook between their respective governments which arose time and again during the Anglo-Iranian dispute. Franks, putting forward the British view, commented that Musaddiq's 'fanatical intransigence' made negotiations impossible. Acheson, on the other hand, voiced the US fear that if the Iranian Government was deprived of oil revenues it might collapse, providing an opening for increased communist influence.[5] He suggested sending a special envoy to Iran and thought that Averell Harriman, who had worked closely with Roosevelt during World War II and had served as US Ambassador to Moscow and to London, might be suitable for the task. The idea of neutral US mediation in the dispute was not one which appealed to the British Foreign Secretary, Morrison. Taking the line which was persistently followed by the British Government during the dispute, he would have liked the USA not to stand in the middle, but to come down firmly on the side of Britain. He wanted the USA to make a categorical statement that Iran should abide by the International Court's interim ruling and that the USA could 'give no sympathy or help to a country which flouts a decision of the world's highest legal authority'.[6] 'What we need now', insisted Morrison, 'is a forthright statement of American support'.[7] That was not, however, forthcoming.

Rather than take sides, President Truman wrote to Musaddiq on 9 July, urging him to give careful consideration to the recommendation of the International Court and offering to send Harriman as his personal representative 'to talk over with you this immediate and pressing situation'.[8] Musaddiq replied that Iran had rejected the International Court's interim ruling, but was prepared to enter into discussions provided the principle of nationalisation was accepted. He said he would welcome a visit from Harriman.[9]

By this time the dispute was having an increasing effect on the Company's staff and operations in southern Iran. Not only had all Company tankers left Abadan, but all British personnel had been withdrawn from the oilfield at Gach Saran and women and children had been evacuated from Abadan where, according to Shepherd, 'the atmosphere has become one of siege'.[10] The British staff, Shepherd reported, were critical of official policy and felt that: 'they were being used as pawns in the game of international politics and only their loyalty to the Company has kept them together as a team for so long. The corporate spirit is still strong and impressive but there is a definite limit to the strains it will stand'.[11] On 4 July, the Consul-General at Khurramshahr wrote that the interference of the provisional board was assuming 'menacing proportions'.[12] On 15 July he added that the Iranians were forbidding the free movement of labour and Company property. Company vehicles were being stopped in the streets by soldiers who were searching both vehicles and occupants, and in Abadan 'a serious incidence of thieving from British houses has begun and the military are taking no action to counteract it'.[13] Other forms of interference ranging from minor irritations to more serious harassment also took place.[14] Moreover, in mid-July a decree was issued, declaring that all contracts of the Company were invalid and offering Iranian oil to all comers on a cash-and-carry basis.[15]

It was against this background that Harriman arrived in Tehran on 15 July with William Rountree of the State Department; Walter Levy, the US oil expert; and Vernon Walters as interpreter.[16] Their initial impressions cannot have been favourable, for they were greeted with Tudeh-inspired mass demonstrations carrying banners and chanting anti-British and anti-American slogans.[17] Violent clashes between Tudeh groups and National Front supporters outside the Majlis buildings resulted in a number of fatalities and several hundred wounded. Martial law was declared the following day.[18] As for Harriman's talks with the Iranian Government, they did not get off to a good start. After lunching with the Shah and seeing Musaddiq, Harriman told Shepherd that his discussions with Musaddiq had been 'completely unfruitful' and that Musaddiq 'did not want to discuss anything with anybody. He blamed the British for Persia's misery'.[19] To make matters worse, on 19 July the Iranians withdrew the residence permit of Seddon, the Company's chief representative in Tehran, and told him that he must leave the country within a week.[20]

Within a few days of his arrival Harriman began, however, to be more optimistic about the possibility of the Iranian Government agreeing to negotiations with a British Government mission. He found that in general the Iranians were sensible, though a few, such as Hassibi, remained 'completely intransigent', unmoved by Levy's repeated attempts to explain the economics of the international oil industry and the need for Iran to work with an established foreign oil company for

80 Iranian women forming part of an anti-British/American demonstration in Tehran, July 1951

the marketing of its oil. In similar vein, Harriman tried to impress upon the Iranians that if they interfered with the Company's complex operations they 'would be cutting their own throats' as they could not expect customers of the Company simply to switch to the NIOC.[21]

On 22 July, after Seddon's permit had been returned, Harriman thought that matters were 'moving in a definitely hopeful direction'.[22] The next day he attended a joint meeting of the Council of Ministers and the mixed oil committee at Musaddiq's house, where he was presented with Iranian proposals which came to be known as the 'Harriman formula'. In essence the formula stated that if the British Government formally accepted the principle of nationalisation the Iranian Government would be prepared to enter into negotiations on the manner in which the nationalisation law was to be carried out. The nationalisation law was explicitly defined as the law of 20 March 1951, which included the key statement that 'all operations for exploration, extraction and exploitation shall be in the hands of the [Iranian] Government'. This wording was reiterated in the Harriman formula.[23]

The British Government's initial reaction on receiving the formula was that it would send a mission to Tehran for negotiations only on the precondition that the Iranians stopped interfering in the Company's operations. The British would not, as the Foreign Office put it, 'accept the present position under which the whole of the Company's operations are being brought to a standstill'.[24] If Iranian interference continued British public opinion would, the Foreign Office went on, look upon the despatch of an official mission as a 'complete surrender' to Iran's 'aggressive and intransigent tactics'.[25]

Harriman, concerned that British insistence on preconditions might wreck his efforts to bring the two sides to the negotiating table, flew to London, accompanied by Levy and Shepherd. He succeeded in persuading the Government to take a more emollient line and on 3 August Britain formally notified Iran that it recognised the principle of nationalisation and would send an official mission to Tehran to negotiate in accordance with the Harriman formula.[26] Instead of insisting on specific preconditions, the British communication to Iran merely referred in vague terms to the need to relieve the 'present atmosphere', to which the Iranians replied in similarly evasive language.[27] The obstacle of British preconditions having thus been circumvented the way was now clear for the despatch of a mission to Tehran for negotiations. It was decided that the mission should be led by Richard Stokes, the Lord Privy Seal, who, Franks told Acheson, was a 'bluff, genial and hearty man'. Acheson later recalled in his memoirs that this description 'proved to be sheer flattery'.[28]

Before the mission set off, discussions were held in London on the line which Stokes should take in his negotiations with the Iranians. One of the participants was Levy, who reported on his discussions in Tehran and caused something of a stir when he gave his views on the kind of oil settlement which he thought the Iranian Government might accept. At a meeting on 29 July, at which Gass was present for the Company, Levy said that in his opinion the Company 'had virtually no bargaining position whatever'.[29] He thought there was no chance of a settlement being reached unless three substantial concessions were made to the Iranians. One was that the British would have to agree to the Company giving up its position as the sole distributor and marketer of Iranian oil so that it would no longer enjoy a monopoly position as the only outlet for Iran's oil. Secondly, to remove the impression that the Company was an instrument for political interference in the internal affairs of Iran, Levy thought that the British Government would have to eliminate, or at least heavily camouflage, its shareholding in the Company. Thirdly, concerning the management of oil operations in Iran, Levy believed that the Iranians would accept a foreign-owned operating company, but not, and this was a vital point, if it merely represented the return of the Company in disguise. The operating company would, Levy suggested, have to be a consortium in which the Company might have a 40 per cent share, the balance being in the hands of other companies.[30] In short, Levy did not think that the Iranian Government would agree to a settlement in which the Company retained its position as the sole operator of the oil industry in southern Iran and the exclusive marketer of the oil it produced there. Such terms had, of course, already been offered by the Jackson mission and rejected by the Iranian Government in June (see chapter 16). At the very least, Levy believed that the Jackson proposals would have to be 'drastically camouflaged and presented in an entirely new way' to have any chance of success.[31]

No sooner had Levy expressed his views than Sir William Strang, the Permanent Under-Secretary at the Foreign Office, wrote to Morrison giving his assessment of the likely reaction of the Company, and especially of Fraser, whose determination to uphold the Company's position in Iran was known in Government circles. Strang was in no doubt that if the concessions suggested by Levy proved to be necessary:

> This will cause a major row with Sir W. Fraser and his Board who are very competent, but very stubborn and not very broad-minded people. There will have to be a kind of revolution in management, and I don't think that Sir W. Fraser is the kind of man either to conduct it or even to recognise its necessity.[32]

At much the same time, Gass, having heard Levy's views on 29 July, wrote to the Foreign Office on the 31st enclosing an 'agenda' for a settlement of the oil dispute. In this document the Company accepted the principle of nationalisation and suggested that its assets in Iran (excluding those used for local distribution and marketing) should be transferred to the NIOC, which would grant the use of the assets to a new subsidiary to be established by the Company. The subsidiary would manage the operations of exploration, production, transportation and refining on behalf of the NIOC and would sell the oil it produced to the Company. Only the Naft-i-Shah oilfield, the Kirmanshah refinery and the Company's facilities for domestic distribution and marketing, all of which served the internal Iranian market, would be transferred to and operated by an organisation established by the Iranian Government. In every other respect, however, the Company would remain the sole operator of the Iranian oil industry in the area covered by the 1933 concession and the sole seller of the oil it produced for export.[33] These proposals were virtually the same as those put forward by the Jackson mission. The Company, in other words, was not offering any significant new concessions.

Meanwhile, Martin Flett of the Treasury drew up a paper assessing the effects which the adoption of Levy's suggested concessions would have on the national interest, the Exchequer and the Company. His conclusions were:

> At first blush it looks as if the sort of scheme for which Mr Levy is groping would not be intolerable either to the national interest or to the Exchequer. It is, however, very difficult indeed to see how such a scheme could be reconciled with the interests of the Company as at present constituted. So far, all our thinking has been on the assumption that the AIOC suitably camouflaged, would retain full responsibility for Persian operations. The whole essence of Mr Levy's approach is that no amount of camouflage will be acceptable and that we must offer the Persians something radically different. There is no doubt whatever that if we could make a camouflaged AIOC acceptable to the Persians this would be the best solution from our point of view.[34]

On 31 July, Flett's paper was discussed at a meeting between Hugh Gaitskell (Chancellor of the Exchequer), P. Noel-Baker (Minister of Fuel and Power), Stokes and various Government officials. Gaitskell made it clear that he was not opposed to Levy's suggestion that the Government might give up its shareholding in the Company, provided that the financial arrangements were satisfactory.[35] However, he was concerned about the effects on the Exchequer of adopting Levy's ideas. On the matter of how Stokes should handle the negotiations with the Iranian Government, it was suggested that it would be better not to mention the *aide-mémoire* containing the Jackson proposals which had been presented to the Iranians in June, but 'rather to dress it up and present its main points in a different order together with trimmings or sweetenings as might be required'.[36] In other words, the Jackson proposals were merely to be camouflaged without incorporating any of Levy's more radical suggestions.

THE STOKES MISSION

Stokes, at the head of a delegation which included Sir Donald Fergusson, the Permanent Secretary of the Ministry of Fuel and Power, and other officials from the Treasury and Foreign Office, arrived in Tehran on 4 August.[37] Harriman and Levy (with Rountree) were already there, having reached Tehran from London on 31 July, on which day the Abadan refinery had to be shut down, its storage

81   Company observers in the Stokes mission leaving for Iran on 3 August 1951. From right to left: W. S. McCann, Joseph Addison, Duncan Anderson, Harold Snow and Edward Elkington

capacity having been filled.[38] For the Company, a small party headed by Elkington was on hand and held a number of discussions with members of the British delegation and their American counterparts. The Company representatives did not, however, take part in any of the official meetings between the British and Iranian negotiators.

On the morning after his arrival Stokes met Musaddiq, who launched into a long harangue about the 'wickedness' of the Company.[39] After joining Shepherd for lunch with the Shah, Stokes saw Musaddiq again in the evening and spoke to him 'more forcibly' about the practical matters which would have to be dealt with in settling the oil dispute. Stokes feared, however, that his 'homily had little effect'.[40]

On 6 August the first official meeting between the British and Iranian delegations took place without noteworthy incident and on 7 August, having got past the introductory phase, Stokes, accompanied by Elkington, visited Abadan where he was taken by members of the NIOC board on a conducted tour of the island. Harriman visited Abadan the same day, but travelled separately and spent most of the

day with the NIOC.[41] Elkington, writing to Fraser, described the Abadan visit in characteristically cavalier language:

> The honours went to Makki, who took charge of the tour of the refinery, and created such confusion that the enormous cavalcade got caught up in knots, circles and figures of eight in the dust, grime and heat of all the slums and villages that could be found on the Island. Luckily, Stokes, who had been in Abadan in 1945, well knew the difference between Company property and that which was not, and expressed himself forcibly on the subject.[42]

The staff at Abadan, Elkington went on, 'gave us a great reception' and were 'full of guts'.[43]

Elkington also gave an informative account of the accommodation arrangements for the negotiators who were gathered in Tehran. They were staying in a palace built by Nasir al-Din Shah, the Qajar who ruled Iran from 1848 until he was assassinated in 1896. Stokes was in a large building and his staff in 'various detached villas which the Shah's harem occupied in the good old days'. Facing the building occupied by Stokes was the palace in which Harriman and his wife were accommodated. The entrances to this 'garden of paradise', as Elkington called it, were heavily guarded by armed police. These arrangements were 'convenient to the extent that Anglo/American relations can proceed unseen by the outside world', but as the telephone exchange was 'practically useless', Elkington had asked Snow, who was one of the Company party, to 'go and live in one of these love-nests, which, good fellow that he is, he has consented to do'. Snow shared his accommodation with two official members of the Stokes mission. In this way, Elkington was able to keep in touch with what was going on in the negotiations, which recommenced on 8 August, after the visit to Abadan.[44]

On 11 August, after further meetings between the British and Iranian delegations, Stokes handed an eight-point memorandum containing proposals for an oil settlement to Javad Bushihri, Minister of Communications, with the intention that Bushihri should sound out

82 (*opposite*) Richard Stokes and Averell Harriman visiting Abadan, 1951. Numbers on the photograph show:
1. Mihdi Bazargan, chairman of the provisional board of the NIOC
2. Averell Harriman
3. Husayn Makki
4. General Kamal, military governor of Abadan
5. Richard Stokes

83   Slum housing on Abadan island, 1951

Musaddiq's reaction to it before Stokes formally submitted it to the Iranian delegation.[45] The preamble to the memorandum stated that it was 'governed by the principle that the Anglo-Iranian Oil Company will cease to exist in Persia and that the Persian Government will acquire full authority over exploration, extraction and exploitation of oil in Persia'.[46] The most important proposals were: 1) the Company would transfer its assets in Iran to the NIOC and receive compensation; 2) a purchasing organisation would be formed to provide an assured outlet for Iranian oil by entering into a long-term contract with the NIOC; 3) the NIOC would be free to sell oil to other buyers provided such sales did not prejudice the interests of the purchasing organisation; 4) the purchasing organisation would 'make available' to the NIOC an operating organisation which would manage oil operations in Iran on behalf of the NIOC; 5) the terms on which the purchasing organisation would buy oil from the NIOC would be so arranged that there would be a 50:50 division of the profits made from Iranian oil.[47]

Despite the wording of the preamble it is apparent, on closer examination, that the eight-point memorandum was really little more than a re-dressed version of the Jackson proposals. Elaborating on the memorandum to the British Government, Stokes made it clear that the British delegation felt it was 'most essential from the point of view of the Company's future, public opinion at home and our balance of payments, to keep effective control of distribution of Persian oil in world markets in the hands of Anglo-Persian [the Company]'.[48] The British negotiators realised that it would be politically impossible to retain the name AIOC for the purchasing company, but Stokes assured the Government in London that 'it is our intention that the [purchasing] company should be 100 per cent AIOC subsidiary or AIOC under a new name, unless the Persians force us to contemplate a *consortium* [emphasis by Stokes]'.[49] As it was the purchasing organisation which was to 'make available' the operating organisation, the eight-point memorandum can be interpreted only as an attempt to keep the operation of the oil industry in Iran and the marketing of Iranian oil in the hands of the Company with no more than a change of name.

In London, Morrison and Gaitskell agreed that Stokes should proceed with the formal presentation of the eight-point memorandum to the Iranians, subject to the condition that on three points 'there should be no weakening'. These were: that the operating company must be in British hands; that the purchasing organisation must be wholly owned by the Company and be managed and controlled from

Britain; and that Iran should not receive more than 50 per cent of the profits from its oil.[50]

While the British Government approved the eight-point memorandum, Musaddiq's first reaction was that it was unacceptable, as Bushihri reported to Stokes on 12 August.[51] Nevertheless, Stokes went ahead and formally presented the memorandum, with minor changes of wording from the earlier draft, to the Iranian delegation on 13 August.[52] The Iranians, however, were swift to reject the proposals, arguing that they failed to conform to the Harriman formula, which had stipulated that the negotiations were to take place on the basis of the nationalisation law of 20 March 1951, in which it was stated, as seen earlier, that oil operations in Iran were to be 'in the hands of' the Iranian Government. Far from achieving those ends, Stokes' proposals would, said the Iranians, 'not only take out of the hands of the Persian Government a substantial part of the powers of management of the oil industry, but ... also revive the former Anglo-Iranian Oil Company in a new form'.[53] There were three features of the proposals to which the Iranians particularly objected. One was that, in their view, the purchasing organisation would have an effective monopoly on sales of Iranian oil because the NIOC's right to sell oil to other buyers was subject to the condition that it was not to prejudice the interests of the purchasing organisation. Secondly, the Iranians found it unacceptable that the purchasing organisation should receive 50 per cent of the profits made from the production of Iranian oil. Thirdly, they felt that the establishment of the operating organisation would be contrary to the principle of nationalisation, constituting a limitation on the sovereign rights of Iran and representing the Company 'under a new guise'. Moreover, the Iranians pointed out that similar proposals had been made by the unsuccessful Jackson mission.[54] There were, argued the Iranians, only three matters for negotiation: the payment of compensation to the Company; arrangements for the continued employment of British technicians; and the terms on which Iranian oil would in future be supplied to Britain.[55]

Though rebuffed by the Iranians, the British, anxious as ever to have the Americans on their side, were given a fillip on 19 August when Harriman attended a meeting of the two delegations and spoke out in support of the eight-point proposals. Then, and in a letter written two days later, he stated that the seizure of foreign-owned property without either paying prompt and adequate compensation or working out new arrangements satisfactory to both parties was 'confiscation rather than nationalisation'.[56] However, his endorsement of the Stokes proposals failed to move the Iranians. On 20 August Stokes and Harriman met with Musaddiq, who still insisted that the only matters for negotiation were the making of arrangements to supply Britain with oil in future; the retention of British staff in Iran; and compensation.[57] Further discussions failed to release the deadlock and on 21 August Stokes withdrew his proposals. On the 23rd he left Iran, emphasising that the negotiations had only been suspended, not broken off.[58] Harriman departed from Tehran the day after. In the meantime, Mason reported that he was making arrangements for the complete withdrawal of British personnel from the oilfields to Abadan, where the number of Britons was to be reduced to a hard core of about 350.[59] By 26 August the evacuation of British staff from the oilfields was complete, their arrival at Abadan heightening the sense of siege felt by the expatriate community which was concentrated there.

EVICTION

Although the Stokes mission had ended in failure, the British could at least draw satisfaction from the public support which Harriman had given to the Stokes proposals when he was in Tehran. Further signs that Harriman had become impatient with the Iranians were given when, stopping in London on his way back to Washington, he told British ministers that he had made it clear to the Iranians that the US Government was 'not prepared to save them from their folly'.[60] Believing that it would be impossible to reach an agreement with Musaddiq so long as he was surrounded by his existing advisers, Harriman thought it would be best to deal with the Shah, who was a 'stable influence and intelligent' and could, in Harriman's opinion, bring about a change of prime minister in Iran, although he would require the support of public opinion and the army.[61]

In the opinions of the British and American negotiators who had been in Tehran it was not, however, only Musaddiq and his advisers who stood in the way of a settlement. So too, they thought, did Fraser and other directors and managers of the Company. Fergusson felt that Fraser, who dominated the Company and was determined not to relinquish its concessionary rights, should cease to be chairman, a view which he said was shared by the Governor of the Bank of England and the National Provincial Bank.[62] Levy was of like mind. After seeing Fraser and other members of the Company's board on 27 August, he told Eric Berthoud at the Foreign Office that Fraser 'had learned absolutely nothing from past events and was completely out of touch

with the position'.[63] The same applied, in Levy's view, to many of the
other directors and senior staff. Levy said that his opinion of Fraser
was shared by Harriman and that the removal of Fraser would help
towards a solution of the Iranian dispute. He concluded by saying that
co-operation between the British and US teams in Tehran had been
excellent, with the result that the previous suspicions of both sides had
been 'very largely dissipated'.[64]

However, over the next few weeks the fragile nature of Anglo-
American unity became apparent as the two countries diverged in their
assessments of the Iranian situation. The British view was that it was
impossible to negotiate an agreement with Musaddiq and that the best
hope of reaching a settlement lay in encouraging the Shah to replace
Musaddiq's Government with a more 'reasonable' one. Any suggestion that Britain was prepared to negotiate with Musaddiq would, it
was thought, help him to remain in power. From this it logically
followed that a settlement was most likely to be reached not by
negotiation with Musaddiq, but by refusing to negotiate with him.
Shepherd, the British Ambassador to Tehran, was a notable exponent
of this hard-line view and was described by Acheson as a 'disciple of
the "whiff of grapeshot" school of diplomacy'. His reports to London
were an important influence on British thinking.[65]

The Americans took a different view. In the aftermath of the Stokes
mission the State Department agreed with the British that it was
impossible to negotiate a settlement with Musaddiq, but was far more
cautious about urging the Shah to replace Musaddiq and nervous
about what might happen in Iran if he went. Given these doubts, the
Americans were much less absolute than the British in their belief that
no agreement could be reached with Musaddiq and were inclined,
when the pressure mounted, to favour continued negotiations in the
hope of reaching a settlement. They tended, in other words, to draw
back from the brink of a full breakdown in negotiations.

The difference in outlook between the British and Americans lay
behind, and helps to explain, a sequence of events in which the same
pattern was repeated, with variations, on several occasions during the
oil dispute. In that pattern, the British tended to favour taking a hard
line with Musaddiq, which meant not making concessions even if this
might result in a breakdown in negotiations. The Americans, on the
other hand, repeatedly used their influence to put pressure on the
British to make concessions to the Iranians in order to keep the
negotiations alive. The British tended to be annoyed by the USA's
preparedness to 'give in' to Musaddiq, rather than take a final stand in

defence of their own position. However, Britain, like Iran, was fully aware of the vital importance of having the support of the all-powerful USA. Consequently, the British were consistently restrained from taking a more extreme course by the need to avoid a rupture in relations with the USA. As for the US Government, it found itself in the awkward position of mediator between two unyielding sides, each of which constantly endeavoured by the usual processes of diplomatic manipulation – most commonly trying to manoeuvre the other side into a position where it appeared to be in the wrong – to win the USA over to its side.

The first round in the escalation of the dispute after the failure of the Stokes mission was opened by Musaddiq in a speech to the Iranian Senate on 5 September. In his speech Musaddiq went through the usual invective against Britain, explained why he could not agree to the Stokes proposals and repeated his counter-proposals that there were only three matters for negotiation with Britain: the retention of British technicians to work in the Iranian oil industry, the payment of compensation to the Company, and the terms of future sales of Iranian oil to Britain. Musaddiq declared that he was prepared to reopen negotiations on these points and said that he intended to give Britain two weeks in which to resume negotiations on the basis he proposed, or to put forward alternative proposals. If a satisfactory answer was not received from Britain within the fortnight, the residence permits of British technicians would be cancelled. In other words, the Company's British staff would be evicted from Iran unless the British Government agreed to negotiate on Musaddiq's terms. Musaddiq then asked the Senate for a vote of confidence, which he won by an overwhelming majority.[66]

The Foreign Office responded the next day with a statement that Musaddiq's speech showed conclusively that no further negotiations with the Iranian Government could produce a result and that the Stokes negotiations, which previously had only been suspended, were now broken off.[67] At the same time the Company issued a notice stating that it was still the owner of the oil produced in southern Iran and warning that if any firm or individual tried to purchase Iranian oil from the Iranian Government the Company would take whatever action might be necessary to protect its rights.[68]

In the meantime, the British Government had for some days been considering the wording of a message for Shepherd to deliver to the Shah, urging him to replace Musaddiq's Government with a more reasonable one and warning that Britain might have to take economic measures which, though ostensibly for the purpose of protecting the British economy from the effects of the oil dispute, were obviously in fact designed to hurt Iran.[69] In discussions with the State Department it became apparent that the USA was deeply disturbed at the idea of putting pressure on the Shah to replace Musaddiq, but was prepared to 'find ways' of not proceeding with the loan to Iran which was under discussion with the Export Import Bank. No objections were raised to Britain taking its own economic measures.[70]

The British Government intended that Shepherd would urge the Shah to replace Musaddiq and inform him of British economic measures at one and the same time. However, the appearance of press articles accurately reporting the economic restrictions which the British were contemplating made it necessary to implement the measures before the Iranians could take evasive action.[71] On 10 September the British Government therefore went ahead and issued an order blocking payments to and from sterling accounts held by Iranians and discontinuing the export to Iran of sugar and iron and steel, scarce commodities.[72] The next day Shepherd saw the Shah and, using wording which had been toned down to meet American sensibilities, said it was clear that steps should be taken to 'summon a more reasonable government'. Shepherd also gave the Shah the official British line that the economic restrictions announced the day before were designed to limit harm to the British economy caused by the oil dispute. However, although Shepherd said that the restrictions were not meant as retaliation or sanctions against Iran, he told the Shah that they might affect the Iranian economy. They could, he went on, be revoked if an acceptable settlement of the oil dispute came into sight.[73] To add to the pressure on Iran, the USA announced that the proposed loan from the Export Import Bank could not be made as Iran now lacked the revenue needed to service it. Musaddiq expressed great indignation and accused the Americans of applying sanctions.[74]

By this time, Musaddiq had declared that he intended to pursue the course of action which he had described to the Senate on 5 September by sending a message to the British via Harriman.[75] Accordingly, in a letter dated 12 September, Musaddiq wrote to Harriman restating that Iran was prepared to negotiate with Britain on the three matters of the employment of British technicians in the Iranian oil industry, compensation for the Company and sales of Iranian oil to Britain. Musaddiq asked Harriman to act as intermediary by transmitting his proposals to the British Government. Musaddiq threatened that if no 'satisfactory conclusion' was reached within fifteen days of the British Government receiving his proposals the residence permits of the Company's remaining British staff in Iran would be cancelled.[76] Harriman replied on 15 September, saying that Musaddiq's proposals represented no advance on those which had been put to Stokes when he was in Tehran. Emphasising that Iran needed to recognise the 'practical business and technical' aspects of the oil industry, which meant, in effect, that Iran could not hope to run its oil industry by itself, Harriman declined to pass Musaddiq's message on to the British Government.[77]

Having failed in his attempt to use Harriman as an intermediary, Musaddiq now tried a different approach. On 19 September, Husayn Ala, the Iranian Minister of Court, called on Shepherd and presented him with a paper which, though it was not dated or signed and was on plain rather than official paper, was said to be a fresh proposal from Musaddiq.[78] In essence, it was an abbreviated repetition of the proposals which Musaddiq had communicated to Harriman seven days earlier, with the significant difference that it excluded the ultimatum threatening to withdraw the residence permits of British staff.[79]

This new initiative by Musaddiq succeeded in driving a wedge between Britain and the USA. It was apparent, by this time, that the Shah lacked the confidence to act on the British suggestion that he should replace Musaddiq's Government. In the words of Shepherd, the Shah was 'more frightened of encouraging a change of government than of letting Musaddiq remain'.[80] Although the prospects of an early change of government were thus diminished, the British Government continued to believe that negotiations with Musaddiq would only help to keep him in power.[81] As for the proposals which Ala had presented to Shepherd, the British Prime Minister, Attlee, was not prepared to reopen negotiations on the basis of a document produced 'in this hole and corner manner'.[82]

The Americans, however, took a different view. Fearful of instability in Iran, they felt that Britain ought not to reject Ala's approach out of hand and that the possibility of entering into further negotiations with Musaddiq should be left open. In the words of Franks, reporting from Washington:

> The State Department had ... shared our views that it was impossible to negotiate with Musaddiq. They were now convinced, however, that the Shah would not dismiss Musaddiq; they were nervous of what might in any case happen after he went. Finally, they believed ... that there was a chance of Musaddiq yielding to pressure and being ready to negotiate constructively. They also had in mind that if it was possible to make a settlement with Musaddiq, it was more likely to 'stick' than a settlement

84   Company staff gathering at the Gymkhana Club, Abadan, before going aboard HMS *Mauritius*, October 1951

with anyone else. For these reasons they thought it desirable not to return a wholly negative answer to these approaches but to leave the way open for them to be developed into something more definite.[83]

Disagreeing with the US assessment of the position, the British Government instructed Shepherd to reject the proposals presented by Ala and to inform the Shah that Britain was anxious to do 'anything possible' to help get rid of Musaddiq.[84] On 22 September, Shepherd duly delivered a note to Ala, turning down the Iranian proposals.[85] It was, however, a few days before he saw the Shah. In the intervening period the escalation of the dispute continued.

On 25 September, the Iranian Government announced that it was instructing the provisional board of the NIOC to inform the Company's British staff in Iran that they must leave the country by 4 October.[86] On 27 September, Iranian troops occupied the Abadan refinery. Also on the 27th, the British Cabinet, restrained by US opposition to military intervention, decided not to use force to hold the refinery and maintain the Company's staff in Abadan.[87] It was decided, instead, to refer the dispute to the Security Council of the

85   The last British staff evacuating Abadan, October 1951

United Nations. That same eventful day, Shepherd saw the Shah and asked him what could be done to stop the expulsion of the British staff. The Shah, to Shepherd's disappointment, said he thought trouble would be avoided and a solution to the dispute made more likely if the British staff were to leave.[88]

On 29 September, the British Government complained to the Security Council that Iran had broken the interim ruling issued by the International Court of Justice on 5 July. The British submitted a draft resolution to the Council, calling on the Iranian Government to comply with the ruling and rescind its expulsion order. As the expulsion was to take place by 4 October, the Council was requested to consider the matter as one of great urgency. The Council responded with alacrity and took up the British complaint on 1 October. However, Iran's Ambassador to the UN requested that the debate be postponed for ten days to allow time for an Iranian representative to travel from Tehran to New York for the proceedings. The Council granted the postponement, which meant that the complaint would not be debated until after the expiry of the deadline for the withdrawal of British staff.[89]

The British were left with no alternative but to evacuate the staff before the deadline imposed by the Iranian Government. Accordingly, the majority of the British personnel at Abadan embarked on the HMS *Mauritius* on 3 October, to be followed by the last of the senior staff the following day.[90] For the moment, the Company was out of Iran.

18

Failed initiatives,
October 1951–August 1953

Having derived about three-quarters of its crude oil and refined products from Iran at the time of nationalisation, the Company was faced
with a major supply crisis when its operations in Iran came to a
standstill. A description of the manner in which the Company dealt
with this problem is reserved for the next volume. It is enough to say
here that the Company quickly managed to obtain increased supplies
of crude oil from other sources. The loss of Iranian crude was thus
rapidly overcome. It was less easy to find substitutes for the huge range
and volume of products produced by the Abadan refinery, but a large
measure of success was nevertheless achieved in procuring alternative
product supplies. Margins may have been reduced by the higher costs
of improvised supply measures, but financially the Company could
weather even a protracted storm, having entered the crisis in a position
of enormous financial strength, with great liquid resources and low
debt, as shown in chapter 10.

As the Company adjusted to the loss of its Iranian supplies, it soon
passed the point where it badly needed to restore the flow of Iranian oil
to keep its business afloat. Fraser nevertheless remained determined
not to give in to Iranian demands for fear that the Company's position
in other oil producing countries would be weakened if it was seen to
make large concessions to Iran. As the Company's supply position
eased there was, therefore, no slackening in Fraser's uncompromising
attitude towards upholding the Company's interests in Iran.

While the Company was broadly successful in making good the loss
of its Iranian oil supplies, Iran was unable to find other buyers for
significant quantities of its oil. The other major international oil
companies, which together with the Company dominated the international oil industry, would not purchase oil from Iran, partly because

they could get ample supplies elsewhere, but more especially because they did not wish to undermine the concessionary system in the Middle East, to which they themselves were heavily committed. As a result, Iran could not sell oil on any scale and lost the revenues from oil, its most important industry. Musaddiq's Government took various measures to try to adjust the economy to the sudden shock of losing its oil income and, according to Homa Katouzian, a biographer of Musaddiq, the strategy of 'non-oil economics' achieved a high degree of success.[1] Nevertheless, there can be little doubt that Iran suffered more from its inability to find customers for its oil than the Company suffered from the loss of its Iranian supplies. Moreover, Iran's problems were compounded by continuing political instability, which Musaddiq was unable to control despite taking emergency powers to increase his own authority.

Fearing that instability in Iran would open the door to communist control, the Truman Administration continued to try to arrange a settlement, to which end the Americans were prepared to go on making concessions to Musaddiq's demands, rather than risk a complete breakdown in negotiations. The British Government favoured taking a harder line with Musaddiq and believed that each successive offer of concessions merely encouraged him to ask for more. The Government was not, however, nearly as uncompromising towards Iran as Fraser, who defended the Company's interests with such force that he came in for increasing criticism from both the US and British Governments, to whom he appeared to be virtually as much of an obstacle to reaching a settlement as Musaddiq. It became increasingly clear that if a settlement was to be reached, one or both of them would have to be removed and for a time it was by no means a foregone conclusion that Fraser would not be the first to go. In the event, however, it was Musaddiq who was deposed, opening the way to a settlement being reached with his successor.

THE McGHEE–MUSADDIQ PROPOSALS, OCTOBER–NOVEMBER 1951

On 8 October, Musaddiq arrived in New York at the head of an Iranian party, having decided that he would personally represent Iran at the hearings of the UN Security Council. Complaining of ill health, he was taken to New York Hospital where he stayed for several days before moving to an hotel.[2] The opening of the Security Council debate was, meanwhile, further postponed until 15 October. By that time the draft resolution which Britain had submitted on 29 September calling on Iran to rescind the expulsion order had been overtaken by events as the expulsion was a *fait accompli*. The British therefore submitted a revised resolution calling for a resumption of negotiations between the two sides in accordance with the principles of the International Court's interim ruling of 5 July.[3]

In the debate which opened on 15 October, Musaddiq argued that neither the International Court nor the Security Council was competent to interfere in the dispute, which he held to be a purely domestic Iranian affair.[4] Four days later, after both sides had been heard, the Council adopted a resolution adjourning the debate until the International Court ruled on its own competence in the matter. This was, in effect, a victory for the Iranians, who had foiled Britain's attempt to win the backing of the Security Council.[5] Moreover, the debate had provided a stage on which Musaddiq was able to make a considerable theatrical impact. As Acheson later recalled, Musaddiq's presentation of the Iranian case before the Council was done 'with great skill and all the drama he had made familiar in the Majlis. Overnight he became a television star.'[6]

Before the Security Council met, Acheson had decided that the State Department, acting in the role of 'honest brokers', should take advantage of Musaddiq's presence in the USA to have talks with him about the oil dispute.[7] Although Acheson said that the Americans would uphold essential British interests in the talks, he made it clear that the State Department was convinced that it would be impossible to reach a settlement which provided for the Company, 'even in disguise', to take part in oil operations in Iran. He assumed, however, that the Company would purchase and distribute the oil produced and refined in Iran.[8] It was on this basis that talks between the State Department and Musaddiq took place in October–November 1951.

The main figure on the US side in the discussions was George McGhee, Assistant Secretary of State for Near Eastern, South Asian and African affairs. He was not new to the Iranian dispute, having met the Company's board, excluding Fraser, in September 1950 and visited Tehran in March 1951 when the oil nationalisation law was being passed. He subsequently met Fraser early in April. In the course of these events McGhee had formed an unfavourable opinion of Fraser and the board, who he regarded as being obdurate and inflexible in their attitude towards Iran's demands.[9] McGhee took a much more sympathetic view of Musaddiq. In the autumn of 1951, during their many and long talks, aggregating some 80 hours in more than 20 meetings, McGhee developed a liking and respect for the Iranian leader, finding him intelligent and humorous.[10] Nevertheless, McGhee found, like everyone else who discussed the oil dispute with Musaddiq, that he was unable to get Musaddiq to understand the 'facts of life about the international oil business'. It appeared to McGhee that Musaddiq simply did not care about operational and technical matters. He insisted that the problem was a purely political one.[11] Moreover, despite his intelligence, Musaddiq was, McGhee found, 'warped by his suspicion of everything British'. Years later, McGhee recognised that Musaddiq's extreme antipathy towards Britain 'probably doomed from the start' US efforts to facilitate a settlement. He also came to think, on reflection, that the British 'probably turned out to be right' in their view that it was impossible to reach an agreement with Musaddiq and that the only hope was to wait for a more amenable successor.[12] These conclusions, however, only came to McGhee later. In the autumn of 1951 he looked upon Musaddiq with warmth and hoped that they could arrive at a settlement.

There were others in the USA who also welcomed Musaddiq's visit and gave it a quintessentially American aura of glamour. After the Security Council debate in New York was over, Musaddiq travelled to Washington amid great publicity on 22 October. The next day he had lunch with President Truman, Acheson and others, before moving to Walter Reed Hospital where he stayed in the presidential suite although the hospital could find nothing wrong with him that could not be put down to his age.[13] According to Acheson, 'Everyone petted and took care of him' as he 'held court in pajamas and bathrobe' and discussed the oil dispute with Acheson and others from the State Department.[14] In common with McGhee, Acheson came to the discussions with a critical opinion of the Company which, he thought, had helped Musaddiq's cause by showing 'unusual and persistent stupidity'.[15] It was only later that his somewhat romanticised view of Musaddiq gave way to a more sober assessment. In his memoirs, Acheson admitted: 'We were, perhaps, slow in realizing that he was essentially a rich, reactionary, feudal-minded Persian inspired by a fanatical hatred of the British and a desire to expel them and all their works from the country regardless of the cost'.[16]

In the days which followed Musaddiq's arrival in Washington events of significance were also taking place in London. On 23 October, the day that Musaddiq lunched with Truman, Fraser came in for renewed criticism from a senior British Government official, this time Sir Edward Bridges, Permanent Secretary of the Treasury. At a meeting on the oil dispute held in his room that day, Bridges said that Fraser was narrow-minded, lacked political insight and should be removed.[17] By this time there was nothing new in such criticisms, but they gained force by repetition and confirmed that Fraser seemed friendless in government circles in both the USA and Britain.

Two days after this latest indictment of Fraser the country's voters, having re-elected Attlee to a second term as Prime Minister in February 1950, went to the polls in the second general election since the end of World War II. The result this time was a defeat for Labour and victory for Churchill, who came back to power at the head of a Conservative Government which included Anthony Eden as Foreign Secretary. What effect would these changes have on British policy towards Iran? Churchill could certainly be expected to take a harder line on the oil dispute than Attlee, whose Government, said Churchill, 'had scuttled and run away from Abadan when a splutter of musketry would have ended the matter'.[18] It might, on the other hand, have been expected that Eden would show a more conciliatory attitude than his predecessor, Morrison, who generally took an aggressively uncompromising approach towards Iran during his time as Foreign Secretary.[19] Unlike Morrison, who knew little of Iran, Eden was not without knowledge of the country, having read Oriental languages with Persian as his main language at Oxford. He was also Under-Secretary at the Foreign Office in 1933 when the Company's concession was re-negotiated, and later paid a number of visits to Iran.[20] In practice, however, the change from Morrison to Eden brought no immediate major change in policy on Iran. Acheson, who had held Morrison in low regard, thought that Eden was a 'great and signal improvement, except on Iran', in which area he perpetuated, in Acheson's view, the adamant attitude which had characterised Morrison's term of office.[21]

Only a few days after Churchill came back to power the State Department put to the British Government the proposals for a settlement of the oil dispute which had been worked out with Musaddiq in New York and Washington. The basic premise of the proposals was stated explicitly: that the Iranian Government would not agree to the Company returning to Iran in any form.[22] Instead, the NIOC would be directly responsible for all aspects of exploration, production and transportation of crude oil in Iran. The Abadan refinery would be sold by the Company to a non-British firm which, it was understood, would be Dutch. The Company would set up a purchasing organisation to acquire very large quantities of Iranian oil on a long-term contract. The NIOC would be free to sell oil in excess of that sold to the

Company's purchasing organisation, provided the terms of sale did not prejudice the interests of the purchasing organisation. All claims and counter-claims to compensation would be cancelled. On the matter of prices, the purchasing organisation would buy products from the Abadan refinery on the basis of costs plus a fixed fee. The price at which the NIOC would sell crude oil was, the State Department found, the most difficult item to agree with Musaddiq. However, the Department was 'reasonably optimistic' that the Iranians would accept a price of $1.10 per barrel.[23] This, thought McGhee, should have produced a result close to 50:50 profit sharing.[24]

On 6 November these proposals were considered at a meeting of the Working Party which had been set up by the British Government to keep developments in Iran under review. Gass, who was the senior Company representative present, was asked for the Company's views and proceeded to raise very strong objections to the scheme. He produced calculations to show that a crude oil price of $1.10 per barrel would approximate not to a 50:50 division of profits, but to a split of about 70:30 in favour of the Iranians. That aside, Gass said that the proposals for the Abadan refinery were 'totally unacceptable'. He had no doubt that the Iranians would apply great pressure on the operator of the refinery to expand social amenities rapidly. The operator, assured of receiving costs plus a fixed fee, would have no incentive to resist such demands. The Company, on the other hand, would be contractually committed to purchase large quantities of products at prices based on costs over which the Company had no control. This, said Gass, 'could well prove entirely unprofitable'. He was also vehemently opposed to the proposal that the Company would get no compensation. He told the meeting:

> In fact, the AIOC was to disappear from Persia for ever without any hope of redress because the Americans chose to accept the outrageous slanders which Mossadeq and his clique had made against the Company. He had torn up an Agreement with the British and he resorted to any lie to prove to the world his action was warranted. If Mossadeq were allowed to get away with this monstrous performance it would have the most disastrous effect on other countries. Other Mossadeqs would arise and what would be left of the fabric of the oil industry to which the Americans professed to attach so much importance if the Mossadeqs were to be allowed to get their way and kick their Concessionaire Companies out and if something in the nature of a 70/30 result were to be labelled 50/50?[25]

Gass thought that the proposals might well be acceptable to the Company if, subject to negotiations on the crude oil price, the

Company were to be the operator of the Abadan refinery. It did not need stating that this was contrary to the fundamental premise which Acheson had put forward: that there was no possibility of a settlement which involved the return of the Company to Iran. The meeting concluded on a note which expressed the Company's irritation with American intervention. It stated that 'The most satisfactory solution from the point of view of AIOC would be for the Americans now to retire from the field and to leave AIOC to handle the matter itself'.[26]

While Gass expressed the Company's views to the Working Party, Eden was in Paris, having travelled there on 4 November for a meeting of the United Nations Assembly, which Acheson also attended. This provided the occasion for them to have a series of discussions about the oil dispute on which, as Eden later recalled, they held 'sharply differing views'.[27] Eden did not accept the US argument that the only alternative to Musaddiq was communist rule. In conformity with the established British line of thinking, he thought that if Musaddiq fell from power, his place might be taken by a more reasonable successor with whom Britain might be able to reach a satisfactory settlement.[28] Briefed from London, Eden rejected the US proposals, which he found unacceptable on the grounds that they would exclude the British from Iran and involved 'confiscation without compensation'. In Eden's view, it would be better to come to no agreement than to reach a 'bad one'.[29] Churchill encouraged him to stand firm with the message: 'I think we should be stubborn even if the temperature rises somewhat for a while'.[30] Unable to secure Britain's agreement to the US proposals, Acheson telephoned McGhee in Washington on 8 November to tell him of the impasse. McGhee passed on the news to Musaddiq, who left Washington to return to Iran via Cairo on 18 November. The McGhee–Musaddiq proposals had come to nothing. Without interlude, however, a new initiative emerged based on the possibility of the International Bank for Reconstruction and Development taking a hand in the oil dispute.

THE INTERNATIONAL BANK'S PROPOSALS, NOVEMBER 1951–MARCH 1952

While Musaddiq was still in Washington it was suggested to him by M. A. H. Isfahani, the Pakistani Ambassador, that the International Bank might be able to help relieve the oil crisis. Musaddiq expressed interest in the idea, and on 10 November Robert Garner, Vice-President of the International Bank, called on him and outlined in general terms the principles on which the Bank might be able to help restart the Iranian oil industry.[31] Garner suggested that the Bank should set up a temporary management to operate Iran's oil industry for up to two years. The management would be responsible to the Bank and be headed by nationals of countries not involved in the dispute. The Bank would also arrange a contract for the sale of Iranian oil to the Company, with some of the sale proceeds being held in escrow pending a final settlement of the dispute. The Bank, Garner made clear, would not adjudicate on the controversy and would be acting as a neutral institution seeking to restart Iran's oil industry on an interim basis so that the flow of oil and revenue could be restored while a permanent settlement was worked out. The International Bank's intervention would be without prejudice to the legal rights of either side in the dispute. Musaddiq agreed to the Bank trying to work out an interim arrangement on the lines proposed. Garner then approached the British Government, which also agreed to the Bank producing specific proposals.[32]

To assist in that task, Gass and Snow travelled to New York on 18 December for consultations with Garner, other Bank officials and Torkild Rieber, the former chairman of the Texas Company, who was employed by the Bank as an adviser.[33] After a series of meetings Garner drew up a letter to Musaddiq, dated 28 December, in which he elaborated on the principles which he had outlined at their earlier meeting in Washington.[34] Hector Prud'homme, the Bank's loan officer, and Rieber then travelled to Iran, arriving there on 31 December, to deliver Garner's letter to Musaddiq and to inspect the oil installations in the south of the country. After having an interview with Musaddiq on 1 January 1952 they visited Abadan and then returned to Tehran. Musaddiq, meanwhile, published a reply to Garner's letter, raising various questions about the Bank's proposals and stating that any intervention by the Bank should be regarded as a 'delegation of authority' from the Iranian Government and that the Bank should act on behalf of the Iranian Government and 'carry out its orders'.[35] This statement was directly contrary to one of the fundamental principles of the Bank's proposals: that it would act as a *neutral* institution and not as the agent of one or other side in the dispute.

In view of the tenor of Musaddiq's reply it was not surprising that Garner, who did not receive it before it was published in the press, told Gass that he was 'not at all pleased' about its 'apparently abusive tone'.[36] On 8 January Gass and Heath Eves met Garner who, having by then received Musaddiq's letter, showed them a copy. Garner told them that he did not intend to reply until Rieber and Prud'homme had returned from Tehran. However, he was clearly of a mind to respond in a compromising tone rather than risk a breakdown of negotiations.[37] Gass, upholding the more uncompromising Company view, 'urged him to stick to his guns and make no concession of principles whatever'.[38] The Bank, he wrote to Fraser, might require 'some stiffening up' in the next round of discussions.[39]

Gass's concern that the Bank's resolve to stand by its proposals might weaken was magnified after Rieber and Prud'homme returned from Tehran to the USA in mid-January and reported on their visit. Rieber thought that the oil installations in southern Iran had been well cared for and were in good condition. However, he had also been impressed by the strength and prevalence of anti-British feeling in Iran and felt that there was no prospect of nationalisation being reversed or, most importantly for Gass, of the Company's restoration in Iran. Rieber did not believe that even the temporary management of the oil industry by the Bank would be acceptable to the Iranians. He therefore suggested that the operation of the Iranian oil industry should be conducted by the Iranians, who should have a sales contract with the Company to provide an outlet for their oil.[40] These suggestions were anathema to Gass, who told Bank officials that such an arrangement 'would mean almost entire capitulation to Mossadeq demands' and that 'we would be conniving at our own expulsion and at the same time providing an outlet for Persian oil'. Concerned about the impact of the Iranian dispute on concessionary agreements with other countries, he added that the relinquishment of managerial control to the Iranians would set a 'deplorable precedent' to neighbouring countries.[41] The Bank 'readily appreciated' Gass's remarks.[42]

In subsequent discussions with the Bank, Gass argued strongly that there were two fundamental aspects of the proposals from which there could be 'no deviation'. One was that effective managerial control of oil operations should be exercised by the Bank. The second was that the exclusion of British nationals from taking part in the Iranian oil industry would be unacceptable.[43] He continued, however, to worry about the attitude of the State Department which, he thought, was prepared to waver and make concessions to Musaddiq because it was 'obsessed with the fear' of closing the door on negotiations. Gass wrote to Fraser: 'as I have said on one occasion to the State Department and innumerable occasions to the Bank, it is far better that there should be no agreement than a bad agreement from the point of view particularly of repercussions on the vast investments elsewhere in the world'.[44] This view was similar to that which Eden had put to Acheson when he rejected the McGhee–Musaddiq proposals in Paris and was still held by the British Government.[45]

While the Bank held discussions on what steps to take next, events in Iran continued to add to the international tension. Of particular note, on 12 January the Iranian Government demanded the closure of all British consulates in the country, claiming that they were used for intelligence purposes and for fostering disturbances.[46] A British note of protest failed to have any effect and the consulates were closed on the 21st. The Iranian Government also announced that it would not accept British diplomatic officials who had previously served in Iran or the British colonies. On these grounds the Iranians refused the appointment of R. M. Hankey as Ambassador in succession to Shepherd, who was being transferred. After Shepherd's departure George Middleton, the British Chargé d'Affaires in Tehran, became the senior British Government representative in Iran.

The International Bank had meanwhile decided to send a second mission to Tehran, where Garner, accompanied by Rieber, Prud'homme and other Bank officials, arrived on 11 February. In discussions with Musaddiq, Garner formed the same impression as others who had come before him: that Musaddiq had an 'unrealistic and purely political approach to the problem'.[47] Garner and his associates explained that if Iranian oil operations were to be restarted in a short time and on a scale sufficient to provide substantial revenues for Iran, a large force of foreign technicians, including some British, would be required. Musaddiq, however, would not agree to the employment of British oil technicians in his country. Since no agreement could be reached on that point, it was temporarily put aside so that discussions could continue on other aspects of the Bank's proposals. It soon, however, became apparent that there were two other issues that separated the two sides. One was Musaddiq's demand that the Bank should state that it was acting on behalf of the Iranian Government in operating the oil industry 'for Iran's account'. This, of course, was a repetition of the point he had made earlier and was open to the same objection that it would contradict the principle of the Bank's neutrality if it acted solely for one party in the dispute. Secondly, the Bank mission was unable to reach agreement with the Iranians on the price which the Company should pay for Iranian oil. Having failed to resolve these differences, Garner, Rieber and Prud'homme left Tehran on 20 February and travelled to London.[48] After their departure Middleton saw Musaddiq and found him as unyielding as ever. In a comment which illustrated his preoccupation with the political rather than the economic aspects of the dispute Musaddiq said that he had to be able to show the Iranian public that all of the revenues from Iran's oil would accrue to Iran. As he told Middleton: 'it was better to receive 100 per cent of 50 cents than 50 per cent of a dollar or even dollars 1.40'.[49] In reporting on that and other remarks, Middleton commented that Musaddiq's 'approach to the problem is in fact an almost purely mystical one'.[50]

After a few days in London, Garner and Rieber went on to Washington and Prud'homme returned to Tehran where he held a series of further discussions with Musaddiq starting on 5 March. However, no progress was made towards a solution of the major problems outstanding and the talks were officially adjourned, but in effect abandoned, on 17 March. Prud'homme returned to Washington shortly afterwards.[51] Another initiative had broken down.

THE IDEA OF A MANAGEMENT COMPANY, APRIL–JULY 1952

Although the International Bank's initiative had failed it had, from the British point of view, a useful result in that it brought the International Bank and, most importantly, the State Department into closer alignment with Britain in their assessments of the political situation in Iran. In particular, the International Bank and the State Department now took the view, held by Britain for some time previously, that it was impossible to reach an agreement with Musaddiq and that the best course was therefore to wait until a more moderate government came to power in Iran before negotiating a settlement of the oil dispute. The timing of such an event could not be accurately predicted, but it was felt that Musaddiq could not remain in power for very much longer, given the economic and political instability in Iran.[52] In the meantime, proposals for a settlement could be drawn up, ready for presentation to whoever succeeded Musaddiq.

The International Bank and the State Department held very similar views on the nature of the proposals which they thought might be acceptable to a future Iranian Government. They both favoured a long-term settlement, rather than an interim arrangement, in which a management company would be established to operate the Iranian oil industry on behalf of the Iranian Government. The Government would be the owner of the oil industry's assets and would enter into a long-term contract for the sale of Iranian oil to the Company. On the composition of the management company, both Garner and the State

Department were emphatic that in their view it would be a political impossibility for the Iranian Government to accept any British participation. The Company would, they thought, most definitely have to be excluded.[53]

In the spring of 1952 the British Government spent much time considering the idea of a management company, in particular the possibility of British participation. By early May the Foreign Office had reached the conclusion that there was no realistic chance of reaching a settlement which would allow the Company to return to Iran to take part in managing the oil industry there. However, while the Foreign Office accepted that the Company would have to be excluded, it still thought that there should be some other form of British participation in the management company. In a note to Eden on 2 May, Strang set out in somewhat vague terms the idea of forming a British company, in which there would be non-British shareholdings, to conduct oil operations in Iran. Strang went on:

> We do not see the possibility of creating such a company unless there is first a radical re-organisation of the AIOC whereby Shell and United States companies take over certain interests. Such a re-organisation is generally recognised to be dependent on a change in the Chairmanship of AIOC.[54]

Although Strang did not elaborate further, it would seem clear that he was proposing a restructuring in which Royal Dutch-Shell and US oil companies would take over Company assets in Iran. Fraser, not surprisingly, stood in the way. On the very day that Strang sent his note to Eden an internal Company report argued 'very much against the suggestion of a complete abdication of our right to participate in whatever internal organisation is eventually decided on in Persia'. There might, the report continued, be a need for a greater 'dilution with foreigners than we had in mind', to which was added the suggestion that the Company might have a 50 per cent interest in the management company (if it was set up), the other 50 per cent being held by non-British interests.[55]

There were thus three distinct views on the possible composition of a management company. One, held by the International Bank and the State Department, was that there could be no British participation at all. Another, held by the Foreign Office, was that the Company would have to be excluded, but that there should be another form of British participation. And, finally, there was the view of the Company, which was not prepared to agree to its own exclusion.

Over the next two months these different strands of opinion showed little sign of convergence. Within the British Government the Foreign Office view prevailed, despite reservations about it in the Ministry of Fuel and Power, and in the third week of June British ministers decided that the idea of a long-term settlement based on a management company which included the greatest possible British interest was acceptable. The Company, however, would be excluded.[56] This was, of course, a major new development in British Government thinking. The idea of a management company which excluded the Company had previously been accepted by the Government in relation to the International Bank's proposals, but they had been for an *interim* arrangement. Now, the British Government was prepared to accept the exclusion of the Company from Iran in a *long-term* settlement.

Fraser, still chairman despite the repeated Government mutterings about getting rid of him, was not prepared to surrender the Company's position so easily. On 15 July he and Gass saw Sir Donald Fergusson of the Ministry of Fuel and Power, who outlined the Government's thoughts on the management company and asked for Fraser's views. Fraser thought that if a new Iranian Government came to power, it would be badly in need of money and would therefore wish to reach a quick settlement of the oil dispute. Feeling that the offer of a financial advance to the Iranian Government would help provide the 'basis of some new bargain', Fraser believed that the best course would be for the Company itself to negotiate with the Iranian Government with the aim of reaching an agreement to get the oil industry restarted quickly. He had no objection to the idea of a management company, but was strongly opposed to the Company being excluded. That, he thought, would have 'the most serious effects' on the Company's position in other Middle East countries and on 'its commercial prestige throughout the world'. On seeing Fergusson's record of the meeting, Eden wrote comments in the margin indicating his absolute disagreement with Fraser's ideas. At the top he wrote: 'Fraser is in cloud cuckoo land'.[57]

MUSADDIQ FALLS AND REBOUNDS, JULY 1952

It was, of course, not only Fraser who was regarded as utterly unrealistic by the British Government. So too was Musaddiq, who travelled to The Hague in June 1952 to represent Iran personally at the hearings held by the International Court of Justice to decide whether or not it had jurisdiction to deal with the oil dispute. At the hearings the Iranians argued that the International Court had no jurisdiction in the dispute, which they held to be a domestic Iranian affair arising from the exercise of national sovereignty. This was consistent with the attitude which the Iranian Government had taken earlier, when it rejected the interim ruling which the International Court issued on 5 July 1951 (see chapter 17). The British, on the other hand, argued that the dispute lay within the Court's jurisdiction.[58] The Court's hearings closed on 23 June, but another month was to pass before its judgement was issued.

In the meantime, Musaddiq left The Hague and returned to Iran where a new Majlis had been elected.[59] Having been re-selected as Prime Minister by the deputies, Musaddiq called on the Shah on 16 July to present the list of his new Cabinet. He said that in addition to being Prime Minister, he also wished to be Minister of War, which would give him control of the armed forces. The Shah, who had the title of Commander-in-Chief of the Armed Forces and had previously always appointed the War Minister, rejected Musaddiq's proposal. Musaddiq promptly handed in his resignation and on 17 July the Shah appointed the pro-British Qavam al-Saltana, who had been Prime Minister three times before, as the new Premier.[60] The moment which the British and Americans had been waiting for had arrived: Musaddiq had fallen from power. 'We are', went a British Government telegram from London to Washington, 'fully aware of the opportunity presented by the fall of Musaddiq and we are determined to make the most of it'.[61]

The Persia (Official) Committee, a group of top-level officials from the Foreign Office, Treasury and Ministry of Fuel and Power which had replaced the Working Party, hurriedly convened on 18 July to consider what action to take in the oil dispute in view of Qavam's accession to power.[62] However, hopes of reaching a settlement with the new Iranian Government were dashed by events in Tehran, where mass demonstrations were held against Qavam. After violent clashes between the army and demonstrators had left many dead or wounded, Qavam resigned on 21 July and Musaddiq was voted back as Prime Minister by the Majlis on the 22nd. This time, the Shah agreed to Musaddiq taking charge of the Ministry of War. Several days later Musaddiq also obtained emergency powers from the Majlis, permitting him to enact legislation without Majlis approval for a period of six months. Meanwhile, on 22 July, the day that Musaddiq regained power, the International Court issued its judgement that it had no jurisdiction in the Iranian oil dispute, which meant that the interim ruling of 5 July 1951 was automatically revoked.[63] These events represented, it should hardly need to be said, a great victory for Musaddiq. As Middleton reported, Musaddiq was 'riding the crest of a wave of popular feeling and success. There are no rival candidates in the field, the Shah has been reduced to impotence and the influence of the army and the mood of its senior officers has been severely shaken'.[64] For the time being, at least, Musaddiq appeared to have consolidated his power and the idea of presenting proposals for a settlement to a successor government could be put aside.

THE CHURCHILL–TRUMAN PROPOSALS,
JULY–OCTOBER 1952

On 25 July, three days after his return to power, Musaddiq sent for Middleton, who arrived to find Musaddiq as usual in bed. According to Middleton, Musaddiq 'launched into his now familiar rehearsal' of the alleged wrongs done to Iran and said he was prepared to seek the agreement of the Majlis to the settlement of compensation by arbitration, in return for which he expected to receive immediate financial aid. Musaddiq made it clear that he would not agree to any form of foreign control over the oil industry in Iran, but that he was willing to make a long-term contract for the sale of Iranian oil to the Company.[65] Middleton was initially inclined not to reject Musaddiq's proposal out of hand.[66] However, when he went to see Musaddiq again on 28 July he was astonished to hear Musaddiq make the 'startling' remark that he had gone a long way in agreeing to examine the *British* proposal for arbitration. The proposal had, Middleton protested, come not from Britain, but from Musaddiq at their last meeting. In the heated discussion which followed, Musaddiq said that he preferred to regard their previous conversation as 'not having taken place'. After the meeting was over Middleton wrote in exasperation that his experience that day had given him the clearest impression yet of Musaddiq's 'complete intransigence'. It was, thought Middleton, 'useless to hope for means of completing negotiations with him ... I am now inclined to think that the only immediate hope is a *coup d'état*'.[67] Middleton was, in short, back to the familiar British view that it was impossible to reach a settlement so long as Musaddiq remained in power.

It looked, briefly, as if the Americans might share that view. After an interview with Musaddiq on 28 July, Loy Henderson, who had succeeded Grady as US Ambassador to Tehran, told Middleton that it had been 'exhausting and depressing'. They agreed that 'Musaddiq has seriously deteriorated and we share doubts about his mental stability

... We cannot ... see any hope of reaching agreement with Musad-diq'.[68] Soon, however, the Americans began once again to diverge from the British line. By the end of July Henderson was reporting to the State Department that Musaddiq now appeared to be in a more stable frame of mind. Henderson went so far as to set out the terms on which he thought Musaddiq might agree to settle the dispute. More importantly, Acheson, feeling that the events of Qavam's brief premiership had weakened the power of the Shah and the army, thought that there was now no alternative to supporting Musaddiq as the only bulwark against communism in Iran. He suggested that the British and US Governments should make joint proposals to Musad-diq, offering immediate financial aid in the form of a US grant on the basis that compensation would be settled by arbitration and nego-tiations promptly started on future arrangements for the distribution of Iranian oil.[69] It was implicit in these proposals that the Iranians would themselves operate the oil industry in Iran. It was an indication of the lengths to which the Americans were prepared to go to reach agreement with Musaddiq that these proposals were, to all intents and purposes, the same as Musaddiq had put to Middleton on 25 July.

Responding to Acheson's proposals on 9 August, Eden set out the British view that the Americans were too ready to look upon Musad-diq as the only alternative to communism in Iran. The British doubted his effectiveness as a barrier to communism and felt that the Iranian army, although humbled by the events of Qavam's four-day premier-ship, might still be an effective counter to communism. As for Ache-son's idea of making joint Anglo-American proposals to Musaddiq, Eden wrote that the British Government was prepared to make a joint offer, but only on stiffer terms than those proposed by Acheson.[70] Acheson, replying to Eden, thought that the terms suggested by Eden were far too severe on Iran and would be 'fatal to agreement'.[71]

Eventually, it was an intervention by Churchill which brought the British and US Governments together. Churchill attached 'the highest importance to the joint approach' and by 23 August was becoming impatient about 'a continuance of the futile parleying which had got us no further in all these months'. 'I do not myself see', he wrote to Truman, 'why two good men asking only what is right and just should not gang up against a third who is doing wrong'.[72] Truman agreed to the joint approach, provided the two Governments were not obliged to follow similar courses of action in the future.[73] The morning after, Churchill cabled: 'Not a minute shall be lost'.[74] The text of the joint proposals was agreed on 26 August.

The Churchill–Truman proposals provided, firstly, for the matter of compensation to be submitted to the International Court of Justice, which was to determine the amount of compensation having regard to all claims and counter-claims of both parties to the dispute. Secondly, the Company and the Iranian Government were to enter into negotiations to make arrangements for the future distribution and marketing of Iranian oil by the Company. If the Iranian Government agreed to these two provisions, then the British Government would relax its financial and trade restrictions on Iran; the US Government would make an immediate payment of $10 million to the Iranian Government; and the Company would arrange to move the oil held in storage in Iran.[75] It should be said that these proposals went significantly further to meet Musaddiq's demands than any previous offer, in that the management and control of the oil industry in Iran would be in the hands of the Iranians. Previous proposals had either provided for the return of the Company in one form or another, as was the case with the terms proffered by Jackson and then Stokes, or for some other form of foreign management, as was the case with the interim arrangements proposed by the International Bank. Now, for the first time, Musaddiq was to be offered a settlement which did not entail foreign management and control of oil operations in his country.

On 27 August Middleton and Henderson called on Musaddiq to deliver the Anglo-American proposals. Their visit turned out to be an occasion for a demonstration of the extremely suspicious and distrustful side of Musaddiq. He gave the joint proposals a reception which, in Middleton's words, was 'not only negative but to an extent hostile'. After reading the terms several times, he entered into a 'long diatribe', the general tenor of which was that the proposals were a 'nefarious snare for the purpose of reimposing upon Persia the 1933 Agreement and bringing into question the Nationalisation Law'. Henderson asked Musaddiq who he thought was laying the trap for him: was it Churchill or Truman? Musaddiq replied that it was the Company. Henderson stated categorically that the Company had nothing to do with the drafting of the part of the proposals which had aroused Musaddiq's suspicions. Musaddiq, however, was in uncompromising mood. He said he would not agree to the Company being free to submit claims for compensation without limitation. This, as he no doubt realised, would have enabled the Company to claim for the loss of future profits caused by the cancellation of the 1933 concession. He argued that the Company's claims should be limited to the loss of its physical property in Iran at the time of nationalisation. He went on to say that he did not want the $10 million grant which was offered by the US Government as part of the proposals. What he wanted was the £49 million which the Company had reserved in its balance sheet for payments which would have become due to Iran if the Supplemental Agreement of 1949 had been ratified by the Iranian Government. Faced with these demands and Musaddiq's general intractability, Middleton and Henderson decided to postpone the official delivery of the joint proposals. They thought that this was preferable to risking an immediate and possibly final break with Musaddiq.[76]

The British and US Governments decided, however, to stand by their proposals, which were officially presented to Musaddiq by Middleton and Henderson on 30 August. Musaddiq said that he would never enter into the type of agreement suggested and that he would give his formal answer to the proposals shortly.[77] Several days later, he issued a statement to the press attacking the proposals.[78] Then, on 16 September, Kazimi, the deputy Prime Minister and Foreign Secretary, read a statement by Musaddiq to the Majlis. It said that the Iranian Government was prepared to submit the matter of compensation to international arbitration only if the following conditions were met: first, compensation would be paid only for the Company's physical property in Iran at the time of nationalisation, the amount to be determined on the basis of any law nationalising any industry in any country which the Company chose; secondly, the arbitrators would examine Iranian counter-claims for losses arising from British measures to prevent sales of Iranian oil during the dispute and from the other British financial and trade restrictions on Iran; thirdly, the Company would pay in advance the £49 million which it had reserved in its balance sheet for payment to Iran under the Supplemental Agreement. Finally, the statement included a threat to break off diplomatic relations with Britain.[79]

Musaddiq set out these terms, without the threat to sever diplomatic relations, in letters to Churchill and Truman on 24 September. In the same letters, he formally rejected the Churchill–Truman proposals, repeating his earlier suspicions that they represented an attempt to revive the 'invalid 1933 Agreement' and were inconsistent with Iran's nationalisation laws. Moreover, he appeared to suspect that the proposals might enable the Company to gain a monopoly on sales of Iranian oil. The British Government, he wrote, 'desired to retain the influence of the former Company under other titles in the same shape and form as before'.[80] At the time of the Stokes mission that might have been true, but since then the British Government had made concessions which made Musaddiq's suspicions seem irrational and exaggerated. As much was said, in diplomatic language, when Eden and Acheson sent separate, but similar, replies to Musaddiq on 5 October. Both said that their joint proposals had recognised Iran's nationalisation of its oil industry and did not seek to revive the 1933 concession or to create a monopoly on sales of Iranian oil. Nor, they went on, did the proposals suggest that there should be foreign management of the Iranian oil industry.[81]

Musaddiq, however, was not satisfied. On 7 October he wrote to Eden and Acheson suggesting that representatives of the Company travel to Tehran within a week for discussions on the terms which Kazimi had put to the Majlis on 16 September and which had been repeated in Musaddiq's letters to Churchill and Truman on 24 September. He added that before the representatives left for Iran the Company should pay £20 million of the £49 million which Musaddiq claimed was due to the Iranian Government. The remaining £29 million was to be paid at the end of the negotiations, for which a maximum period of three weeks was envisaged.[82] On receiving Musaddiq's letters, the British and US Governments showed their familiar tendency to diverge. The British were disinclined to make concessions which went further than the Churchill–Truman proposals, even if this meant risking the collapse of negotiations with Musaddiq. The Americans did not wish to take that risk. Anxious to reach some settlement, any settlement, they were prepared to make more concessions to try to secure Musaddiq's agreement.[83] From the sidelines, Fraser remained immovable in his view that a firm stand should be taken with Musaddiq. Responding to the American idea that the Iranians might pay compensation in free oil, he told the Persia (Official) Committee on 13 October:

> The suggestion now appeared to be that an agreement should be reached that Iran should pay compensation in free oil. Iran would then receive financial help, and the oil would start to flow. It was apparently suggested that the Company should then assist the Iranians by buying their oil. The Company, he thought, could give no such undertaking. Their ability to market the Iranian oil was one of their few remaining assets in the area and they were not prepared to pledge it away. They would buy Iranian oil as ordinary customers if they wanted it, but not as a favour to Iran... He thought, in fact, that there was little hope of a final deal with Dr Musaddiq on lines which both he and the Company would accept, nor was he confident that Dr Musaddiq would loyally carry out any agreement which might be reached.[84]

Despite American suggestions that further concessions should be made, the British Government decided to stand firm and on 14 October Middleton handed Musaddiq a note conveying the British rejection of his terms. Dealing with the main elements in turn, the British Government insisted that claims for compensation should be submitted without limitation, meaning that the Company would be able to claim not only for the loss of its physical property in Iran, but also for the loss of future profits caused by the termination of the 1933 concession. It would, the British Government pointed out, be for the arbitrators to decide on the merits of the claims. With regard to Musaddiq's argument that Iran should be able to claim against the Company for blocking sales of Iranian oil, the British Government contended that the Company had merely exercised its legal right to stop the sale of oil which it regarded as its property. Finally, on the question of the £49 million which Musaddiq claimed was owed to Iran by the Company, the British Government pointed out that this sum would have been due if the Supplemental Agreement had come into effect. However, as the Iranian Government had rejected the Supplemental Agreement, it could hardly lay claim to the financial benefits which it would have provided.[85] Musaddiq struck back almost immediately, announcing on Radio Tehran on 16 October that he intended to sever diplomatic relations with Britain.[86]

Over the next few days there were further Anglo-American discussions, in which the Americans again suggested making concessions which went further than the joint Churchill–Truman offer. Fraser's view remained characteristically clear-cut. At a meeting of the Persia (Official) Committee on 21 October he said that 'it was no use going on haggling with Persia, since the Persians would never accept any agreement which would satisfy the Company's legitimate claims'.[87] That sentiment was evidently reciprocally held by Musaddiq, who broke off diplomatic relations with Britain the next day. By the end of October the British embassy staff had left Tehran and their Iranian counterparts had left London.

As yet another set of proposals lay in ruins, there remained, broadly speaking, two schools of thought on the best approach to take towards Musaddiq. One, held by the US Government, was that there was a chance of reaching a settlement by offering still further concessions to Iran. The other, held by the British Government, and most emphatically by Fraser, was that it was precisely because he thought that further concessions could be gained, that Musaddiq had refused to come to terms. For those who held this point of view, it logically followed that the best chance of reaching a settlement was to be had not by making further concessions, but by taking a firm stand. The tension between these two approaches remained as the search for a settlement continued after the failure of the Churchill–Truman proposals.

TRUMAN'S OUTGOING INITIATIVES, NOVEMBER 1952–JANUARY 1953

In the last two months of 1952 Anglo-American discussions on the oil dispute continued in Washington, London and at the meeting of Foreign Ministers in Paris, where Eden and Acheson had talks about Iran.[88] The discussions in general followed the pattern, by now well-established, of the Americans being much more inclined to make concessions to Musaddiq than the British, who preferred to take an uncompromising stand. In the middle of December it looked, briefly, as if the Anglo-American differences might even result in a public split which, as Eden put it, would 'cause immense harm in the Middle East and elsewhere'. An open break was, however, averted and it was agreed that Henderson, who had travelled to Washington, should present new proposals to Musaddiq after he returned to Tehran, whither he departed on 20 December.

While the Anglo-American discussions on Iran were taking place, the US presidential elections of autumn 1952 resulted in victory for General Dwight Eisenhower, a Republican. He was to take over from Truman on 20 January 1953, with John Foster Dulles as his Secretary of State, replacing Acheson. These political events in the USA injected a new element of urgency into the negotiations with Musaddiq, with whom it was hoped to reach agreement before the change of US administrations took effect.

After returning to Tehran from Washington, Henderson called on Musaddiq on 25 December and outlined the terms of a proposed settlement. The four main elements were: first, that the matter of compensation would be settled by international arbitration; secondly, that upon the signing of an agreement to settle compensation by arbitration, the US Government would make an advance of $100 million to Iran to help meet her immediate financial needs until substantial revenues began to come in from oil; thirdly, in order to provide Iran with an assured future income from oil, a long-term sales contract would be negotiated between the Iranian Government and the Company. However, the Company would not have exclusive monopoly rights to Iranian oil. Finally, Henderson made it clear that these proposals would leave the Iranians in control of their own oil industry inside Iran. These terms were similar to the earlier Churchill–Truman offer, the most notable difference being that the proposed US financial aid had been increased from $10 million to $100 million, a substantial inducement to the Iranian Government to accept the proposals. Also notable, for future reference, is that Musaddiq told Henderson that he would be willing to settle compensation on the basis of any English law nationalising any industry which the Company selected.[89] This was later to become a significant issue.

The proposals which Henderson put to Musaddiq on 25 December were regarded by the State Department as a package, in which the individual parts were not intended to be separately negotiable. Musaddiq, however, soon sought changes which would have affected the Company. When Henderson saw him again on 31 December, Musaddiq said that he did not want to commit himself to enter into negotiations with the Company for a long-term sales contract. He would, he said, make a decision on the channels through which Iranian oil exports would be brought to market only *after* the matter of compensation had been settled by arbitration and *after* Iran had received the $100 million offered by the USA. Henderson was 'somewhat taken aback' by this development as Musaddiq had previously said that he would not object to negotiating a sales contract with the Company, provided it would not have a monopoly over Iranian oil exports. However, Musaddiq, having had a change of heart, now stuck to his new position, which he justified on the grounds that if the Company was the main purchaser of Iran's oil, there was a danger that Britain might at some time or other try to put pressure on Iran by stopping the Company's purchases of Iranian oil. Henderson tried to persuade Musaddiq not to insist on the new demand he had introduced into the negotiations and pointed out that Musaddiq's attitude would only strengthen the arguments of those who felt that he changed his mind so often that it was impossible to treat with him.[90] Musaddiq was unmoved and told Henderson in a further 'difficult and exhausting' interview on 2 January 1953 that he would not give the Company the 'tremendous power' which it would have over the economy of Iran if it had a long-term contract to buy the bulk of the country's oil production. However, he said that he would be prepared to negotiate a long-term contract with an international consortium, in which the Company could participate.

On the matter of compensation, Musaddiq again said that he was willing to have it settled on the basis of any English law of national-isation which the Company found acceptable.[91] As the British Coal Nationalisation Act of 1946 allowed for compensation to be paid for loss of future profits, the British Government believed that the way was open for the Company to claim compensation not just for its physical property in Iran at the time of nationalisation, but also for the future profits lost as a result of the early termination of the 1933 concession. Henderson pointed this out to Musaddiq, who raised no objection, provided that the terms of reference for arbitration made no explicit statement that the Company would be entitled to claim for loss of future profits. He seemed, in other words, to be prepared to accept terms which would allow the Company to claim compensation for loss of future profits, but he could not, for political reasons, be *seen* to agree to these terms. In the words of a British Government telegram to Washington:

> We have agreed on a formula which without mentioning the United Kingdom Coal Act will require the Court to apply the principles on which compensation under that Act was assessed. We are thus assured of being able to claim compensation for loss of profits under the Concession. Dr Musaddiq has told Mr Henderson that he fully under-stands the United Kingdom Coal Act allowed for future profits and has stated that he accepts the fact provided that it is not clear on the surface that he has done so.[92]

It was on that understanding that the British Government reluctantly agreed to accept Musaddiq's demand that the purchasing organisation for the bulk of Iran's oil exports should be a consortium, which would include the Company.[93]

While the Company was being increasingly marginalised in the proposals which were being drawn up for the future of the Iranian oil industry, the mood in government circles in London and Washington in the second week of January was one of optimism that a settlement could be clinched before 20 January, when the change of US presi-dency was to come into effect. As Acheson, glad of British flexibility on the matter of the consortium, wrote on 13 January: 'I believe we are close to far-reaching and helpful developments. I am appreciative of Eden's constructive and statesmanlike attitude without which no pro-gress would have been possible'.[94] His expectant tone was reasonable enough, given that Musaddiq's main demands seemed to have been met by the British. There remained, however, plenty of scope for suspense about Musaddiq's next move: would he now agree to a settlement or would he raise further objections to the terms which were offered?

The answer to that question began to unfold on 15 January, when Henderson presented Musaddiq with the written draft agreements which were, as Henderson put it, 'the result of an enormous amount of effort' by the British and US Governments. They provided for: firstly, compensation to be settled by international arbitration on the basis of any English law of nationalisation selected by the Company, meaning, in effect, the Coal Nationalisation Act of 1946; secondly, the USA to make a payment of $100 million to Iran against future deliveries of oil to the Defence Materials Procurement Agency ($50 million was to be paid on the entry into force of the arbitration agreement and the rest in instalments); thirdly, the Iranian Government to negotiate a long-term sales contract with an international consortium in which the Company would have a share.

Henderson's meeting with Musaddiq lasted for seven hours and obviously, from Henderson's description, went very badly indeed. For example, on the question of compensation, Musaddiq

> desired it to be understood that he would not enter an agreement with AIOC. That company did not exist from the point of view of Iran. It would be necessary for the agreement to be with officials of 'former Company AIOC' ... I [Henderson] reminded him that the AIOC is a company which is registered in the UK, conducts business in a number of regions of the world, has hundreds of tankers and a large number of refineries and oil wells which can be seen with the naked eye. I expressed the fear that he would cause Iran and himself to appear ridiculous in front of the entire world if he were to insist on phrasing which indicated that in his view AIOC no longer existed.[95]

Musaddiq then went on to raise other objections, the majority of which were, according to Henderson, 'so petty as not to warrant detailed mention'. Musaddiq insisted on redrafting, in Henderson's presence, various parts of the proposals, which he made 'so confusing' that they became 'practically meaningless'. The paragraph on the terms of reference for arbitration 'brought his emotions to the peak'. Eventually, Musaddiq said that he would have to study the proposals privately before making a response.[96]

After a further meeting on 17 January, Henderson saw Musaddiq again on the 19th, when their conversation was taken up mainly with the difficult matter of trying to agree on the wording of the terms of reference for arbitration. After the meeting, Henderson recorded his frustrations in dealing with Musaddiq:

Of the conversations I have had with Musaddiq so far, the one this afternoon ... was the most discouraging. In some of our prior conversations, he has evidenced some reasonableness while in others he has shown himself illogical, confused, and stubborn. Even when the position seemed hopeless in the past, I have been able at least to some degree, to remedy it and to keep the conversations on what I have hoped has been a forward course. I must confess, however, that after our conversation this afternoon, my hopes of ever persuading him to accept a settlement which we would consider reasonable and fair are beginning to wane.

Henderson went on:

This is not to accuse Musaddiq with duplicity or with not desiring basically to reach a reasonable settlement. My concern rests more with his apparent inability to continue complicated negotiations in one direction for any length of time. He appears to become confused, to change his mind, to forget, to pay heed to advisers who pander to his suspicions, to think up new ideas, or not to explain frankly what is in back of his mind.[97]

The next day, Eisenhower succeeded Truman as US President, with the Iranian dispute still unresolved.

LAST EFFORTS WITH MUSADDIQ, JANUARY–JULY 1953

With Eisenhower in the White House, Eden sent a long telegram setting out the British Government's view on the Iranian dispute to Sir Roger Makins, who had replaced Franks as the British Ambassador to Washington, asking him to bring it to the attention of Dulles. In his telegram, Eden argued forcefully that many concessions had been made to Musaddiq. For example, by the end of 1952 the US Government was offering Iran financial aid of $100 million, compared with only $10 million in the Churchill–Truman proposals a few months earlier. Moreover, the British Government had conceded that the Iranian Government's long-term sales contract for oil exports could be with an international consortium in which the Company would only be one of the participants. Eden felt that the British could not make more concessions and urged the new US Administration to take a firm stand with Musaddiq. Eden wrote:

If we are to go back to him with redrafts which go some way to meet his objections, we shall merely encourage him as in the past, to believe that by maintaining pressure and striking emotional attitudes, he will prevail upon the United States Government to reduce HMG eventually to make further concessions even to the point of capitulation to his full demands.[98]

Eden was sure, he went on, that the right approach was to confront Musaddiq again with the proposals of 15 January.[99]

The State Department, responding to Eden's views, told Makins that in its opinion the proposals of 15 January did not have a 'sanctity in themselves'. The Department hoped that the British Government would consider making changes to the proposals to meet Musaddiq's objections without, it was added, relinquishing ground on the matters which were of the greatest importance to Britain. Makins then took the matter further in discussions with Henry Byroade, who was McGhee's successor as Assistant Secretary of State for Near Eastern, South Asian and African affairs. Repeating Eden's comments that if further concessions were made to Musaddiq he would only be encouraged to ask for still more, Makins pressed Byroade for an assurance that the State Department would not press Britain to give more ground. Byroade replied, no doubt to Makins' comfort, that the matter had been discussed at length in the State Department, who felt that 'we were very close to the point where no further moves could be made ... the next round of discussions in Tehran would probably be decisive'.[100]

That round commenced on 28 January, when Henderson saw Musaddiq, who made it clear that he was willing to agree to the Company claiming compensation for the loss of its physical assets in Iran, but not for the loss of future profits. As has been seen, Henderson was under the impression that this matter had already been resolved on the basis that the Company would claim compensation in accordance with the principles of the British Coal Nationalisation Act, which allowed for claims for loss of future profits. Henderson reminded Musaddiq of four previous conversations at which:

> He had explicitly told me that he would be willing ... to settle compensation on basis any British law nationalising a British industry acceptable to the former company; and that in answer to questions put by me he had stated that such basis for settlement would be agreeable to him even though British law selected would provide for payment to former owners of compensation for loss of future profits ... He said he did not remember making such statement to me. We had discussed many things.[101]

Henderson warned Musaddiq that if he insisted on limiting claims for compensation to the Company's physical assets in Iran, the negotiations would probably be terminated. Characteristically, Musaddiq then came up with yet more counter-proposals: he said he might agree to compensation for loss of future profits if the British specified the maximum amount they were claiming; alternatively, he might agree to an arrangement whereby Iran would pay 25 per cent of the proceeds from oil exports for a limited term of years as compensation. His main concern, in making these counter-proposals, was to put an upper limit on the amount of compensation so that it would not be an unbearable burden for Iran.[102]

Anglo-American discussions on Musaddiq's counter-proposals followed a familiar pattern. The State Department took a generally sympathetic view of Musaddiq's desire to limit the amount of compensation so that Iran would not be overburdened in paying it off. The British Government, on the other hand, was not inclined to make concessions and preferred to stick to the proposals which had been put to Musaddiq on 15 January.[103] A reconciliation between these different views was achieved by a compromise, suggested by the British, under which the amount of compensation would be decided, without a pre-set maximum, by the International Court of Justice. In paying the compensation, the Iranian Government would be required to make payments in cash only to the extent of 25 per cent of the proceeds from oil exports. To limit the period of repayment, it was provided that if in any year Iran's cash payments were less than 5 per cent of the compensation due, then Iran would deliver to the Company free oil equal in value to the shortfall in the cash payment. The effect of this formula would be to limit Iran's cash payments to 25 per cent of the proceeds from her oil exports for 20 years. If that was not enough to repay the compensation in full, the balance would be paid in free oil.[104]

The State Department accepted this proposal as a 'reasonable final position' for Henderson to put to Musaddiq. Dulles thought that the US and British Governments had now reached the limit of concessions to Musaddiq. The Americans, reported Makins, 'regarded this as the last round and ... if Musaddiq did not accept, Henderson would disengage'. Moreover, 'if a break came on the basis of this final proposal, they [the Americans] would join us in defending it as firm and reasonable'.[105] Eden was very glad that the USA and Britain were in such close agreement.[106] The British Government had at last succeeded in achieving the vital goals of forging a common front with the US Government and winning US commitment to a set of proposals on the understanding that no more concessions would be made to Musaddiq.

On 20 February, Henderson presented Musaddiq with the revised proposals, which were the same as those he had put forward on 15 January except that they included the new formula for compensation. Henderson told Musaddiq that if he did not show the conciliatory attitude which had been demonstrated by the British Government, the USA would see no purpose in continuing the negotiations. Musaddiq, however, still objected to the terms of reference for arbitration on compensation, which provided for the International Court to apply the principles of the British Coal Nationalisation Act, meaning that the Company would be able to claim for loss of future profits. Henderson pointed out that it was Musaddiq himself who had first suggested that compensation should be determined in accordance with any British law of nationalisation which was acceptable to the Company. After further discussion, the meeting terminated on the understanding that Musaddiq would consult his advisers before giving a definite reply to the proposals.[107] Henderson saw Musaddiq again three days later, when they went over the same area of disagreement, without result.[108]

Further talks between Henderson and Musaddiq were held up by a new eruption of political instability in Iran. In January, Musaddiq had succeeded in obtaining an extension of his emergency powers to rule by decree, despite oppposition from former allies and supporters, notably Kashani, Makki and Baqa'i, who argued that the extension of Musaddiq's special powers was unconstitutional and would end in dictatorship.[109] The extension of Musaddiq's powers did not, however, bring political stability to the country. In February, tension between Musaddiq and the Shah came to a head when, on the 28th, a pro-Shah crowd organised by Kashani marched to Musaddiq's house in Tehran and tried to break in. Musaddiq, in bed at the time, escaped over the back wall in his pyjamas and went straight to the Majlis where he told the deputies that members of the Shah's court had tried to assassinate him.[110]

It was not until the atmosphere in Tehran had calmed down that Henderson met Musaddiq again on 4 March, when Musaddiq said that the proposals of 20 February were still being studied and that the political upheaval had prevented the drafting of a reply. In a typically distracted outburst, he went on to state, as he had done many times in the oil dispute, that the Company's 1933 concession was invalid. The 1933 concession had, he said, been negotiated by a 'tyrannical Government which did not represent the people of Iran and consequently the agreement could not be regarded as binding on Iran'.[111] This was not the first time that an Iranian Government had repudiated a concessionary agreement with the Company entered into by a previous regime. As was seen in chapter 2, the same had happened in 1932, when Riza Shah cancelled the D'Arcy concession of 1901. As Musaddiq took increasingly dictatorial powers, it did not seem to occur to him that an agreement reached directly with him might also later be repudiated by a successor regime. The Company, on the other hand, had suffered the experience before and was alert to the possibility of a repetition. As Gass had told the Persia (Official) Committee in January 1953, the Company wished the matter of compensation to be settled by international arbitration and not by direct negotiation with Musaddiq because

> A future Persian Government might well choose to repudiate any compensation agreement negotiated directly with Dr Musaddiq on the grounds that he had dictatorial powers and had used them against the wishes of the people; the award ought therefore to have, if possible, the sanctity of a judgement by the International Court.[112]

After the meeting of 4 March, further talks between Henderson and Musaddiq followed, but none reversed the march towards the final breakdown of negotiations. On 7 March a communiqué was issued in Washington, stating that the US Government regarded the proposals of 20 February as reasonable and fair; on the 9th and 18th, Henderson had more talks with Musaddiq, without constructive result; on the 20th, Musaddiq made a broadcast speech rejecting the proposals of 20 February and making a counter-offer; on 4 April, he again met Henderson, who said that it would 'serve no useful purpose to start discussing the oil dispute again'; on 8 May, Musaddiq came up with a new suggestion, on which the British Government quickly poured cold water, that Eisenhower might personally arbitrate on the matter of compensation.[113] As a last resort, Musaddiq wrote to Eisenhower on 8 May, appealing for financial aid. Eisenhower replied in a letter which was delivered to Musaddiq by Henderson on 3 July. In the letter, Eisenhower turned down Musaddiq's request for aid on the grounds that it would not be fair to spend US taxpayers' money assisting Iran, which could have access to funds from the sale of oil if a reasonable agreement was reached on compensation. Eisenhower went on to say that the US Government did not regard it as reasonable that compensation should be limited to the loss of the Company's physical assets in Iran.[114] With that letter, the door for negotiations with Musaddiq was finally sealed.

MUSADDIQ'S OVERTHROW, AUGUST 1953

By the time that Musaddiq received Eisenhower's letter early in July, the US and British Governments were well advanced in making preparations to give covert support to a coup to oust Musaddiq. The coup, which has been described in detail in published accounts by some of the participants, was put into operation on 15 August and resulted, after several days of confusion, in Musaddiq having to flee from his house before giving himself up for arrest on 20 August. He was succeeded as Prime Minister by General Fazlullah Zahidi, who formed a new Government on 21 August. The next day, the Shah, who had fled from Iran during the coup, returned to Tehran.[115] Having helped to overthrow Musaddiq, the Americans and the British could now try to negotiate a settlement of the oil dispute with his successor.

CONCLUSION

Before the Company was evicted from Iran, the terms offered to Musaddiq by the Jackson and Stokes missions had been based on the idea that the Company would return to Iran as the sole operator of the oil industry, while at the same time retaining its position as the international distributor and marketer of Iranian oil. However, from the time of the Company's eviction up to Musaddiq's overthrow, a series of offers was made to the Iranian Government, conceding more ground at each step. In that process, the idea of the Company recovering its exclusive position in Iran was steadily eroded: first, the notion of a managing agency replaced the thought of the Company being the sole operator of the Iranian oil industry; then came the Churchill–Truman proposals, which would have left the Iranians themselves to operate the industry in Iran without any foreign management, whilst leaving international distribution and marketing in the hands of the Company; then came the proposals of 20 February 1953, under which a consortium, in which the Company was but one of the participants, would have been responsible for the distribution and marketing of Iranian oil exports.

From the Company's point of view, the most important feature of these developments was that its own role in the future of the Iranian oil industry was increasingly marginalised as the British Government was consistently pressured by the USA to make concessions to Musaddiq. Fraser, who was determined not to give ground on the Company's essential interests, refused to accept this process without demur. His uncompromising resolution made him appear at times obstructive to a settlement, so much so that he became an increasingly unpopular figure not only in Iran, but also in government circles in the USA and Britain, where senior civil servants such as Bridges and Strang repeatedly raised the notion that he should be removed from the chairmanship of the Company.

If Fraser was thought to be impeding the negotiation of a settlement, Musaddiq was felt to be a still greater obstruction, at least by the British, as he rejected proposal after proposal despite considerable concessions being made to meet his demands. Eventually, however, the limit was reached and the Americans came round to the British view that the offer of concessions to Musaddiq merely encouraged him to seek still more, so that there seemed no end to the dispute. Having failed to resolve the dispute by negotiation, the two western Governments resorted to a coup, which, by overthrowing Musaddiq, did some later damage to relations between the western powers and Iran.[116]

Although the coup was no doubt meant to facilitate a settlement of the oil dispute and to restore economic and political stability to Iran, its purpose was not to defend the Company's position in Iran. As has been seen, the Company's exclusion from the future management of the Iranian oil industry had already been offered to Musaddiq, who had turned it down. Having succeeded in pushing the British and Americans to the limit of their flexibility, Musaddiq made the negotiating mistake of demanding still more, missing his chance of clinching a settlement at the optimum moment. Had he capitalised on his success, the future might have been different.

19

The consortium, August 1953–October 1954

HOOVER STEPS IN

Although Musaddiq was out of office, he was not yet out of the limelight. Indicted on charges of treason, he continued to uphold his political beliefs at his trial before a military court until he was sentenced to three years' imprisonment on 21 December 1953.[1] In the meantime, with Musaddiq still commanding widespread support and nationalist sentiment running high in Iran, there was no possibility of Zahidi's Government simply repealing the oil nationalisation laws and returning to the *status quo ante*. Nor, however, were the western powers prepared to continue holding out the concessions which had been offered to Musaddiq in the proposals of 20 February 1953. Between these limits, a settlement would have to be found which preserved at least the form of nationalisation, if not its substance.

In the search for a settlement the pattern of international diplomacy assumed the familiar form of a triangular relationship between Britain, the USA and Iran, with the USA acting as an intermediary, trying to form a bridge between the other two countries. This arrangement was not to the liking of the British Government and the Company, who repeatedly stated that their first aim was to restore diplomatic relations with Iran so that negotiations for a settlement could be held directly between the parties to the dispute. Britain's insistence on that point reflected the feeling of unease, shared by the Government and the Company, that an intermediary, which meant in effect the USA, might take an initiative in negotiating a settlement on terms which did not fully uphold British interests.

For its part, the US Government's main foreign policy concern remained the containment of communism, to which end it sought to buttress Zahidi's new Government by granting it $45 million of emergency aid on 5 September 1953. Later that month, Herbert Hoover Junior, son of the former US President, was appointed as special adviser to Dulles on the Iranian oil problem.[2] His appointment marked the commencement of a new round of negotiations between the three countries.

At the beginning of October the idea was put forward that Hoover should visit Tehran. Eden, wary of Hoover acting as an intermediary, telegraphed Dulles expressing his strong feeling that the immediate objective should be to restore diplomatic relations between Britain and Iran. Negotiations through intermediaries could not, Eden went on, be a wholly satisfactory substitute for direct contact. Although he accepted the suggestion that Hoover should go to Tehran, Eden requested that he should concern himself only with making it clear that Britain wished to resume diplomatic relations as soon as possible, assessing the political situation in Iran as it affected the oil dispute, explaining the problems involved in bringing Iranian oil back into world markets, and eliciting ideas from the Iranians about a possible oil settlement.[3] Hoover was not, in other words, to put forward proposals for a settlement.

On his way to Tehran in mid-October, Hoover stopped in Amsterdam, where he met Gass and Jackson. They gave him the Company's view that 'Having been the party most hurt, we should like to play if possible the biggest part in operating the oil industry in Persia and in disposing of Persian oil'.[4] In short, they wished the Company to be restored to its former pre-eminence in the Iranian oil industry as far as that was possible. That, as will be seen, remained the view of the board, dominated by Fraser, throughout the negotiations that followed.

The Company's views were not, however, shared by the Iranian Government, as Hoover discovered after he arrived in Tehran on 17 October. Over the next few days he and Henderson had discussions with the Shah, Zahidi, other ministers and a technical commission which Zahidi had set up to advise his Government on the oil problem. In those discussions the US representatives urged the Iranians to resume diplomatic relations with Britain. They also presented a memorandum which set out the difficulties of disposing of Iranian oil in the world's oil markets now that the shortage of oil caused by the stoppage of the Iranian oil industry had been made good from other sources. Emphasising that there was a world surplus of oil, the memorandum stated that the oil companies could not be expected to pay more for

Iranian oil than they paid for oil from other Middle East countries. Moreover, the memorandum went on, in other oil producing countries the oil companies were allowed to be in charge of production and refining.

Zahidi, however, had to think not only of the economic aspects of a settlement, but also of public sentiment and political considerations in Iran. He thought it would be difficult for Iran to accept a settlement in which the oil industry in Iran was under foreign control. He also saw no reason why Iran should have to pay compensation if agreement was reached on a 50:50 division of profits. As for the future role of the Company, he made it clear that 'it would be too much for his Government or any other to agree to a settlement that included AIOC's re-entry into Iran, either openly or in disguised form'. Nor, he said, should the Company be allowed a dominant position in any consortium formed to purchase Iranian oil and dispose of it in world markets.[5] Before leaving, Hoover was presented with an unsigned memorandum giving the Iranians' views.[6]

Meanwhile, on 29 October Fraser and Gass met with Sir Pierson Dixon of the Foreign Office to discuss the line they should take with Hoover, who was to come to London after he left Tehran. They agreed on the stance, consistent with the earlier British view, that the first step should be to re-establish diplomatic relations between Britain and Iran, whereupon direct negotiations between the two countries could commence. Negotiations through intermediaries were, it was still felt, unsatisfactory. On the nature of a future settlement, Fraser thought that the attitude of the Iranians should be ascertained by direct contact before a decision was taken on whether to try for a settlement which involved the Company returning to Iran on its own, or as a participant in a consortium. He thought it was essential that the Company should issue the invitations to other oil companies in the event of it being decided to take the idea of a consortium further.[7] He was determined, in other words, that the Company should play the leading role.

Hoover arrived in London on 4 November. Beginning that afternoon, and continuing over the next few days, he held a series of meetings with senior Government officials and representatives of the Company. Fraser stuck to the line that the Company wished to have complete first-hand information on the situation in Iran before it would agree to negotiations on any basis other than the Company going back to Iran on its own.[8] Hoover was not averse to the principle of the Company's restoration in Iran. Indeed, he reported that the US Government and the US oil companies thought that the best solution to the oil dispute would be for the Company to go back into Iran as sole operator of the oil industry there.[9] However, he did not believe that the Iranians would accept a settlement on those lines.[10] He could see no prospect of the Company's unpopularity in Iran being eradicated for a long time and said that the Iranians felt 'as a first principle, that the AIOC should not control the oil industry'.[11] Hoover also presented the unsigned memorandum which the Iranians had given to him in Tehran. Apart from the usual dose of rhetoric against the Company, it suggested that the NIOC should be responsible for the management of oil operations in Iran and that an international consortium, in which the Company would have only a minority interest, should be formed to purchase Iranian oil and market it internationally. Compensation to the Company should, the memorandum continued, be paid by the other members of the consortium.[12]

By way of response to the Iranian memorandum it was decided, with Hoover's agreement, that Henderson should be asked to inform the Iranian Government that the British could not accept the Iranian proposals and that it was essential to restore diplomatic relations so that direct negotiations could take place between the the two countries.[13] Hoover then departed for New York, where he stayed for little more than a fortnight before the next round in his shuttle diplomacy.

On 27 November, he returned to London for further discussions. The day before his arrival, the Persia (Official) Committee met, with Fraser and Gass present, to discuss the line which should be taken with Hoover. They agreed to reiterate the views which they had expressed before, i.e. that they favoured a resumption of diplomatic relations followed by direct negotiations between Britain and Iran. On the idea of forming a consortium to manage the oil industry in Iran, Fraser said that he did not wish to discuss such a scheme with anyone outside the British Government until it was clear that no other form of settlement was possible.[14] The next day, Hoover met the Persia (Official) Committee and put forward the US assessment of the position. He was pessimistic about the prospects of an early resumption of diplomatic relations between Britain and Iran, doubting whether the Iranian Government could risk such a step while Musaddiq was still on trial and attracting public attention. Time, he believed, was short as the emergency aid which the USA had granted to Iran would not last much longer and the negotiation of a settlement would not be instant. For these reasons Hoover wanted to move speedily and open negotiations with the Iranians on the basis of a settlement involving a consortium on his next visit to Tehran.[15] That was, of course, exactly what the

British wished to avoid, for it amounted to an immediate opening of
negotiations through an intermediary while British diplomatic relations with Iran remained broken; and, moreover, it meant accepting
the idea of a consortium, which the British were not yet conivinced was
necessary. Eden, concerned, met Hoover on 30 November. He said
that his objective was to resume diplomatic relations with Iran. Once
that had been achieved, he would wait for a report from the British
representative in Iran before asking the Cabinet to take a decision on
how to settle the oil dispute. Hoover agreed not to go to Tehran for the
time being, but asked Eden what the British Government would do if,
after the resumption of diplomatic relations, its representative in Iran
reported that the Iranians would not accept the return of the
Company. He thought that the British should prepare for such an
event by considering alternatives to the restoration of the Company in
Iran. To that end, he wished to have discussions, of a purely hypothetical nature, with the major oil companies on the sort of settlement
that might be acceptable if it became clear that the Iranians would
reject the return of the Company on its own.[16]

Later that same day, the Persia (Official) Committee met to consider
Hoover's request. Fraser insisted, as ever, that he could not abandon
the idea of restoring the Company's exclusive position in Iran until he
had a first hand report on the situation there from a representative of
the Company or, failing that, from a representative of the British
Government. He pointed out that 'British prestige in the Middle East
was at stake and we could not afford to have a consortium forced on
us'. Giving nothing of substance away, he agreed to talk to Hoover
about the formation of a consortium, but only on the basis that 'he
could not commit himself in any way until he knew the facts at first
hand about the situation in Persia'. He also agreed to hold informal
talks on the idea of a consortium with the other major oil companies.[17]

On 3 December Fraser accordingly wrote to the six other major
international oil companies which, together with the Company, made
up the 'Seven Sisters' of the oil industry, inviting them to come to
London at their earliest convenience for discussions of a hypothetical
nature on the formation of a consortium. The companies concerned
were Royal Dutch-Shell and the five US oil majors: Standard Oil (NJ),
Socony, Socal, the Texas Company and Gulf. A few days later an
invitation was also sent to the French company, Compagnie Française
des Pétroles (CFP). In his letters, Fraser made it plain that in his
opinion, and Hoover's, the ideal solution to the oil dispute would be
for the Company to return to Iran alone. With a characteristic lack of

decorum, he made it clear that he was only issuing the invitations to discuss a consortium on the insistence of Hoover. He would have preferred, he wrote, to wait for the resumption of diplomatic relations between Britain and Iran and then to open direct negotiations with the Iranian Government.[18]

While Fraser showed evident displeasure at being pushed into talks about a consortium by the US Government, he could at least draw cheer from another development in the dispute: the announcement on 5 December of a resumption in diplomatic relations between Britain and Iran. The way was at last clear for Britain to send her own representatives to Iran to assess the situation there at first hand and to prepare the way for negotiations to settle the oil dispute.

THE FORMATION OF THE CONSORTIUM

Pending a decision on the choice of Ambassador to send to Iran, the Foreign Office announced on 7 December that Denis Wright, who had been head of the Economic Relations Department of the Foreign Office since 1951, would be the new Chargé d'Affaires in Tehran, where he would be the top-ranking British official until a new Ambassador was appointed.

While Wright prepared to go to Tehran, the oil companies which Fraser had invited to discuss the formation of a consortium sent representatives to London where they held a series of meetings, with Hoover present, in mid-December. In the words of Winthrop Aldrich, the US Ambassador to London, it was the 'largest and most influential group of private companies and respective executives ever gathered together' – no overstatement, considering the combined economic might of the major international oil companies.[19]

As chairman of the meetings, Fraser played a leading role, reiterating the point that talks about a settlement had to be on a strictly hypothetical basis until Wright reached Tehran and reported on the prospects of the Company returning to Iran on its own. The Company, he made clear, wished to be restored to its former position in Iran. If this were found to be impossible it should, said Fraser, have the largest possible share of a consortium. He also said that the Company should receive compensation from the Iranian Government as well as payment from the other participants in a consortium, who would, in effect, be buying their shares from the Company.[20] It was decided that a working group should be set up to continue the discussions in London early in the new year and that a technical mission should visit Abadan to examine the condition of the installations there.[21] In deciding what course to take next much, however, depended on Wright's report on the situation in Iran.

Wright arrived in Tehran on 21 December, charged with the task of assessing the attitudes of the Iranian Government and people to a settlement of the oil dispute. The instructions which he received before leaving London set out in detail the lines of argument which he was to present in trying to steer the Iranians towards making a direct settlement with the Company, which, his briefing stated, 'would be the most satisfactory outcome'.[22] Any British hopes that the Company might be able to recover its exclusive position in the Iranian oil industry were soon, however, shaken. On 29 December Wright had his first meeting with Abdullah Intizam, the Iranian Foreign Minister, and asked whether it would be possible for the Company to go back to Iran. Intizam replied that: 'he was certain that public opinion in Persia would not tolerate the return of the Company and any attempt by the Persian Government to bring them back would greatly embarrass the Government and poison Anglo-Persian relations again'.[23] That view was confirmed by Zahidi, who saw Wright on 6 January 1954 and said that the restoration of the Company in Iran 'would be impossible. The Government could not face being accused of preparing the way for the return of the AIOC'.[24] The next day, Wright reported to the Foreign Office that he had taken every possible opportunity of sounding out opinion on the possibility of the Company returning to Iran. With the sole exception of the Pakistani Chargé d'Affaires, whom Wright thought was 'not particularly bright', everyone with whom he had discussed the subject was convinced that any attempt to restore the Company to Iran on its own would be 'doomed to failure'. There still seemed, Wright thought, to be 'much latent support for Musaddiq throughout the country. No Persian Government in the foreseeable future can afford to ignore the nationalism which he stirred up'. It was Wright's 'firm conviction' that if the Iranian Government tried to bring back the Company, it would be 'courting disaster'. He therefore recommended that the British Government should give up the idea of the Company returning in sole charge of the Iranian oil industry and inform the Iranian Government that Britain would be prepared to accept a settlement based on the formation of a consortium which would be responsible for the production and marketing of Iranian oil, provided the Company had a major share in it.[25]

Wright's report and recommendations marked a turning point in British policy, influencing the Government to abandon hopes of restoring the Company's exclusive position in Iran and to work instead towards the formation of a multinational consortium to manage oil operations in Iran. This change of policy was agreed by the Cabinet on receipt of Wright's report.[26] It had yet, however, to be endorsed by the Company, and Fraser was not easily swayed. On 8 January he saw Eden, who explained the position and asked Fraser to resume discussions about the formation of a consortium with the other oil companies.[27] Fraser, unyielding as ever, said that he thought it was too early to conclude that it was impossible for the Company to be restored to its former position in Iran. He suggested that the Iranian Government should be asked to agree to a visit from representatives of the Company, who would form their own judgement about the prospects of the Company's return. Wright was asked for his views on Fraser's suggestion and replied that he thought it would be 'most undesirable' to put Fraser's proposition to the Iranian Government.[28]

The Persia (Official) Committee, meeting on the morning of 11 January, heeded Wright's advice and decided that the Foreign Office should draft a message for Wright to communicate to the Iranian Government, stating that as the Iranians did not appear to favour a settlement based on the return of the Company alone, inter-company discussions on the formation of a consortium, in which the Company would have an 'appropriate share', were to be resumed.[29] However, before that message was given to the Iranians there were two matters which came first: one was to gain the Company's agreement to give up the idea of its 100 per cent return to Iran; and the other was to wait for Hoover to get clearance from the US authorities for the US oil companies to participate in talks on a consortium without running foul of the US anti-trust laws. Concurrently, another matter of key importance was: what share should the Company have in a consortium?

That question was considered at a second meeting of the Persia (Official) Committee on 11 January when it was made clear, with Gass, Snow and Addison present for the Company, that a consortium in which the Company's share was materially less than 50 per cent would be unacceptable to both the British Government and the Company.[30] Three days later the Company held a special board meeting at which the directors accepted the idea of a settlement based on a consortium in which the Company would have a 50 per cent interest.[31] The board's decision was at once communicated to Eden, who requested Makins to show the State Department the text of the statement which had been drafted for Wright to make to the Iranian Government and to explain that the British Government thought the 'appropriate share' for the Company to have in a consortium was 50 per cent.[32]

A week later, on 21 January, the anti-trust obstacle to the resumption of talks on a consortium was removed by the US National Security Council, which authorised the US oil companies to go ahead.[33] That hurdle having been cleared, Wright was able, on 23 January, to give Intizam the British Government's message that it had, in effect, given up the idea of the Company's 100 per cent return to Iran and that inter-company talks on a consortium were to be resumed. The message was deliberately vague about the extent of the Company's participation in a consortium, sticking to the wording, agreed earlier, that the Company would have an 'appropriate share'.[34] The Company and the three Governments of Britain, the USA and Iran were now all on the same wavelength about one fundamental aspect of a settlement: it would be based on a consortium and not on the Company recovering its exclusive position in Iran.

Action to resume inter-company discussions on a consortium was taken in the last week of January, when Hoover paid another visit to London, where he met the Middle East Oil Committee, successor to the Persia (Official) Committee, on 25 January.[35] The next day he lunched at Britannic House and agreed with Fraser the wording of a letter inviting the seven other international oil majors, including CFP, to resume talks on a consortium in London. All of the companies replied, agreeing with Fraser's suggestions that an inter-company working group should meet in London; that a technical committee should be set up and send a technical mission to Iran to inspect the oil installations; and, at a higher level, that the principal representatives of the companies should also meet for further discussions on the consortium.[36]

During February, a series of meetings of the inter-company working group was duly held and a technical mission was sent to Iran to examine the state of the oil installations.[37] It found that the Iranians had maintained the oilfields and Abadan refinery to a high standard and that operations could be restarted with little difficulty at an early date.[38]

In the meantime, high-level discussions on the consortium were dominated by an issue which was of fundamental importance to the Company: what share would it have in the consortium? As has been seen, the British Government and the Company had agreed that a share materially less than 50 per cent would not be acceptable. That view had been communicated to the US Government, but not in as many words to the Iranians, who had been given the more vague wording that the Company ought to have an 'appropriate share'. The

Americans, however, did not believe that the Iranian Government would agree to a settlement in which the Company's share was as much as 50 per cent.[39] This was confirmed by Hoover, who criticised Fraser's attitude at a meeting of the Middle East Oil Committee on 29 January. Hoover said that both the US Government and the US oil companies thought it was 'completely unrealistic' to seek a 50 per cent share for the Company.[40] He also reported that the US Government was disturbed by British inflexibility in this matter, fearing that the generally favourable prospects of a settlement might be spoilt because of an 'unreasonably rigid attitude' about the Company's share in the consortium. Moreover, although the Americans accepted that the Company ought to receive a consideration from the other oil companies buying shares in the consortium, Hoover did not think that the Iranian Government should have to pay compensation to the Company.[41]

After the weekend of 30/31 January, Fraser, Jackson and Gass attended a meeting of the Middle East Oil Committee on Monday, 1 February when Fraser refuted Hoover's statements, saying that none of the US oil companies had made any suggestion to him that the Company's share in the consortium should be less than 50 per cent. Fraser told the Committee that, in the Company's view, 'it was a matter of the greatest importance to British prestige in the Middle East that their share in the consortium should be as high as possible short of control. They could not countenance any suggestion that their share should be reduced because of alleged misdeeds in the past. They were taking the lead in forming the consortium, although it might be commercially more advantageous to sell their interests, but they were willing to give up control and to accept a share of 49 per cent.'[42] With Hoover and Fraser in disagreement, Sir Harold Caccia, who had succeeded Dixon as Deputy Under-Secretary at the Foreign Office, reported that Hoover had told him earlier that day that 'someone' – unspecified – had suggested a split of 40 per cent for the Company, 40 per cent for the US companies and 20 per cent for Royal Dutch-Shell.[43]

After Hoover returned to the USA on 7 February, the US Government continued to argue that it would not be possible to reach a settlement in which the Company held 50 per cent of the consortium. In a State Department *aide-mémoire* handed to Makins on 19 February it was suggested that the Company might have 35 per cent, the US companies 35 per cent and Royal Dutch-Shell 30 per cent. The *aide-mémoire* also made it clear that the State Department would not advise the US companies to go to London for further talks until agreement had been reached on the apportionment of shares in the consortium.[44]

In London, Fraser, Jackson and Gass discussed the problem with Caccia on 22 February. Later that day Eden telegraphed Makins, saying that he would agree to a Company share of 44 per cent in the consortium, with the remainder being split equally between the seven other companies which had taken part in the London talks. Eden wrote that the British Government would 'in no circumstances' agree to the split proposed by the State Department.[45] On 23 February Makins met Dulles, with Hoover and Byroade present, and handed over a note based on Eden's instructions.[46] Two days later the US Government put forward new proposals, suggesting that the shares in the consortium should be 40 per cent for the Company, 40 per cent for the US companies, 18 per cent for Royal Dutch-Shell and 2 per cent for CFP. The State Department made it clear that this offer of a 40 per cent share for the Company was 'final and had been confirmed at the highest level'.[47] The British Government was satisfied that it would not be able to get better terms for the Company, which had no option but to accept the US proposal. As Fraser explained at a Company board meeting on 1 March:

> The State Department had said categorically that any proposal to give AIOC more than 40 per cent interest would be unacceptable and that that decision had been confirmed at the highest level. The Foreign Secretary had stated that there was no likelihood in these circumstances of securing a modification of this decision and had indicated – at the same time making it clear that there was no thought of putting pressure on the AIOC Board – that Her Majesty's Government were very anxious to see the Persian situation cleared up if it were at all possible in view of present international conditions.[48]

After considerable discussion, the directors agreed that although they were 'exceedingly disappointed' by the terms they were 'prepared reluctantly' to accept a 40 per cent share in the consortium for the Company.[49] The Foreign Office informed the State Department of the board's decision, emphasising that the British Government was not prepared to see the Company's share reduced further. In reply the State Department agreed that the Company's 40 per cent share was not a matter for further negotiation.[50]

The Company's share in the consortium having been settled, the way was open for the principals of the participating companies to gather again in London for further discussions. Between 9 and 14 March a series of meetings took place in the course of which it became apparent that the Company and the US oil majors were far apart, first on the sum which was to be paid to the Company by the other members of the consortium in consideration for their shares and, secondly, on the amount of compensation which the Company was to claim from the Iranian Government. As regards the consideration to be paid by the other companies, Fraser asked for an initial payment of £120 million followed by further payments per ton of oil produced which, at prevailing exchange rates and production forecasts, would have amounted to some £140 million over twenty years. The Company also expected the other members of the consortium to pay about £20 million for their shares of stores and oil stocks. In all, therefore, the Company wanted a consideration, much of which would be payable in future years, of £280 million.[51] The US companies, on the other hand, suggested a payment of £32.4 million in the first year of operations, followed by payments per ton of oil exported in future years until a further £138.6 million had been paid. There should, they argued, be no payment for stores and stocks.[52] The total consideration suggested by the US companies was thus £171 million, most of it to be paid from production in future years. On compensation from the Iranian Government, Fraser suggested that the Company should receive 110 million tons of free oil over twenty years, which at current prices (but not costs) would add up to about £530 million over the period.[53] The differences between the Company and the US majors were thought by Royal Dutch-Shell to be of an 'alarming magnitude' and there was no sign of compromise from either side as strongly worded memoranda were exchanged in an atmosphere of obvious acrimony.[54]

In Washington, Dulles, hearing of the deadlock between the companies, sent for Makins urgently on 17 March and spoke to him 'most earnestly' about the crisis which had developed in the inter-company talks. Dulles placed the blame squarely on the Company, whose financial claims were, he said, 'utterly unrealistic'. They were, he went on, not only considered unreasonable by the US oil companies, but would be 'completely unacceptable' to the Iranian Government. Dulles warned that the US Government was not prepared to urge the US companies to refrain from breaking off negotiations, as was their intention, unless there was a 'drastic change' in the Company's attitude.[55] His stern message, passed on to Eden by Makins, was:

> The United States had been trying to do two things in Persia: to establish a stable, friendly and reasonably strong country capable of resisting Communist penetration, and to preserve the rights of the

foreign investor and concessionaire against expropriation. Now, if it was impossible to secure both objectives, it might be necessary 'to concentrate on the former, and salvage what was possible'.

More generally, it had been Mr. Dulles' policy to work closely with us in the Middle East; to leave us in the lead and, in general, to defer to our judgement. But the outstanding problems did not seem to get settled, and the United States Administration observed a tendency on our part to 'overstay the market' ... if we could not agree about the oil problem, he believed the United States Administration would have more often to take their own line and rely on their own judgement in dealing with Middle Eastern countries and problems. He did not want to do this, more especially as he realized the ill-effects that this would have over the whole field of our relations ... But the position taken by the AIOC was completely unrealistic, and he felt that a turning point had been reached, not only in the oil dispute, but in the policy of Anglo-American solidarity in Middle Eastern affairs.[56]

As this message makes clear, Fraser had taken Anglo-American harmony to the brink of destruction by his hard bargaining. The oil dispute had taken him far beyond the normal bounds of commerce to the point where his actions quite literally threatened to upset Britain's most important foreign alliance. Makins thought that if the negotiations broke down there would be 'no sympathy or support in any American heart for what will universally be regarded as the obstinacy and unreasonableness of the AIOC ... nobody here believes that Her Majesty's Government could not bring her influence to bear on the Company'.[57]

Eden intervened to retrieve the situation. In reply to Dulles, he presented calculations to show that when future payments to the Company were discounted to give their present value (by a technique known as discounted cash flow which has become the accepted method for rigorous financial analysis), the real financial worth of the amounts claimed by the Company was much less than that indicated by a simple aggregation of future payments. Moreover, distinguishing between prices and costs, he pointed out that the Company could obtain 110 million tons of oil from Kuwait at a cost of very much less than £530 million.[58] Although Eden was on sound economic ground in presenting these arguments, he also tried to placate the Americans by the expedient which came to him more naturally than it did to Fraser: he made concessions. With regard to the consideration which the other participants in the consortium were to pay to the Company, he told Fraser to come to a compromise.[59] As for the compensation to be paid by the Iranian Government, Eden said that this would be negotiated by the British Government and not by the Company. The amount proposed by the British Government was a payment of about £100 million spread over twenty years.[60] In other words, Eden took the negotiation of compensation out of the Company's hands and greatly reduced the amount which was claimed.

On 19 March Fraser explained the situation at a Company board meeting, at which the directors agreed to try to reach a compromise on the consideration to be paid by the other companies for their shares in the consortium.[61] At meetings later that day the other companies put forward revised proposals under which the Company stood to receive £32.4 million in the first year after the re-commencement of Iranian oil exports, plus payments per ton of oil exported from Iran by the other consortium participants until £182 million had been paid. The total consideration would therefore be £214.4 million.[62] On 20 March the Company's board met again to consider the revised proposals. Fraser explained that both Eden and R. A. Butler, the Chancellor of the Exchequer, had indicated that although they thought the offer was unsatisfactory, it represented the best that the Company was likely to get. The board decided to accept the proposals subject to a satisfactory settlement being reached on the other issues.[63] Further disappointment was, however, to follow on the amount of compensation to be sought from the Iranian Government.

After the board had reached its decision Eden telegraphed Makins, pointing out that the Company, having already agreed to a share of only 40 per cent in the consortium, had now accepted the 'extremely low' offer of consideration from the other participants in the consortium. In the circumstances, anything less than net compensation of about £100 million from the Iranian Government would, wrote Eden, be 'derisory'. In instructing Makins to convey these points to the State Department, Eden requested him to emphasise that the Company had made heavy sacrifices to meet the US Government's views. The British, he went on, now felt entitled to ask the US Government for support on the crucial matter of compensation. If that support was not forthcoming: 'it will not be clear to any ordinary person what is the difference between a policy of Anglo-United States solidarity and the United States administration taking its own line.'[64] The State Department, however, continued to try and make the terms more palatable to the Iranians. Instead of agreeing that the Company should receive compensation of about £100 million, the Department suggested that this figure should be the *maximum* to be paid.[65] That principle was incorporated in an understanding on compensation which was reached between the US and British Governments.[66] At a Company board meeting on 30 March Fraser reported that the inter-governmental understanding provided for the Company to receive net compensation (after allowing for Iranian counter-claims) between a minimum of zero and a maximum of £100 million. The amount would include payment by Iran for the Naft-i-Shah oilfield, the Kirmanshah refinery and the Company's distribution facilities in Iran, all of which were to be taken over and run by the Iranians. However, the Company's loss of future profits and the value of its other assets in southern Iran – which though formally nationalised would in fact be operated for profit by the consortium – were to be excluded from claims for compensation on the grounds that they were covered by the consideration which the other participants were to pay to the Company for their shares in the consortium. Fraser said he was 'extremely disappointed by the final terms of the understanding', but could think of no alternative which would be acceptable to the Governments. The directors, with no other option before them, agreed that the sum to be paid for compensation should be in accordance with the inter-governmental understanding.[67]

That settled, the inter-company talks in London came to a close with the signing of a memorandum of understanding on 9 April. The memorandum provided for the formation of a consortium in which the shares would be: 40 per cent for the Company; 14 per cent for Royal Dutch-Shell; 8 per cent each for the five US companies of Standard Oil (NJ), Socony, Socal, Texas and Gulf; and 6 per cent for CFP.[68] This arrangement was later modified in April 1955 when each of the US companies gave up one-eighth of its holding so that a 5 per cent share could be made available for nine smaller US oil companies to hold through the joint organisation which they formed for the purpose, the Iricon Agency.[69]

FINAL NEGOTIATIONS IN TEHRAN

After agreement had been reached in London on the formation of the consortium, negotiations moved to Tehran, where discussions were held between representatives of the consortium and the Iranian Government on arrangements for the future operation of the Iranian oil industry. At the same time, the matter of compensation was dealt with at inter-governmental level, the main British negotiator being Sir Roger Stevens, who had been appointed Ambassador to Tehran.

The consortium negotiating team, consisting of Orville Harden of Standard Oil (NJ), John Loudon of Royal Dutch-Shell and Snow of the Company, arrived in Tehran on 11 April. Snow was accompanied by his Company colleagues, J. Addison, D. Anderson, P. T. Cox, K. Le Page and J. M. Pattinson. Hoover had arrived earlier and with Henderson had already seen Zahidi, Intizam and Ali Amini, the Finance Minister, who was to be the head of the Iranian negotiating team. Other members of that team were Murtiza Bayat, chairman of the NIOC and a former Prime Minister, and Nuri Isfandiari, who had previously been an Imperial Delegate in London. The main consultant to the Iranian Government was Torkild Rieber, who had advised the World Bank in its negotiations with Musaddiq in 1952. In addition to these leading figures there was a large supporting cast of assistants, interpreters and secretaries and an audience of onlookers including representatives of the consortium companies who came to Iran as observers.

The next five weeks were taken up with intensive discussions on the issues to be settled in terminating the dispute. Between 20 April and 10 May a series of five meetings took place on compensation, in which the British side, led by Stevens, and the Iranians, led by Amini, exchanged claims and counter-claims without agreement being reached.[70] Parallel with the talks on compensation, between 14 April and 18 May sixteen meetings were held between the consortium negotiators, headed by Harden, and Amini's team on the myriad of matters, too numerous to catalogue, to be settled in reaching agreement on the future operation of the Iranian oil industry.[71] Two matters, in particular, stood out as problems. The most important was the question of who was to have effective managerial control over oil operations. The consortium negotiators insisted that the consortium must have effective control, but the Iranians felt that they could not be seen to agree to an arrangement which smacked of a return to the pre-nationalisation concessionary days. They suggested as a possible solution that the consortium might carry out producing and refining operations as an agent of the NIOC. The consortium negotiators agreed to put that idea to the participating companies.[72] Secondly, there were differences about the nationality of the operating companies which the consortium proposed to establish to conduct operations in Iran. The Iranians felt unable to accept the suggestion that the companies should be British for fear, once again, of seeming to acquiesce in the revival of the pre-nationalisation concession. They wanted the operating companies to have Iranian nationality, to which the consortium negotiators had objections.[73] These matters, together with the question of compensation, had not been settled by 19 May, when the consortium negotiators left Tehran for London.

86 Consortium negotiators talking with Iranian ministers in Tehran, 12
April 1954. From right to left: John Loudon, Harold Snow, Orville Harden,
Abdullah Intizam and Ali Amini

Their arrival in London was followed by a new round of inter-
company meetings which resulted in the negotiators being authorised
to agree with the Iranians that the consortium might operate the
Iranian oil industry as, in effect, an agent of the Iranian Government,
which would be the owner of the assets operated by the consortium.[74]
This arrangement was designed to preserve the principle of nationalisation whilst giving the consortium effective control of oil operations.
As regards organisational structure and the nationality of the operating companies, it was agreed that the consortium should establish
three principal companies: a holding company incorporated in
England and owned directly by the members of the consortium; and
two operating companies of neutral (Dutch) nationality, which would
conduct the operations of production and refining respectively. In
addition, a service company was to be formed in Britain to procure
supplies for the operating companies.[75]

On 16 June Hoover left London for Tehran, to be followed three
days later by the consortium negotiators, now led by Howard Page of
Standard Oil (NJ) in place of Harden, who was not well.[76] On 21 June

Page, with Loudon and Snow, presented Intizam with the proposals that had been agreed in London.[77] Beginning the next day and ending on 24 August, a series of thirty-three meetings was held between the two sides led by Page and Amini respectively.[78] Concurrently, between 30 June and 3 August, seventeen meetings were held on compensation, with Stevens taking the lead for the British.[79] The outcome of the compensation meetings was a settlement, initialled by the two sides on 4 August, under which the Company was to receive net compensation of £25 million payable over ten years beginning on 1 January 1957.[80] These terms were publicly announced on 5 August.[81] On that same day, Page and Amini issued a joint statement outlining the agreed basis on which oil operations were to be restarted in Iran.[82]

Over the next few weeks Page and Snow shuttled back and forth between Tehran and London for discussions with the numerous interests and advisers involved in bringing the agreement on principles to a final conclusion.[83] Eventually, the oil agreement was signed in Iran at 2.00 am on 19 September and flown by chartered plane to The Hague, where it was signed by Royal Dutch-Shell, CFP and the two operating companies later that day. Between 1.30 and 3.00 am the next day, it was signed by Fraser at the Berkeley Arms, an hotel near London Heathrow Airport. On the 21st, the US oil companies signed and Amini submitted the agreement to the Majlis. A month later to the day, the Majlis gave its approval by a vote of 113 in favour, 5 against and 1 abstention. On 28 October, the Iranian Senate followed suit by 41 votes in favour, 4 against and 4 abstentions. On the 29th the Shah gave his royal assent.[84]

In its essential features, the agreement provided for a consortium holding company, Iranian Oil Participants Ltd (IOP), to be incorporated in England and have its headquarters there. The IOP was to be the parent of two wholly owned operating companies, incorporated under the laws of the Netherlands, which would operate the oil industry in southern Iran. They were the Iraanse Aardolie Exploratie en Productie Maatschappij (Iranian Oil Exploration and Producing Company), which was to undertake exploration and production; and the Iraanse Aardolie Raffinage Maatschappij (Iranian Oil Refining Company), which was to undertake refining. The operating companies were to be registered in Iran, to have their headquarters there, and to include two Iranian directors on their boards. They were to be the sole operators of the oilfields and Abadan refinery, which they were, in effect, to manage on behalf of the NIOC, which was to be the owner of the assets. Another consortium company, Iranian Oil Services Ltd, was

87 Howard Page and Ali Amini at the conclusion of the oil agreement, September 1954

to be incorporated in England with its headquarters in London. Its function was to provide the operating companies with supplies, engineering services and non-Iranian staff.

Apart from being the owner of the oil industry in southern Iran, the NIOC was also to be responsible for providing infrastructure such as industrial training, public transport, road maintenance, housing, medical care and social amenities. In addition it was not only to own, but also to manage the Naft-i-Shah oilfield, Kirmanshah refinery and internal distribution facilities, all of which were concerned with supplying Iran's home market. Profits made from the oil operations under the consortium's control were to be divided equally between the consortium and the Iranian Government, preserving the principle of 50:50 profit sharing which had become the norm in the Middle East.

As to duration, the agreement was to last for twenty-five years, with provision for three five-year extensions, giving a maximum duration of forty years. However, each of the extensions was conditional on a reduction in the area covered by the agreement (initially about 100,000 square miles), so that in the last five-year period it would be about half the size of the original area.

On the matter of compensation, as has been seen, the Company was to receive a net sum of £25 million from the Iranian Government in ten equal instalments starting on 1 January 1957. The Company was also to receive a sum of £32.4 million from the other consortium participants in the first year of operations, plus a further payment per ton of crude oil produced until $510 million (£182 million at prevailing exchange rates) had been paid. Finally, the Company's share in the consortium was the 40 per cent which had been negotiated earlier in 1954.[85]

No sooner had these terms received the Shah's assent than the Company tanker, the *British Advocate*, berthed at Abadan on 29 October and began loading the first cargo of Iranian oil under the consortium agreement.[86] The Iranian oil industry was back in operation.

CONCLUSION

The terms which had been offered to Musaddiq in February 1953 had gone a long way to meet his demands and would have excluded the Company from taking part in oil operations in Iran, leaving it with only a share in a consortium formed to purchase and distribute Iranian oil exports. However, with the fall of Musaddiq and the accession to power of Zahidi, the British and Americans were at last able to negotiate with an Iranian Government which was much more prepared to come to a compromise than Musaddiq had been. Having gone to their limits in making concessions to Musaddiq, the western powers now looked upon the proposals of February 1953 as a once-only offer which, having been rejected, was no longer available.

Fraser, seeing the opportunity to re-assert the Company's interests, hoped that the Company could be restored to its former position of having sole charge of the oil industry in southern Iran. However, the US and British Governments, influenced by the reports of Hoover and Wright respectively, did not believe that it would be politically possible for Zahidi's Government to agree to the Company's return to Iran as the exclusive operator of the oil installations. It was, of course, impossible for Fraser to flout the combined authority of the US and British Governments, and he and the rest of the Company's board had no alternative but to agree to participate in a consortium, which would operate the Iranian oil industry. At that stage, the British Government and the board shared the view that the Company's share in the consortium ought to be 50 per cent, or marginally less if that was necessary. As had happened so often throughout the oil dispute, the US

Government took a different view, being prepared to sacrifice the Company's interests in order to make the terms of a settlement more acceptable to the Iranians. On this occasion, the Company was, in effect, forced by pressure from the USA to agree to a share of 40 per cent in the consortium.

Fraser, contesting every inch of ground, then sought a consideration from the other members of the consortium and compensation from the Iranian Government on terms which were considered so unreasonable by the Americans that the US Government threatened to break away from the Anglo-American alliance in its policies in the Middle East. In seeking to uphold the Company's interests, Fraser was on the verge of causing a split between the two closest western allies. Eden intervened to mend the damage, taking the negotiation of compensation out of the Company's hands and subsequently agreeing to a sum which left Fraser and the board again feeling disappointed. It might seem that the Company, having suffered this series of disappointments, did badly out of the final settlement. Yet most writers on the oil dispute have tended to regard its outcome as a defeat for Iran.[87] Why should that be so?

In the end, the Iranian Government had gained very little of substance out of the dispute. The NIOC was, it is true, the owner of the oil industry in the south of the country, but with the consortium exercising effective managerial control over operations, the vesting of ownership with the NIOC was really no more than a token gesture of nationalisation. Moreover, with the profits from Iranian oil exports being split 50:50 between the Iranian Government and the consortium, Iran gained nothing more than other Middle East oil producing countries had obtained without going through more than three years of hiatus. In fact, having to pay compensation as well as divide the profits made from Iranian oil equally with the consortium, Iran retained less of the profits made from its oil than other producing countries did from theirs. In short, Iran won little from the crisis.

On the other side, the Company also appeared to have lost from the dispute. It had, in effect, given up 60 per cent of its interests in Iran in return for a consideration from the other members of the consortium which dissatisfied Fraser and the board, plus compensation from the Iranian Government which was also regarded as inadequate. How could such a result be regarded as a victory? The answer is twofold.

First, the Company was so successful in making good the loss of its Iranian oil supplies by turning to other sources that a 40 per cent share in the Iranian oil industry was at least as much as it needed for its business. Indeed, the Company's 100 per cent return to Iran would probably have been a commercial embarrassment in that it would have added greatly to the surplus of supply over demand which had been a persistent and growing problem in the Company's business for some time, as was seen in earlier chapters. A 40 per cent share in the consortium was, therefore, very satisfactory for the Company on commercial grounds.

Secondly, there was the matter of the Company's wider position in the international oil industry. Fraser and the board were, of course, aware that the Company did not desperately need Iranian oil and, as has been mentioned, they had come to look upon the dispute primarily as a matter of prestige, which the Company could not afford to lose in case its position in other oil producing countries was thereby undermined. On that score, the Company's honour was satisfied not merely by its 40 per cent share in the consortium, but, more importantly, by the manner in which the seemingly inexorable advance towards its complete exclusion from Iran had been put into reverse during the crisis. For that, the Company paradoxically owed a large debt of gratitude to Musaddiq who, having played with great effect on US fears of communism to wring concession after concession out of the British, was offered terms in February 1953 which would have left the Iranians in charge of their own oil industry and the Company with no more than a share in a consortium responsible for the international marketing of Iranian oil exports. At that point Musaddiq made the great mistake of failing to realise that he had extracted all the concessions he could get. Having driven the USA and Britain to their limits, he asked for still more and precipitated his downfall, opening the way for new negotiations with Zahidi's Government. Fraser seized on the opportunity to recover lost ground and upheld the Company's interests with characteristic vigour. He could not, as he hoped, secure the complete reinstatement of the Company in Iran. However, the final terms of the settlement represented a great recovery from a situation which in early 1953 had looked a virtually lost cause for the Company. From the Company's point of view, Fraser, so often criticised by Government ministers and officials for his stiff resistance to the exclusion of the Company from Iran, was vindicated.

20

Conclusion

By any measure, the Company achieved a tremendous expansion in the scale of its operations in the 1930s and 1940s. Its crude oil production rose from some 6 million tons in 1930 to more than 40 million tons in 1950, an annual compound rate of growth of about 10 per cent. Most of that crude oil was processed at Company refineries, whose combined throughputs increased from about 6 million tons in 1930 to over 30 million tons in 1950, a rate of expansion which, high as it was, could not match the burgeoning production of the Company's prolific oilfields. As was seen in chapter 11, the Company therefore turned increasingly to crude oil sales to dispose of its output in the postwar years.

Most, but not all, of the growth in crude oil production and refining took place in Iran, which remained, throughout the period up to 1951, overwhelmingly the most important centre of operations. In 1930 the Company's oil exports from Iran came from just two oilfields, at Masjid i-Suleiman and Haft Kel. By 1950, however, the Company had seven producing oilfields in Iran. These accounted for about three-quarters of the Company's crude oil production, the remainder coming from the Company's producing interests in Iraq, Kuwait, Qatar and Britain, all of which came onstream in the period covered by this volume. In crude oil refining, the Company acquired or constructed, as whole- or part-owner, seven new refineries in the 1930s and 1940s. They were at Lavera and Dunkirk in France, Kirmanshah in Iran, Haifa in Palestine, Venice in Italy, the Eurotank refinery in Germany and a refinery in Kuwait. Adding these refineries to those which had come into operation earlier, the Company wholly or partly owned twelve refineries by the end of 1950, with others already planned or soon to be acquired at Antwerp, the Isle of Grain in Kent and

Hamburg. Abadan, far and away the most important, was the largest refinery in the world and Britain's greatest single overseas investment. Conversely, the Company was easily the largest foreign investor in Iran, where the number of its employees and contractors rose from about 31,000 in 1930 to 80,000 in the late 1940s; and capital expenditure increased from some £2.3 million in 1930 to £17.6 million in 1949.

The financial rewards which the Company earned looked to be growing even faster than its physical operations. Between 1930 and 1950 consolidated pre-tax profits grew from approximately £6.5 million to nearly £86 million, a remarkable average annual growth rate of close to 14 per cent. Figures such as these would seem to give substance to the impression, popularised by Anthony Sampson in his book *The Seven Sisters*, that the international oil majors, of which the Company was one, were able to reap huge rewards from the oil industry over which they exercised virtually untrammelled dominion through joint production companies such as the IPC, restrictive contracts such as the Red Line Agreement and marketing cartels, most notably the famous Achnacarry accord.[1] Accredited with virtually mystical powers, the major oil companies were, according to Sampson, 'invulnerable to the laws of supply and demand' and seemed 'often enough like private governments, to which the Western nations had deliberately abdicated part of their diplomacy'.[2]

The picture of smooth, unruffled dominance over market forces and nation states is not, however, altogether convincing on closer inspection of the Company's fortunes in the period, starting with its profits. With perfect data, much light could be thrown on the Company's performance by examining the real worth and sources of its profits. In this case, unfortunately, the data are seriously deficient for such a purpose, not least because the Company did not produce consolidated accounts, aggregating the results of all its subsidiaries in a single statement, until later in its history. A *post hoc* consolidation of the accounts, done for the purposes of this volume, cannot possibly compensate fully for the shortcomings of the original data, but it does at least help to elucidate the financial results of the Company and its subsidiaries as a group. Allowing for the inaccuracies which inevitably accompany such an exercise, some conclusions may be drawn.

To that end, some summary financial data are presented in table 20.1, covering the period 1929–50 except for the abnormal years of World War II. Column (a) confirms the growth in pre-tax profits from £6.2 million in 1929 to £85.7 million in 1950, giving the impression, noted above, of dazzling financial results. An adjustment for price inflation is, however, required to gain a more realistic measure of the Company's profits. The adjusted data, displayed in column (b) of table 20.1, show a considerably less spectacular peak in pre-tax profits of £33.8 million in 1950. That figure, it should be noted, was more than double the pre-tax profit made in 1949. The reason was, very largely, the devaluation of sterling from US $4.03:£1 to US $2.80:£1 in September 1949. At a constant dollar/sterling exchange rate the real pre-tax profit in 1950 would have been more like £23.5 million. Thus, by adjusting for price inflation and currency devaluation, the Company's pre-tax profit in 1950 is cut from a heady figure of £85.7 million to a more sober £23.5 million.

The rise in pre-tax profits from £6.2 million in 1929 to the adjusted figure of £23.5 million in 1950 was still, it should be said, very substantial, representing an average annual rate of growth of about 6.5 per cent. How was that achieved? A large part of the answer is again given in table 20.1, which shows in column (c) the great increase in the Company's physical sales of oil from 5.8 million tons in 1929 to 39.8 million tons in 1950. The pre-tax profit per ton, unadjusted for price inflation and currency devaluation, is shown as an index in column (d) and gives the flattering impression that profit margins doubled between 1929 and 1950. However, the data in column (e), which are adjusted for inflation and devaluation, give a different picture. As can be seen, between 1929 and 1950 the Company's adjusted pre-tax profit margins per ton of sales fell by nearly half, from an index of 100 in 1929 to 55 in 1950. The main cause was a fall in real crude oil prices, shown in columns (f) and (g).

This brief statistical excursion has not been without purpose: it has given a broad measure of the growth in the Company's pre-tax profits after making appropriate adjustments; and it has indicated roughly how those profits were made. Despite the development of the alkylation process for the manufacture of aviation spirit, the installation of cracking plant to produce larger volumes and higher grades of motor spirit and the entry into petrochemicals (described in the next volume), the Company's growing profits were not fundamentally based on the development and commercial exploitation of a new product or products. Nor, despite continuing improvements in production techniques, were they founded on one or more cost-reducing technological breakthroughs. Mergers and acquisitions took place in, for example, the Company's postwar expansion into European markets, but they were relatively few in number and minor in scale. None of these possible sources of increased profits accounted for very much com-

Table 20.1 *The Company's profit margins, 1929–1938 and 1946–1950*

	(a) Pre-tax profits at current prices (£millions)	(b) Pre-tax profits at constant (1929) prices (£millions)	(c) Sales tonnage (million long tons)	(d) Index of pre-tax profit per ton at current prices (1929=100)	(e) Index of pre-tax profit per ton at constant prices (1929=100)	(f) Crude oil at current prices (US $ per barrel)	(g) Index of crude oil at constant prices (1929=100)
1929	6.2	6.2	5.8	100	100	1.27	100
1930	6.5	7.1	6.0	101	111	1.29	103
1931	3.6	4.2	5.7	59	69	0.65	60
1932	3.5	4.1	6.2	53	62	0.87	81
1933	3.1	3.8	6.6	44	54	0.67	65
1934	5.3	6.1	7.4	67	76	1.00	91
1935	5.3	6.0	8.0	62	70	0.97	86
1936	7.8	8.2	8.8	83	87	1.09	91
1937	9.8	9.0	10.4	88	81	1.18	86
1938	8.7	8.0	10.1	81	73	1.13	82
1946	28.9	15.3	20.4	133	70	1.06	44
1947	37.3	17.1	24.0	145	66	1.55	56
1948	51.0	21.0	29.2	163	67	2.11	69
1949	41.2	16.7	33.2	116	47	1.84	59
1950	85.7	33.8(23.5)*	39.8	201	79(55)*	1.71	53

* Figures in parentheses are at constant sterling/dollar exchange rate.

Sources: BP consolidated balance sheets (see Appendix 1); G. Jenkins, *Oil Economists' Handbook* (5th edn London, 1989), p. 3. The price index used in calculating profits at constant prices is that for plant and machinery in C. H. Feinstein, *National Income, Expenditure and Output in the United Kingdom, 1855–1965* (Cambridge, 1972), table 63.

pared with the dominant thrust behind the Company's expansion: the production and sale of ever greater quantities of oil, albeit at falling margins.

The rapid growth of the Company's sales and its rising profits were made possible by economic conditions which were fundamentally extremely favourable. The underlying trend was one of steeply rising demand for oil, which increasingly displaced coal as an energy source and was an essential fuel for a new generation of consumers for whom road transport became the norm.[3] Although world oil supply grew still more rapidly than demand, resulting in falling prices and pressure on margins, the growth in oil consumption provided the Company with great opportunities to increase its sales of the very low-cost Middle East oil which it possessed in such abundant quantities.

While favourable underlying economic conditions enabled the Company to expand, there was another, more compelling factor forcing the pace: the Company's overriding desire to retain its concessions in producing countries, especially Iran. As was seen in chapter 11, Fraser laid it down as the basis of the Company's postwar policy that the rate of growth of production in Iran should at least keep pace with the growth in world production. That policy was founded on the realisation that failure to preserve Iran's position in the international oil industry would cause the Iranian Government to become dissatisfied with the Company, whose concession would then be at risk. A rapid rate of growth was, therefore, seen as a prerequisite for the retention of the concession. For the same reason, it was explicitly laid down by Fraser that the Company should aim to achieve steady, constant growth, free of large fluctuations. For concessionary reasons, the Company therefore sought not only to expand its sales rapidly, not minding falling margins, but also to avoid the turbulence of peaks and troughs. The goal, in short, was one of smooth, stable growth.

In setting that aim, Fraser would no doubt have been aware that the Company had repeatedly found that it was unable to influence its economic and political environment sufficiently to achieve the goal of stability. The Achnacarry Agreement setting out the basis of the 'As Is' principles was a case in point, failing to insulate the Company from the great economic depression of the early 1930s. In fact, when the 'As Is' principles were at their zenith, the Company was making less return on its capital than at any other time between 1929 and 1951. In a clear demonstration of the linkage between economic conditions and concessionary relations, the slide in profits brought about by the great depression resulted in a fall in the amount of royalties payable to Iran, which in turn triggered the cancellation of the D'Arcy concession by Riza Shah in 1932. After the negotiation of the 1933 concession the same pattern of events was repeated in the late 1930s, when the economic downturn of 1938–9 made Riza Shah dissatisfied with the Company's performance. On that occasion, the possibility of the concession being cancelled was averted by Cadman on his visit to Tehran in 1939. However, it was not long before the wartime fall in Iranian oil exports, and hence revenues, again prompted the Shah to threaten action against the concession, an event which was avoided by the 'make-up' payments agreement of 1940. The pattern was unmistakable: economic fluctuations, whatever their causes, resulted in concessionary instability.

It was not, of course, only the Iranian Government that suffered a loss of revenues when the Company went through economic downturns. So too did the Company. On such occasions it administered the well-tried remedies of severe cost-cutting. Between 1930 and 1932, during the great depression, the number of Company employees and contractors in Iran was cut by more than half, from some 31,000 to about 15,000. Between the same dates, capital expenditure in Iran was reduced from more than £2.3 million to a mere £74,000. Meanwhile, the sum paid out as dividends to shareholders was brought down from approximately £3.6 million to £1.7 million. A decade later, in the early 1940s, the same measures had to be taken. The number of employees and contractors in Iran, having risen to 49,000 in 1938, was reduced to 29,000 in 1940 as Iranian oil exports were cut back after the outbreak of war. At the same time, capital expenditure in Iran, having climbed to £10.4 million in 1938, was pruned to £3 million in 1940. The shareholders once again saw their dividends fall as the sum paid out was reduced from £6.1 million in 1937 to £2.1 million in each of 1939 and 1940. These movements, and the subsequent recovery during and after the war, are plotted in figure 20.1, which stops short of the greatest upheaval of all: the Iranian crisis of 1951–4.

These fluctuations call into question the glib assumption that the Company and the other oil majors were masters of the environment in which they operated, using their vast size and resources to control events to their own advantage. It would seem that the Company's main success lay less in achieving the stability which it would have liked than in coping with its unstable and uncertain environment. In that situation, there were three characteristics of the Company which most obviously helped it to weather the various shocks to which it was subjected. One, already described, was the flexibility which the

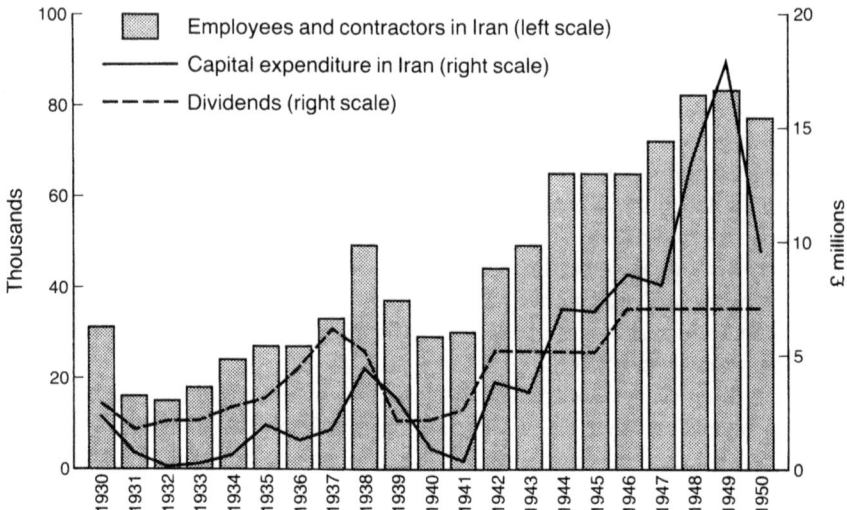

Figure 20.1 Fluctuations in the Company's investment, employment and dividends, 1930–1950

Company showed in retrenching deeply and rapidly when conditions were adverse. A second was the Company's large scale and diversity of interests, the advantage of which was most clearly demonstrated in the crisis of 1951–4, when the Company, having been deprived of its Iranian oil supplies, was able to turn to other sources to sustain its business. A third factor which contributed importantly to the Company's resilience was its financial robustness, the product of a consistent policy of maintaining an exceptionally strong balance sheet. This may have looked very conservative and unadventurous in expansive phases of the cycle, but when downturns came, the Company's high liquidity and virtual absence of debt ensured that it would not be plunged straight away into financial difficulties.

In financial policy, as in other economic matters, there was, on the whole, a smooth transition from Cadman to Fraser. The same could not, however, be said about the spheres of government relations and international affairs, in which the Company was involved to a most unusual degree, even by the standards of international oil companies, during this period. Why did matters of state and diplomacy require so much attention, diverting management from concentrating on the economic functioning of the business? To some degree, this was the consequence of the growth in state intervention in the British economy, which, as was seen in chapter 12, reached new peacetime levels in the postwar years. In that respect, the Company's experience of government regulation and interference would have been shared by other businesses. However, the interest of the Government in the Company's affairs went much further than that. From its formation in 1909 the Company had, in effect, been marked out as a special case, largely because it was dealing with a strategic commodity which was of great importance to the defence of the realm. It was primarily for that reason that the British Government had acquired its majority shareholding in 1914. Although the Government undertook not to interfere with the commercial operations of the Company, its position as a majority shareholder inevitably tended to politicise the Company, which, whether it was true or not, was widely seen as an arm of the British Government. On balance, it was not in the Company's interest to be closely associated with a declining imperial power in an age of rising nationalism. It might have helped in the negotiations for the concession in Kuwait, but on the other hand the Government's shareholding had the effect of excluding the Company from oil-rich Venezuela. As for the diplomatic support which the Company had from the British Government in the two major concessionary crises of 1932–3 and 1951–4, that surely would have been given as a matter of national interest, regardless of the Government's shareholding.

Most importantly, the Government's shareholding contributed to the Iranian impression that the Company was a state within a state in Iran, where the Company was repeatedly accused of interfering in the country's internal affairs. That damaging image was exacerbated by the Company's involvement in matters which extended beyond narrowly commercial activities. There was very little in the way of amenities and infrastructure in the southern part of the country, where the Company's oilfields and the Abadan refinery were located. As a result, the Company had to be self-sufficient, importing virtually all of its needs and establishing its own facilities. Townships grew up, utilities were provided, education promoted, communications and transport links inaugurated and housing constructed under Company auspices. The result was the creation of an enclave community, which appeared to the Iranians as a form of private government. Moreover, in the Company's early years the area in which it operated was under the control of local chieftains such as the Bakhtiari Khans and the Shaykh of Muhammara, with whom the Company had to make its own arrangements for security and other matters. One of the consequences was that the Company was accused of having interfered in local affairs, over which the central government in Tehran had little authority.

The Company was so closely involved in government relations and international affairs that its chairmen were required to deal with issues which were far outside the realms of activity normally associated with business. Cadman found this no hardship. By background and temperament he moved easily between the worlds of business and government, displaying consummate diplomatic skills in his dealings with representatives of governments, whether they were British officials and ministers or the autocratic Riza Shah. After successfully negotiating the 1933 concession, Cadman even volunteered to act as a kind of special envoy to Iran in negotiating a general Anglo-Iranian treaty, an idea which, in the event, the British Government turned down.

Fraser was a different character altogether. A complete oilman, he had proved himself under Cadman to be generally capable of dealing with the running of the Company's business. However, after Cadman died and Fraser became chairman, he had to spend more time on political and diplomatic matters, in which he lacked the skills of his predecessor. Moreover, he had the ill luck that his chairmanship happened to coincide with a high watermark of government involvement in the affairs of the Company. A forthright personality, his autocratic, abrasive and uncompromising temperament made him unpopular with British officials and ministers from the very beginning of his chairmanship.

Fraser's relations with the British Government, already strained, were tested virtually to breaking point during the nationalisation crisis of 1951–4. In the eyes of civil servants and politicians, he took a narrow view of the Company's interests and held to it so determinedly that he seemed at times to be an obstacle to the reaching of a settlement. To his critics, both then and later, this has been regarded as a failing of Fraser's, who might, it has been argued, have headed off the nationalisation crisis if he had been more willing to make concessions to Iran. For example, Acheson thought that the confrontation with Iran might have been avoided if more heed had been paid to the example of Aramco in Saudi Arabia, 'graciously granting what it no longer had the power to withhold'.[4] On the other hand, it could be said that without Fraser's fearlessly staunch defence of the Company's interests, the Company would have been 'sold out' by the western powers for the sake of bringing the dispute to a swift conclusion, assuaging US fears that a protracted crisis might open the door to communist control of Iran. Seen in such a light, the Iranian dispute

was not a case of western governments abdicating matters of international diplomacy to the Company: it was exactly the opposite – a
case of the Company being treated as a pawn, to be sacrificed if
necessary for the sake of the USA's overriding concern with the global
containment of communism.

Each view of Fraser contains elements of truth. The Supplemental
Agreement did indeed look like a measure which offered too little too
late after Aramco and Saudi Arabia reached their 50:50 profit sharing
agreement at the end of 1950. However, once the Supplemental Agreement had been rejected and Musaddiq had come to power what
concessions could anyone have made which would have satisifed him,
short of giving in completely to all of his demands? As it was, he
extracted concession after concession in the many rounds of negotiations before the western powers eventually concluded that it was
impossible to reach agreement with him. In short, if Fraser was uncompromising so too, in at least equal measure, was Musaddiq. That, of
course, was the essence of the problem: the parties to the dispute were
so far apart that there was no compromise settlement which would
have been regarded as satisfactory by the opposite sides.

The same difficulty besets assessment of such a controversial figure
as Fraser – there is no middle ground on which to rest a final judgement which would be universally accepted by such disparate groups as
US politicians and officials, their British counterparts, Company shareholders (other than the British Government) and Iranian nationalists:
not to mention historians of various persuasions. It can, however, be
said with certainty that Fraser acted in what he perceived to be the
interests of the *Company*. If those interests did not conform with the
international strategic concerns of the USA, or other powers, that did
not alter Fraser's conception of the Company's cause, to which he
remained utterly committed.

Although Fraser and the board were disappointed with some of the
terms of the final settlement, the conclusion of the nationalisation
dispute was unmistakably a watershed in the Company's development.
The Company may have held a 40 per cent interest in the consortium,
but Iran was no longer the dominant centre of its operations, investment and employment, or the main plank of its forward planning.
Having been forced to diversify its sources of oil supply during the
crisis, the Company had developed a wider, more balanced spread of
international interests, which gave it a much more multinational character than before. To these changes of substance was added a change
of name when, on 16 December 1954, an extraordinary general

88   Changing the plaque at the entrance to Britannic House, Finsbury Circus, London to show the name The British Petroleum Company instead of Anglo-Iranian Oil Company, December 1954

meeting of shareholders agreed that the Company should be known not as the Anglo-Iranian Oil Company, but as British Petroleum. That decision symbolised the passing of the Anglo-Iranian years and the opening of a new era: an epoch in the Company's history.

APPENDIX I

A note on statistics and their sources

FINANCIAL STATISTICS: THE CONSOLIDATION OF ACCOUNTS

In this volume the term 'the Company' has been used with a certain amount of latitude to describe a group of companies consisting of a parent company, the Anglo-Persian Oil Company (APOC) – renamed the Anglo-Iranian Oil Company (AIOC) in 1935 – which was not only a holding company, but also traded in its own right; and numerous subsidiaries and associates such as the D'Arcy Exploration Company (engaged in exploration and production), the British Tanker Company (engaged in shipping) and various overseas marketing subsidiaries and associates.

In such a group, no assessment of the financial results and position of the group as a whole is possible without consolidated accounts which combine the assets and liabilities, revenues and expenditures of the constituent companies into a single annual balance sheet and profit and loss account. However, in the case of the group collectively described as the Company, the accounts of the parent, APOC/AIOC, were published separately from its subsidiaries until 1947, when consolidated balance sheets began to be published, but excluding most of the marketing subsidiaries and the Tanker Insurance Company. In order to give a consistent overall picture of the Company's finances for the purpose of this history, the annual balance sheets of the constituent companies were consolidated for each year. In a few minor cases the balance sheets of subsidiaries were not available and one or two of the overseas subsidiaries drew up their accounts in a form which was not ideal for the purposes of consolidation. The statistics cannot, therefore, be taken as exact, but it is considered that they are sufficiently accurate to give a useful indication of the Company's financial results and position.

In compiling them, and in extracting data on, for example, capital expenditure, use was made of the series of schedules which were used as the basis for the contemporary preparation of the accounts. They can be found in the file series BP CRO 4P 6001–7086 and contain far greater detail than appears in the published accounts.

CRUDE OIL PRODUCTION, REFINERY THROUGHPUTS AND EMPLOYMENT

Statistics on crude oil production and refinery throughputs were compiled as series for the whole period covered by this volume from sources which it would be tedious to give separately for each table, graph etc. Unless other sources are specifically mentioned at the bottom of a table or graph the data has been drawn from two sets of sources: the final accounts schedules in the file series BP CRO 4P 6001–7086; and the monthly management reports to the board. Both are kept in the BP archives. The same remarks apply to the statistics on numbers employed, except that the sources were BP 54374–8; 54381; 71487; and 96438.

APPENDIX 2

Notes and sources
to the graphs and diagrams

Figure 4.1

Note: Return on capital employed represents pre-tax profits as percentage of average net assets for the year.

Sources: G. Jenkins, *Oil Economists' Handbook*, 5th edn (London, 1989), p. 3; BP consolidated balance sheets (see Appendix 1).

Figure 4.2

Notes: 1. Includes company tanker bunkers and refinery works fuel.
2. Fuel oil in 1938 includes Abadan bunkers as in 1937.

Source: BP 109194.

Figure 4.3

Note: Arabia and Baluchistan included with Iran.

Sources: BP 109194; 77799.

Figure 4.4

Note: Interpolated data used for fuel oil sales in 1929.

Source: BP 109194.

Figure 4.5

Note: Fuel oil includes bunkers up to 1930.

Source: BP 77799.

Figure 4.6

Source: BP 77799.

Figure 6.1

Source: Jenkins, *Oil Economists' Handbook*, p. 96.

Figure 11.1

Note: Irano Products sold products such as lubricating oil extracts, slack
           waxes, shale resins, powdered shale and kerosene extracts from the
           Llandarcy, Pumpherston and Grangemouth refineries.
Source: BP 72322.

Figure 11.2

Notes: 1. This graph is not comparable with figure 4.2 owing to changes in
           classification in the source data. The main change is that in figure 4.2
           fuel oil includes quantities of diesel oil, which are classified as gas
           and diesel oil in this graph.
           2. Data on the breakdown of product sales is not available for
           1946–7.
Sources: BP 102729; 102749; 102760; 102771; 109194.

Figure 11.3

Notes: 1. This graph is not comparable with figure 4.4, in which fuel oil
           sales include diesel oil, which comes under gas and diesel oil in this
           graph.
           2. Excludes marine bunkers and sales to the armed services.
Sources: BP 109194; 102729; 102738; 102760; 102771.

Figure 11.4

Note: 1. Owing to inconsistencies in the source data, this graph incor-
           porates adjustments to UK motor spirit sales in 1946–7 and to
           bitumen sales in the Consolidated area in 1946.
           2. UK sales in this graph are not comparable with those in figure
           11.3, which includes sales to other local distributors not shown here.
Sources: BP 102729; 102771; 109188; 109190; 77799.

Figure 11.5

Source: BP 77799.

Figure 11.6

Source: BP 77799.

Figure 14.1

Notes: 1. Capital expenditure in 1942–50 includes £6.3 million spent on aviation spirit plant which was recoverable from the British Government.

2. The price index used for the calculation of capital expenditure at constant (1930) prices is that for plant and machinery in C. H. Feinstein, *National Income, Expenditure and Output in the United Kingdom, 1855–1965* (Cambridge, 1972), table 63.

Sources: See Appendix 1.

Figure 14.2

Note: Tons converted from barrels at 7.45 barrels = 1 long ton.

Source: DeGolyer and MacNaughton, *Twentieth Century Petroleum Statistics* (Dallas, Texas, 1990).

Figure 14.3

Note: Masjid i-Suleiman includes minor production from the Lali field from 1947 onwards.

Sources: See Appendix 1.

APPENDIX 3
Draft Achnacarry Agreement, 18 August 1928

Certain politicians, with the support of a portion of the press, have endeavoured to create in the public mind the opinion that the petroleum industry operates solely under a policy of greed and has itself initiated methods of wanton extravagance. This contention is absolutely unjustified, ignoring as it does the problem of the oil industry.

Since its inception the oil industry has looked forward with apprehension to the gradual depletion and final exhaustion of its supplies of crude oil. The temporary shortage of supplies that existed in certain countries during the great war further accentuated this fear and caused vast sums of good money to be expended to locate and develop reserves in all parts of the world where petroleum potentialities appeared, as well as in accumulating large reserve stocks above ground.

Now the situation has changed. An adequate supply for a long time to come is assured. This is the result of the application of science to the petroleum industry. More effective methods of handling crude have been developed so that the yield of gasoline from a given amount of crude has been enormously increased. On the other hand, methods of consumption are being made more economical; high compression motors are being developed; diesel engines take the place of steam engines and fuel oil is being utilized more efficiently and more economically. All of these developments will have a marked tendency to slow down the rate of increase in consumption.

Excessive competition has resulted in the tremendous overproduction of to-day, when over the world the shut-in production amounts to approximately 60% of the production actually going into consumption. In other branches of the business, over-competition has had a similar result, so that it may be fairly said that money has been poured into manufacturing and marketing facilities so prodigally that those now available are far in excess of those required to handle efficiently the present world's consumption.

Up to the present each large unit has tried to take care of its own over-production and tried to increase its sales at the expense of someone else. The effect has been destructive rather than constructive competition, resulting in much higher operating costs. Certainly no company can expect to obtain an increased outlet for its own production when all companies have a surplus for which they are desirous of securing a market. Naturally each is determined to hold his share of the business, and the result is non-compensatory returns. There is only a certain consumption to be supplied.

The petroleum industry has not of late years earned a return on its investment sufficient to enable it to continue to carry in the future the burden and responsibilities placed upon it in the public's interest, and it would seem impossible that it can do so unless present conditions are changed. Recognizing this, economies must be effected, waste must be eliminated, the expensive duplication of facilities curtailed, and the following sets out the more important principles for bringing this about in all countries other than the domestic market in the U.S. and imports into the U.S. –

1. The acceptance by the units of their present volume of business and their proportion of any future increase in consumption.
2. As existing facilities are amply sufficient to meet the present consumption these should be made available to producers on terms which shall be based on the principle of paying for the use of these facilities an amount which shall be less than that which it would have cost such producer had he created these facilities for his exclusive use, but not less than the demonstrated costs to the owner of the facilities.
3. Only such facilities to be added as are necessary to supply the public with its increased requirements of petroleum products in the most efficient manner. The procedure now prevailing of producers duplicating facilities to enable them to offer their own products regardless of the fact that such duplication is neither necessary to supply consumption nor creates an increase in consumption should be abandoned.
4. Production shall retain the advantage of its geographical situation, it being recognized that the value[s] of the basic products of uniform specification are the same at all points of origin or shipment and that this gives to each producing area an advantage in supplying consumption in the territory geographically tributary thereto, which should be retained by the production in that area.
5. With the object of securing maximum efficiency and economy in transportation, supplies shall be drawn from the nearest producing area.
6. To the extent that production is in excess of the consumption in its geographical area then such excess becomes surplus production which can only be dealt with in one of two ways; either the producer to shut in such surplus production, or offer it at a price which will make it competitive with production from another geographical area.
7. The best interests of the public as well as the petroleum industry will be served through the discouragement of the adoption of any measures the effect of which would be to materially increase costs with consequent reduction in consumption.

8. If these principles are followed, the result will be a stabilization of the world's market outside of the U.S. domestic market which will be in the interest of all.

To give effect to the above it will be necessary for the groups to adopt a uniform policy, which in each country of production will have to be considered separately based on the particular conditions existing in such country. The policy to be followed for all exports from countries of production shall be broadly on the following lines:–

1. An arrangement to cover all exports of petroleum and its products, with the exception of lubricating oils, paraffin wax and specialty products, which shall be the subject of further consideration, and excepting all exports made to the U.S.
2. A proportion of the exports of each group to be ascertained on the following basis:–
   (a) The quantities of each product the groups have delivered in each country during the period from ..... to ..... shall be added, and the proportion of each group's deliveries of the total shall be ascertained for each product and each country. The proportion of the total deliveries of each product so ascertained for each group shall be the quota each group shall be entitled to supply of the total imports of all groups in the country in question.
   (b) The percentage shall furthermore be ascertained which the total of each group's deliveries in all countries bears for each product to the total of all of the groups' deliveries.
   (c) If during any one year the total of all the quantities supplied by a group as under (a) is less than the quantity based on the group's percentage of the total supplies of all groups as under (b), then the group shall have the right to offer to the over-supplying groups the difference, and the over-supplying groups shall be obliged to purchase from the under-supplying group the quantity of a product so offered in proportion to their over-supply, at the Gulf price basis.

EXAMPLE.

Assume that groups X, Y and Z had in the countries I, II, III and IV the following basic deliveries for the computation of their quota as under the arrangement:

	X	Y	Z
Country I	20	380	0
,, II	0	0	100
,, III	200	25	0
,, IV	50	60	10
	270	465	110

then the groups' percentages as under (b) above would be:

$$X - 270/845 \text{ or } 32\%$$
$$Y - 465/845 \text{ or } 55\%$$
$$Z - 110/845 \text{ or } 13\%$$

Now, assume that in country "I" consumption increased by 100%; in country "II" by 200%; in "III" it remained stable and in country "IV" it increased by 10%. Then we should have the following picture of the deliveries of the various groups in these countries:–

	X	Y	Z
Country I	40	760	0
„ II	0	0	300
„ III	200	25	0
„ IV	55	66	11
The groups would have delivered the following quantities ...	295	851	311

The quantity based on the percentage as under (b) above would have been of the total of 1,457.

For X	32%	–	466
X actually delivered			295
and therefore under-supplied			171
For Y	55%	–	801
Y actually supplied			851
and therefore over-supplied			50
For Z	13%	–	190
Z actually supplied			311
and therefore over-supplied			121

Following the above example, "Y" and "Z" would be obliged to purchase from "X" in proportion to their over-supply a quantity of 171 at the Gulf price basis.

3. If an under-supplying group does not exercise its right to offer to the over-supplying group or groups on the basis of (b) of No. 2 the under-supplied quantities, then this group shall forfeit the right to do so thereafter, under the arrangement.

4. Any group that does not supply its full quota as under (a) of No. 2 through force majeure shall forfeit its right to supply a quota proportion equal to the proportion of its under-supplies, this forfeit to last after the

case of force majeure ceased to exist for a period equal to that during which the party under-supplied.

5. Any party that does not supply for a period of six months its full quota as under (a) of No. 2 for any reason but force majeure shall lose of its quota right, as under (a) and (b) above, the same proportion as it under-supplied and this part of its proportion shall be allotted to the other group or groups in proportion to their quota.

6. The groups to agree on a standardization of quality for the various products.

(Note: it will be necessary to work out detailed specifications therefor.)

7. For the practical carrying out of the arrangement, the groups will form an association in which each group shall be interested in proportion to its quota as under (a) of No. 2, and which shall be managed by a working representative of each group, and each group shall have the right to recall at any time its representative, replacing same by another. Each representative shall have the right to name a substitute in his stead.

8. The Association shall make arrangements with the marketing organisations owned by each group or by each member of each group and/or with the parent companies that have agreements with marketing organizations, such arrangements to provide that each marketing organization shall maintain its quota of deliveries irrespective of source of supply and obligate itself to purchase from the groups exclusively all its imports of petroleum products that fall under the scope of the arrangement. Such arrangements furthermore to provide that each marketing organization keep the Association currently informed of the demand it expects to have of each product. The Association will obligate itself to supply the demand of the marketing organizations and not to sell and deliver to others in the countries in which the marketing organizations operate at prices below those quoted at the time to the marketing organizations.

9. The Association will inform the groups as far in advance as possible and keep the groups currently informed of the total demand for the various products it will have as under No. 8.

10. The Association will allocate to each group its quota of each product and will direct shipments to the geographically most favorably situated area.

11. Each group shall have the right to transport the quantities to which its quota entitled it, to the ports allocated to same as under No. 10, by its own vessels. Transportation facilities for quantities a group cannot so transport or may elect not to so transport shall be furnished by the Association and not by the group, and before chartering outside tonnage for such purpose the Association shall charter from the other groups tonnage owned by them which they may offer to the Association at freight rates at which the Association is able to charter outside tonnage.

12. The Association shall prepare for six months in advance, on the basis of

current freight rates, statements of relative freight rates from each port of shipment to each port of import.

13. Each group shall be paid f. o. b. port of shipment for each product on the basis of the Gulf price; or, if the goods are supplied c. i. f. port of import, the marketing organizations shall pay to each group this price plus the freight rate scheduled for the port of import under the statement prepared by the Association.

    (Note: The U.S. Gulf prices shall be the basis until further notice by the Association).

14. Any group having a surplus of any product or products on hand and desirous of selling such surplus over and above the quantities which it is entitled to supply under its quota shall offer same to the other groups at a price below that posted by the Association for the product or products in question. Each of the groups shall then have the right to purchase the quantities so offered in proportion to their quota and if one group declines to so purchase and the other be willing to purchase then the willing group may purchase all of the quantities so offered. The quantities so purchased shall then be supplied by the purchasing group or groups under this agreement at the posted price against their quota.

E X A M P L E

Let us assume AP is entitled to supply 100,000 tons, and this quantity is shipped from Persia to Italy instead of from Persia to the AP's own marketing organisation in Great Britain, this marketing organisation receiving instead 100,000 tons from the Shell from the U.S. Then, assuming other figures, we would have the following picture:

1. AP's selling organisation in Great Britain is supplied by the Shell with 100,000 tons from the U.S., paying therefor to Shell 10c per gallon and 20/– freight.
2. AP's selling organisation realises from the sale and delivery of these 100,000 tons a return of £150,000.
3. AP supplies to Italy 100,000 tons and receives therefor 10c per gallon and a freight rate of 18/– per ton.

Then, AP has realised for its 100,000 tons of products a net return of £150,000, less local marketing expenses in Great Britain, less cost of transportation from U.S. to Great Britain instead of realising for this quantity £150,000 less the same marketing expenses and less cost of transportation from Persia to Great Britain. Consequently AP saves the difference in cost between transportation from U.S. to Great Britain as against transportation from Persia to Great Britain.

The fact that AP's marketing organisation may have paid to the Shell a higher freight rate for the transportation of the 100,000 tons from the U.S. to Great Britain than self cost does not in the least change this, since AP will receive for the transportation from Persia to Italy a proportionately equally higher freight rate. Whatever excess it may have paid over self cost through its British marketing organisation for Atlantic maritime transportation to Great Britain will be returned to AP by the proportionately equally high excess over self cost AP will receive for the transportation from Persia to Italy. On the other hand, if the Shell sells through its Italian marketing organisation the 100,000 tons of products supplied by AP to Italy then the Shell will realise for the 100,000 tons supplied to AP in Great Britain the Italian price less the freight from Persia to Italy instead of the Italian price less the freight from the United States to Italy, thereby making an additional profit by saving the difference in freight from Persia to Italy as against U.S. to Italy.

Source: BP 106331. (Note: Emphases and style as in original.)

Notes

INTRODUCTION

1 For descriptions of the development of the international oil industry in the early twentieth century, see J. G. Clark, *The Political Economy of World Energy: A Twentieth-Century Perspective* (Hemel Hempstead, 1990), chs. 1–3; G. Jones, *The State and the Emergence of the British Oil Industry* (London, 1981); D. Yergin, *The Prize: The Epic Quest for Oil, Money and Power* (New York, 1991), chs. 4–12. On BP, see R. W. Ferrier, *The History of The British Petroleum Company: Volume 1, The Developing Years 1901–1932* (Cambridge, 1982).

2 On the Burmah Oil Company, see T. A. B. Corley, *A History of the Burmah Oil Company, 1886–1966*, 2 vols. (London, 1983 and 1988).

3 Jones, *The State*, p. 225.

4 For other examples of businessmen experienced in government, see W. J. Reader, 'Imperial Chemical Industries and the State, 1926–1945', in B. Supple (ed.), *Essays in British Business History* (Oxford, 1977), p. 231.

1 MANAGEMENT AND FINANCE, 1928–1939

1 BP 104123, Anglo-Persian Oil Company (APOC) 18th ordinary general meeting, 2 November 1927.

2 *Ibid.*, APOC 19th ordinary general meeting, 6 November 1928.

3 *Ibid.*, APOC 21st ordinary general meeting, 17 June 1930.

4 For a biography of Cadman, see J. Rowland and Basil (Second Baron) Cadman, *Ambassador for Oil: The Life of John, First Baron Cadman* (London, 1960).

5 On Cadman's earlier managerial reforms and the division of responsibilities, see R. W. Ferrier, *The History of The British Petroleum Company: Volume 1, The Developing Years, 1901–1932* (Cambridge, 1982), pp. 333–41, and 'The early management organisation of British Petroleum and Sir John Cadman', in L. Hannah (ed.), *Management Strategy and Business Development* (London, 1976), pp. 130–47.

6 A. D. Chandler, *Strategy and Structure: Chapters in the History of the Industrial Enterprise* (Cambridge, Mass., 1962), ch. 4.
7 Ferrier, *History*, pp. 333–41.
8 BP 104123, APOC 18th ordinary general meeting, 2 November 1927.
9 *Ibid.*, APOC 21st ordinary general meeting, 17 June 1930.
10 *Ibid.*, Proceedings at annual general meeting, 12 June 1934.
11 BP 106231, Minutes of Anglo-Iranian Oil Company (AIOC) board meeting on 20 June 1938.
12 *Ibid.*
13 Ferrier, *History*, ch. 13, especially pp. 624–8.
14 For example, see H. Katouzian, *The Political Economy of Modern Iran, 1926–1979* (New York, 1981), p. 118.

2 A NEW CONCESSION AND A 'FRESH' START IN IRAN, 1932–1939

1 For a detailed account of the grant of the D'Arcy concession and the Armitage-Smith Agreement, see R. W. Ferrier, *The History of The British Petroleum Company: Volume 1, The Developing Years, 1901–1932* (Cambridge, 1982), ch. 1 and pp. 365–71. For a briefer summary, see Ferrier, 'The development of the Iranian oil industry', in H. Amirsadeghi (ed.), *Twentieth-Century Iran* (London, 1977), pp. 93–128.
2 On the condition of Iran in the early twentieth century, see, for example, Amirsadeghi (ed.), *Iran*; M. Reza Ghods, *Iran in the Twentieth Century, A Political History* (London, 1989); G. Lenczowski (ed.), *Iran under the Pahlavis* (Stanford, 1978).
3 For the heroic version, see, for example, the introduction to Lenczowski (ed.), *Iran*, and L. P. Elwell-Sutton, 'Reza Shah the Great: Founder of the Pahlavi dynasty' in *ibid.*, pp. 1–50. For an unfavourable assessment of Riza Shah, see, for example, Ghods, *Iran*, ch. 6.
4 W. Knapp, '1921–1941: The period of Riza Shah', in Amirsadeghi (ed.), *Iran*, p. 28.
5 F. Diba, *Mohammad Mossadegh, A Political Biography* (London, 1986), ch. 4; F. Azimi, 'The reconciliation of politics and ethics, nationalism and democracy: an overview of the political career of Dr Muhammad Musaddiq', in J. A. Bill and W. R. Louis (eds.), *Musaddiq, Iranian Nationalism and Oil* (Austin, Texas, 1988), pp. 47–68.
6 Knapp, 'The period of Riza Shah', in Amirsadeghi (ed.), *Iran*, p. 24.
7 *Ibid.*, p. 25.
8 Ghods, *Iran*, p. 110.
9 *Ibid.*, pp. 105 and 110; Knapp, 'The period of Riza Shah', in Amirsadeghi (ed.), *Iran*, pp. 49–50.
10 For descriptions of Riza Shah's reforms, see, for example, the essays in Lenczowski (ed.), *Iran*; Ghods, *Iran*, ch. 6; Knapp, 'The period of Riza Shah', in Amirsadeghi (ed.), *Iran*, pp. 23–51.

11 On the negotiations from 1928 to 1932, see Ferrier, *History*, ch. 13.

12 PRO FO 371/16080, Hoare to Simon, 19 December 1932 and Mihdi Quli Khan Hidayat, *Khatirat va Khatarat* (Recollections and Adversity) (Tehran, 1950), p. 504.

13 BP 96487, Jacks to Cadman, 2 December 1932.

14 On Riza Shah's attitude to the British Government after his accession to power, see Nance F. Kittner, 'Issues in Anglo-Persian diplomatic relations, 1921–1933', PhD thesis, School of Oriental and African Studies, University of London, 1980. For an Iranian view, see R. K. Ramazani, *The Foreign Policy of Iran, 1500–1941: A Developing Nation in World Affairs* (Charlottesville, Virginia, 1966).

15 J. M. Pattinson in a personal reminiscence on 6 February 1982 to R. W. Ferrier remarked that to the Shah's annoyance it was the Company which provided, at the request of Iranian officials, the car in which he travelled and that he had not realised that the only practical way across the River Karun at that point was on a ferry operated by the Company. Nobody had queried Riza Shah's itinerary which called for a river crossing where there was no bridge.

16 BP 69264, Translation of an article from the 'Iswestja' of 2 December 1932.

17 *The Times*, 15 December 1932.

18 BP 69266, Isa Khan to Taqizadeh, 28 November 1932.

19 *Ibid.*, Taqizadeh to Isa Khan, 1 December 1932.

20 PRO FO 371/16078, Jacks' telegram, 28 November 1932.

21 BP 96487, Fraser to Jacks, 28 November 1932.

22 *Ibid.*, Jacks to Fraser, 2 December 1932.

23 *Ibid.*, Taqizadeh and Fateh interview, 2 December 1932; *ibid.*, Jacks to Cadman, 3 January 1933. Taqizadeh later denied any personal responsibility for the negotiation of the 1933 concession because 'all important matters in this country are decided by the great man of the time, who wanted the slate cleaned and to start afresh', as he explained to the Majlis in 1949.

24 *Ibid.*, Jacks to Fraser, 2 December 1932.

25 BP 70455, Meeting at FO, 29 November 1932.

26 PRO FO 371/16078, Beckett legal submission, 29 November and *The Times*, 30 November 1932.

27 BP 70455, Fraser to Jacks, 30 November 1932.

28 *Ibid.*, Cadman to Fraser, 29 and 30 November.

29 PRO FO 371/16078, Hoare to FO, 29 November 1932.

30 *Ibid.*, Vansittart minute, 29 November 1932; *ibid.*, Hearn note, meeting at FO, 30 November 1932.

31 *Ibid.*, Hoare to FO, 29 November 1932.

32 PRO CAB 64/32, 30 November 1932.

33 PRO FO 371/16078, Hoare to FO, 2 December 1932.

34 *Ibid.*, Vansittart note, 3 December 1932.

35 BP 70455, Fraser and Hearn note, 4 December 1932.

36 BP 69267, Note on meeting at FO, 5 December 1932.
37 *Ibid.*, Note on meeting at FO, 6 December 1932.
38 PRO FO 371/16078, Rendel note, 6 December 1932.
39 *Ibid.*, Hoare to FO, 5 December 1932.
40 *Ibid.*, Rendel note, 6 December 1932.
41 BP 70455, Fraser note, 7 December 1932.
42 *Parl. Debates*, House of Commons, vol. 272, cols. 1790–3, 8 December 1932.
43 *Ibid.*
44 PRO FO 371/16079, Hoare to Simon, 14 December 1932.
45 *Ibid.*, Vansittart Cabinet submission, 13 December 1932; PRO CAB 67/32, 14 December 1932.
46 BP 69267, FO to Company, 15 December 1932.
47 PRO FO 371/16079, Rendel note, 15 December 1932.
48 BP 69404, Company to Jacks, 15 December 1932.
49 BP 96487, Cadman to Jacks, 23 December 1932.
50 BP 88373, Cadman note on visit from Ansari, 16 December 1932.
51 *Ibid.*, Cadman note on visit from Ansari, 19 December 1932.
52 League of Nations, *Official Journal*, 13th year, 19 December 1932, pp. 2298–305.
53 *Ibid.*, pp. 1989–90.
54 PRO FO 371/16080, Malkin minute, 12 January 1933.
55 *Ibid.*
56 *Ibid.*, Malkin minute, 2 January 1933.
57 *Ibid.*, Vansittart to Hoare, 12 January 1933.
58 *Ibid.*, Tehran to FO, 14 December 1932.
59 For text, see PRO FO 371/16935, Provisional minutes of third meeting, 17th session, Council of League of Nations, 26 January 1933.
60 BP 88374, Simon's speech, 26 January 1933.
61 BP 96659, Cadman's Geneva notes, 26 January 1933.
62 *Ibid.*, 3 February 1933.
63 *Ibid.*, 4 February 1933.
64 BP 69388, Paris notes, February 1933.
65 BP 69363, Lefroy note on Paris meeting, 10 February 1933.
66 BP 88375, Jacks to Cadman, 11 February 1933.
67 BP 96487, Cadman to Jacks, 23 February 1933.
68 BP 88376, Lloyd note, 'The D'Arcy concession', 6 January 1933.
69 BP 88375, Cargill to Whigham, 13 February 1933.
70 BP 96487, Young to Hearn, 5 April 1933.
71 On this phase of the negotiations, see, generally, BP 96659, Cadman diary.
72 BP 88377, Young to Hearn, 20 April 1933.
73 BP 88372, Cadman to Simon, 19 April 1933.
74 BP 96659, Cadman diary.
75 *Ibid.*
76 *Ibid.*

77 Thomas Herbert, *A Discription of the Persian Monarchy* (London, 1634), p. 98.
78 BP 96659, Cadman diary.
79 *Ibid.*
80 *Ibid.*
81 *Ibid.*
82 A detailed account of the itinerary is in *ibid.*
83 PRO FO 371/16938, Hoare to FO, 25 May 1933; *ibid.*, Hoare to Simon, 3 June 1933.
84 For a published account of the terms of the 1933 concession, see B. Shwadran, *The Middle East, Oil and the Great Powers*, 2nd edn (New York, 1959), pp. 52–5.
85 BP 85908, Cadman to Elkington, 9 June 1933; *ibid.*, Cadman to Riza Shah, 23 June 1933.
86 *Ibid.*, Cadman to Elkington, 12 January 1934.
87 PRO FO 371/16961, Mallet to Simon, 17 June 1933.
88 *Ibid.*, Hoare to Simon, 3 June 1933.
89 See, generally, PRO FO 371/17909, Persia, annual report, 1933, pp. 33–44.
90 BP 84880, Elkington's correspondence with Lt.-Col. Loch (Political Resident in the Persian Gulf); *The Times*, 30 June, 4 July and 16 September 1933.
91 BP 85909, Cadman to Riza Shah, 31 October 1933.
92 *Ibid.*, Riza Shah to Cadman, 3 November 1933.
93 *Ibid.*, Cadman to MacDonald, 8 November 1933.
94 *Ibid.*, Cadman to MacDonald, 12 November 1933.
95 *Ibid.*, MacDonald to Cadman, 1 December 1933.
96 *The Times*, 28 November 1933.
97 *Ibid.*, 7 December 1933; L. Lockhart and Rose L. Greaves, 'The record of the British Petroleum Company Ltd' (BP internal record, vol. 1, 1968), p. 114.
98 *Ibid.*, p. 115.
99 BP 85909, Cadman to Simon, 9 December 1933; *ibid.*, Cadman to Eden, 21 December 1933; *ibid.*, Eden to Cadman, 22 December 1933; *ibid.*, Cadman to Fisher, 29 December 1933.
100 *Ibid.*, Cadman to Fisher, 29 December 1933.
101 *Ibid.*
102 *Ibid.*, unsigned and undated note.
103 BP 85907, Cadman to Baldwin, 29 April 1929.
104 BP 85909, Cadman to Elkington, 5 January 1934.
105 PRO CAB 23/78, Cabinet minute, 15 January 1934.
106 *Ibid.*, Cabinet minute, 24 January 1934. See, also, PRO FO 416/92, Simon to Hoare, 25 January 1934.
107 There is little serious study of economic growth in Iran during this period, but see A. Banani, *The Modernisation of Iran 1921–1941* (Stan-

ford, 1961); J. Bharier, *Economic Development in Iran 1900–1970* (Oxford, 1971); and C. Issawi, 'The Iranian economy, 1925–1975: fifty years of economic development', in Lenczowski (ed.), *Iran*, pp. 129–33. Two detailed monographs are useful: W. Floor, *Industrialization in Iran, 1900–1941* and *Labour union, law and conditions in Iran (1900–1941)* in University of Durham, Centre for Middle Eastern and Islamic Studies, Occasional Papers, series 2, no. 23 (1984) and no. 26 (1985).

108 PRO FO 416/94, Butler to Seymour, 27 June 1936, Economic report for Iran, January to May 1936; *ibid.*, Urquhard, Memorandum on political affairs in Azerbaijan, 31 August 1936.

109 D. N. Wilber, *Riza Shah Pahlavi: The Resurrection and Reconstruction of Iran* (New York, 1975), p. 177.

110 See, for example, PRO FO 416/92, Hoare to Simon, 12 January 1933; *ibid.*, Hoare to Simon, 25 February and 22 April 1933; *ibid.*, Mallet to Simon, 12 August 1933; *ibid.*, Mallet to Hoare, 6 and 21 October and 17 November 1933.

111 PRO FO 371/16939, Mallet to Simon, 9 September 1933, Stanley Simmons enclosure, Memorandum on Japanese trading activities in Persia. On India, see PRO FO 416/92, Hoare to Simon, 11 December 1933.

112 PRO FO 416/92, Butler to Eden, 11 July 1936.

113 PRO FO 416/94, Butler to Eden, 11 July 1936 and enclosure, Stanley Simmons' memorandum respecting Anglo-Iranian trade relations, 3 July 1936. ICI had shown interest in Iran, PRO FO 371/16939, Department of Overseas Trade to FO, 20 July 1933; PRO FO 416/94, Simmons interview with Badr, 6 July 1936.

114 PRO FO 416/94, Knatchbull-Hugessen to Samuel Hoare, 30 October 1935. Bulletins of the Bank Melli gave indications on company formations and capital investments; *ibid.*, Knatchbull-Hugessen to Eden, 27 June 1935.

115 *Ibid.*, Butler to Eden, 25 July 1936 and enclosure Stanley Simmons, German economic penetration, 24 July 1936.

116 PRO FO 416/95, Seymour to Eden, 15 February 1937.

117 *Ibid.*, Seymour to Eden, 11 March 1937.

118 BP 104123, Minutes of general meeting, 11 June 1936.

119 The Company's decision was conveyed to the Iranian Government in BP 71074, Fraser to Davar, 30 July 1936.

120 Lockhart and Greaves, 'Record', vol. 1, p. 133.

121 BP 84881, Cadman to Riza Shah, 28 November 1938.

122 *Ibid.*, Riza Shah to Cadman, 18 December 1938.

123 *Ibid.*, Cadman to Riza Shah, 5 January 1939.

124 *Ibid.*, Riza Shah to Cadman, 25 January 1939.

125 *Ibid.*, Cadman to Riza Shah, 30 January 1933.

126 BP 85904, Gass to Cadman, 19 February 1939.

127 BP 85905, Gass memorandum, 16 May 1939.

128 *Ibid.*

129 *Ibid.*
130 BP 84881, Cadman diary, visit to Tehran, May–June 1939.
131 *Ibid.*
132 *Ibid.*
133 *Ibid.*
134 *Ibid.*
135 BP 84881, Cadman to Chamberlain, 10 June 1939.
136 *Ibid.*, Cadman diary, visit to Tehran, May–June 1939.

3 OPERATIONS AND EMPLOYMENT IN IRAN, 1928–1939

1 An exceptionally informative account of the Company's exploration activities in Iran by a former chief geologist is P. T. Cox, 'Exploration for oil in the D'Arcy concession area of Iran', (BP internal paper, 1968).
2 F. Fesharaki, *Development of the Iranian Oil Industry* (New York, 1976), p. 27. British Petroleum Company, *Our Industry*, 5th edn (London, 1977), p. 502. S. H. Longrigg, *Oil in the Middle East: Its Discovery and Development*, 3rd edn (London, 1968), pp. 61–3. Where discovery dates given in the above publications do not agree, the dates given by Fesharaki and BP are those which appear in the text.
3 This account of Kirmanshah owes much to the explanations of A. W. M. Robertson, a former manager of the refinery, who played an important role during the advance of the British forces entering Iran from Iraq on 25 August 1941 up the Partaq Pass under General Sir William Slim. On 'a very brave man', see Field Marshal Sir William Slim, *Unofficial History* (London, 1959), pp. 191–5. On Kirmanshah, see also Longrigg, *Oil*, pp. 60–1.
4 BP 67627, Jameson report on visit to Iran, 1938, p. 31.
5 *Ibid.*, p. 32.
6 PRO FO 371/13784, Clive to Henderson, 28 November 1929.
7 This controversial incident has acquired some significance in the history of labour relations and the development of communism in Iran, though its impact upon the Company has generally been exaggerated. For different accounts, see G. Lenczowski, *Russia and the West in Iran, 1918–1948: A Study in Big-Power Rivalry* (New York, 1949), pp. 110–18 and 138–41; E. Abrahamian 'The strengths and weaknesses of the labour movement in Iran', in M. E. Bonine and N. R. Keddie (eds.), *Continuity and Change in Modern Iran* (New York, 1981); W. Floor, *Labour Union Law and Conditions in Iran (1900–1941)* in University of Durham, Centre for Middle Eastern and Islamic Studies, Occasional Papers, series 2, no. 26 (1985); H. Ladjevardi, *Labour Unions and Autocracy in Iran* (New York, 1985), pp. 20–2; and consular reports and legation despatches in PRO FO 371/13783 and 13784.
8 BP 59010, Elkington to Medlicott, 2 May 1929; *ibid.*, Abadan to London, 4 May 1929.

9 *Ibid.*, Abadan to London, 6 May 1929.

10 *Ibid.*, Elkington to Medlicott, 8 May 1929; *ibid.*, Abadan to London, 7 May 1929; *ibid.*, Elkington to Cadman, 9 May 1929.

11 PRO FO 371/13783, Clive to FO, 9 May 1929.

12 *Ibid.*, Clive to FO, 7 May 1929; *ibid.*, Clive to Sir Austen Chamberlain, 16 May 1929; BP 59010, London to Tehran, 5 May 1929 (83); *ibid.*, London to Tehran, 17 May 1929 (95).

13 PRO FO 371/13784, Clive to Austen Chamberlain, 1 June 1929; BP 59010, Elkington, An appreciation of the political situation in Khuzistan with special reference to the present unrest, 17 June 1929; *ibid.*, Abadan to Tehran, 20 May 1929.

14 PRO FO 371/13783, Note, Mirza Hussayn Movaqqar, 23 May 1929.

15 BP 59010, Jacks to Elkington, 7 June 1929.

16 *Ibid.*, Abadan to London, 16 May 1929; PRO FO 371/13783, Fletcher to Barratt, 18 May 1929.

17 PRO FO 371/13783, Clive to FO, 7 May 1929; BP 59010, Tehran to Abadan, 18 May 1929.

18 BP 59010, Clive to Cadman, 30 May 1929.

19 *Ibid.*, Gass interviews with Timurtash, 29 and 30 May 1929; PRO FO 371/13784, Fletcher to Barratt, 1 June 1929.

20 BP 59010, Timurtash to Greenhouse, 27 May 1929.

21 *Ibid.*, Cadman to Timurtash, 28 May 1929.

22 Fletcher, the acting British Vice Consul at Muhammara, was worried about the emotion which might be engendered at the religious observances. PRO FO 371/13784, Fletcher to Barratt, 1, 8, 15 and 29 June and 6 July 1929 and *ibid.*, Fletcher to Clive, 10 June 1929.

23 BP 59010, Elkington to Medlicott, 5 July 1929.

24 *Ibid.*, Greenhouse to London, 3 August 1929; *ibid.*, Greenhouse to Cadman, 11 August 1929. It had even been suggested that Colonel T. E. Lawrence had been stirring up trouble for Iran from Iraq, *ibid.*, London to Tehran, 26 July 1929.

25 According to D. N. Wilber: 'Riza Shah was in a savage mood' throughout the summer of 1929. See *Riza Shah Pahlavi: The Resurrection and Reconstruction of Iran* (New York, 1975), p. 132.

26 BP 59011, Comparative rates of pay of the Ulen Company (railway contractors) and the Company, 1929.

27 *Ibid.*, Elkington, A report on the relations between the Company and its labour and suggestions for improvement of existing system, 1 June 1929; *ibid.*, Elkington, Labour welfare, 1 June 1929.

28 *Ibid.*, Elkington to Medlicott, 11 July 1929.

29 BP 59010, Elkington to London, 22 May 1929.

30 BP 59011, Elkington to Medlicott, 11 July 1929.

31 *Ibid.*, Medlicott to Elkington, 25 July 1929.

32 *Ibid.*, Tehran to London, 22 October 1928; *ibid.*, Abadan to London, 24 and 25 October and 3 November 1929; *ibid.*, London to Abadan, 28

October 1929.

33 *Ibid.*, Abadan to London, 3 November 1929 and London to Abadan, 5 November 1929.

34 *Ibid.*, Abadan to London, 22 November 1929.

35 *Ibid.*, Abadan to London, 9 January 1930.

36 *Ibid.*, Cadman to Jacks, 18 December 1929.

37 *Ibid.*

38 BP 53967, Fields Staff Department Compendium, 1928.

39 BP 72015, Bell to Cadman, Staff policy in Persia, 27 July 1932.

40 BP 68040, Elkington to Jameson, 21 December 1935.

41 BP 80902, Weekly expenditure in Iran of the average married man in grades C and D, 1935.

42 *Ibid.*

43 A full breakdown of staff posts and salaries in 1931 is in BP 66948.

44 BP 59011, View of Elkington, Minutes of meeting held at Britannic House, 11 February 1930.

45 BP 67627, Jameson report on visit to Iran, June 1938.

46 *Ibid.*

47 BP 70268, Ministry of Finance, 31 May 1933.

48 *Ibid.*, Fraser to Minister of Finance [Taqizadeh], 12 July 1933.

49 *Ibid.*, Elkington to Fraser, 22 October 1933; *ibid.*, Fraser to Jacks, 23 and 24 October 1933.

50 *Ibid.*, Fraser to Elkington, 3 November 1933.

51 *Ibid.*, Fraser to Jacks, 16 November 1933.

52 *Ibid.*

53 *Ibid.*, Fraser to Elkington, 24 November 1933.

54 *Ibid.*, Note of meeting at Abadan, 27 November 1933.

55 *Ibid.*, Jacks to Fraser, 15 December 1933.

56 BP 70267, A general plan for the training of Persian nationals for the Persian oil industry, 8 February 1934.

57 *Ibid.*, Jacks to Fraser, 18 May 1934.

58 *Ibid.*, Jacks to Fraser, 21 May 1934.

59 *Ibid.*, Fraser to Jacks, 31 May 1934.

60 BP 70266, Jacks to Davar, 5 June 1934; *ibid.*, Jacks to Fraser, 9 June 1934.

61 *Ibid.*, Jacks to Fraser, 9 July 1934.

62 BP 70266, Davar to Jacks, 26 August 1934.

63 L. Lockhart and Rose L. Greaves, 'The record of the British Petroleum Company Ltd' (BP internal record, 1968), vol. 1, p. 79.

64 *Ibid.*

65 BP 70266, Idelson opinion, 16 January and 12 February 1935.

66 Lockhart and Greaves, 'Record', vol. 1, p. 79.

67 *Ibid.*, pp. 79–80.

68 For Gass's drafts, see BP 700266, Gass note, Article 16, 20 April 1935; *ibid.*, Gass to Elkington, 14 May 1935; *ibid.*, Gass, draft Article 16, 17 May 1935. For comments, see, for example, BP 70266, Elkington to Gass, May 1935;

ibid., Gass to Elkington, 31 May 1935; *ibid.*, Mylles comments to Gass on automation, n. d.; *ibid.*, Elkington to Gass, 5 June 1935; *ibid.*, Gass to Elkington, 5 June 1935; *ibid.*, Gass to Elkington, 18 June and 4, 15, 19, 31 July and 28 August 1935. On automation, see also BP 68040, Jameson to Baylis, 28 August 1935.

69 BP 70269, Gass to Elkington, 18 and 25 October 1935.

70 BP 84877, Davar to Cadman, 7 November 1935; *ibid.*, Cadman to Davar, 29 November 1935.

71 BP 67623, Fraser, Report on visit to Iran, spring 1936.

72 *Ibid.*

73 BP 6S559, General Plan, 1936.

74 *Ibid.*

75 BP 67627, Jameson, Report on visit to Iran, 1938.

76 *Ibid.*

77 BP 54366 and BP 54368, General manager's reports for 1925 and 1926.

78 BP 70268, Elkington to Fraser, 4 December 1933.

79 BP 70267, Outline of Company training schemes, 8 February 1934.

80 BP 54380, General manager's annual report for 1934.

81 BP 70267, Outline of Company training schemes, 8 February 1934.

82 BP 54379, General manager's annual report, 1935.

83 BP 65559, General Plan, 1936.

84 On wastage, see BP 15922, The problem of wastage in training, 1948.

85 BP 71487, AIOC record of activities in the war years.

86 BP 71182, Hawker, Note on Shah's visit to Abadan, 31 March 1940.

87 BP 70268, Elkington to Fraser, 7 November 1933.

88 BP 67590, Jameson, Diary of visit to Iran, 27 February 1934.

89 BP 72348, Mylles to Gass, 7 June 1934.

90 BP 71880, Pattinson to Rice, 29 June 1944.

91 BP 59011, Elkington, Labour welfare, 17 June 1929.

92 BP 67590, Wilson brief, 22 February 1934.

93 BP 67623, Jameson, Report on visit to Iran, 1936.

94 BP 68835, Gobey to Pattinson, 14 August and 17 September 1938.

95 Much of the information on leisure activities is derived from R. W. Ferrier's personal observations of the Abadan area and countryside or conversations with former members of the Company, British and Iranian, who served in Khuzistan, or from articles which appeared in the pages of *The Naft* magazine.

4 CO-OPERATION IN THE MARKETS, 1928–1939

1 See, for example, D. H. MacGregor *et al.*; 'Problems of rationalisation', *Economic Journal,* vol. 40 (1930), pp. 1351–68; A. Mond, *Industry and Politics* (London, 1927); L. Urwick, *The Meaning of Rationalisation* (London, 1929); J. Stamp, *Criticism and Other Addresses* (London, 1931). For more recent accounts of the rationalisation movement, see L. Hannah,

The Rise of the Corporate Economy (London, 1976), ch. 3, and 'The rationalization movement – a revolution in business policy', *Journal of Business Policy*, vol. 1 (1970–1), pp. 35–41.

2 For a contemporary account of the control of competition in Britain, see A. F. Lucas, *Industrial Reconstruction and the Control of Competition* (London, 1937).

3 BP 37143, Fraser to Eady (enclosing note on India), 20 December 1944. See also R. W. Ferrier, *The History of The British Petroleum Company: Volume 1, The Developing Years, 1901–1932* (Cambridge, 1982), p. 511.

4 Ferrier, *History*, pp. 510–12.

5 BP 106331, Short journal of events, Cadman to Teagle, 22 February 1928.

6 BP B15 4469, Watson to Fraser, 17 August 1928; *ibid.*, Watson to Fraser, 18 August 1928.

7 *Oil and Gas Journal*, 20 September 1928. Cited in 82nd Congress, 2nd Session, Senate Small Business Committee, Staff report of the Federal Trade Commission (FTC), *The International Petroleum Cartel* (1952), p. 199 and in J. M. Blair, *The Control of Oil* (New York, 1976), p. 55.

8 For comparison, see extracts from the Achnacarry Agreement cited in FTC, *International Petroleum Cartel*, p. 199 ff.

9 *Ibid.*, pp. 218–19.

10 *Wall Street Journal*, 7 December 1928; FTC, *International Petroleum Cartel*, p. 218.

11 *Ibid.*, pp. 221–4.

12 BP B15 4469, Export Association memorandum in Godber to Fraser, 15 February 1930.

13 BP 110641, Teagle to Cadman, 16 February 1929; FTC, *International Petroleum Cartel*, pp. 211–13; Blair, *Control of Oil*, pp. 156–7.

14 BP 110641, Teagle to Fraser, 5 April 1929.

15 FTC, *International Petroleum Cartel*, p. 213.

16 British Petroleum Company, *Our Industry*, 5th edn (London, 1977), p. 580.

17 FTC, *International Petroleum Cartel*, pp. 221 and 225–6.

18 BP 110641.

19 BP 110645, Jackson to Cadman, 28 March 1930 and 19 May 1930.

20 BP 108309, Watson memorandum, '"As Is" in the Oil Industry', 16 October 1930; *ibid.*, Watson memorandum of discussions, 22 October 1930.

21 FTC, *International Petroleum Cartel*, p. 224.

22 *Ibid.*, pp. 228–35 and 241–9.

23 On Russian oil, see J. A. Bowden, '"That's the Spirit": Russian Oil Products Ltd (ROP) and the British oil market, 1924–1939', *Journal of European Economic History*, vol. 17, no. 3 (1989), pp. 641–63. See also V. A. Shishkin, 'Soviet oil exports between the two world wars', in R. W. Ferrier and A. Fursenko (eds.), *Oil in the World Economy* (London, 1989), pp. 74–82.

24 Bowden, '"That's the Spirit"'.
25 BP 110645, Jackson to Cadman, 6 May 1931; *ibid.*, Jackson to Cadman, 24 April 1931.
26 FTC, *International Petroleum Cartel*, pp. 239–40.
27 BP 110645, Jackson to Cadman, 6 May 1932.
28 FTC, *International Petroleum Cartel*, p. 240.
29 Shishkin, 'Soviet oil exports', p. 79.
30 On the Romanian oil industry, see M. Pearton, *Oil and the Romanian State* (Oxford, 1971), particularly in this context, pp. 156–201.
31 FTC, *International Petroleum Cartel*, p. 236.
32 *Ibid.*, p. 237; Blair, *Control of Oil*, p. 158.
33 FTC, *International Petroleum Cartel*, p. 238.
34 BP 63281, Fraser to Jackson, 4 November 1932.
35 *Ibid.*, Fraser to Jackson, 5 January 1933.
36 *Ibid.*, Kessler to New York, 12 January 1933.
37 BP 86000, Jackson to Cadman, 20 January 1933.
38 BP 85986, Heath Eves to Fraser, 17 March 1933; *ibid.*, Jackson to Lloyd, 22 March 1933; *ibid.*, Lloyd to Jackson, 23 March 1933; *ibid.*, Heath Eves to Jackson, 7 and 10 April 1933. See also *The Times*, 29 March 1933; *Financial Times*, 6 April 1933; FTC, *International Petroleum Cartel*, pp. 249–50.
39 FTC, *International Petroleum Cartel*, pp. 250–51.
40 G.D. Nash, *United States Oil Policy, 1890–1964: Business and Government in Twentieth Century America* (Pittsburgh, 1968), p. 111.
41 FTC, *International Petroleum Cartel*, p. 253.
42 BP, *Our Industry*, p. 580.
43 BP 91058, Minutes of meetings, 30 April, 3 and 4 May 1933.
44 *Ibid.*, Minutes of meetings, 30 April, 3 and 4 May 1933; BP 72276, Draft Memorandum of Principles. See also FTC, *International Petroleum Cartel*, pp. 255–65.
45 BP 110644, Minutes of meetings held at Shell-Mex House.
46 BP 63983, Monetary discussions, 30 August 1935; BP 110644, Notes on discussions re UK, 14 August 1935; *ibid.*, Revision of 'As Is' quotas, 23 September 1935; BP 91058, Minutes of September meeting, September 1936.
47 BP 96602, The new Shell/Standard-Vacuum Agreement, Memorandum to Snow, 17 December 1936.
48 BP B15 4469, Memorandum on Australia, 29 August 1928.
49 The passage on Italy is drawn from BP 37143, Memorandum on Italy, Morris to Fraser, 16 May 1945; BP 37141, Morris, Memorandum on Italy, 3 September 1943.
50 BP 71529, Cadman diary, 13 December 1928.
51 BP 68878, Heath Eves meeting with Rieber, February 1929.
52 *Ibid.*, Cadman to Holmes, 28 May 1929.
53 BP 110645, Jackson to Cadman, 21 November 1930; *ibid.*, Jackson to Cadman, 3 October 1930.

54 BP 72534, Jackson to Cadman, 21 April 1931; BP 71383, Jackson to Cadman, 23 April 1931; BP 110645, Jackson to Cadman, 4 January 1932.
55 BP 71528, Cadman diary, 8 and 11 April 1930. On the earlier ideas for co-operation, see Ferrier, *History*, pp. 475–6.
56 BP 71528, Cadman diary, 16 June 1930.
57 *Ibid.*, 30 June 1930.
58 *Ibid.*, 3 and 4 July 1930.
59 *Ibid.*, Cadman to Teagle, 8 July 1930.
60 *Ibid.*, Cadman diary, 8 July 1930.
61 *Ibid.*, 10 July 1930.
62 *Ibid.*, Cadman to Jacks, 23 July 1930.
63 *Ibid.*, Cadman diary, 27 August 1930.
64 *Ibid.*, 3 September 1930.
65 *Ibid.*, 8 and 9 September 1930.
66 BP 110645, Jackson to Cadman, 17 April and 20 October 1931. The Cleveland Petroleum Products Company was the British subsidiary of the Pan American Petroleum and Transport Company, partly owned by the Standard Oil Company of Indiana.
67 *Ibid.*, Jackson to Cadman, 16 October 1931.
68 *Ibid.*, Jackson to Cadman, 26 October and 6 November 1931.
69 *Ibid.*, Jackson to Cadman, 6 November 1931.
70 Henrietta M. Larson, Evelyn H. Knowlton and C. S. Popple, *History of Standard Oil Company (New Jersey): New Horizons, 1927–1950* (New York, 1971), pp. 45–50.
71 Calculated from data in BP 109194.
72 BP 68318, Meeting of Achnacarry partners, 10 February 1932; *ibid.*, Germany refinery proposals.
73 *Ibid.*, 'As Is' meeting, 13 September 1933.
74 *Ibid.*, Krauss memorandum, 11 December 1933.
75 The deterioration in Olex's position was noted in *ibid.*, Medlicott to Krauss and Barstow, 8 June 1933. On the financial position of the Company in Germany, see *ibid.*, Figures for 'Olex' DBPG, June 1933.
76 *Ibid.*, Morris note, 15 September 1934; *ibid.*, 'Olex' DBPG receipt of petroleum products, 1936–7.
77 BP 106231, Minutes of AIOC board meeting on 28 September 1937.
78 BP 104123, AIOC annual general meeting, 21 June 1937.

5 CONCESSIONARY INTERESTS OUTSIDE IRAN, 1928–1939

1 BP 45758, 1918 surveys in Colombia.
2 R. W. Ferrier, *The History of The British Petroleum Company: Volume 1, The Developing Years, 1901–1932* (Cambridge, 1982), pp. 555–8.
3 BP 71198, Hearn memorandum, 23 September 1927.
4 *Ibid.*, De Böckh to Cadman, 23 September 1927.
5 BP 78501, Colombian Petroleum Laws, 1919–31.

6 BP 71198 and 59936, Hearn report on Uraba survey and suggestion _re_ future, 13 July 1928.

7 BP B49 (Sec), D'Arcy Exploration Company (DEC) minutes, 13 December 1928.

8 BP 71968, Memorandum on Colombian oil legislation, undated.

9 BP 70594, Fraser to Cadman, 20 February 1929.

10 _Ibid._, Cadman to Fraser, 20 February 1929.

11 BP 45758, Law 160, 1936.

12 _Ibid._, Cox report, 1937.

13 BP 45737, Extract from Venezuelan Petroleum Law, 18 July 1925.

14 PRO FO 371/11110, Snow's minute, 12 April 1926.

15 BP 45737, Brown to Lloyd, 16 June 1926; _ibid._, Rodrigues' opinion, 23 November 1926.

16 BP 44213, Case to Counsel to advise, February 1927.

17 _Ibid._, Minutes of directors' meeting, 11 April 1927.

18 BP 44227, Cull and Co. to Wilson, 20 April 1930, enclosure, Caracas Petroleum Corporation.

19 _Ibid._, Cull and Co. to Fraser, 12 February 1934, enclosure, Orinoco Oilfields Ltd.

20 BP 45747 on legal matters; BP 44227 on meetings between the companies, November 1933 to April 1934.

21 BP 44227, Meeting, 25 January 1934.

22 _Ibid._, Wyllie note _re_ Venezuela, 29 January 1934.

23 BP 45747, Brown to Lefroy, 25 April 1934.

24 _Ibid._, 20 May 1934.

25 _Ibid._, 23 May 1934.

26 BP 44210, Ultramar proposition, 1 February 1938.

27 _Ibid._, Hearn to Meyer, 8 March 1938.

28 The English translation of the 1899 agreement is printed in A. H. T. Chisholm, _The First Kuwait Oil Concession Agreement: A Record of the Negotiations 1911–1934_ (London, 1975), p. 85. Brief descriptions of the establishment of British hegemony over the Gulf shaykhdoms can be found in Chatham House Study Group, _British Interests in the Mediterranean and Middle East_ (London, 1958), p. 31; G. E. Kirk, _A Short History of the Middle East_, 3rd edn (London, 1955), pp. 88–9; Sir Reader Bullard, _Britain and the Middle East_ (London, 1951), pp. 47 and 53; G. W. Stocking, _Middle East Oil: A Study in Political and Economic Controversy_ (London, 1970), p. 109; T. E. Ward, _Negotiations for Oil Concessions in Bahrain, El Hasa, The Neutral Zone, Qatar and Kuwait_ (printed for private circulation, 1965), pp. 62–3.

29 Chisholm, _Kuwait_, pp. 3 and 89.

30 Ferrier, _History_, pp. 561–70.

31 Shaykh Mubaruk died in 1915 and was succeeded by Shaykh Jaber. He died in 1917 and was succeeded by Shaykh Salim who died in 1921, to be succeeded by Shaykh Ahmad. All were members of the Sabah dynasty.

32 The text of the nationality clause is printed in Chisholm, *Kuwait*, p. 16.
33 From 1820 onwards Britain had established domination over Bahrain through a series of treaties. For a brief description, see G. Lenczowski, *The Middle East in World Affairs*, 4th edn (New York, 1980), pp. 669–70.
34 Chisholm, *Kuwait*, p. 18.
35 *Ibid.*
36 *Ibid.*, p. 22.
37 *Ibid.*, p. 26.
38 The English text of the concession document is printed in *ibid.*, pp. 242–9.
39 *Ibid.*, pp. 80–1; S. H. Longrigg, *Oil in the Middle East: Its Discovery and Development*, 3rd edn (London, 1968), pp. 112–13.
40 For general accounts of economic, social and political conditions in Iraq in the period, see Edith and E. F. Penrose, *Iraq: International Relations and National Development* (London, 1978), chs. 1–4; S. H. Longrigg, *Iraq, 1900 to 1950: A Political, Social and Economic History* (London, 1953); G. Lenczowski, *The Middle East*, ch. 7.
41 Longrigg, *Iraq*, p. 174, and *Oil*, pp. 43 and 66; Penrose and Penrose, *Iraq*, pp. 76–7.
42 Longrigg, *Oil*, pp. 66–7.
43 Longrigg, *Oil*, p. 69, and *Iraq*, p. 175; Penrose and Penrose, *Iraq*, pp. 64–5.
44 Longrigg, *Oil*, pp. 70–2.
45 In relation to the TPC, see, for example, Penrose and Penrose, *Iraq*, pp. 57–60; I. H. Anderson, *Aramco, the United States and Saudi Arabia: A Study of the Dynamics of Foreign Oil Policy, 1933–1950* (Princeton, 1981), pp. 13–19.
46 Longrigg, *Oil*, pp. 69–70; Penrose and Penrose, *Iraq*, p. 68.
47 CFP agreed to purchase Gulbenkian's share of TPC oil at a fair price. See Penrose and Penrose, *Iraq*, p. 68.
48 *Ibid.*, p. 69.
49 For descriptions of the dispute, see, for example, Longrigg, *Iraq*, pp. 144–8 and 152–8; Penrose and Penrose, *Iraq*, pp. 48–51.
50 Penrose and Penrose, *Iraq*, p. 69.
51 BP 63291, Meeting, 13 March 1928; *ibid.*, Ritchie to IPC, 28 March 1928.
52 PRO FO 371/13028, Dobbs to Colonial Office, 18 April 1928.
53 Penrose and Penrose, *Iraq*, p. 69. For evidence of the appeal of the railway scheme to King Faisal, PRO FO 371/13028, Dobbs to Colonial Office, 25 May 1928.
54 *Ibid.*, Dobbs to Colonial Office, 18 and 19 May 1928.
55 *Ibid.*, Ritchie discussion at the Colonial Office, 21 May 1928.
56 PRO FO 371/13029, Shuckburgh's interview with Ritchie, 11 June 1928.
57 Records of the Turkish Petroleum Company, later renamed Iraq Petroleum Company (referred to as IPC) C22H, Ritchie to TPC, 5 July 1928.

58 *Ibid.*, TPC board meeting, 12 July 1928.
59 Penrose and Penrose, *Iraq*, pp. 69–70; Longrigg, *Oil*, pp. 73–4.
60 BP 70221, Skliros to Cadman, 21 November 1930.
61 *Ibid.*, Cadman diary, 23 February 1931.
62 Penrose and Penrose, *Iraq*, p. 71; Longrigg, *Oil*, pp. 74–5.
63 Longrigg, *Oil*, p. 76; Penrose and Penrose, *Iraq*, p. 139. For a description of the pipeline, see, generally, IPC, *An Account of the Construction of the Pipeline of the IPC Ltd* (London, 1934).
64 Longrigg, *Oil*, p. 77.
65 *Ibid.*
66 *Ibid.*, p. 78.
67 *Ibid.*
68 *Ibid.*, pp. 89–90.
69 *Ibid.*, p. 75; Penrose and Penrose, *Iraq*, p. 71.
70 Longrigg, *Oil*, p. 82; Penrose and Penrose, *Iraq*, p. 140.
71 Longrigg, *Oil*, p. 83.
72 Longrigg, *Oil*, pp. 107–9.
73 *Ibid.*, pp. 105–6 and 137.
74 *Ibid.*, pp. 113–16.
75 *Ibid.*, pp. 90–3.
76 Ferrier, *History*, ch. 12.

6 THE BRITISH GOVERNMENT AND OIL, 1928–1939

1 R. W. Ferrier, *The History of The British Petroleum Company: Volume 1, The Developing Years, 1901–1932* (Cambridge, 1982), ch. 5. See also G. Jones, *The State and the Emergence of the British Oil Industry* (London, 1981).
2 Jones, *The State*; B. S. McBeth, *British Oil Policy, 1919–1939* (London, 1985), ch. 2.
3 *Ibid.*
4 D. J. Payton-Smith, *Oil: A Study of Wartime Policy and Administration* (London, 1971), p. 40.
5 PRO FO 371/281, Board of Trade memorandum, 22 December 1928.
6 PRO FO 370/301, FO minute, 19 January 1929.
7 Payton-Smith, *Oil*, p. 40.
8 'The attempt to secure Britain's oil independence failed miserably', McBeth, *British Oil Policy*, p. 148.
9 McBeth, *British Oil Policy*, especially pp. 121–6; Payton-Smith, *Oil*, pp. 17–18.
10 PRO CAB 50/18, Fraser memorandum, The Scottish shale oil industry, 5 July 1937.
11 PRO POWE 33/543, Cadman to Redemayne, 27 and 28 May 1919; *ibid.*, Jones to Cadman, 5 and 20 June 1919.

12 *Ibid.*, Clarke memorandum, Scottish shale oil industry, 7 October 1921; *ibid.*, Fraser memorandum, 21 October 1921; PRO CAB 50/18, CID Committee, Oil from coal, 6 May 1937.

13 P. T. Cox to R. W. Ferrier, personal communication.

14 BP 44191, Cadman to Board of Trade, 6 April 1929.

15 McBeth, *British Oil Policy*, p. 126.

16 Payton-Smith, *Oil*, p. 17.

17 BP 87219, AIOC record of activities in the war years.

18 *Ibid.*

19 Cited in McBeth, *British Oil Policy*, p. 91.

20 On ICI generally during this period, see W. J. Reader's classic history, *Imperial Chemical Industries, A History: Volume 2, The First Quarter-Century, 1926–1952* (London, 1975). ICI's interest in oil from coal is described in ch. 10.

21 For a brief description of their involvement, see McBeth, *British Oil Policy*, pp. 104–6.

22 Reader, *ICI*, pp. 166–70.

23 Payton-Smith, *Oil*, p. 22.

24 *Ibid.*; Reader, *ICI*, pp. 170–81 and 263–5.

25 Committee of Imperial Defence, *Report of Sub-Committee on Oil from Coal* (Cmd 5665, 1938). See also documents in BP 72488.

26 BP 72488, Memorandum on the present views of the AIOC on synthetic processes, n.d. On the Jowitt Committee, see Payton-Smith, *Oil*, p. 94.

27 Payton-Smith, *Oil*, pp. 54–5; Reader, *ICI*, pp. 265–6.

28 Payton-Smith, *Oil*, pp. 56–7 and 273–5.

29 PRO CAB 27, CID minutes, 16 February 1928.

30 *Ibid.*, Churchill to Government directors, 28 February 1928.

31 PRO T 161/5151, Starling to Grieve, 1 January 1932.

32 *Ibid.*, Grieve minute, 6 January 1932.

33 *Parl. Debates*, House of Commons, vol. 219, cols. 128–31, 25 June 1928, Gillett.

34 *Ibid.*, vol. 219 , col. 225, 26 June 1928, Kenworthy.

35 *Ibid.*, Churchill.

36 *Ibid.*, col. 226, 26 June 1928, Harris.

37 *Ibid.*, cols. 692–3, 28 June 1928, Kenworthy.

38 *Ibid.*, cols. 692–4, 28 June 1928, Churchill; see also PRO T 161/293, Treasury brief for Anglo-Persian Oil Company debate, 3 July 1928.

39 *Parl. Debates.*, House of Commons, vol. 226, cols. 558 and 560, 7 March 1929, Grattan-Doyle.

40 *Ibid.*, cols. 559–60, 7 March 1929, Baldwin. The Prime Minister was further questioned about oil prices on 12, 19 and 21 March.

41 *Prices of Petroleum Products* (Cmd 3296, 1929).

42 PRO T 161/293, Memorandum by Petroleum Department on oil companies' statement, 21 March 1929.

43 *Ibid.*, 'Petrol prices', 21 March 1929.

7 RESEARCH AND TECHNICAL PROGRESS, 1928–1939

1 R. W. Ferrier, *The History of The British Petroleum Company: Volume 1, The Developing Years, 1901–1932* (Cambridge, 1982) especially pp. 274–9 and 397–460.

2 The activities of Sunbury were described in technical reports and monthly reports, the earliest of which are from 1917. See Sunbury Technical Records (hereinafter referred to as STR) (TB311).

3 STR 003.30 (1) (TB1678), Dunstan, Biennial report on chemical research and development, 1929–30, 31 August 1931. See also *The Naft*, September 1931, no. 5, pp. 18–19.

4 *The Naft* magazine carried a coverage of these events over the years.

5 STR 056.1(f) (TB806), Correspondence and minutes, 1923–37. The new research advisers in 1937 were Professor I. M. Heilbron (Manchester University), Professor A. C. G. Egerton (Imperial College) and Professor E. K. Rideal (Cambridge University). In October 1938 the Committee was augmented by Professor Robert Robinson (Oxford) and in autumn 1939 Sir Frank Smith, formerly secretary of the Royal Society and secretary of the Department of Scientific and Industrial Research, was appointed adviser on Scientific Research and Development.

6 Much of the information from the Company's investigatory work in Iran percolated into contemporary technical understanding from publications by members of the staff. See, for example, D. Comins, 'Gas saturation pressure of crude under reservoir conditions as a factor in the efficient operation of oilfields', *Proceedings of First World Petroleum Congress*, vol. 1 (1933), pp. 458–66; M. C. Seamark, 'The drilling and control of high pressure wells', *ibid.*, pp. 354–60; C. A. P. Southwell, 'Scientific unit control', *ibid.*, pp. 304–9; L. A. Pym, 'The measurement of gas-oil ratios and saturation pressures and their interpretation', *ibid.*, pp. 452–7; D. T. Jones, 'The surface tension and specific gravity of crude oil under reservoir conditions', *ibid.*, pp. 467–72; M. W. Strong, 'The significance of underground temperatures', *ibid.*, pp. 124–8; C. J. May and A. Laird, 'The efficiency of flowing wells', *Journal of the Institute of Petroleum Technologists*, vol. 20 (1934), pp. 214–47; L. A. Pym, 'Bottom hole pressure measurement', in A. E. Dunstan, A. W. Nash, B. T. Brooks and H. Tizard (eds.), *The Science of Petroleum, Volume 1* (Oxford, 1938), pp. 508–15.

7 J. H. Jones, 'A seismic method of prospecting', *Proceedings of First World Petroleum Congress*, vol. 1 (1933), pp. 169–73; and 'The refraction method of seismic prospecting' in Dunstan, Nash, Brooks and Tizard (eds.), *Science of Petroleum, Volume 1*, pp. 382–6.

8 STR V500.1 (TB553), Gas – general correspondence, 1917–48; STR V500.33 (TB1252), Gas – general Sunbury reports, 1921–46.

9 BP 42762, Porosity and limestone productivity, correspondence, 1925–41.

10 BP 53812, C. J. May, Report no. CO/1937, 28 February 1929.

11 BP 67587, Report on questions discussed by deputy director (J. A. Jameson) at Fields and Ahwaz, March/April 1930.

12 BP 42762, Porosity and main limestone productivity, correspondence, 1925–41; STR (TB889), Geological correspondence, 1931–2.

13 BP 42762, Porosity and main limestone productivity, correspondence, 1925–41.

14 *Ibid.*

15 BP 53878, H. S. Gibson, 'The production of oil from the M-i-S Field', Production report no. 16, 1933.

16 BP 42762, P. T. Cox, 'Asmari limestone investigation', Progress report C CRL/B.11/No.150, 8 September 1934 and Clark memorandum, 27 September 1934.

17 On these developments, see BP 85999, National Oil Refineries, 1917–27; STR Q420.1 (TB757), Thermal cracking correspondence and reports; STR Q420.37 and (TB311), Research monthly reports 1925–9; also, *The Naft*, January 1930, no. 1, pp. 21–6.

18 BP 68925, Kellogg cracking plant, 1929–30.

19 STR (TB311), Research monthly reports, 1929–33; STR B129.33 (TB66), Inhibitors and inhibition, Sunbury reports, 1932–58; STR B129.1 (TB924), Inhibitors and inhibition, Sunbury correspondence, 1931–52; STR G221.33 (TB457), Refining treatment, general – Sunbury reports, 1923–47; STR library, W. H. Thomas, 'The stabilisation of cracked gasoline', 7th annual AIOC chemical conference, June 1932; M. E. Kelly, 'The use of inhibitors in Abadan with particular reference to gum and colour formation', 14th annual AIOC chemical conference, July 1939.

20 STR (TB311), Research monthly reports, 1932–3.

21 STR library, H. W. Rigden, 'Re-running of acid-treated pressure distillate in Abadan with particular reference to corrosion problems', 8th annual AIOC chemical conference, July 1933.

22 BP 64468–9 and 64477–8, Abadan development reports, 4 March 1931 to 18 February 1934.

23 STR Q425.33 (TB255), Catalytic cracking, Sunbury reports, 1937–9; STR Q425.1 (TB549), Catalytic cracking, correspondence, 1937–44; STR Q425.33a (TB309), D. A. Howes 'Decarbonisation – Summary of work carried out December 1937 to April 1938'.

24 STR Q425.a (TB573), Catalytic cracking, Llandarcy progress reports, February–September 1939.

25 STR 017.1 (TB600), CRA Agreement.

26 Sunbury library pamphlet no. 5667, 'The ideal motor spirit', 1929; *The Naft*, September 1929, no. 5, pp. 3–4 and 23–4.

27 G. S. Gibb and Evelyn H. Knowlton, *History of Standard Oil Company (New Jersey): Volume 2, The Resurgent Years, 1911–1927* (New York, 1956), pp. 539–44.

28 BP 68942, Ethyl – general, 1928–31; BP 62422, Tetraethyl lead and Ethyl fluid, February 1928–October 1931.

29 *The Naft*, May 1931, no. 3, p. 4 for an illustration.

30 BP 102537, Management committee report to the board, May 1931.

31 STR B120.1 (TB1292), Benzines, general correspondence, 1929–37.

32 In the late 1920s and 1930s *The Naft* magazine described and illustrated these records in details. There was much trade press comment as in W. B. Rowntree, 'Trends and developments in racing fuels', *The Autocar*, 25 September 1933, pp. 390–2.

33 There is much documentation on the subject of the evolution and application of engine test procedures in the late 1920s and 1930s. Stansfield himself left a personal memoir issued in June 1947 on 'Engine research, Sunbury, 1926–47'. Relevant technical archives are STR B124.33 (TB452), Motor fuels reports, 1928–37; STR 0381.33 (TB78), Engine indicators report, 1926–43; STR 0380.33 (TB77), Engines, general reports, 1926–36; STR 0388.33 (TB224), Engine testing and correlation reports, 1927–44; STR 0380.1 (TB886), Engines, general correspondence, 1927–51. Also, 'Motoring press representatives visit Sunbury', *The Naft*, September 1929, pp. 23–4; 'New engine research laboratory at Sunbury', *ibid.*, November 1929, pp. 32–3.

34 R. Stansfield and R. E. H. Carpenter, 'The strobophonometer', *Journal of the Institute of Petroleum Technologists*, vol. 18 (1932), pp. 513–25.

35 STR (TB791), Special dossier of Sunbury reports, 1923–35; STR 0382.1 (TB886), Petrol engines – general correspondence, 1929–54.

36 STR B122.1 (TB592), Aviation fuel components, correspondence, 1928–39.

37 BP 95244, Chairman's folder no. 583, Aviation spirit. See also STR B121.33 (TB220), Aviation fuels, Sunbury reports, 1927–38.

38 BP 64478, Abadan laboratory report no. 4287, 5 February 1934; BP 64483, Abadan laboratory report no. 4515, 20 May 1934.

39 'The great international air race', *The Naft*, September 1934, no. 6, pp. 11–14.

40 The production of alcohols and chemical intermediates from ethylene and propylene also seemed promising. The Company was for a short time attracted to this idea, but with the improvement in oil trading conditions in 1933 it concentrated on its traditional business and the possibility of diversifying into petrochemical activities was dropped until the later talks with ICI and the more successful collaboration with the Distillers Company from 1947, which will be examined in volume 3.

41 STR T480.32 (TB889), Pyrolysis – general Abadan reports, 1927–9; STR T480.1 (TB1294), Benzole pyrolysis, correspondence, 1933–9; STR library, D. A. Howes, 'Pyrolysis and polymerisation processes', 11th annual AIOC chemical conference, July 1936; STR T485.33 (TB82), Catalytic polymerisation, Sunbury report, 1935–9; STR T485.1 (TB625), Catalytic polymerisation correspondence, 1935–9; STR library, J. W. T. Jones, 'Catalytic polymerisation', 13th annual AIOC chemical conference, July 1938; L. C. Strang, 'Catalytic polymerisation', 14th annual AIOC chemical conference, July 1939.

42 STR library, S. F. Birch, 'Current organic research', 11th annual AIOC chemical conference, July 1936; and, 'Acid condensation', 13th annual AIOC chemical conference, July 1938; STR W540.33a (TB309), and, 'Report on acid condensation of olefins and paraffins', March 1939.

43 For an authoritative summary of the discovery, see J. H. D. Hooper, 'The alkylation process – a golden anniversary', *Chemistry and Industry*, 20 October 1986, pp. 683–6.

44 See J. H. D. Hooper, 'Alkylation's 40th birthday', *BP Shield* (July 1976).

45 STR W540.33 (TB122), Alkylation, Sunbury reports, 1936–43.

46 British Patent No. 479345 of 31 January 1938, 'Improvements relating to the production of motor oil', in the names of S. F. Birch and A. E. Dunstan. Tait was very unhappy at the omission of his name.

47 STR 056.1(f) (TB806), Correspondence and minutes, 1923–37, 31 December 1937.

48 STR W540.33(a) (TB309), Howes, 'Report on acid condensation of olefins and paraffins', March 1939. The process was operable, but the design of the plant was not satisfactory.

49 STR 056.1(f) (TB806 and 513), Correspondence and minutes, 1938–46.

50 For example, in STR library, R. K. Speirs, 'Acid condensation in Abadan', 14th annual AIOC chemical conference, July 1939.

51 Thus S. F. Birch, A. E. Dunstan, F. A. Fidler, F. B. Pim and T. Tait, 'Saturated high octane fuels without hydrogenation', *Journal of the Institute of Petroleum Technologists*, vol. 24 (1938), pp. 303–20; 'Condensation of olefins with isoparaffins', *Oil and Gas Journal*, vol. 37, no. 6 (1938), pp. 49, 52–5 and 58; and 'High octane isoparaffinic fuels', *Industrial and Engineering Chemistry*, vol. 31 (1939), pp. 884–91 and 1079–82.

52 Howes personal communication to Hooper, 1976.

53 G. Egloff and G. Hulla, *The Alkylation of Alkanes – Patents*, vol. 1 (New York, 1948).

54 BP 80742, Patents and processes, 11 June 1940; STR 010.1 (TB6000) CRA Agreement – correspondence.

8 WAR, 1939–1945

1 BP 25553, AIOC record of activities in the war years, chs. 1–3.

2 This description of wartime organisation is based on BP 71480, AIOC record of activities in the war years, ch. 4; D. J. Payton-Smith, *Oil: A Study of Wartime Policy and Administration* (London, 1971), pp. 43–4, 77–93 and 114–18. For a list of Company employees on wartime committees, see BP 25553, AIOC record of activities in the war years, chs. 1–3.

3 BP 25557, AIOC record of activities in the war years, ch. 7; BP 71483, AIOC record of activities in the war years, chs. 10–14.

4 According to Sir Edward Bridges of the Treasury, Sir Horace Wilson, who was the head of the Civil Service when Fraser became chairman, 'tried to prevent Sir William Fraser from being appointed Chairman of the Anglo-

Iranian'. See PRO T 273/360, Bridges note of meeting with Sir Thomas Gardiner, 18 July 1950.

5 BP 106232, Minutes of AIOC board meeting on 30 June 1942.

6 BP 106233, Minutes of AIOC board meeting on 29 June 1943.

7 BP 104123, 34th ordinary general meeting of the AIOC, 21 September 1943.

8 BP 106232, Minutes of AIOC board meeting on 28 October 1941; BP 106233, Minutes of AIOC board meetings on 23 May 1944 and 27 February 1945.

9 *Ibid.*, Minutes of AIOC board meeting on 21 December 1943.

10 This section on shipping and the pattern of trade is based on BP 71480, 87219 and 71483, containing various chapters of the 'AIOC record of activities in the war years'. Also, Payton-Smith, *Oil*, pp. 127, 159–67, 177–8, 231, 343, 380.

11 S. H. Longrigg, *Oil in the Middle East: Its Discovery and Development*, 3rd edn (London, 1968), p. 137.

12 *Ibid.*, p. 117.

13 Edith and E. F. Penrose, *Iraq: International Relations and National Development* (London, 1978), pp. 103–4.

14 BP 71488, AIOC record of activities in the war years.

15 *Ibid.*

16 Penrose and Penrose, *Iraq*, p. 145; Longrigg, *Oil*, p. 121.

17 BP 71480 and 87219, AIOC record of activities in the war years.

18 BP 87219 and 71483, AIOC record of activities in the war years.

19 BP 87219, AIOC record of activities in the war years.

20 *Ibid.*

21 *Ibid.*

22 See H. Longhurst, *Adventure in Oil: The Story of British Petroleum* (London, 1959), ch. 12. For a much more critical view of PLUTO, see Payton-Smith, *Oil*, pp. 332–5, 410–13 and 445–9.

23 BP 71483, AIOC record of activities in the war years.

24 *Ibid.*

25 On the interest in planning at Courtaulds, see D. C. Coleman, *Courtaulds: An Economic and Social History, Volume 3, Crisis and Change, 1940–1965* (Oxford, 1980), pp. 6–7.

26 BP 68947, Hubbard memorandum on Central Planning Department, 12 October 1953.

27 *Ibid.*

9 TRANSITION IN IRAN, 1938–1947

1 BP 84881, Cadman to Riza Shah, 2 October 1939.

2 BP 71066, Rice to Pattinson, 26 November 1939.

3 BP 85906, Jameson to Cadman, 23 February 1940.

4 BP 84882, Elkington to Gass, 20 February 1940.

5 PRO FO 416/98, Seymour minute, 30 March 1940; *ibid.*, Baggally to Muqaddam, 23 April 1940; *ibid.*, Halifax to Bullard, 24 April 1940.
6 BP 71066, Rice to Pattinson, 22 June 1940.
7 BP 69457, Tehran to London, 2 July 1940.
8 PRO FO 416/98, Bullard to Halifax, 2 July 1940.
9 BP 69457, Tehran to London, 7 July 1940.
10 *Ibid.*, Tehran to London, 16 July 1940.
11 *Ibid.*, London to Tehran, 21 July 1940. Tonnage figures from BP 29818, Chairman's statement to stockholders, 6 October 1945. It should be noted that the figures for royalty oil, which was defined as oil sold for consumption in Iran or exported, differ from those given later in this chapter for production, which included oil used as refinery fuel.
12 BP 69457, London to Tehran, 31 July 1940.
13 *Ibid.*, Tehran to London, 3 August 1940.
14 *Ibid.*
15 PRO FO 416/98, Halifax to Bullard, 6 August 1940.
16 BP 69457, London to Tehran, 6 August 1940.
17 *Ibid.*, Tehran to London, 21 August 1940; BP 72188, Unsigned note on differences between the Iranian Government and the AIOC Ltd regarding royalty payments, 27 January 1943.
18 For reference to the agreement, see BP 25553, AIOC record of activities in the war years.
19 BP 72188, Fraser note, 26 January 1943.
20 Record notes of the discussions are in BP 72188.
21 *Ibid.*, Iliff to Rice, 25 February 1943.
22 BP 106232, Minutes of AIOC board meeting on 25 May 1943.
23 PRO FO 416/99, Bullard to Eden, Memorandum on Iranian foreign trade, 1939–40.
24 *Ibid.*, Bullard to Eden, 27 December 1940.
25 PRO FO 416/98, Bullard to Halifax, 4 January 1941.
26 PRO FO 416/99, Bullard to FO, May 1941.
27 *Ibid.*, Eden to Bullard, 30 June 1941.
28 *Ibid.*, Eden to Cripps, 11 July 1941.
29 PRO FO 371/27196, Minutes of Middle East Ministerial Committee, 11 July 1941.
30 PRO FO 416/99, Eden to Cripps, 19 July 1941.
31 See, in this connection, PRO FO 371/27196, Seymour note, Persian economic pressure, 21 July 1941.
32 *Ibid.*, Minutes of Chief of Staff Committee, 28 July 1941; *ibid.*, War Cabinet conclusion, 28 July 1941.
33 *Ibid.*, Eden to Cripps, 29 July 1941.
34 *Ibid.*, WO to C in C Middle East, 6 August 1941; *ibid.*, Eden to Bullard, 18 August 1941.
35 For detailed accounts of the military operation so far as it affected Company staff, see BP 25553 and 71484, AIOC record of activities in the

war years. Also, H. Longhurst, *Adventure in Oil: The Story of British Petroleum* (London, 1959), pp. 99–104.

36 L. P. Elwell-Sutton, 'Reza Shah the Great: Founder of the Pahlavi dynasty', in G. Lenczowski (ed.), *Iran under the Pahlavis* (Stanford, 1978), pp. 1–2.

37 R. K. Ramazani, *Iran's Foreign Policy, 1941–1973* (Charlottesville, Virginia, 1975), p. 52; Rose Greaves, '1942–1976: The reign of Muhammad Reza Shah', in Hossein Amirsadeghi (ed.), *Twentieth-Century Iran* (London, 1977), p. 53.

38 BP 4308, The case of the AIOC Ltd's concession.

39 BP 25553, AIOC record of activities in the war years.

40 *Ibid.*

41 For a description of some of the defensive precautions, see *ibid.*

42 S.H. Longrigg, *Oil in the Middle East: Its Discovery and Development*, 3rd edn (London, 1968), p. 127.

43 *Ibid.*

44 *Ibid.*, p. 128.

45 Sunbury Technical Records (hereinafter referred to as STR), P416.30(a) (TB1009), Aviation spirit components, Job 303, Alkylation unit, Abadan.

46 STR B121.32 (TB640), Aviation fuels, Abadan reports, 1931–40 and STR T480.1 (TB1294), Benzole pyrolysis correspondence.

47 R. Stansfield, 'Aero engine work in the laboratory', 14th annual AIOC chemical conference, July 1939.

48 BP 71480, AIOC record of activities in the war years.

49 *Ibid.* and D. J. Payton-Smith, *Oil: A Study of Wartime Policy and Administration* (London, 1971), p. 271.

50 Payton-Smith, *Oil*, pp. 271–2.

51 STR T482.33(b) (TB236), Sunbury isomerisation reports, July 1940–October 1942.

52 S. F. Birch, P. Docksey and J. H. Dove, 'Process development and production of isohexane and isoheptane and aviation fuel components', *Journal of the Institute of Petroleum Technologists*, vol. 32 (1946), pp. 167–99.

53 Payton-Smith, *Oil*, pp. 271 and 384.

54 BP 71066, Pattinson to Rice, 24 November 1942.

55 BP 72358, Pattinson to Jameson, 4 and 12 July 1944.

56 *Ibid.*, Jameson to Pattinson, 28 July 1944.

57 BP 25553, AIOC record of activities in the war years.

58 *Ibid.* On price inflation, see for example, C. Issawi, 'The Iranian economy, 1925–1975: fifty years of economic development', in Lenczowski (ed.), *Iran*, p. 134.

59 BP 71487, AIOC record of activities in the war years.

60 *Ibid.*

61 *Ibid.*

62 BP 25553, AIOC record of activities in the war years.

63 *Ibid.*

64 BP 71487, AIOC record of activities in the war years.

65 BP 53007, Jameson to Dr C. F. R. Harrison, 28 August 1944.

66 On the emergence of these various groups, parties and personalities, see R. M. Savory, 'Social development in Iran during the Pahlavi era', in Lenczowski (ed.), *Iran*, p. 100; Lenczowski, 'Political process and institutions in Iran: the second Pahlavi kingship', in *ibid.*, pp. 441 and 447; M. Reza Ghods, *Iran in the Twentieth Century, A Political History* (London, 1989), pp. 124–33; Greaves, '1942–1976: The reign of Muhammad Riza Shah', in Amirsadeghi (ed.), *Iran*, p. 54; F. Azimi, 'The reconciliation of politics and ethics, nationalism and democracy: an overview of the political career of Dr Muhammad Musaddiq', in J. A. Bill and W. R. Louis (eds.), *Musaddiq, Iranian Nationalism and Oil* (Austin, Texas, 1988), p. 51.

67 Cited in Greaves, '1942–1976: The reign of Muhammad Riza Shah', in Amirsadeghi (ed.), *Iran*, p. 57.

68 On the extent of Allied influence, see, for example, *ibid.*, pp. 54–6.

69 PRO FO 371/35128, Peterson to Brown, 27 September 1943; *ibid.*, Brown to Peterson, 13 October 1943; *ibid.*, Bullard to FO, 22 October 1943.

70 *Ibid.*, Bullard to FO, 10 and 13 November 1943; PRO POWE 33/117, Bullard to FO, 15 November 1943.

71 PRO FO 371/35128, Legh-Jones to Godber, 17 November 1943.

72 PRO POWE 33/317, Boyle to Godber, 24 February 1944; *ibid.*, Bullard to FO, 14 and 20 March, 1944.

73 Royal Institute for International Affairs, *Survey of International Affairs: The Middle East in the War* (Oxford, 1954), p. 474.

74 PRO POWE 33/317, Van Hasselt to Godber, 13 April 1944; *ibid.*, Bullard to FO, 8 May 1944.

75 E. Abrahamian, *Iran between Two Revolutions* (Princeton, 1982), p. 210.

76 PRO FO 371/40241, Bullard to FO, 2 and 3 October 1944.

77 Sir Clarmont Skrine, *World War in Iran* (London, 1962), p. 227.

78 PRO FO 371/40241, Bullard to FO, 9 October 1944.

79 *Ibid.*, Bullard to FO, 26, 28 and 30 October, 1944; for text of press conference, see *ibid.*, Soviet *Monitor*.

80 Ramazani, *Iran's Foreign Policy*, p. 100.

81 PRO FO 371/40242, Bullard to FO, 10 November 1944.

82 *Ibid.*, Bullard to FO, 24 November; PRO FO 371/3276, Bullard to Eden, 23 January 1945, Report on Persian affairs, October–December 1944.

83 R.W. Ferrier, 'The development of the Iranian oil industry', in Amirsadeghi (ed.), *Iran*, p. 103.

84 PRO FO 371/40242, Bullard to FO, 4 December 1944.

85 B. Shwadran, *The Middle East, Oil and the Great Powers*, 2nd edn (New York, 1959), p. 68; Ghods, *Iran*, p. 136. On the shifting factions which brought about Bayat's fall from office, see Greaves, '1942–1976: The reign of Muhammad Riza Shah', in Amirsadeghi (ed.), *Iran*, pp. 59–60.

86 Ghods, *Iran*, pp. 138–9.
87 *Ibid.*, p. 141.
88 *Ibid.*, p. 143; Shwadran, *The Middle East*, p. 69.
89 Shwadran, *The Middle East*, pp. 70–1.
90 Ghods, *Iran*, p. 148; Shwadran, *The Middle East*, p. 75.
91 On Soviet denials of the linkage, see Shwadran, *The Middle East*, pp. 75–6.
92 Ghods, *Iran*, p. 148.
93 Shwadran, *The Middle East*, p. 76.
94 PRO FO 371/52671, Farquhar to FO, 29 and 20 March 1946.
95 On the Security Council's deliberations, see Shwadran, *The Middle East*, pp. 70–8.
96 *Ibid.*, pp. 78–9.
97 Greaves, '1942–1976: The reign of Muhammad Riza Shah', in Amir-sadeghi (ed.), *Iran*, pp. 61–2.
98 Ghods, *Iran*, pp. 155–6.
99 Greaves, '1942–1976: The reign of Muhammad Riza Shah', in Amir-sadeghi (ed.), *Iran*, p. 63.
100 Shwadran, *The Middle East*, p. 80.
101 Azimi, 'The reconciliation of politics and ethics, nationalism and democracy: an overview of the political career of Dr Muhammad Musaddiq', in Bill and Louis (eds.), *Musaddiq*, p. 52; H. Ladjevardi, 'Constitutional government and reform under Musaddiq', in *ibid.*, p. 71.
102 The full text of the single article law can be found in Shwadran, *The Middle East*, pp. 80–1.
103 PRO FO 371/61974, Cresswell to FO, 22 October 1947.
104 *Ibid.*, Pyman minute, 25 October 1947.

10 MANAGEMENT AND FINANCE, 1946–1951

1 BP 106233, Minutes of AIOC board meeting on 30 July 1946.
2 BP 106234, Minutes of AIOC board meeting on 26 April 1949.
3 *Ibid.*, Minutes of AIOC board meeting on 28 March 1950.
4 BP 70288, Mayhew, Departmental establishments, 28 January 1948.
5 BP 72322, Mayhew memorandum, Head Office establishments, 23 March 1948.
6 BP 53922, Report of the committee to review Head Office organisation and procedure, 1 January 1949.
7 BP 66874, Extracts from a report by Harold Whitehead & Partners, n. d.
8 *Ibid.*
9 R.W. Ferrier, *The History of The British Petroleum Company: Volume 1, The Developing Years, 1901–1932* (Cambridge, 1982), pp. 336–41.

11 THE SEARCH FOR MARKET OUTLETS, 1943–1951

1 For illustrations of these trends, see for example, BP 9210, AIOC inter-managerial conference, Diagrams accompanying an address by Coxon, April 1951.
2 See, for example, BP 67593.
3 BP 67593, Planning Committee, 19 October 1943.
4 BP 69275, Fraser memorandum to the board, 25 November 1943.
5 *Ibid.*
6 BP 106600, Joint statement of aims, 30 September 1942.
7 BP 43853, Morris to Legh-Jones, 12 August 1943.
8 *Ibid.*, Astley-Bell to Morris, 17 August 1943.
9 *Ibid.*, Morris to Fraser, 1 September 1943.
10 BP 59341, unsigned draft memorandum of proposals, 3 January 1941.
11 *Ibid.*, Smith to Snow, 18 December 1944; PRO T 273/359, Bridges to Padmore and Cripps, 26 April 1945.
12 PRO T 273/359, Eady to Padmore, 27 February 1945.
13 *Ibid.*, Bridges note, 11 May 1945.
14 *Ibid.*, 25 May 1945.
15 *Ibid.*, Eady to Bridges, 27 June 1945.
16 *Ibid.*, 2 August 1945.
17 *Ibid.*, Bridges note, 5 September 1945.
18 BP 106617, unsigned note on AIOC policy, 18 June 1946.
19 BP 66261, Note, Snow, 14 June 1946.
20 For reference to the meeting, see BP 66261, Morris to Fraser, 2 July 1946.
21 BP 66261, Note, Snow, 14 June 1946.
22 *Ibid.*
23 BP 66874, Snow to Morris, 7 November 1946.
24 *Ibid.*
25 BP 67593, Eighth meeting of Planning Committee, 5 August 1943; *ibid.*, Coxon to Fraser, 3 September 1943; *ibid.*, Central Planning Department, Utilisation of petroleum products, 1 December 1945; *ibid.*, Central Planning Department, Utilisation of petroleum products, 19 February 1946.
26 *Ibid.*, Central Planning Department, Utilisation of petroleum products, 17 May 1946.
27 BP 72322, Note on departmental organisation, 2 May 1947; BP 66874, Snow, Departmental organisation, 14 November 1947.
28 BP 66874, Snow, Departmental organisation, 31 May 1949.
29 *Ibid.*, MacWilliam to Snow, 29 December 1950; *ibid.*, Harrison to Snow, 29 December 1950.
30 *Ibid.*, MacWilliam to Snow, 29 December 1950.
31 *Ibid.*, MacWilliam to Snow, 15 January 1951.
32 *Ibid.*, Snow to Morris, 16 January 1951.
33 *Ibid.*, Morris manuscript note.

34 BP 69275, Central Planning Department, Future refining requirements, 28 October 1946; *ibid.*, Central Planning Department, Future refining policy, 20 December 1946.

35 *Ibid.*

36 *Ibid.*, Central Planning Department, Future refining policy, 20 December 1946; *ibid.*, Coxon to Fraser, 9 September 1948.

37 BP 72329, Central Planning Department, New refinery projects for the UK, 10 June 1947; BP 67593, Extract from first report on co-ordination of oil refinery expansion in the OEEC countries, 1949.

38 BP 69275, Central Planning Department, Future refining policy, 20 December 1946.

39 *Ibid.*, Central Planning Department, Future refining requirements, 28 October 1946; *ibid.*, Central Planning Department, Future refining policy, 20 December 1946.

40 BP 72329, Central Planning Department, New refinery projects for the UK, 10 June 1947.

41 *Ibid.*, Central Planning Department, New refinery projects for the UK, 10 June 1947; *ibid.*, Coxon to Fraser, 9 September 1948.

42 BP 69275, Central Planning Department, Future refining policy, 20 December 1946; *ibid.*, Planning Committee minutes, 5 February 1947.

43 BP 72329, Coxon to Jameson, 27 June 1947; *ibid.*, Central Planning Department, New refinery projects for the UK, 10 June 1947.

44 *Ibid.*, Coxon to Fraser, 9 September 1948.

45 Potted descriptions of the refinery can be found in BP 30875; 30903; 30902; 30906.

46 BP 61671, AIOC to subsidiary companies, 9 November 1947.

47 BP 28204, Fraser's statement to shareholders, 20 June 1950; BP 102771, Management report for AIOC board meeting on 28 February 1952.

48 BP 71379, Distribution information memorandum no. 24, 15 August 1950.

49 BP 69501, Report on Germany, 24 January 1947.

50 BP 71379, Distribution information memorandum no. 22, 14 April 1950.

51 *Ibid.*, No. 22, 14 April 1950; no. 24, 15 August 1950; no. 25, 6 November 1950.

52 *Ibid.*, No. 22, 14 April 1950; no. 24, 15 August 1950; no. 25, 6 November 1950; no. 26, 29 January 1951.

53 BP 67903, Snow to Anderson, 18 February 1949; BP 62436, Watts memorandum to Co-ordination Committee, 18 April 1962.

54 BP 71379, Distribution information memorandum no. 28, 14 May 1952; BP 67903, Crude petroleum supply agreement between Compagnie Française Belge des Pétroles SA (Petrofina) and AIOC, 22 January 1951.

55 BP 71379, Distribution information memorandum no. 25, 6 November 1950.

56 BP 67664, European market review, February 1951.

57 *Ibid.*

58 *Ibid.*

59 *Ibid.*; BP 71379, Distribution information memorandum no. 18, 31 August 1949.
60 BP 67664, European market review, February 1951.
61 *Ibid.*
62 BP 71379, Distribution information memorandum no. 25, 6 November 1950.
63 *Ibid.*, No. 24, 15 August 1950; BP 67664, European market review, February 1951.
64 BP 67664, European market review, February 1951.
65 BP 66869, O'Brien to Snow, 26 January 1948; *ibid.*, Snow record note on Greece, 4 February 1948; *ibid.*, unsigned note on Steaua Agencies Ltd, 13 January 1948; BP 71379, Distribution information memorandum no. 14, 25 March 1949 and no. 18, 31 August 1949.
66 BP 67664, European market review, February 1951.
67 *Ibid.*
68 BP 70985, Jackson to Fraser, 29 July and 7 December 1944. BP 70986, Fraser to Jackson, 2 January, 4 January and 8 October 1945; *ibid.*, Jackson to Fraser, 9 January, 17 January, 6 March and 3 December 1945.
69 BP 70987, Jackson to Fraser, 13 August 1946.
70 BP 66821, Fraser to Jackson, 10 October 1946.
71 BP 26878, Standard Oil Company (New Jersey) and Middle Eastern oil production. A background memorandum on company policies and actions, March 1947.
72 BP 66821, Fraser to Jackson, 10 October 1946.
73 *Ibid.*, AIOC offtake of crude oil and products, 4 November 1946.
74 BP 26882, Fraser to Jackson, 8 October 1946.
75 BP 66821, Fraser to Jackson, 10 October 1946.
76 BP 26882, Jackson to Fraser, 15 October 1946.
77 *Ibid.*
78 *Ibid.*, Fraser to Jackson, 8 October 1946; *ibid.*, Jackson to Fraser, 10 October 1946.
79 *Ibid.*, Jackson to Fraser, 30 October 1946.
80 BP 66823, Heads of Terms made between Standard Oil (NJ) and AIOC, 20 December 1946; BP 66821, Jackson to Fraser, 20 December 1946.
81 BP 66823, Jackson to Harden, 20 December 1946.
82 BP 26936, Anglo-Iranian and Standard Oil Company, agreement dated 25 September 1947; BP 62375, Anglo-Iranian and Socony-Vacuum Oil Company, agreement dated 25 September 1947.
83 *Ibid.* The quantities were subject to a tolerance of plus or minus 10 per cent at the buyers' option.
84 BP 26878, Fraser to Jackson, 25 February 1947.
85 BP 71340, Snow to Jackson, 20 May 1947; *ibid.*, Jackson to Fraser, 3 June 1947; *ibid.*, Snow note on Socony inquiry for additional crude, 20 June 1947.
86 BP 71340, Snow to Jackson, 20 May 1947.

87 *Ibid.*
88 BP 71340, Snow note on Socony inquiry for additional crude, 20 June 1947.
89 *Ibid.*
90 On the negotiations, see generally, BP 71340. The legal agreement is in BP 64736.
91 BP 64734, Anglo-Iranian, Standard Oil, Socony-Vacuum and Middle East Pipelines Ltd, Pipeline agreement dated 23 March 1948.
92 82nd Congress, 2nd Session, Senate Small Business Committee, Staff report of the Federal Trade Commission (FTC), *The International Petroleum Cartel* (1952), p. 156.
93 *Ibid.* See also documents in BP 66822 and 26878.
94 BP 66822, Jackson to Snow, 4 May 1948; BP 26879, Snow to Jackson, 24 May 1948. See also BP 26878, Snow to Jackson, 5 July 1948.
95 The relevant agreements are in BP 54006 and 59376.

12 THE COMPANY AND THE BRITISH GOVERNMENT 1946–1951

1 See generally, R. N. Gardner, *Sterling-Dollar Diplomacy: Anglo-American Collaboration in the Reconstruction of Multilateral Trade* (Oxford, 1956). Also, A. S. Milward, *The Reconstruction of Western Europe, 1945–1951* (London, 1984).
2 On the postwar British economy, see generally, Sir Alec Cairncross, *Years of Recovery: British Economic Policy, 1945–1951* (London, 1985); G. D. N. Worswick and P. H. Ady (eds.), *The British Economy, 1945–1950* (Oxford, 1952); J. C. R. Dow, *The Management of the British Economy, 1945–1960* (Cambridge, 1964).
3 For an example of the Company's difficulties in guessing how long it might take to obtain an authorisation, see BP 72328, Central Planning Department memorandum on refineries, 30 November 1950.
4 For example, BP 66887, Snow record note of meeting with Petroleum Division, 10 October 1949; *ibid.*, Snow to Morris, 2 February 1949.
5 PRO BT 230/121, Helmore to Miss Goodwin, 19 January 1949.
6 BP 72337, Jameson to Jackson, 30 July 1948.
7 PRO T 236/219, Bank of England memorandum on oil policy, 15 January 1945.
8 PRO T 236/1314, Cabinet Investment Programme Committee meeting, 5 September 1947.
9 *Ibid.*
10 PRO T 236/1314, Memorandum on overseas steel programme for the oil companies, 29 October 1947.
11 *Ibid.*, Bridges memorandum, 26 November 1947.
12 *Ibid.*, Flett to Rowe-Dutton, 8 December 1947.

13 *Ibid.*, Cabinet Production Committee, 'Oil', memorandum by the Minister of Fuel and Power, 10 December 1947.
14 BP 106618, Central Planning Department, Steel requirement to implement the AIOC programme for 1952–3, 17 June 1948.
15 *Ibid.*, Wright to Coxon, 21 May 1948.
16 A full record of the working party's meetings is in BP 66889.
17 PRO T 236/2612, Morris to Fraser, 2 October 1947; *ibid.*, Fraser to Eady, 3 October 1947.
18 For an exposition of Bank policy, see PRO T 236/2612, O'Brien to Brooks, 17 October 1947.
19 PRO T 236/2612, Morris to Fraser, 2 October 1947.
20 *Ibid.*, Rudd memorandum on AIOC European marketing subsidiaries, 23 December 1947.
21 BP 66869, Snow record note of meeting with O'Brien, 4 February 1948.
22 *Ibid.*, O'Brien to Snow, 26 January 1948.
23 BP 71379, Distribution information memorandum no. 14, 25 March 1949.
24 *Ibid.*
25 PRO T 236/3872, Davis to Legh-Jones, 15 March 1950.
26 BP 66894, Record note of meeting at Ministry of Fuel and Power, 11 December 1946.
27 PRO T 236/2142, Record note of meeting at the Petroleum Division between Fergusson, Butler, Berthoud and Fraser, 17 June 1947.
28 *Ibid.*
29 *Ibid.*
30 *Ibid.*
31 *Ibid.*, Berthoud to Rowe-Dutton, 18 June 1947.
32 *Ibid.*, Rowe-Dutton to Clarke and Eady, 23 June 1947.
33 PRO T 236/1314, Flett to Clarke and Rowe-Dutton, 7 January 1948; PRO T 236/2882, Wolfson to Rudd, 23 March 1949; PRO T 236/2612, Rudd to Kelly, 2 February 1950.
34 *Parl. Debates*, House of Commons, vol. 421, cols. 1778–9, 8 April 1946, Shinwell.
35 PRO POWE 33/1623, Agnew to Starling, 12 April 1946.
36 *Ibid.*, Prime Minister's personal minute, no. 216/46, to Minister of Supply, 21 June 1946.
37 *Ibid.*, Prime Minister's personal minute, no. 228/46, to Minister of Fuel and Power, 30 June 1946.
38 *Ibid.*, Minister of Fuel and Power to Prime Minister, 8 July 1946.
39 *Ibid.*, Press release no. 101, 1 August 1946.
40 B. Pimlott, *Hugh Dalton* (London, 1985), p. 477.
41 *Parl. Debates*, House of Commons, vol. 427, cols. 916–30, 16 October 1946, Shinwell.
42 PRO POWE 33/1623, Petroleum board to Butler, 25 October 1946.
43 Pimlott, *Dalton*, p. 478.

44 *Ibid.*; see also Cairncross, *Years of Recovery*, ch. 13.
45 Pimlott, *Dalton*, pp. 478–9.
46 Cairncross, *Years of Recovery*, p. 383.
47 PRO POWE 33/1624.
48 *Ibid.*
49 *Ibid.*, Wilson to Ellis, 25 March 1947.
50 PRO POWE 33/1625, Fergusson's minute, 7 August 1947.
51 *Ibid.*, Prime Minister's minute to Minister of Fuel and Power, 31 July 1947.
52 BP 66897, Unsigned notes on meeting at the Petroleum Division on 15 August 1947; *ibid.*, Unsigned notes on meeting held at the Petroleum Division on 15 August 1947 on the subject of petroleum supplies to the UK.
53 *Ibid.*, Unsigned notes on meeting at the Petroleum Division on 11 August 1947; *ibid.*, Richardson, Estimated UK stock position at 25 January 1948; *ibid.*, Snow record note, 2 September 1947.
54 *Ibid.*
55 BP 66897, Snow, British supply position 1948 – fuel and gas oil, 19 September 1947.
56 *Ibid.*, Snow record note, 29 September 1947.
57 *Ibid.*, Snow, UK forward stock position – fuel oil, 10 October 1947.
58 BP 66896, Snow to Morris, 19 November 1947.
59 BP 66897, Snow, Supply position – first quarter 1948, 26 November 1947.
60 *Ibid.*, Snow to Butler, 1 December 1947.
61 *Ibid.*, Snow, Supply position – first quarter 1948, 9 December 1947.
62 BP 66895, Unsigned note of meeting held at Ministry of Fuel and Power on 19 December 1947 to discuss the oil position.
63 *Ibid.*, Fraser to Murphy, 30 December 1947.
64 See documents in PRO POWE 33/1626.
65 PRO POWE 33/1659, Kelf-Cohen minute, 25 October 1950.
66 Henrietta Larson, Evelyn Knowlton and C. S. Popple, *History of Standard Oil Company (New Jersey): New Horizons, 1927–1950* (New York, 1971), pp. 701–3.
67 D. S. Painter, 'Oil and the Marshall Plan', *Business History Review*, vol. 58, no. 3 (1984), p. 362.
68 Larson *et al.*, *New Horizons*, pp. 701–3.
69 *Ibid.*, pp. 705–6.
70 BP 66896, Snow to Fraser, 5 July 1949.
71 *Ibid.*
72 BP 96429, Snow record note, 2 August 1949.
73 *Ibid.*, Stockwell to Snow, 16 August 1949; *ibid.*, Snow record note, 18 October 1949; *ibid.*, Snow to Fraser, 28 November 1949. Also, Painter, 'Oil and the Marshall Plan', p. 380.
74 BP 96429, Snow to Fraser, 12 January 1950.
75 Painter, 'Oil and the Marshall Plan', pp. 380–1; Larson *et al.*, *New Hori-*

zons, pp. 706–10. On US protests, see also, for example, *New York Times*, 20 and 21 December 1949, 1 February 1950.

76 Painter, 'Oil and the Marshall Plan', p. 382; Larson *et al.*, *New Horizons*, pp. 710–13.

77 BP 66896, Fraser to Fergusson, 13 November 1950.

78 Dow, *Management of the British Economy*, pp. 27–8; A. R. Ilersic, *Government Finance and Fiscal Policy in Post-War Britain* (London, 1955), pp. 92–3; I. M. D. Little, 'Fiscal policy' in Worswick and Ady (eds.), *British Economy*, pp. 169 and 171.

79 Cairncross, *Years of Recovery*, pp. 230, 405, 408; Dow, *Management of the British Economy*, pp. 34–6, 61; G. D. N. Worswick, 'Personal income policy' in Worswick and Ady (eds.), *British Economy*, pp. 328–31.

80 *Economist*, 4 August 1951, p. 297. Cited in Dow, *Management of the British Economy*, p. 61.

81 PRO T 273/360, Bridges note of meeting with Gardiner, 18 July 1950.

82 PRO T 273/359, Eady note, 19 February 1945.

83 *Ibid.*

84 *Ibid.*, Eady to Bridges, 2 August 1945.

85 PRO T 236/219, Unsigned Ministry of Fuel and Power note to Blaker, August 1945.

86 PRO T 273/360, Bridges note of meeting with Gardiner, 18 July 1950.

87 *Ibid.*, Bridges note of meeting with Gardiner, 27 October 1950.

88 *Ibid.*

89 *Ibid.*, Bridges note of meeting with Gardiner, 12 February 1951.

90 *Ibid.*, Bridges note of meeting with Gardiner, 27 October 1950.

91 *Ibid.*

13 POLITICS AND JOINT INTERESTS IN THE MIDDLE EAST, 1946–1951

1 M. B. Stoff, *Oil, War and American Security: The Search for a National Policy on Foreign Oil, 1941–1947* (New Haven, 1980), pp. 135–6.

2 *Ibid.*

3 For full accounts of these events and what follows in this section, see generally, Stoff, *Oil, War and American Security*; and 'The Anglo-American oil agreement and the wartime search for foreign oil policy', *Business History Review*, vol. 55, no. 1 (1981), pp. 59–74; I. H. Anderson, *Aramco, the United States and Saudi Arabia: A Study of the Dynamics of Foreign Oil Policy, 1933–1950* (Princeton, 1981); A. D. Miller, *Search for Security: Saudi Arabian Oil and American Foreign Policy, 1939–1949* (Chapel Hill, 1980).

4 The British delegation was led by Sir William Brown, head of the Petroleum Division of the Ministry of Fuel and Power. The other members were John Le Rougetel of the Foreign Office (later Ambassador to Iran, 1946–50; Frederic Harmer, a young Treasury official who was later to

become a British Government director of the Company; Commodore A. W. Clarke, representing the three service departments; Frederick Starling and Victor Butler of the Petroleum Division; and Sir Frederick Godber of Royal Dutch-Shell.

5 Cited in Stoff, *Oil, War and American Security*, p. 167.

6 The text of the Anglo-American Petroleum Agreement is printed in Anderson, *Aramco, the United States and Saudi Arabia*, appendix B, pp. 218–23.

7 Anderson, *Aramco, the United States and Saudi Arabia*, especially ch. 6.

8 *Ibid.*, pp. 178–97.

9 S. H. Longrigg, *Oil in the Middle East: Its Discovery and Development*, 3rd edn (London, 1968), pp. 118 and 174. During the war Gulbenkian, being resident in France, was similarly classified as an enemy alien, but he was reinstated as a British national in 1943 when he moved to Lisbon in Portugal. His claim for compensation for loss of IPC oil was made in 1943 and settled in May 1945.

10 Iraq Petroleum Company records (hereinafter referred to as IPC) 3B 2003, Minutes of meeting, 16 September 1946.

11 *Ibid.*, Minutes of meetings, 23, 24 and 27 September 1946.

12 For legal opinions, see, for example, IPC 3B 2003, Opinion of Sir David Maxwell Fyfe, H. G. Robertson and Vladimir Idelson, 26 November 1946; *ibid.*, Walter Monckton's opinion, 13 January 1947.

13 IPC 3B 1132, Copy of writ, 13 December 1946.

14 IPC 3B 2003, Whishaw to Fraser, 14 January 1947.

15 IPC 3B 2003, Sheets to Shephard, 15 February 1947; *ibid.*, Sheets to New York, 14 May 1947.

16 Documents on the negotiations with Gulbenkian are in IPC 3B 2003–4.

17 Edith and E. F. Penrose, *Iraq: International Relations and National Development* (London, 1978), p. 147.

18 S.H. Longrigg, *Iraq, 1900–1950: A Political, Social and Economic History* (London, 1953), p. 374; and *Oil*, p. 180; Penrose and Penrose, *Iraq*, pp. 147–8.

19 For a detailed account, see Longrigg, *Iraq*, ch. 10; also, Penrose and Penrose, *Iraq*, ch. 12.

20 This account of events is drawn largely from Longrigg, *Iraq*, ch. 10; Penrose and Penrose, *Iraq*, ch. 5; G. Lenczowski, *The Middle East in World Affairs*, 4th edn (New York, 1980), chs. 7 and 10.

21 Longrigg, *Oil*, pp. 180, 248 and 252; Lenczowski, *Middle East*, p. 277; Penrose and Penrose, *Iraq*, p. 148.

22 Longrigg, *Oil*, p. 176.

23 Documents on the negotiations may be found in PRO FO 371/75178–9 and 82464–9. For published reference to the conditional offer of a £3 million loan, see Longrigg, *Iraq*, pp. 356–7.

24 PRO FO 371/82465, EQ 1531/53.

25 PRO FO 371/82468, State Department memorandum, June 1950 and *ibid.*, Trevelyan to Furlonge, 7 July 1950.

26 Longrigg, *Oil*, p. 190; and *Iraq*, p. 376.
27 Longrigg, *Oil*, p. 191; and *Iraq*, p. 377; Penrose and Penrose, *Iraq*, p. 158; Lenczowski, *Middle East*, p. 279.
28 Longrigg, *Oil*, pp. 191–2; and *Iraq*, p. 377; Penrose and Penrose, *Iraq*, pp. 158–9.
29 Lenczowski, *Middle East*, pp. 278–9; Longrigg, *Oil*, p. 193; and *Iraq*, p. 377; Penrose and Penrose, *Iraq*, p. 160.
30 Longrigg, *Oil*, pp. 176–8.
31 *Ibid.*, pp. 180–2.
32 *Ibid.*, pp. 183–4.
33 *Ibid.*, pp. 185–6.
34 *Ibid.*, p. 227.
35 *Ibid.*, p. 231.
36 This account of oil development in Kuwait is based on Longrigg, *Oil*, pp. 137 and 221–5.

14 OPERATIONS AND EMPLOYMENT IN IRAN, 1946–1951

1 S. H. Longrigg, *Oil in the Middle East: Its Discovery and Development*, 3rd edn (London, 1968), pp. 123, 128 and 149–50.
2 Sunbury Technical Records (hereinafter referred to as STR) J283.33(ad) (TB624), Laboratory apparatus, Fields laboratory. On the compilation of crude oil dossiers and handbooks and the evaluation of crude oils, see STR A101.1 (TB1295), Crude oils, general, correspondence, 1926–47 and STR A110.1 (TB563), Crude Oil Advisory Group minutes, 1948–51.
3 For a technical introduction to oil production methods see British Petroleum Company, *Our Industry*, 5th edn (London, 1977), ch. 7. For earlier descriptions of technical aspects of production in the period covered by this volume, see BP 2047–56, 93591 and 90448, 'Petroleum Engineering compendium of the Iranian oilfields'; various papers in *Proceedings of First World Petroleum Congress* (1933); C. J. May and A. Laird, 'The efficiency of flowing wells', *Journal of the Institute of Petroleum Technologists*, vol. 20 (1934), pp. 124–8; H. S. Gibson, 'The production of oil from the fields of south-western Iran', *ibid.*, vol. 34 (1948), pp. 374–402. Some relevant documents are also in BP 67587, 42762, 53878.
4 The fire caught at a well being drilled by Rig 20 on 1 May 1951. The American oil fire specialist, Myron Kinley, was called in to deal with the blaze, though it took him several weeks to extinguish the flames and cap the well. For a fuller account, see H. Longhurst, *Adventure in Oil* (London, 1959), ch. 14.
5 Longrigg, *Oil*, p. 48.
6 BP 59011, Jameson to Elkington, 7 November 1929.
7 BP 68040, Elkington to Jameson, 28 January 1935.
8 BP 16250, Memorandum of October 1943, 1 December 1943.
9 *Ibid.*, Hazhir to Company, 3 May 1947.

10 BP 68936, Meeting on General Plan, 17 October 1947. See also BP 16250, Meeting on social conditions in Iran, 30 September 1947.

11 BP 68936, Gass, Meetings on 25, 26, 29 and 30 November 1947. See also BP 9252, Northcroft to Gass, 14 December 1947 regarding Iranian suspicion 'as to whether the Company has in fact been doing everything possible in regard to Iranianisation' and *ibid.*, Northcroft to Gass, 19 January 1948.

12 BP 9252, Gass memorandum, General Plan, 5 January 1948; *ibid.*, Idelson to Gass, 5 February 1948; *ibid.*, Joint opinion, 23 February 1948: 'We are therefore of the opinion that the Plan is not inconsistent with the Concession'.

13 BP 8334, Chisholm note, Present situation in Iran, 22 January 1948: 'Intensified efforts are being made to weld healthily together our Iranian, and non-Iranian employees' and BP 9252, Gass to Northcroft, 13 January 1948 on 'the sincerity of our intentions'.

14 On the terms and conditions of employment in oil companies operating in the Middle East, see BP 79663, Minutes of meetings at Britannic House of oil company representatives, 23, 24 and 25 September 1947.

15 BP 9252, Gass to Gobey, June 1948; *ibid.*, Records of meetings between Gobey and Pirnia; *ibid.*, Gobey to Gass, 24 August 1948.

16 *Ibid.*, Gobey to Gass, 24 August 1948: 'Settlement of the General Plan prior to outcome of other and wider discussions was not Government intention'.

17 BP 16250, Mylles to Elkington, September 1947.

18 BP 15918, Report from Iran on education and training, 1947.

19 BP 71880, Elkington memorandum to Finance Committee, 19 November 1948.

20 *Ibid.*, Northcroft to Morrison, 15 March 1948.

21 Ministry of Education statistics cited in J. Murray, *Iran Today: An Economic and Descriptive Survey* (Tehran, 1950).

22 BP 69296, Jameson to Aspinall, 9 December 1948.

23 Overseas Consultants Inc., *Report on Seven Year Development Plan for the Plan Organization of the Imperial Government of Iran* (New York, 1949), vol. 1.

24 *Ibid.*

25 BP 54363, General manager's annual report, 1923.

26 See, for example, BP 41113, Notes on meeting at Abadan, 22 March 1936; *ibid.*, Visit of Minister of Education to Abadan, 13 April 1938; *ibid.*, Pattinson to Rice, 15 February 1943; *ibid.*, Elkington to Pattinson, 16 April 1943; *ibid.*, Hawker, Note on educational policy, 28 July 1943; BP 71874, Elkington to Jones, 6 April 1945; *ibid.*, Jameson, Budget proposal for Abadan primary school, 19 September 1945; *ibid.*, Gass to Abadan, 10 December 1947.

27 BP 68192, Note on educational facilities, 3 April 1951.

28 BP 70268, Meeting with Jahangir, 20 October 1933.

29 BP 72616, Elkington to Gass, 28 September 1935.

30 BP 9252, Notes of 3rd meeting at Ministry of Finance, 8 August 1948.

31 BP 71066, Pattinson to Rice, 18 August 1941.
32 BP 67116, Summary of Persianization, 1951.
33 BP 69296, Jameson to Aspinall, 9 December 1948.
34 BP 71066, Pattinson to Rice, 18 August 1941.
35 PRO FO 371/45461, Bullard, Amenities for Persian employees of the AIOC, 15 September 1945.
36 BP 41355, Leggett report on visit to Iran, October 1946; BP 3B 5122, Notes of meetings held during visit for Leggett; PRO FO 371/61984, Hird, Labour conditions – Anglo-Iranian Oil Co. – Persia, 31 December 1946.
37 BP 71874, Jones to Elkington, 20 June 1947.
38 On Mylles' opposition to the co-operative proposal, see BP 16250, Mylles to Elkington, n. d. September and 19 November 1947.
39 BP 9252, Gass, Memorandum on the General Plan, 5 January 1948.
40 BP 69517, Note of meetings, 25 and 29 January 1949.
41 PRO FO 371/82402, Meeting in Mr Gee's room on 28 September 1950.
42 BP 68186, Note on housing as at 31 December 1950.
43 *Ibid.*
44 International Labour Organisation, *Labour Conditions in the Oil Industry in Iran* (Geneva, 1950), p. 31.
45 *Ibid.*, p. 58.
46 *British Medical Journal*, vol. 1 (1951), p. 101.
47 BP 71837, The Anglo-Iranian Oil Company and Iran, July 1951; International Court of Justice (ICJ), Pleadings, *Anglo-Iranian Oil Co. Case, United Kingdom v. Iran* (Leiden, 1952), p. 206.
48 On these regulations, see BP 71069, Mylles to Pattinson, 19 January 1944.
49 BP 71069, Pattinson to Jameson, 12 January 1944.
50 PRO FO 371/40158, Ward report, 12 January 1944.
51 *Ibid.*, Meeting at Ministry of Labour, 12 January 1944; PRO FO 371/40159, Picton and Barber report, Investigation into difficulties between the Anglo-Iranian Oil Company Limited and British staff employed as shift operators at the Company's oil refinery in Abadan, 13 April 1944.
52 PRO FO 371/45483, Hudson Davies, Report on operations of AIOC in Iran, spring 1945.
53 BP 72358, Mylles to Pattinson, 22 November 1944.
54 PRO FO 371/45483, Report of Hudson Davies and Picton.
55 PRO LAB 13/39, Picton note, 2 August 1945; *ibid.*, Meeting to discuss Hudson Davies report, 15 September 1945.
56 BP 70998, Fraser, Visit to Iran, October–December 1945.
57 BP 16249, Kazeruni memorandum, 21 November 1945.
58 BP 71069, Pattinson to Elkington, 25 March 1944; BP 72358, Pattinson to Elkington, 7 November 1944; BP 72127, Discussions following chairman's visit to Iran, October–December 1945.
59 BP 16249, Elkington to Jones, 14 January 1946.
60 BP 16249, Jones to Elkington, 13 and 25 March, 1946; *ibid.*, Elkington, Tudeh Party activities amongst Anglo-Iranian Oil Co. labour, March–

May 1946; *ibid.*, Rice, Some impressions of Iran, March 1946.

61 R. A. Chisholm was appointed for this purpose, BP 16249, Elkington to Northcroft, 8 April 1946 and *ibid.*, Jones to Elkington, 13 May 1946.

62 BP 16256, Elkington to Cooke, 3 June 1946.

63 PRO FO 371/52719, Le Rougetel to FO, 16 July 1946. Dr Jowdat, Tudeh Party representative, said of the strike that 'there were no industrial grounds for it and the motive must be entirely political'. See also for local impressions, *ibid.*, Trott to Tehran, 17, 18 and 20 July 1946 and the comments of Sir Clarmont Skrine, Consul General at Ahwaz, in *ibid.*, Le Rougetel to Howe, 17 July 1946. British Cabinet reaction is in PRO FO 371/52718, Cabinet conclusion, 16 July 1946. Qavam's views and the action of the local authorities are in PRO FO 371/52720, Le Rougetel to FO, 22 July 1946, and 371/52721, Le Rougetel to FO, 1 August 1946.

64 PRO FO 371/52723, Audsley to Overton, 29 July 1946; *ibid.*, Notes on visits to Persia, 31 May to 8 June and 24 June to 9 July 1946. See also PRO FO 371/52770, Audsley report in Le Rougetel to Bevin, 13 July 1946. Philip Noel-Baker, Minister of State at the Foreign Office believed, as might be expected from his political inclinations, that the Company should have given in to the demands, PRO FO 371/52721, Noel-Baker minute, Persia and the Anglo-Iranian Oil Company, to Bevin, 17 July 1946. However, Ahmad Aramesh, Acting Minister of Commerce, told trade union leaders that the demand for high wages 'would drive many factories out of existence', *ibid.*, Le Rougetel to FO, 30 July 1946.

65 PRO FO 371/52726, Bevin, Cabinet memorandum, Labour conditions in the Anglo-Iranian Oil Company, 10 October 1946; *ibid.*, Howe minute to Bevin, 21 October 1946; *ibid.*, Note of FO and Ministry of Fuel and Power, Persia, mid-October 1946. On negotiations in Tehran, PRO FO 371/52726, Howe minute to Sargent, Negotiations for minimum wage in Persia, 14 October 1946; *ibid.*, Le Rougetel to FO, 11 October 1946; *ibid.*, Le Rougetel to Baxter, 16 October 1946 and Le Rougetel to Bevin 23 and 29 October 1946; speeches of Firuz and Leggett on 11 October in *Ittila'at*, 13 October 1946.

66 The case for trade unionism has been forcefully presented by H. Ladje-vardi, *Labour Unions and Autocracy in Iran* (New York, 1985).

67 The main Company records on the disturbances of March–April 1951 are BP 68907–8, 68913, 68920. Once the preliminary actions had taken place reports were made by Embassy officials. See PRO FO 371/91455–7. Telegrams between Drake and Elkington on the conditions for ending the strike are mostly in BP 71148.

15   THE SUPPLEMENTAL AGREEMENT, 1947–1950

1 For biographies of Musaddiq, see F. Diba, *Mohammad Mossadegh, A Political Biography*, (London, 1986); H. Katouzian, *Musaddiq and the Struggle for Power in Iran* (London, 1990).

2 For a recent commentary on the Seven Year Plan in the context of Iranian development planning, see Frances Bostock and G. Jones, *Power and Planning in Iran: Ebtehaj and Economic Development under the Shah* (London, 1989). See also Overseas Consultants Inc., *Report on Seven Year Development Plan for the Plan Organization of the Imperial Government of Iran*, 5 vols. (New York, 1949).

3 PRO FO 371/62047, Le Rougetel to FO, 27 June 1947.

4 BP 80924, Seddon to Rice, 29 October 1947.

5 *Ibid.*, Gass to Fraser, 27 November 1947.

6 PRO FO 371/68704, Le Rougetel to Burrows, 2 December 1947.

7 PRO FO 371/61992, Le Rougetel to FO, 6 December 1947.

8 Royal Institute for International Affairs (RIIA), *Survey of International Affairs: The Middle East 1945–1950* (Oxford, 1954), p. 89.

9 BP 9257, Gass to Fraser, 3 December 1947.

10 PRO FO 371/61992, Le Rougetel to Burrows, 16 December 1947; see also Le Rougetel's judgement on him in *ibid.*, Le Rougetel to Bevin, Political situation in Persia, 8 December 1947; see also E. Abrahamian, *Iran Between Two Revolutions* (Princeton, 1982), pp. 240–50.

11 PRO FO 371/68704, Le Rougetel to FO, 30 December 1947; *ibid.*, Le Rougetel to FO, 3 January 1948.

12 *Ibid.*, 7 January 1948.

13 RIIA, *The Middle East*, pp. 90–2.

14 PRO FO 371/68706, Cresswell to FO, 22 June 1948.

15 See his proclamation on 14 June 1948 against Hazhir in which 'general strikes, mass demonstrations, and public protests were a duty', SD 891.6363/6, Wiley to SD, 15 June 1948. A useful introduction to Kashani's life is Yann Richard, 'Ayatollah Kashani: Precursor of the Islamic Republic', in N. Keddie (ed.), *Religion and Politics in Iran* (New Haven, 1983), pp. 101–24.

16 Rose Greaves, '1942–1976: The reign of Muhammad Riza Shah', in H. Amirsadeghi, *Twentieth-Century Iran* (London, 1977), p. 69.

17 PRO FO 371/68704, Le Rougetel to Bevin, 14 January 1948.

18 PRO FO 371/68706, Cresswell to FO, 24 June 1948.

19 *Ibid.*, 30 June 1948; he would negotiate 'in a spirit of friendliness and understanding'; *ibid.*, Cresswell to Bevin, 23 June 1948.

20 *Ibid.*, Cresswell to FO, 24 June 1948.

21 PRO FO 371/68731, Cripps to Fraser, 27 May 1948.

22 BP 101099, Fraser to Imami, 1 June.

23 PRO FO 371/68731, Berthoud to Sargent, 27 August 1948.

24 *Ibid.*, Sargent to Cripps, 30 August 1948; *ibid.*, Berthoud to Sargent, 3 September; *ibid.*, Sargent to Cripps, 4 September 1948.

25 *Ibid.*, Miss Loughanne note, AIOC to Young, 27 August 1948; *ibid.*, Clinton-Thomas note, AIOC, 1 September 1948.

26 *Ibid.*, Cripps to Bevin, 14 September 1948.

27 On Pirnia's role in the Iranian oil industry, see Husayn Pirnia, *Dah Sal*

Kushish dar Rah-i Hifz va Bast-i Huquq-i Iran dar Naft (Ten Years Struggle to Preserve and Expand Iranian Rights in Oil) (Tehran, 1952). Pirnia was the eldest son of the former Prime Minister, Mushir al-Dowla, educated at the Ecole Normale in Paris and at Cambridge University, where he submitted a doctoral thesis on finance.

28 BP 95576, Northcroft to Gass, 24 August 1948.

29 BP 71181, 25-point memorandum received on 28 September 1948.

30 Minutes of Gass's discussions in Tehran are in BP 71181.

31 PRO FO 371/68709, Burrows to Le Rougetel, 3 December 1948.

32 PRO FO 371/68732, Le Rougetel to Bevin, 21 December 1948.

33 *Ibid.*, Le Rougetel to Bevin, 30 November 1940.

34 *Ibid.*, Le Rougetel to Bevin, 21 December 1948 and *ibid.*, Wright to Butler, 21 December 1948.

35 *Ibid.*, Wright to Le Rougetel, 21 December 1948.

36 PRO FO 371/75495, Le Rougetel to FO, 12 January 1949.

37 *Ibid.*, Tehran to FO, 13 January 1949.

38 *Ibid.*, Clinton-Thomas minute, AIOC, 25 January 1949.

39 *Ibid.*, Bevin to Le Rougetel, 29 January 1949.

40 *Ibid.*, Bevin to Cripps, 18 February 1949.

41 *The Times*, 11 January 1949.

42 PRO FO 248/1489, Le Rougetel to FO, 31 January 1949; see also Northcroft's comments in BP 80924.

43 PRO FO 248/1489, Le Rougetel to Bevin, 2 February 1949 and enclosure of Taqizadeh's speech in the Majlis.

44 M. Reza Ghods, *Iran in the Twentieth Century, A Political History* (London, 1989), p. 179; Greaves, '1942–1976: The reign of Muhammad Riza Shah', in Amirsadeghi (ed.), *Iran*, pp. 69–70.

45 For the preliminary information and impressions of Gass, see BP 3B 5076, Tehran to London, 3 February 1949; *ibid.*, Gass to Fraser, 10 February 1949; BP 3B 5077, Gass to Fraser, 16 February 1949; particularly useful, BP 3B 5073, Gass diary.

46 See Northcroft advice on tackling the task as soon as possible, BP 3B 5077, Northcroft to Gass, 30 January 1949.

47 BP 3B 5086, Meeting, 13 February 1949.

48 BP 3B 5077, Gass to Fraser, 23 February 1949.

49 *Ibid.*, Fraser to Gass, 16 March 1949.

50 BP 3B 5086, Meeting, 14 February 1949.

51 BP 3B 5073, Gass diary, 24 February 1949. See BP 3B 5086 for records of the meetings.

52 BP 3B 5086, Meeting, 21 February 1949; BP 3B 5076, Gass to Fraser, 22 February 1949.

53 BP 3B 5086, Meeting, 23 February 1949; BP 3B 5076, Gass to Fraser, 24 February 1949.

54 BP 3B 5086, Meetings, 26, 27 and 28 February and 1 March 1949; BP 3B 5073, Gass diary, *passim*; BP 3B 5077, Gass to Fraser, 1 and 2 March 1949;

BP 3B 5076, Gass to Fraser, 27 and 28 February 1949.

55 BP 3B 5086, Meeting with Sa'id, 9 March 1949.

56 BP 3B 5076, Gass to Fraser, 10 March 1949.

57 *Ibid.*, Fraser to Gass, 11 March 1949.

58 BP 3B 5086, Gass meeting with Sa'id, 19 March 1949; BP 3B 5076, Gass to Fraser, 19 March 1949.

59 PRO FO 371/75496, Bevin to Le Rougetel, 22 March 1949; *ibid.*, Clinton-Thomas note, 22 March 1949; *ibid.*, Berthoud note to Bevin, 22 March 1949.

60 PRO FO 371/75495, Bevin note to Strang *et al.*, 24 March 1949.

61 PRO FO 371/75496, 6 April 1949 and BP 3B 5073, Gass diary, 6 April 1949.

62 PRO FO 371/75496, Chadwick note, 8 April 1949; BP 3B 5073, Gass diary, 8 April 1949.

63 PRO FO 371/75496, Chadwick note, 13 April 1949; BP 3B 5073, Gass diary, 12 April 1949.

64 PRO FO 371/75496, Memorandum, Anglo-Iranian Oil Company, n. d. (probably 17 April 1949).

65 *Ibid.*, Wright minute, 18 April 1949.

66 Records of the meetings are to be found in BP 3B 5073 and 3B 5086.

67 BP 3B 5086, Meeting, 26 April 1949; BP 3B 5073, Gass diary, 26 April 1949.

68 BP 3B 5086, Meeting with Sa'id, 28 April 1949.

69 BP 3B 5073, Gass diary, 2 May 1949.

70 BP 3B 5086, Meeting with Sa'id, 7 May 1949; BP 3B 5073, Gass diary, 7 May 1949.

71 BP 3B 5086, Meeting with Sa'id, 10 May 1949.

72 BP 3B 5073, Gass diary, 10 May 1949.

73 *Ibid.*, 11 May 1949.

74 *Ibid.*, Meeting with Sa'id, 11 May 1949.

75 BP 3B 5073, Gass diary, 17 May 1949.

76 PRO FO 248/1489, Bevin to Le Rougetel, 18 May 1949.

77 *Ibid.*, FO to Le Rougetel, 21 May 1949; BP 3B 5073, Gass diary, 19 May 1949.

78 PRO FO 248/1489, Le Rougetel to FO, 16 June 1949.

79 BP 3B 5073, Gass diary, 20 June 1949.

80 PRO FO 371/75458, Bevin to Le Rougetel, 1 July 1949.

81 BP 3B 5073, Gass diary, 3 July 1949.

82 *Ibid.*, 5 July 1949.

83 PRO FO 248/1489, Le Rougetel to FO, 5 July 1949.

84 BP 3B 5073, Gass diary, 6 July 1949.

85 BP 3B 5076, Gass to Fraser, 9 July 1949; BP 3B 5073, Gass diary, 8 July 1949.

86 BP 3B 5076, Gass to Fraser, 10 July 1949.

87 PRO FO 248/1489, Le Rougetel to FO, 11 July 1949.

88 BP 3B 5073, Gass diary, 9 and 10 July 1949; PRO FO 248/1489, Wheeler to Le Rougetel, 9 July 1949.

89 PRO FO 248/1489, Le Rougetel to FO, 11 July 1949.

90 BP 3B 5073, Gass diary, 11 July 1949.

91 *Ibid.*, 12 July 1949; BP 3B 5076, Gass to Fraser, 13 July 1949.

92 PRO FO 248/1489, Le Rougetel to FO, 15 July 1949; BP 3B 5073, Gass diary, 13 and 14 July 1949; BP 3B 5086, Gass to Fraser, 14 July 1949.

93 PRO FO 248/1489, Le Rougetel to FO, 15 July 1949.

94 BP 3B 5076, Gass to Fraser, 15 July 1949; BP 3B 5073, Gass diary, 14 July 1949.

95 BP 3B 5073, Gass diary, 17 July 1949.

96 B. Shwadran, *The Middle East, Oil and the Great Powers*, 2nd edn (New York, 1959), p. 105.

97 PRO FO 371/75498, Le Rougetel to FO, 17–20 July 1949.

98 PRO FO 248/1489, Le Rougetel to Attlee, 1 August 1949.

99 BP Z 0196, Hobson to Rice, 9 and 10 August 1949.

100 *Ibid.*, Hobson to Rice, 6 September 1949.

101 International Court of Justice (ICJ), Pleadings, *Anglo-Iranian Oil Co. Case, United Kingdom v. Iran* (Leiden, 1952), p. 218.

102 BP Z 0196, Hobson to Rice, 17 September 1949; *ibid.*, Rice to Hobson, 23 September 1949; PRO FO 371/75499, Lawford to FO, 18 September 1949; *ibid.*, FO to Tehran, 28 September 1949.

103 BP Z 0196, Hobson to Rice, 26 September 1949; PRO FO 371/75499, Lawford to FO, 26 September 1949.

104 BP 90819, Minutes of informal discussions, 20 October 1949; BP 3B 5074, Gass, London diary, 20 October 1949.

105 PRO FO 371/75500, Iranian *mémoire* to Secretary of State, 18 October; *ibid.*, Chadwick to Young, 19 October 1949.

106 *Ibid.*, Bevin to Le Rougetel, 26 October 1949.

107 For a first-hand American account of the Shah's visit to Washington, see G. McGhee, *Envoy to the Middle World: Adventures in Diplomacy* (New York, 1983), pp. 62–71.

108 Diba, *Mossadegh*, p. 95.

109 Abrahamian, *Iran Between Two Revolutions*, pp. 250–61.

110 BP Z 0197, Northcroft to Rice, 6 and 8 April 1950.

111 *Ibid.*, Gass to Northcroft, 14 April 1950; *ibid.*, Hawley to Bank of England, 12 April 1950; *ibid.*, Gass to Northcroft, 15 and 17 April 1950.

112 ICJ, Pleadings, *Anglo-Iranian Oil Co. Case*, p. 218.

113 Abrahamian, *Iran Between Two Revolutions*, p. 263; BP Z 0197, Northcroft to Rice, 21 and 22 June 1950.

114 BP Z 0198, Notes on Majlis committee elected to study the Supplemental Agreement Bill, 2 July 1950.

115 McGhee, *Envoy*, p. 325.

116 R. K. Ramazani, *Iran's Foreign Policy, 1941–1973* (Charlottesville, Virginia, 1975), pp. 156–7; H. Katouzian, 'Oil boycott and the political

economy: Musaddiq and the strategy of non-oil economics', in J. A. Bill and W. R. Louis (eds.), *Musaddiq, Iranian Nationalism and Oil* (Austin, Texas, 1988), p. 216.

117 PRO FO 371/82374, Shepherd to FO, 26 June 1950.

118 BP Z 0198, Northcroft to Rice, 8 July 1950; SD 888.2553/7–1350, Tehran to Secretary of State, 13 July 1950.

119 W. R. Louis, *The British Empire in the Middle East, 1945–1951* (Oxford, 1984), p. 644; PRO FO 371/82375, Barnet note, 2 August 1950; *ibid.*, Furlonge note, 11 August; PRO FO 371/82383, Wright minute, 4 August 1950; *ibid.*, Young to Furlonge, 8 August 1950.

120 SD 888.2553/7–2450, SD to London, 7 August 1950; SD 888.2553/8–950, Tehran to Secretary of State, 9 August 1950.

121 PRO FO 371/82375, Bevin to Franks, 12 August 1950.

122 BP Z 0198, Gass note, 28 September 1950; *ibid.*, Northcroft to Rice, 23 August 1950.

123 *Ibid.*, Northcroft to Rice, 4 September 1950; ICJ, Pleadings, *Anglo-Iranian Oil Co. Case*, p. 218.

124 BP Z 0198, Northcroft to Rice, 31 August, 2 and 3 September 1950.

125 ICJ, Pleadings, *Anglo-Iranian Oil Co. Case*, p. 218.

126 PRO FO 371/82343, Shepherd to FO, 2 October 1950.

127 BP Z 0198, Northcroft to Rice, 5 October 1950.

128 *Ibid.*, Northcroft to Rice, 14 and 15 October 1950.

129 *Ibid.*, Rice to Northcroft, 13 October 1950; *ibid.*, Northcroft to Rice, 19 October 1950.

130 Ramazani, *Iran's Foreign Policy*, pp. 190–1.

131 PRO FO 371/82376, Wright minute, 16 October 1950. Cited in Louis, *British Empire*, p. 645.

132 PRO FO 371/82376, Wright minute, 23 October 1950. Cited in Louis, *British Empire*, p. 645.

133 PRO FO 371/82376, Wright minute, 24 October 1950. Cited in Louis, *British Empire*, p. 646. Louis argues with elegant cynicism that Fraser's new-found emollience was based on his anticipation of the announcement of the 50:50 profit sharing agreement between Aramco and the Saudi Arabian Government at the end of 1950. This conflicts with the Foreign Office quotation which Louis cites on p. 597, complaining that the British had no advance warning of the 50:50 agreement.

134 BP Z 0200, Northcroft to Rice, 4 November 1950.

135 Husayn Makki, *Kitab-i siah* (Tehran, 1954), summarised in Ramazani, *Iran's Foreign Policy*, pp. 189–94.

136 BP Z 0200, Northcroft to Rice, 4 November 1950.

137 *Ibid.*, Northcroft to Rice, 26 November 1950.

138 SD 888.2553/12–2150 and 12–2250, Tehran to SD, 21 December 1950.

139 The best accounts of the negotiation of the Saudi 50:50 agreement are in I. H. Anderson, *Aramco, the United States and Saudi Arabia: A Study of the Dynamics of Foreign Oil Policy, 1933–1950* (Princeton, 1981), and

'The American oil industry and the fifty-fifty agreement of 1950', in Bill and Louis (eds.), *Musaddiq*, pp. 143–63.

140 On the rumours and the Company's lack of knowledge, see BP Z 0197, Northcroft to Rice, 27 June 1950; *ibid.*, Rice to Northcroft, 30 June 1950; BP Z 0198, Northcroft to Rice, 9 October 1950; *ibid.*, Rice to Northcroft, 11 October 1950; *ibid.*, Northcroft to Rice, 29 November 1950; BP Z 0200, Heath Eves to Jackson, 5 and 7 December 1950; *ibid.*, Northcroft to Jackson, 11 December 1950; *ibid.*, Rice to Northcroft, 15 December 1950; *ibid.*, Northcroft to Rice, 17 December 1950; *ibid.*, Jackson to Long, 22 December 1950; *ibid.*, Long to Jackson, 26 December 1950; PRO FO 371/91759, Fry, Note on Anglo-American relations in Saudi Arabia, 7 February 1951, cited in Louis, *British Empire*, p. 597. The archival evidence on the Company's lack of knowledge about the terms being negotiated by Aramco contrasts with McGhee's assertion that the Company 'in its anguish, accused Aramco of not having given them advance warning of the fifty-fifty agreement. Actually we had given the AIOC three months warning' in McGhee, *Envoy*, p. 325.

141 For a published appreciation of these points, see Louis, *British Empire*, pp. 598–9.

142 BP 29812, AIOC annual report, 1950, p. 7.

16 NATIONALISATION, JANUARY–JUNE 1951

1 PRO FO 371/91522, Fry minute, 6 February, 1951.

2 Cited in W. R. Louis, *The British Empire in the Middle East, 1945–1951* (Oxford, 1984), p. 650.

3 PRO FO 371/91527, Berthoud minute, 18 April 1951; PRO FO 371/91528, Franks to Strang, 21 April 1951.

4 M. Reza Ghods, *Iran in the Twentieth Century, A Political History* (London, 1989), pp. 182–4.

5 J. A. Bill, 'America, Iran and the politics of intervention, 1951–1953', in J. A. Bill and W. R. Louis (eds.), *Musaddiq, Iranian Nationalism and Oil* (Austin, Texas, 1988), p. 278.

6 Louis, *British Empire*, p. 680.

7 *Ibid.*, p. 644.

8 PRO FO 371/91459, Shepherd to Furlonge, 6 May 1951. Cited in Louis, *British Empire*, p. 652.

9 See, for example, the essays in Bill and Louis (eds.), *Musaddiq*; Louis, *British Empire*, especially pp. 651–4; F. Diba, *Mohammad Mossadegh, A Political Biography* (London, 1986); H. Katouzian, *Musaddiq and the Struggle for Power in Iran* (London, 1990); M. Elm, *Oil, Power and Principle: Iran's Oil Nationalization and its Aftermath* (New York, 1992).

10 BP Z 0200, Le Page to Northcroft, 11 January 1951.

11 *Ibid.*, Le Page to Rice, 13 January 1951.

12 On the widespread support commanded by religious leaders, see for

example, Ann Lambton, 'The impact of the West on Persia', *International Affairs*, vol. 33, no. 1 (January 1957), p. 24.

13 For some details of these meetings, see Louis, *British Empire*, pp. 649–51.

14 BP 29811, AIOC annual report, 1951, p. 8.

15 BP 3B 5072 and Z 0200, Northcroft note, January 1951.

16 BP Z 0200, Northcroft to Rice, 4 February 1951.

17 B. Shwadran, *The Middle East, Oil and the Great Powers*, 2nd edn (New York, 1959), p. 106.

18 *Ittila'at*, 22 February 1951.

19 BP Z 0200, Northcroft to Rice, 22 and 24 February 1951.

20 Rose Greaves, '1942–1976: The reign of Muhammad Riza Shah', in H. Amirsadeghi (ed.), *Twentieth-Century Iran* (London, 1977), p. 74.

21 *Parl. Debates*, House of Commons, vol. 484, col. 1261, 21 February 1951.

22 BP Z 0200, Northcroft to Rice, 25 February to 3 March 1951.

23 BP Z 0203, Seddon to Northcroft, 6 March 1951; *ibid.*, Glennie note with enclosures, 5 March 1951.

24 *Ibid.*, Northcroft to Rice, 4 and 5 March 1951.

25 *Ibid.*, Northcroft to Rice, 4 March 1951 and Razmara's speech, *Ittila'at*, 4 March 1951.

26 *Ibid.*, Northcroft to Rice, 6 March 1951.

27 Greaves, '1942–1976: The reign of Muhammad Riza Shah', in Amirsadeghi (ed.), *Iran*, p. 74.

28 BP Z 0203, Northcroft to Seddon, 7 March 1951; PRO FO 371/91452, Shepherd to FO, 7 March 1951.

29 R. K. Ramazani, *Iran's Foreign Policy, 1941–1973* (Charlottesville, Virginia, 1975), p. 197.

30 BP Z 0203, Northcroft to Gass, 7 March 1951. See also, Fu'ad Ruhani, *Tarikh-i Milli Shudan-i San'at-i Naft-i Iran* (A History of the National-isation of Oil in Iran) (Tehran, 1973).

31 *Correspondence between His Majesty's Government in the United Kingdom and the Persian Government, and Related Documents concerning the Oil Industry in Persia, February to September 1951* (Cmd 8425, 1951), p. 25.

32 BP Z 0203, Northcroft to Seddon, 12 March 1951.

33 Cmd 8425, pp. 25–7.

34 Shwadran, *The Middle East*, p. 106.

35 *Ibid.*

36 PRO FO 371/91524, Morrison minute, 15 March 1951; *ibid.*, Ross note, 19 March 1951; *ibid.*, Berthoud note on Persian oil, 16 March 1951.

37 PRO FO 371/91525, Strang note, 19 March 1951.

38 PRO FO 371/91524, FO to MoD, 20 March 1951; PRO FO 371/91525, Minutes of meeting at FO, 20 March 1951.

39 PRO FO 371/91471, Franks to FO, 16 and 17 April 1951.

40 British correspondence on the talks is in PRO FO 371/91471. For US documentation, see SD 888.2553/4–1751.

41 PRO FO 371/91528, Furlonge note, 21 April 1951.
42 PRO FO 371/91529, Strang to Franks, 10 April 1951.
43 Cmd 8425, pp. 28–9; PRO FO 371/91528, Shepherd to FO, 27 April 1951.
44 Greaves, '1942–1976: The reign of Muhammad Riza Shah', in Amir-sadeghi (ed.), *Iran*, pp. 75–7.
45 For correspondence and memoranda about the strikes, see BP 33732, 68908 and 71148.
46 BP 67241, Appendix I to Fields withdrawal report, August 1951.
47 BP 71148, Elkington to Drake, 23 April 1951.
48 *Parl. Debates*, House of Commons, vol. 487, cols. 1011–12, 1 May 1951.
49 *Ibid.*, col. 146, 2 May 1951.
50 Cmd 8425, pp. 31–2.
51 *Ibid.*, pp. 32–3.
52 *Ibid.*, pp. 33–4.
53 SD Secret file CS/S Discussion of US oil company reps. with McGhee, 14 May 1951.
54 Statement by the Department of State in Y. Alexander and A. Nanes (eds.), *The United States and Iran: A Documentary History* (Frederick, Maryland, 1982), pp. 215–26.
55 *The Times*, 22 May 1951.
56 Cmd 8425, pp. 34–6.
57 BP Z 0168; Cmd 8425, p. 36.
58 Cmd 8425, p. 37.
59 *Ibid.*, pp. 37–8.
60 *Ibid.*, pp. 38–9.
61 BP 100557, Shepherd to FO, 30 May 1951.
62 Louis, *British Empire*, p. 662.
63 PRO FO 371/91538, Brief for Cabinet meeting prepared by Furlonge, 28 May 1951.
64 *Parl. Debates*, House of Commons, vol. 488, col. 42, 29 May 1951, Herbert Morrison.
65 Cmd 8425, pp. 39–41.
66 PRO FO 371/91542, Truman to Musaddiq, 31 May 1951; PRO FO 371/91543, Truman to Attlee, 31 May 1951. Truman's letter to Attlee, addressed simply to the Prime Minister, was also sent to Grady, who mistakenly (it was claimed) forwarded it to Musaddiq.
67 Shwadran, *The Middle East*, pp. 113–14.
68 BP Z 0209, Seddon to Drake, 7 June 1951.
69 BP 101108, Drake to Elkington, 6 June 1951.
70 BP Z 0209, Drake to Seddon, 10 and 11 June 1951; *ibid.*, Green to Seddon, 12 June 1951.
71 BP 100171, Drake's general management (commercial no. 2) file, 19 June–31 August, Diary of events (document 194b).
72 *Ibid.*

73 PRO FO 371/91545, Shepherd to FO, 14 June 1951; BP 3B 5070, Minutes of first meeting, 14 June 1951; PRO FO 371/91546, Logan note, Persian oil, 16 June 1951; *ibid.*, Shepherd to FO, 15 June 1951.

74 BP 100557, FO to Tehran, 16 June 1951.

75 BP 3B 5070, Fateh note of meeting with Musaddiq, 18 June 1951.

76 Cmd 8425, p. 42.

77 BP 3B 5070, *Aide-mémoire*, 19 June 1951.

78 *Daily Telegraph*, 20 June 1951.

79 BP 100171, Drake's general management (commercial no. 2) file, 19 June–31 August 1951, Minutes of a meeting in Khurramshahr, 19 June 1951.

80 *Ibid.*, Drake to Jackson, telegram 1, 19 June 1951.

81 International Court of Justice (ICJ), Pleadings, *Anglo-Iranian Oil Co. Case, United Kingdom v. Iran* (Leiden, 1952), p. 404.

82 *The Times* and *New York Times*, 21 June 1951.

83 *Parl. Debates*, House of Commons, vol. 489, cols. 522–3, Herbert Morrison, 20 June 1951.

84 PRO FO 371/91554, Strang note, 20 June 1951.

85 PRO FO 371/91556, Strang note, 23 June 1951.

86 ICJ, Pleadings, *Anglo-Iranian Oil Co. Case*, p. 404.

87 BP 100171, Drake's general management (commercial no. 2) file, 19 June–31 August 1951, Drake to Fraser, telegrams 3157–9, 21 June 1951.

88 *Ibid.*, Proclamation, 20–21 June 1951.

89 *Ibid.*, Jackson to Drake, 21 June 1951.

90 *Ibid.*, Minutes of a meeting at Khurramshahr, 21 June 1951.

91 *Ibid.*, Drake to Fraser, telegram 4, 22 June 1951.

92 *Ibid.*, Drake to Fraser, telegram 3160, 21 June 1951.

93 *Ibid.*, Provisional board to Drake, 23 June 1951.

94 *Ibid.*, Minutes of a meeting at Khurramshahr, 25 June 1951.

95 *Ibid.*, Drake to Fraser, telegram 3178, 25 June 1951.

96 *Ibid.*, Mason to Fraser, telegram 3179, 25 June 1951.

97 *Parl. Debates*, House of Commons, vol. 488, col. 30, 30 May 1951; Royal Institute for International Affairs (RIIA), *Survey of International Affairs, 1951* (Oxford, 1954), p. 311.

98 *Parl. Debates*, House of Commons, vol. 489, cols. 1184–6, 26 June 1951.

99 Mohammad Riza Shah Pahlavi, *Mission for My Country* (London, 1961), p. 91.

100 RIIA, *Survey*, p. 317.

101 BP 100171, Drake's general management (commercial no. 2) file, Note, L. H. Baxter, interpreting, 28 June 1951.

102 Cmd 8425, pp. 42–4.

103 PRO FO 371/91464, Shepherd note, A comparison between Persian and Asian nationalism in general, 2 October 1951. Cited in Louis, *British Empire*, p. 640.

17 EVICTION, JUNE–OCTOBER 1951

1 International Court of Justice (ICJ), Pleadings, *Anglo-Iranian Oil Co. Case, United Kingdom v. Iran* (Leiden, 1952), p. 424.
2 *Correspondence between His Majesty's Government in the United Kingdom and the Persian Government and Related Documents concerning the Oil Industry in Persia, February 1951 to September 1951* (Cmd 8425, 1951), p. 47.
3 *Ibid.*, pp. 45–9, ICJ interim order, 5 July 1951.
4 *Ibid.*, p. 51, Shepherd to Kazimi, 7 July 1951; *ibid.*, p. 52, Musaddiq to Shepherd, 7 September 1951; ICJ, Pleadings, *Anglo-Iranian Oil Co. Case*, p. 718.
5 PRO FO 371/91555, Franks to FO, 4 July 1951. See also D. Acheson, *Present at the Creation: My Years in the State Department* (New York, 1969), pp. 507–8.
6 PRO FO 371/91555, Morrison to Acheson, 7 July 1951.
7 *Ibid.*, Morrison to Franks, 7 July 1951.
8 Y. Alexander and A. Nanes (eds.), *The United States and Iran: A Documentary History* (Frederick, Maryland, 1980), pp. 218–19.
9 *Ibid.*, p. 220.
10 BP 5E 4082, Shepherd to FO, 4 July 1951.
11 *Ibid.*
12 *Ibid.*, Khurramshahr to Tehran, 4 July 1951.
13 *Ibid.*, Khurramshahr to Tehran, 15 July 1951.
14 See, for example, BP 100171, Drake's general management (commercial no. 2) file, 19 June–31 August 1951, Incidents Abadan, 10/11 July 1951 (document 154) and Abadan Sitrep, 12 July 1951 (document 155).
15 A. W. Ford, *The Anglo-Iranian Oil Dispute of 1951–1952* (Berkeley, 1954), p. 99.
16 For a published account of the Harriman mission and subsequent events described in this chapter see, M. Elm, *Oil, Power and Principle: Iran's Oil Nationalization and its Aftermath* (New York, 1992), ch. 8.
17 BP 5E 4082, Shepherd to FO, 16 July 1951.
18 *Ibid.*
19 *Ibid.*, Shepherd to FO, 17 July 1951.
20 *Ibid.*, Shepherd to FO, 19 July 1951.
21 PRO FO 371/91567, Shepherd to FO, 20 July 1951.
22 BP 5E 4082, Shepherd to FO, 22 July 1951.
23 Cmd 8425, pp. 52–3, Harriman formula, 23 July 1951.
24 BP 5E 4082, FO to Tehran, 26 July 1951.
25 *Ibid.*
26 Cmd 8425, p. 53, Middleton to Kazimi, 3 August 1951.
27 *Ibid.*
28 Acheson, *Present at the Creation*, p. 509.

29 PRO FO 371/91575, Ramsbotham record of meeting in the Ministry of Fuel and Power on 29 July 1951.

30 *Ibid.*

31 *Ibid.*

32 PRO FO 371/91573, Strang to Morrison, 29 July 1951.

33 PRO FO 371/91575, Gass to Bowker, 31 July 1951 with enclosure, Agenda for a working basis of co-operation between the Iranian Government and Anglo-Iranian.

34 *Ibid.*, Flett note, 31 July 1951.

35 *Ibid.*, Ramsbotham record of meeting in the room of the Chancellor of the Exchequer on 31 July 1951.

36 *Ibid.*

37 The officials who accompanied Stokes were Sir Donald Fergusson, Permanent Secretary of the Ministry of Fuel and Power; M. T. Flett of the Treasury; P. Ramsbotham of the Foreign Office; and Dr W. Nuttal, technical adviser of the Ministry of Fuel and Power.

38 BP 43859, Northcroft diary, 1 August 1951. A sulphur dioxide plant remained in operation at Abadan, treating kerosene.

39 BP 100387, Stokes to London, telegram 1038, 5 August 1951.

40 *Ibid.*, Stokes to London, telegram 1039, 5 August 1951.

41 *Ibid.*, Stokes to London, 8 August 1951; BP 43859, Northcroft diary, 8 August 1951.

42 BP 43859, Elkington to Fraser, 10 August 1951.

43 *Ibid.*

44 *Ibid.*

45 BP 100387, Stokes to London, telegram 1076, 11 August 1951.

46 *Ibid.*

47 *Ibid.*, Stokes to London, telegram 1077, 11 August 1951.

48 *Ibid.*, Stokes to London, telegram 1085, 12 August 1951.

49 *Ibid.*, Stokes to London, telegram 1082, 12 August 1951.

50 *Ibid.*, FO to Tehran, 13 August 1951.

51 *Ibid.*, Stokes to London, telegram 1085, 12 August 1951.

52 *Ibid.*, Stokes to London, 13 August 1951. For the text of the eight-point memorandum, see Cmd 8425, pp. 54–5.

53 Cmd 8425, pp. 55–7, Reply of the Persian delegation to the proposals of the British delegation, 18 August 1951.

54 *Ibid.*

55 BP 100387, Tehran to London, 15 August 1951.

56 *Ibid.*, Stokes telegram, 19 August 1951; BP 43859, Harriman to Musaddiq, 21 August 1951.

57 BP 43859, Northcroft diary, 22 August 1951.

58 For a detailed narrative, see Northcroft's diary of events in BP 43859 and telegrams in BP 100387. The letter in which Stokes withdrew his proposals is in Cmd 8425, p. 57.

59 BP 43859, Northcroft diary, 24 August 1951.

60 PRO FO 371/91581, FO to Shepherd, 28 August 1951. On Harriman's views on the oil dispute at this time, see also SD 888.2553/8–2551, Harriman to Truman, 22 August 1951; SD 888.2553/8–2451, Harriman to Acheson, 24 August 1951.
61 PRO FO 371/91581, FO to Shepherd, 28 August 1951.
62 PRO FO 371/91621, Note on AIOC management, 28 August 1951.
63 PRO FO 371/91584, Berthoud record of conversations, 28 August 1951.
64 *Ibid.*
65 Acheson, *Present at the Creation*, p. 509.
66 BP 100652, Shepherd to FO, 6 September 1951; BP 43859, Northcroft diary, 6 September 1951.
67 BP 100652, FO to Shepherd, 6 September 1951; BP 43859, Northcroft diary, 7 September 1951.
68 BP 100652, London to Athens, 5 September 1951.
69 BP 100387, FO to Shepherd, 30 August 1951.
70 Of special importance in the consultations between Britain and the USA on these matters was a meeting between Morrison, Harriman and Acheson on 11 September. See BP 100652, Morrison to FO, 11 September 1951.
71 BP 100652, FO to Tehran, 10 September 1951.
72 *Ibid.*, London to Rangoon, 10 September 1951.
73 *Ibid.*, Shepherd to FO, 12 September 1951.
74 *Ibid.*, Washington to London, 12 September 1951.
75 BP 43859, Northcroft diary, 10 September 1951.
76 Cmd 8425, pp. 60–1, Musaddiq to Harriman, 12 September 1951.
77 *Ibid.*, pp. 62–4, Harriman to Musaddiq, 15 September 1951.
78 BP 100652, Shepherd to FO, 20 September 1951.
79 Cmd 8425, pp. 64–5, Text of the document hand by Ala to Shepherd on 19 September 1951.
80 BP 100652, Shepherd to FO, 18 September 1951. There were also other occasions on which Shepherd reported that the Shah was too timorous and vacillating to change the Government. See *ibid.*, Shepherd to FO, 12 September 1951; *ibid.*, Shepherd to Strang, 15 September 1951; *ibid.*, Shepherd to FO, 19 September 1951.
81 See, for example, BP 100652, Shepherd to FO, 10 September 1951; PRO FO 371/91589, Middleton to FO, 18 September 1951.
82 PRO FO 371/91591, Younger note, The Persian situation, 21 September 1951.
83 BP 100652, Franks to FO, 21 September 1951.
84 *Ibid.*, FO to Shepherd, 22 September 1951.
85 Cmd 8425, pp. 65–6, Shepherd to Ala, 22 September 1951.
86 BP 100652, Shepherd to FO, 25 September 1951; BP 43859, Northcroft diary, 26 September 1951.
87 W. R. Louis, *The British Empire in the Middle East 1945–1951* (Oxford, 1984), pp. 686–8.
88 BP 100652, Shepherd to FO, 27 September 1951.

89 For a published account of these events, see B. Shwadran, *The Middle East, Oil and the Great Powers*, 3rd edn (New York, 1973), p. 104.

90 BP 100572, Khurramshahr to Tehran, 3 October 1951; BP 43859, Northcroft diary, 4 October 1951.

18 FAILED INITIATIVES, OCTOBER 1951–AUGUST 1953

1 H. Katouzian, *Musaddiq and the Struggle for Power in Iran* (London, 1990), ch. 11. Also Katouzian, 'Oil boycott and the political economy: Musaddiq and the strategy of non-oil economics' in J. A. Bill and W. R. Louis (eds.), *Musaddiq, Iranian Nationalism and Oil* (Austin, Texas, 1988).

2 G. McGhee, *Envoy to the Middle World: Adventures in Diplomacy* (New York, 1983), p. 388.

3 B. Shwadran, *The Middle East, Oil and the Great Powers*, 3rd edn (New York, 1973), p. 105.

4 *Ibid.*

5 *Ibid.*, pp. 106–7.

6 D. Acheson, *Present at the Creation: My Years in the State Department* (New York, 1969), p. 510.

7 BP 100572, Franks to FO, 11 October 1951.

8 *Ibid.*, Franks to FO, 11 and 17 October 1951.

9 McGhee, *Envoy*, ch. 27 esp. pp. 323–4 and 333.

10 *Ibid.*, ch. 31 esp. p. 390.

11 *Ibid.*, pp. 390–1.

12 *Ibid.*, pp. 390–2.

13 *Ibid.*, pp. 398–9.

14 Acheson, *Present at the Creation*, p. 510.

15 *Ibid.*, p. 501.

16 *Ibid.*, p. 504.

17 PRO FO 371/91607, Minutes of meeting in Bridges' room, 23 October 1951.

18 Cited in M. Elm, *Oil, Power and Principle: Iran's Oil Nationalization and its Aftermath* (New York, 1992), p. 199.

19 See, for example, W. R. Louis, *The British Empire in the Middle East 1945–1951* (Oxford, 1984), pp. 632–89.

20 A. Eden, *Full Circle: The Memoirs of Anthony Eden* (London, 1960), pp. 189–92.

21 Acheson, *Present at the Creation*, p. 511.

22 BP 100572, Washington to FO, 30 October 1951.

23 *Ibid.*

24 McGhee, *Envoy*, p. 402.

25 BP 100572, Seddon note on meeting at the Treasury on 6 November 1951.

26 *Ibid.* A Treasury note recording the same meeting is also in BP 100572.

27 Eden, *Full Circle*, p. 200.

28 *Ibid.*, p. 201.
29 *Ibid.*
30 PRO FO 371/91610, Churchill to Eden, 8 November 1951.
31 BP 100571, Statement by International Bank on its negotiations with the Persian Government, 4 April 1952.
32 *Ibid.*
33 BP 71836, Gass to Fraser, 2 January 1952.
34 BP 101074, Garner to Musaddiq, 28 December 1951.
35 *Ibid.*, Musaddiq to Garner, 3 January 1952.
36 *Ibid.*, Snow record note of telephone conversation between Fraser and Gass on 4 January 1952.
37 *Ibid.*, Gass to Fraser, 11 January 1952.
38 *Ibid.*
39 *Ibid.*
40 *Ibid.*, Gass to Jackson, 22 January 1952.
41 *Ibid.*, Gass to Jackson, 23 January 1952.
42 *Ibid.*
43 *Ibid.*, Gass to Fraser, telegrams 118 and 120, 26 January 1952.
44 *Ibid.*, Gass to Fraser, 2 February 1952.
45 See, for example, BP 100571, London to Washington, 11 February 1952.
46 Elm, *Oil*, p. 256.
47 BP 100571, Tehran to London, 19 February 1952.
48 *Ibid.*, Statement by International Bank on its negotiations with the Persian Government, 4 April 1952.
49 *Ibid.*, Middleton to London, 25 February 1952.
50 *Ibid.*
51 *Ibid.*, Statement by International Bank on its negotiations with the Persian Government, 4 April 1952.
52 PRO FO 371/98687, Petroleum Division note of discussion on Anglo-Persian relations held at the Ministry of Fuel and Power on 17 March 1952; *ibid.*, Middleton to Ross, 17 March 1952; *ibid.*, Franks to FO, 31 March 1952; *ibid.*, Ramsbotham minute, 9 April 1952; PRO FO 371/98688, Logan note on Persia, US State Department's views, 16 April 1952.
53 PRO FO 371/98687, Franks to FO, 31 March 1952; PRO FO 371/98688, Ross, Next steps in Persia, 23 April 1952; *ibid.*, Franks to FO, 29 and 30 April 1952; *ibid.*, Logan minute, 3 May 1952; PRO FO 371/98690, Makins note on Persian oil dispute, 21 June 1952.
54 PRO FO 371/98689, Strang to Eden, 2 May 1952.
55 BP 100571, Note on Washington telegrams received on 2 May 1952.
56 For a blow-by-blow documentary coverage of these events, see PRO FO 371/98690.
57 *Ibid.*, Fergusson note of interview with Fraser and Gass on 15 July 1952.
58 International Court of Justice (ICJ), Pleadings, *Anglo-Iranian Oil Co. Case, United Kingdom v. Iran* (Leiden, 1952).
59 Elm, *Oil*, pp. 239–40.

60 *Ibid.*, p. 241; BP 100571, Tehran to London, 17 July 1952.
61 BP 100571, London to Washington, 18 July 1952.
62 BP 101912, Minutes of Persia (Official) Committee meeting on 18 July 1952.
63 Elm, *Oil*, pp. 243–7; Shwadran, *The Middle East*, p. 110.
64 PRO FO 371/98691, Middleton to FO, 27 July 1952.
65 BP 100571, Middleton to FO, 25 July 1952.
66 *Ibid.*, Middleton to FO, 27 July 1952.
67 *Ibid.*, Middleton to FO, No. 45, 28 July 1952.
68 *Ibid.*, Middleton to FO, No. 46, 28 July 1952.
69 *Ibid.*, Franks to FO, 31 July 1952.
70 BP 46596, Eden to Acheson, 9 August 1952.
71 *Ibid.*, Acheson to Eden, 13 August 1952.
72 PRO FO 371/98693, Churchill to Truman, 23 August 1952.
73 *Ibid.*, Truman to Churchill, 24 August 1952.
74 *Ibid.*, Churchill to Truman, 25 August 1952.
75 The text of the Churchill–Truman proposals is in *Correspondence between Her Majesty's Government in the United Kingdom and the Iranian Government, and Related Documents, concerning the Joint Anglo-American Proposals for Settlement of the Oil Dispute, August 1952 to October 1952* (Cmd 8677, 1952), p. 3.
76 BP 46596, Note on meeting with Musaddiq on 27 August drafted jointly by Middleton and Henderson.
77 *Ibid.*, Joint telegram drafted by Middleton and Henderson, n. d.; PRO FO 371/98695, Middleton to FO, 30 August 1952.
78 BP 46596, Tehran to London, 7 September 1952.
79 *Ibid.*, Tehran to London, 17 September 1952.
80 Cmd 8677, pp. 4–8, Musaddiq to Churchill, 24 September 1952. The same letter was sent to Truman. See *ibid.*, p. 9.
81 *Ibid.*, p. 9, Eden to Musaddiq, 5 October 1952; *ibid.*, pp. 9–10, Acheson to Musaddiq, 5 October 1952.
82 *Ibid.*, pp. 10–11, Musaddiq to Eden, 7 October 1952; *ibid.*, pp. 11–12, Musaddiq to Acheson, 7 October 1952.
83 BP 46596, Washington to London, Nos. 98–100, 9 October 1952; *ibid.*, FO to Washington, 11 October 1952; *ibid.*, Washington to London, 11 October 1952; *ibid.*, Glennie to Gass, 13 October 1952; *ibid.*, Washington to London, 12 October 1952; *ibid.*, London to Washington, 12 October 1952.
84 BP 101912, Minutes of Persia (Official) Committee meeting on 13 October 1952.
85 Cmd 8677, pp. 12–14, Note handed to Musaddiq by Middleton on 14 October 1952.
86 Shwadran, *The Middle East*, p. 112.
87 BP 101912, Minutes of Persia (Official) Committee meeting on 21 October 1952.

88 Documents on the Anglo-American discussions at this time can be found in BP 46596.
89 BP 100570, Henderson to SD giving the draft for a proposed Persian communiqué, 6 January 1953.
90 BP 46596 and 100570, Henderson to SD reporting a further conversation with Musaddiq on 31 December 1952.
91 BP 100570, Henderson to SD reporting a conversation with Musaddiq on 2 January 1953.
92 *Ibid.*, London to Washington, 14 January 1953.
93 *Ibid.*, Draft telegram London to Washington, 8 January 1953; *ibid.*, London to Washington, 14 January 1953.
94 *Ibid.*, Acheson to Gifford and Byroade, 13 January 1953.
95 *Ibid.*, Message from Henderson, 17 January 1953.
96 *Ibid.*
97 *Ibid.*, Henderson to SD reporting his interview with Musaddiq on 19 January 1953.
98 PRO FO 371/104610, Eden to Makins, 23 January 1953.
99 *Ibid.*
100 BP 100570, Washington to London, 27 January 1953.
101 PRO FO 371/104611, Henderson to SD, 28 January 1953.
102 *Ibid.*
103 The views of the State Department and the British Government are documented in BP 100570.
104 *Ibid.*, London to Washington, 18 February 1953.
105 *Ibid.*, Washington to London, 18 February 1953.
106 *Ibid.*, London to Washington, 19 February 1953.
107 *Ibid.*, Paraphrase of telegram from Henderson giving a full report of his conversation with Musaddiq on 20 February 1953.
108 *Ibid.*, Washington to London, 24 February 1953.
109 Elm, *Oil*, p. 294; Shwadran, *The Middle East*, p. 113; Rose L. Greaves, 'The record of the British Petroleum Company Ltd' (BP internal record, vol. 3, 1970), p. 71.
110 Elm, *Oil*, p. 295.
111 BP 100570, Paraphrases of telegram from Henderson reporting his interview with Musaddiq on 4 March 1953.
112 BP 101911, Minutes of Persia (Official) Committee meeting on 30 January 1953.
113 These events are documented in BP 100570.
114 Elm, *Oil*, pp. 290–1; Shwadran, *The Middle East*, pp. 114–15.
115 For accounts of the coup by participants see, C. M. Woodhouse, *Something Ventured* (London, 1982), pp. 105–35; Kermit Roosevelt, *Countercoup: The Struggle for Control of Iran* (New York, 1979). For later comment by historians, see J. A. Bill, 'America, Iran and the politics of intervention, 1951–1953' in Bill and Louis (eds.), *Musaddiq*; J. A. Bill, *The Eagle and the Lion: The Tragedy of American-Iranian Relations* (New Haven, 1988); Elm, *Oil*, ch. 20. Elm states on pp. 293 and 297 that

the Company initiated the coup, giving as evidence an article by Robert Scheer in the *Los Angeles Times*, 29 March 1979. Scheer's article was based on an interview with Kermit Roosevelt, who was in charge of the operations of the US Central Intelligence Agency in the Middle East at the time of the coup. However, Roosevelt has proved to be an unreliable witness. His *Countercoup* (1979), which does not mention the Company, is described by J. A. Bill as 'very careless in its detail and must be used with extreme care' in Bill and Louis (eds.), *Musaddiq*, p. 294, reference no. 64. In an earlier edition of *Countercoup*, withdrawn from publication, Roosevelt had stated that the Company was involved, but he later admitted to Sir Denis Wright (the British Chargé d'Affaires on the resumption of diplomatic relations with Iran in 1953 and later Ambassador to Tehran) that he had 'mentioned AIOC rather than embarrass M16'. Wright himself believed that the Company was not involved. These remarks are based on: Bamberg interview with Wright, 25 August 1992; BP 112228, Wright to Elm, 20 August 1992.

116 See, Elm, *Oil*, ch. 22; Bill, 'America, Iran and the politics of intervention', in Bill and Louis (eds.), *Musaddiq*.

19 THE CONSORTIUM, AUGUST 1953–OCTOBER 1954

1 After his release from prison Musaddiq was confined to his estate at Ahmadabad until his death in 1967. On his trial, see H. Katouzian, *Musaddiq and the Struggle for Power in Iran* (London, 1990), ch. 14.

2 B. Shwadran, *The Middle East, Oil and the Great Powers*, 3rd edn (New York, 1973), p. 143.

3 A. Eden, *Full Circle: The Memoirs of Anthony Eden* (London, 1960), pp. 215–16.

4 PRO CAB 134/1149, Persia (Official) Committee, 23 October 1953.

5 SD 888.2553/10–2253, Tehran to Secretary of State, 22 October 1953; SD 888.2553/10–2453, Tehran to Secretary of State, 24 October 1953; SD 888.2553/10–2653, Tehran to Secretary of State, 26 October 1953; SD 888.2553/10–2953, Tehran to Secretary of State, 29 October 1953. Records of the meetings and the memorandum presented by Hoover can also be found in BP 90653 and 100711.

6 SD 888.2553/11–253, Memorandum, Iranian oil, Tehran to Secretary of State, 2 November 1953.

7 PRO FO 371/104642, Belgrave minute, Persia, 2 November 1953.

8 PRO CAB 134/1148, Minutes of Persia (Official) Committee meeting on 4 November 1953.

9 *Ibid.*, Minutes of Persia (Official) Committee meeting on 7 November 1953.

10 *Ibid.*

11 *Ibid.*, Minutes of Persia (Official) Committee on 4 November 1953.

12 *Ibid.*

13 *Ibid.*, Minutes of Persia (Official) Committee meeting on 7 November, Annex A.

14 *Ibid.*, Minutes of Persia (Official) Committee meeting on 26 November 1953.
15 *Ibid.*, Minutes of Persia (Official) Committee meeting on 27 November 1953.
16 *Ibid.*, Minutes of Persia (Official) Committee meeting on 30 November 1953.
17 *Ibid.*
18 BP 58246, Letter of 3 December 1953 sent by Fraser to other oil majors and CFP.
19 SD 888.2553/12–1253, Aldrich to Dulles, 12 December 1953.
20 BP 90653, Unsigned note on discussions between the oil companies, n. d.; *ibid.*, Minutes of Persia (Official) Committee meeting on 18 December 1953.
21 BP 94794, Persia, Meeting No. 1 on 14 December 1953; *ibid.*, Persia, Meeting No. 2 on 14 December 1953; *ibid.*, Meeting No. 3 on 15 December 1953; *ibid.*, Meeting No. 4 on 17 December 1953.
22 PRO FO 371/104587, FO to Wright, 17 December 1953.
23 BP 100711, Wright to London, 29 December 1953.
24 BP 79673, Wright to London, 6 January 1954.
25 *Ibid.*, Wright to London, 7 January 1954.
26 PRO CAB 134/1148, Minutes of Persia (Official) Committee meeting on 8 January 1954.
27 *Ibid.*
28 *Ibid.*, Minutes of Persia (Official) Committee meeting on morning of 11 January 1954.
29 *Ibid.*
30 *Ibid.*, Minutes of Persia (Official) Committee meeting on afternoon of 11 January 1954.
31 BP 106235, Minutes of AIOC board meeting on 14 January 1954.
32 PRO FO 371/110046, FO to Washington, 13 and 14 January 1954; *ibid.*, Dixon to Gass, 15 January 1954.
33 *Ibid.*, Makins to FO, 21 January 1954.
34 *Ibid.*, FO to Wright, 22 January 1954; *ibid.*, Outward telegram from Commonwealth Relations Office, 27 January 1954.
35 BP 101904, Meeting of Middle East Oil Committee on 25 January 1954. The Middle East Oil Committee's terms of reference were 'to keep under review Government policy in regard to oil from the Middle East'. PRO CAB 134/1084, Middle East Oil Committee, 15 January 1954.
36 BP 79673, Letter of 26 January 1954 sent by Fraser to other oil majors and CFP; BP 5E 4066, Gass to Butler, 26 January 1954; *ibid.*, Hopwood to Fraser, 27 January 1954; *ibid.*, Follis to Fraser, 27 January 1954; *ibid.*, Jennings to Fraser, 28 January 1954; *ibid.*, Swensrud to Fraser, 28 January 1954; *ibid.*, Leach to Fraser, 28 January 1954; *ibid.*, Rathbone to Fraser, 29 January 1954.
37 Records of the working party meetings are in BP 5E 4066 and BP 94794.

38 PRO FO 371/110046, Unsigned note, 10 March 1954. The full reports of the technical mission are in BP 4591.

39 PRO FO 371/110046, Belgrave record of conversation on Persian oil with Moline, 15 January 1954; *ibid.*, Makins to FO, 19 January 1954; *ibid.*, Dixon note for oral statement in Cabinet, 20 January 1954; *ibid.*, Dixon note, 25 January 1954.

40 BP 101904, Middle East Oil Committee meeting on 29 January 1954.

41 *Ibid.*, Middle East Oil Committee – Persian oil, 29 January 1954.

42 *Ibid.*, Discussion on points arising from the meeting with Hoover on 29 January 1954.

43 *Ibid.*

44 PRO FO 371/110047, Makins to FO, 19 February 1954.

45 *Ibid.*, Eden to Makins, 22 February 1954.

46 *Ibid.*, Makins to FO, 23 February 1954.

47 *Ibid.*, Makins to FO, 25 February 1954.

48 BP 106235, Minutes of AIOC board meeting on 1 March 1954.

49 *Ibid.*

50 PRO FO 371/110047, FO to Washington, 2 March 1954; PRO FO 371/110048, Makins to FO, 4 March 1954.

51 BP 79661, Iran, Basis for the settlement with Anglo-Iranian, 12 March 1954.

52 *Ibid.*, Iran, American Group's views on 'Basis for the settlement with the Anglo-Iranian', 12 March 1954.

53 *Ibid.*, Iran, Basis for the settlement with Anglo-Iranian, 12 March 1954; *ibid.*, Iran, American Group's further views on 'Basis for the settlement with Anglo-Iranian', 16 March 1954.

54 *Ibid.*, Iran, Basis for the settlement with Anglo-Iranian, 14 and 17 March 1954; *ibid.*, Iran, American Group's further views on 'Basis for the settlement with Anglo-Iranian', 16 March 1954.

55 PRO FO 371/110048, Makins to FO, telegram 447, 17 March 1954.

56 *Ibid.*, Makins to FO, telegram 448, 17 March 1954.

57 *Ibid.*, Makins to FO, telegram 449, 17 March 1954.

58 *Ibid.*, FO to Washington, 18 March 1954. For an explanation of discounted cash flow techniques, see for example, S. Lumby, *Investment Appraisal*, 2nd edn (Wokingham, England, 1984), ch. 4.

59 Elm, *Oil*, pp. 317–18.

60 PRO FO 371/110048, FO to Washington, telegrams 1024 and 1025, 18 March 1954.

61 BP 106235, Minutes of AIOC board meeting on 19 March 1954.

62 *Ibid.*, Minutes of AIOC board meeting on 20 March 1954.

63 *Ibid.*

64 PRO FO 371/110049, Eden to Makins, 20 March 1954.

65 *Ibid.*, FO to Washington, telegrams 1121 and 1122, 23 March 1954.

66 *Ibid.*, FO to Washington, 26 March 1954.

67 BP 106235, Minutes of AIOC board meeting on 30 March 1954.

68 BP 79660, Memorandum of Understanding, 9 April 1954.
69 The new entrants were Richfield Oil Corporation; American Independent Oil Co., itself a consortium of ten US independents; Standard of Ohio; Getty Oil Company; Signal Oil and Gas Company; Atlantic Refining; Hancock Oil Company; and San Jacinto Petroleum Corporation. S. H. Longrigg, *Oil in the Middle East: Its Discovery and Development*, 3rd edn (London, 1968), pp. 278–9.
70 Minutes of the meetings are in BP 79671.
71 A full record of the meetings is in BP 94788, with supporting documents in BP 100720.
72 *Ibid.*
73 *Ibid.*
74 Minutes of the inter-company meetings are in BP 94794.
75 BP 94794, Record of decisions taken by Group principals at meetings held in London, 16 June 1954.
76 BP 100547, Diary of events.
77 *Ibid.*
78 Minutes of the meetings are in BP 94787 and 101896.
79 Minutes of the meetings are in BP 79671.
80 BP 79674, *Aide-mémoire* on compensation settlement as initialled in Tehran on 4 August 1954.
81 BP 58303, Statement by AIOC issued in London on 5 August 1954.
82 *Ibid.*, Copy of joint statement by Persian Government negotiators and oil consortium negotiators issued in Tehran on 5 August 1954.
83 See, generally, BP 100547, Diary of events.
84 *Ibid.*
85 BP 112226, Fraser's circular to shareholders, October 1954.
86 BP 100547, Diary of events.
87 Elm, *Oil*, pp. 330–1.

20   CONCLUSION

1 A. Sampson, *The Seven Sisters: The Great Oil Companies and the World They Made* (London, 1975). As another example of this view, see J. M. Blair, *The Control of Oil* (New York, 1976).
2 Sampson, *Seven Sisters*, pp. 24–5.
3 For a general description of the rise of oil as an energy source, see J. G. Clark, *The Political Economy of World Energy: A Twentieth Century Perspective* (Hemel Hempstead, 1990), chs. 3 and 4.
4 D. Acheson, *Present at the Creation: My Years in the State Department* (New York, 1969), p. 506. For similar views, see for example, G. McGhee, *Envoy to the Middle World: Adventures in Diplomacy* (New York, 1983), ch. 27, and the collection of essays in J. A. Bill and W. R. Louis (eds.), *Musaddiq, Iranian Nationalism and Oil* (Austin, Texas, 1988).

Biographical details of important personalities

Abraham, William Ernest Victor (1897–1980).
Geologist, Burmah Oil Company, 1920–37. London office, 1937–40. Served in army, 1940–5, rising to rank of Maj.-General. Director, Burmah Oil Company, 1945–55; Managing Director, 1948–55. Director, AIOC, 1953–5.

Acheson, Dean (1893–1971).
Private secretary to Louis D. Brandeis, Associate Justice, US Supreme Court, 1919–21. Member of law firm of Covington, Burling & Rublee, 1921–33. Under-Secretary, US Treasury, 1933. Assistant Secretary, State Department, 1941; Under-Secretary, 1947; Secretary of State, 1949–53.

Addison, Joseph (1912–1994)
Law graduate, articled to Linklaters and Paines, 1933–6; solicitor, Linklaters and Paines, 1936–9. Army (rising to Major), 1939–45. Joined AIOC, Legal Branch of Concessions Department, 1946. Transferred to Distribution Department, 1950. Observer with Stokes mission to Tehran, August 1951. Manager of Persian Department, 1954. Member of AIOC delegation to Tehran for the consortium negotiations, April-September 1954. General Manager, Iranian Oil Participants Ltd, 1955–71. Awarded CBE.

Ahmad, al-Jaber as-Sabah (1885–1950) (Ahmad I).
Succeeded his uncle as ruler of Kuwait in 1921 and survived a constitutional crisis to retain absolute control until he died after an illness in January 1950.

Ala, Mirza Husayn Khan (1882–1964).
Iranian politician. Educated Westminster School. Barrister, Inner Temple. Minister of Public Works and Agriculture, Iran, 1918.

Minister, USA, 1921–5; Paris, 1929. Joint Managing Director, Bank Melli, 1933–4. Minister, London, 1934–6. President, Bank Melli, 1941. Minister of Court, 1942, 1950–5, 1957–63. Ambassador, USA, 1945–50. Foreign Minister, 1950. Prime Minister, March–April 1951 and 1955–7. Senator, 1963–4.

Alanbrooke, Field Marshal; Alan Francis Brooke (1883–1963).
Joined Royal Field Artillery, 1902. Brigade-Major, 1915. Held numerous staff appointments in 1920s and 1930s. Chief of Imperial General Staff, 1941–6. Field Marshal, 1944. Government Director, AIOC, 1946–54. Baron, 1945. 1st Viscount Alanbrooke of Brookeborough, 1946.

Aliabadi, Abdul Husayn (1914–   ).
Iranian lawyer and judge. Educated Tehran and Paris Universities. Professor of law. Provisional board member of the National Iranian Oil Company, 1951. Director-General, Prime Minister's office. Adviser to Minister of Justice. Public Prosecutor General, 1964.

Amini, Ali (1905–1992).
Iranian politician. Educated University of Paris. Member of Majlis. Minister of National Economy. Minister of Finance, 1953. Leader of Iranian delegation at consortium negotiations. Minister of Justice. Ambassador to USA, 1956–8. Prime Minister, 1961–2.

Attlee, Clement Richard (1883–1967).
MP, 1922–55. Leader of Labour Party, 1935–55. Leader of Opposition, 1935–40. Lord Privy Seal, 1940–2. Secretary of State for Dominions, 1942–3. Lord President of the Council, 1943–5. Deputy Prime Minister, 1942–5. Leader of Opposition, 1945. Prime Minister, 1945–51. Leader of Opposition, 1951–5. 1st Earl Attlee, 1955.

Barnes, Hugh Shakespear (1853–1940).
Lt.-Governor of Burma, 1903–5. Council of India, 1905–13. Director, Imperial Bank of Persia; Chairman, 1916–37. Director, APOC/AIOC, 1909–40. Knighted, 1903.

Barstow, George Lewis (1874–1966).
Controller of Supply Services, Treasury, 1919–27. Director, Prudential Assurance Company Ltd, 1928; Chairman, 1941. Government Director, APOC/AIOC, 1927–46. Knighted, 1920.

Bayat, Murtiza Quli (b. 1887).
Iranian landowner and politician. Majlis Deputy for the whole period of Riza Shah's rule. Prime Minister, November 1944–

April 1945. Provisional board member, National Iranian Oil Company, 1951.

Bazargan, Mihdi (1905–1995).

Iranian politician. Degree in thermodynamics, Paris University. Professor, Technical College, Tehran University, then Dean. A strong supporter of Musaddiq and a leading member of the National Front. Provisional board member, National Iranian Oil Company, 1951, then Managing Director. Leader of Freedom Movement Party.

Beckett, William Eric (1896–1966).

Called to Bar 1922. Assistant Legal Adviser, Foreign Office, 1925; Second Legal Adviser, 1929–45; Legal Adviser, 1945–53. Knighted, 1948.

Berthoud, Eric Alfred (1900–89).

Served with APOC/AIOC in France, Holland and Germany, 1926–39. Wartime service in Egypt and Austria. Under-Secretary, Petroleum Division, Ministry of Fuel and Power, 1946–8. Assistant Under-Secretary, Foreign Office, 1948–52. Ambassador, Denmark, 1952–6; Poland, 1956–60. Knighted, 1954.

Bevin, Ernest (1881–1951).

General Secretary, Transport and General Workers' Union, 1921–40. MP, 1940–51. Minister of Labour and National Service, 1940–5. Foreign Secretary, 1945–51. Lord Privy Seal, 1951.

Bradbury, John Swanwick (1872–1950).

Insurance Commissioner, 1911–13. Joint Permanent Secretary, Treasury, 1913–19. Delegate to Reparation Commission, Paris, 1919–25. Government Director, APOC, 1925–7. Knighted, 1913. Created 1st Baron Bradbury of Winsford, 1925.

Bridges, Edward (1892–1969)

Joined Treasury, 1919. Secretary to the Cabinet, 1938–46. Permanent Secretary, Treasury, 1919–56. Fellow, All Souls College, Oxford, 1920–7 and 1954–8. Knighted, 1939. Created 1st Baron, 1957.

Bullard, Reader William (1885–1976).

Consular Service, 1909–30. Consul-General, Moscow, 1930; Leningrad, 1931–4; Jedda, 1936–9. Minister (later Ambassador) Tehran, 1939–46. Knighted, 1936.

Butler, Victor Spencer (1900–69).

Under-Secretary, Ministry of Fuel and Power, 1946–54. Senior executive, Shell Petroleum Company, 1955–61.

Caccia, Harold Anthony (1905–90).
> Foreign Office, 1929. Served in Peking, 1935. Assistant Private Secretary to Foreign Secretary, 1936. Wartime service in Greece, North Africa, Italy. Assistant Under-Secretary, Foreign Office, 1946; Deputy Under-Secretary, 1949, 1954–6. Ambassador, Austria, 1951–4; USA, 1956–61. Permanent Under-Secretary of State, 1962–5. Head of Diplomatic Service, 1964–5. Knighted, 1950. Created Baron Caccia of Abernant, 1965.

Cadman, John (1877–1941).
> Chief Inspector of Mines, Trinidad and Tobago, 1904–8. Professor of Mining and Petroleum Technology, Birmingham University, 1908–20. Served on the Admiralty Commission that reported on the oil prospects of Iran, 1913. Consulting Petroleum Adviser, Colonial Office, and member of the Advisory Council for Scientific and Industrial Research, 1916. Director of the newly created Petroleum Executive, 1917 and Chairman of the Inter-Allied Petroleum Council, 1918. Appointed Technical Adviser to APOC, 1921; Managing Director, 1923; Chairman, APOC/AIOC, 1927–41. Knighted, 1918. Created 1st Baron Cadman of Silverdale, 1937.

Cargill, John Traill (1867–1954).
> Director, Burmah Oil Company, 1902–43; Chairman, 1904–43. Chairman, Concessions Syndicate Ltd, 1905. Senior Partner, Finlay, Fleming & Company, Rangoon. Director, APOC/AIOC, 1909–43. Baronet, 1920.

Churchill, Winston Leonard Spencer (1874–1965).
> MP, 1900–22 and 1924–64. President, Board of Trade, 1908–10. Home Secretary, 1910–11. First Lord of Admiralty, 1911–15. Chancellor of Duchy of Lancaster, 1915. Minister of Munitions, 1917–19. Secretary of State for War and Air, 1919–21. Secretary of State for the Colonies, 1921–2. Chancellor of Exchequer, 1924–9. First Lord of Admiralty, 1939–40. Prime Minister, 1940–5. Leader of Opposition, 1945–51. Prime Minster, 1951–5. Knighted, 1953.

Clive, Robert Henry (1877–1948).
> Foreign Office, 1902. Minister, Tehran, 1926–31. Ambassador to Holy See, 1933–4; Japan, 1934–7; Belgium, 1937–9. Knighted, 1927.

Cox, Peter Thomas (1902–73).
> Geologist, APOC, 1924. General Fields Manager, 1948–51. Exploration Manager and Chief Geologist, AIOC, 1953. Managing Director, BP Exploration Company, 1957–66.

Cripps, Richard Stafford (1889–1952).
> Called to Bar 1913. MP 1931–50. Ambassador, USSR, 1940–2.
> Minister of Aircraft Production, 1942–5. President, Board of
> Trade, 1945–7. Minister for Economic Affairs, 1947. Chancellor
> of the Exchequer, 1947–50. Knighted, 1930.

Davar, Ali Akbar (1887–1937).
> Iranian Prosecutor General, 1911. Director of Public Education,
> 1921. Deputy for Varamin, 1922. Formed Radical Party, 1925.
> Minister of Public Works, 1925–6. Minister of Justice, 1927–33.
> Minister of Finance, 1933–7. Committed suicide, 1937.

Deterding, Henri Wilhelm August (1866–1939).
> Joined Royal Dutch Oil Company, 1896. General Manager,
> 1900. Managing Director, Asiatic Petroleum Company, 1903.
> Managing Director, Royal Dutch-Shell, 1907–36. Honorary
> knighthood, 1920.

Dixon, Pierson John (1904–65).
> Foreign Office, 1929. Principal Private Secretary to Foreign Secre-
> tary, 1943–8. Ambassador, Czechoslovakia, 1948–50. Deputy
> Under-Secretary of State, Foreign Office, 1950–4. Permanent
> Representative to the UN, 1954–60. Ambassador, France,
> 1960–4. Knighted, 1950.

Drake, Arthur Eric Courtney (1910–).
> Joined APOC as Assistant Accountant and worked in London,
> 1935–7. Sent out to Iran as a Senior Accountant, 1937. Appointed
> Commercial Superintendent at Abadan, 1944. General Manager,
> Iran and Iraq, 1950–1, at the time of the nationalisation. AIOC
> Representative in New York, 1952–4. Returned to London and
> became Director, 1958; Deputy Chairman, 1962; Chairman,
> 1969–75. Knighted, 1970.

Dulles, John Foster (1888–1959).
> Lawyer. Member of US delegation to UN, 1945–50. Secretary of
> State, 1953–61.

Eady, Crawfurd Wilfrid Griffin (1890–1962).
> Civil Service, 1913. Under-Secretary of State, Home Office,
> 1938–40. Chairman of Board of Customs and Excise, 1941–2.
> Joint Second Secretary, Treasury, 1942–52. Knighted, 1939.

Eden, Robert Anthony (1897–1977).
> MP, 1923–57. Parliamentary Private Secretary to Foreign Secre-
> tary, 1926–9. Under-Secretary, Foreign Office, 1931–3. Lord
> Privy Seal, 1933–5. Foreign Secretary, 1935–8. Secretary of State
> for Dominions, 1939–40. Secretary of State for War, 1940.
> Foreign Secretary, 1940–5. Deputy Leader of Opposition,

1945–51. Foreign Secretary, 1951–5. Prime Minister, 1955–7. Knighted, 1954. 1st Earl of Avon, 1961.

Eisenhower, General Dwight David (1890–1969).

Supreme Commander, Allied Expeditionary Force in Western Europe, 1944–5. Chief of Staff, US Army, 1945–8. President of the USA, 1953–61.

Elkington, Edward H. O. (1890–1964).

Indian Army and Consular Service, 1910–21. APOC, 1921. General Manager, APOC/AIOC, Iran and Iraq, 1929–37. Deputy Director, Production Department, 1946; Director, 1948–56.

Fallah, Riza (1910–   ).

Iranian petroleum expert. Studied oil engineering at Birmingham University. Principal, Abadan Technical Institute. Director, Technical and International Affairs, National Iranian Oil Company. General Manager, Abadan refinery, following nationalisation. Deputy Chairman and General Managing Director, 1974.

Fateh, Mustafa (b. 1899).

Iranian economist, oil expert and banker. Educated John Hopkins and Columbia Universities. The Company's most senior Iranian employee, having joined in 1924. Manager (Distribution); Assistant General Manager (Administration), 1947; Adviser and Consultant on Employee Relations, 1950. Chairman, Bank of Tehran.

Fergusson, John Donald Balfour (1891–1963).

Treasury, 1919. Private Secretary to successive Chancellors of the Exchequer, 1920–36. Permanent Secretary, Ministry of Agriculture, 1936–45. Permanent Secretary, Ministry of Fuel and Power, 1945–52. Knighted, 1937.

Fisher, Norman Fenwick Warren (1879–1948).

Permanent Secretary, Treasury, and Head of Civil Service, 1919–39. Director, AIOC, 1941–8. Knighted, 1919.

Franks, Oliver Shewell (1905–1992).

Professor of Moral Philosophy, Glasgow, 1937–9. Ministry of Supply, 1939–46. Provost, Queen's College, Oxford, 1946–8. Ambassador, USA, 1948–52. Director, Lloyd's Bank, 1953–75; Schroders, 1969–84. Knighted, 1946. Created Baron Franks of Headington, 1962.

Fraser, William Milligan (1888–1970).

Joined his father's Pumpherston Oil Company in Scottish shale oil industry, 1909. Became Director, 1913; Managing Director, 1915. Chairman in the USA of the wartime Inter-Allied Petroleum Conference Specifications Commission, 1918. Headed the negotiations for the amalgamation of the shale oil companies into Scottish Oils Ltd. and their purchase by APOC, 1919. Managing Director, APOC/AIOC, 1923–56; Deputy Chairman, 1928–41; Chairman, 1941–56. Knighted, 1939. Created 1st Baron Strathalmond of Pumpherston, 1955.

Furughi, Mirza Muhammad Ali Khan (1873–1942).
Minister of Justice, Iran, 1911–15. Foreign Minister, 1923–4, 1930–3. Minister of Finance, 1924–5. Minister of War, 1925–7. Prime Minister, 1941–2.

Gardiner, Thomas Robert (1883–1964).
Post Office, 1906. Controller, London Postal Service, 1926–34; Director-General, 1936–45. Government Director, AIOC, 1950–3. Knighted, 1936.

Gass, Neville Archibald (1893–1965).
Joined APOC, 1919 after war service. Worked for Strick, Scott & Company, APOC's managing agents, 1919–23, dealing with refinery and distribution affairs. Transferred to Abadan, 1923, and became Personal Assistant to J. Jameson, the Joint General Manager. Appointed Assistant General Manager, Commercial Operations, 1926. Deputy General Manager, 1930. Returned to join London Head Office, 1934. Appointed Managing Director, AIOC, 1939; Deputy Chairman, 1956; Chairman, 1957–60. Knighted, 1958.

Gibson, Horace Stephen (1897–1963).
Petroleum Engineer, APOC, 1922. Production Manager, Masjid i-Suleiman, 1932; General Fields Manager, 1945–9; Head of AIOC Research Station, Kirklington Hall, 1949; Managing Director, Iraq Petroleum Company, 1950. Knighted, 1956.

Gobey, George Nigel Stafford (1901–76).
Appointed Commercial Superintendent, Abadan, 1935. Manager, Personnel, 1937. Assistant General Manager (Administration), 1938. General Manager, Iran and Iraq, 1949–50.

Godber, Frederick (1888–1976).
Asiatic Petroleum Company, 1904. Director, Shell Union Oil Corporation, 1922; Director, Shell Transport and Trading Company, 1928; Managing Director, 1934. Chairman, Shell Union Oil, 1937–46. Chairman, Shell Transport and Trading Company, 1946–61. Knighted, 1942. Created 1st Baron Godber of Mayfield, 1956.

Grady, Henry Francis (1882–1957).
US businessman and diplomat. Assistant Secretary of State, 1939–41. Ambassador, India, 1947–8; Greece, 1948–50; Iran, 1950–1.

Greenway, Charles (1857–1934).
Joined Shaw, Wallace & Company, India, 1893; Partner, 1897. Senior Partner, R. G. Shaw & Company. Managing Director, Lloyd, Scott & Company, 1910. Director, APOC, 1909–34; Managing Director, 1910–19; Chairman, 1914–27; President, 1927–34. Baronet, 1919. Created 1st Baron Greenway of Stanbridge Earls, 1927.

Gulbenkian, Calouste Sarkis (1869–55).
British subject, 1902. Director, National Bank of Turkey, 1910. Director, Turkish Petroleum Company, 1912–29. Director, Iraq Petroleum Company, 1929–33.

Gulshayan, Abbas Quli (1902–   ).
LLB, Paris University. Attorney-General, then Director-General, Minister of Justice. Finance Minister, Minister of National Economy. Mayor of Tehran. Senator in 1949 when the Gass–Gulshayan Supplemental Agreement met with stiff opposition in the Majlis.

Hakimi, Ibrahim (1870–1959).
Physician to Qajar Court. Entered politics 1908. Member of Majlis and then many Cabinets variously as Minister of Education, Finance, Foreign Affairs and Justice. Prime Minister, 1945 (twice) and 1947. President of Senate, 1954–5.

Harden, Orville (1894–1957).
Member of the Co-ordination Committee, Standard Oil (New Jersey), 1928–34; Director, 1929; Supervisor of Latin American affiliates, 1934–6; Vice-President and Vice-Chairman of the Executive Committee, 1935; Contact Director for overseas marketing, 1943. Director, Aramco. First leader of consortium delegation to Iran, 1954.

Harmer, Frederic (1905–1995)
Treasury, 1939. Assistant Secretary, 1943–5. In Washington, March–June 1944 and September–December 1945 for Anglo-American economic and financial negotiations. Resigned from government service, December 1945. Deputy Chairman, P&O, 1957–70. Government Director, AIOC/BP, 1953–70. Knighted, 1968.

Harper, Kenneth Brand (1891–1961).

Joined Finlay Fleming & Company, Rangoon, 1913. General Manager for the East, 1929–36. Served on India's Council of State in Delhi and participated in the Burma round-table conference in London, 1931–2. London office, 1936–7. Director, Burmah Oil Company, 1937–57; Managing Director, 1948–51; Chairman, 1948–57. Director, AIOC/BP, 1947–57. Director, Shell, 1948–61. Knighted, 1936.

Harriman, William Averell (1891–1986).

US government service, 1932. President Roosevelt's special representative, London, 1941. Ambassador, USSR, 1943–6; UK, 1946. President Truman's special envoy to Iranian Government, 1951.

Hazhir, Abdul Husayn (b. 1895).

Iranian politician. Graduate of School of Political Science. Served in Foreign Ministry and joined Russian Embassy as an interpreter. Rejoined Government to find favour under Davar. Head of Industrial and Agricultural Bank. Minister of Commerce under Furughi. Finance Minister. Minister of Court. Prime Minister, June–November, 1948.

Hearn, Arthur Charles (1877–1952).

Assistant Director of Stores, Admiralty. Chairman, Anglo-French Commission to Romania, 1919–20. Joined APOC, 1920. Deputy Director, 1925; Director, 1927–38. Knighted, 1945.

Heath Eves, Hubert (1883–1961).

Burmah Oil Company, 1909. Head of Distribution Department, APOC, 1921; Director 1924. Chairman of the wartime Tanker Tonnage Committee. Deputy Chairman, AIOC, 1941–50. Knighted, 1946.

Henderson, Loy Wesley (1892–1988).

Lawyer. Served in Eastern Europe for Department of State, 1930. Second Secretary US Embassy, Moscow, 1934; First Secretary, 1935; Chargé d'Affaires, 1935–8. First Secretary, Department of State, 1938. Counsellor, USSR, 1942. Minister, Baghdad, 1943–5. Director, Near-Eastern, South Asian and African Affairs, 1945–8. Ambassador, India, 1948–51; Iran, 1951–5.

Hikmat, Ali Asghar (b. 1892).

Iranian politician. Held many education and government posts including Chancellor, Tehran University, 1935; Head of Iran Literary Academy; Professor in Faculty of Letters, 1940; Minister of Interior, 1940; Minister of Health, 1942; Minister of Justice, 1944; Foreign Minister, 1948–52; Ambassador, India, 1952.

Holmes, Frank (Francis) (1874–1947).

Worked as mining engineer in various places including his native New Zealand before World War I. Joined British forces, arranging supplies for troops in Mesopotamia and elsewhere. After war became involved in concession-seeking in Middle East and played a major part in the negotiations for the Kuwait concession, becoming the Shaykh's representative in London after the concession was signed in 1934.

Hoover, Herbert Clark (1903–69).

Mining engineer. President, United Geophysical Company, Inc., 1935–52. Consultant to governments of Venezuela, Iran, Brazil, Peru, etc., 1940–53. Special Adviser to Secretary of State, 1953. Under-Secretary of State, 1954–7.

Idelson, Vladimir Robert (d. 1954).

PhD, Berlin. Called to Russian Bar, 1906. Russian Treasury, 1917. International lawyer in London, 1919–26. Called to English Bar, 1926. British subject, 1930. Took silk, 1943.

Intizam, Abdullah (1907–).

Held senior diplomatic posts Iranian Foreign Ministry in The Hague, Washington, Warsaw, Berne, Prague and Stuttgart. Chief Secretary, Iranian delegation to the United Nations. Foreign Minister under Ala (1951) and Zahidi (1953–5). Director, later Chairman, of National Iranian Oil Company from 1957.

Jacks, Thomas Lavington (1884–1966).

Oil Assistant, Strick, Scott & Company, Muhammara, 1909–13; Assistant Manager, 1917–20. Joint General Manager, APOC, 1923–5; Resident Director, Iran, 1926–35.

Jackson, Basil Rawdon (1892–1957).

APOC, 1921. Production Department, 1921–9; APOC/AIOC representative in USA, 1929–34 and 1939–48. D'Arcy Exploration Company, 1935–9. Managing Director, AIOC, 1948; Deputy Chairman, 1950; Chairman, 1956–7.

Jameson, James Alexander (1885–1961).

Engineer, APOC, 1910. General Manager, Iran, 1926–8; Deputy Director and General Manager of Production, 1927–39; Director, 1939–52.

Jones, Ivor Maurice (n.d.).

Joined APOC as accountant at Muhammara, 1921. Employed by the Khanaqin and Rafidain oil companies, subsidiaries of APOC. Appointed APOC's Manager in Baghdad, 1933. Returned to Abadan as Commercial Manager, 1934; Assistant General Manager, 1937; General Manager, Iran and Iraq, 1945–9.

Kashani, Ayatullah Abul Qasim (1882–1962).
  Iranian religious leader and politician. Majlis Deputy for Tehran, 1949. Speaker of Majlis, 1952–3. Lived in enforced obscurity after 1953 coup.
Kazimi, Baqer (b. 1891).
  Iranian politician. Degree in political science, Washington. Iranian Foreign Ministry, 1911. Held several government posts including Ambassador to France, and was Foreign Minister and Deputy Prime Minister under Musaddiq.
Keswick, William (1903–90).
  Worked in China with Jardine Matheson, 1931–41, becoming 'Taipan' (head of the firm), Hong Kong, 1934–5; Shanghai, 1935–40. Chairman, Shanghai Municipal Council, 1938–41. Director, Matheson & Company, 1943–75; Chairman, 1949–66. Director, Hudson's Bay Company, 1943–72; Governor, 1952–65. Director, AIOC/BP, 1950–73. Knighted, 1972.
Lees, George Martin (1898–1955).
  Served in Mesopotamia during World War I. Political Officer, Kurdistan, 1919–20. Studied geology at Imperial College, London, 1920. Joined APOC as geologist, 1921. Carried out surveys in Persia and Oman, 1921–5. Studied at University of Vienna, PhD, 1928. APOC/AIOC Chief Geologist, 1930–53. Fellow of the Royal Society, 1948.
Le Rougetel, John Helier (1894–1975).
  Foreign Office, 1920. Served in Austria, Canada, China, Hungary, Japan, Netherlands, Romania, USSR. Ambassador, Iran, 1946–50; Belgium, 1950–1. High Commissioner, South Africa, 1951–5. Knighted, 1946.
Levy, Walter James (1911–).
  LLD from Kiel, 1932. Assistant Editor, Petroleum Press Bureau, London, 1936–41. Member, US Enemy Oil Committee, 1942–4. Special Assistant, Office of Intelligence Research, Department of State, 1945–8. Chief of Oil Branch, Economic Co-operation Administration, 1948–9. Formed consulting firm in 1949 advising World Bank and governments on oil policy. Consultant, Policy Planning Staff, Department of State, 1952–3.
Lloyd, Geoffrey William (1902–84).
  Parliamentary Under-Secretary of State, Home Office, 1935–9. Secretary for Mines, 1939–40. Secretary for Petroleum, 1940–2. Parliamentary Secretary for Petroleum, Ministry of Fuel and Power, 1942–5. Minister, Petroleum Warfare Department,

1940–5. Minister of Fuel and Power, 1951–5. Created Baron Geoffrey-Lloyd, 1974.

Lloyd, John Buck (1874–1952).
Senior Partner, Shaw, Wallace & Company, India, 1904. Opened Muhammara office for Lloyd, Scott & Company, 1909. Director, Strick, Scott & Company, 1909–22. Director, APOC/AIOC, 1919–46. Knighted, 1928.

Lund, Frederick William (1874–1965).
Director, British Steamship Investment Trust. Director, APOC/AIOC, 1917–50.

McGhee, George Crews (1912–   ).
Geologist and independent oil producer from 1940. Co-ordinator of aid to Greece and Turkey, 1947–9. Special Assistant to Secretary of State, 1949. Assistant Secretary for Near Eastern, South Asian and African affairs, 1949–51. Ambassador, Turkey, 1951–3. Director, Middle East Institute, 1953–8. Ambassador, Germany, 1963–8.

Makins, Roger Mellor (1904–   ).
Called to Bar, 1927. Foreign Office, 1928. Minister, Washington, 1945–7. Assistant Under-Secretary, Foreign Office, 1947–8. Deputy Under-Secretary, 1948–52. Ambassador, USA, 1953–6. Joint Permanent Secretary, Treasury, 1956–9. Chairman, UK Atomic Energy Authority, 1960–4; Hill Samuel, 1966–70. Knighted, 1949. Created 1st Baron Sherfield, 1964.

Makki, Husayn (1913–   ).
Iranian politician. Resigned from the army after the Allied occupation and joined the Ministry of Roads, 1943. Several minor public posts until elected to Fifteenth Majlis. He oppposed the Supplemental Oil Agreement, supporting Musaddiq, but in the last term of Musaddiq's office began to oppose him and was an effective element in his downfall. He held office under General Zahidi, Musaddiq's successor.

Mallet, Victor Alexander Louis (1893–1969).
Third Secretary in Tehran, 1919; Chargé d'Affaires 1933, 1935. Counsellor and Chargé d'Affaires, Washington, 1936–9. Minister, Stockholm, 1940–5. Ambassador, Spain, 1945–6; Italy, 1947–53. Knighted, 1944.

Mansur, Ali (b. 1888).
Iranian politician. Graduate of School of Political Science. Under-Secretary, Foreign Ministry, 1919. Under-Secretary, Ministry of Interior, 1920. Governor of Azerbaijan, 1926–31.

Minister of Interior, 1931–3. Prime Minister, 1940–1. Subsequently Governor-General of Khurasan and then Azerbaijan. Head of Seven-Year Plan Organisation. Prime Minister, March–June 1950.

Matin-Daftari, Ahmad (n.d.).
Iranian politician. Musaddiq's grand-nephew and son-in-law. Professor of Law. Minister of Justice. Prime Minister, 1935–40 under Riza Shah. Arrested for suspected pro-German activities by the Allies.

Middleton, George Humphrey (1910–).
Consular Service, 1933. Foreign Office, 1943. Counsellor then Chargé d'Affaires, Tehran, 1951 and 1952. Deputy High Commissioner, India, 1953–6. Ambassador, Lebanon, 1956–8. Political Resident, Persian Gulf, 1958–61. Ambassador, Argentina, 1961–4; United Arab Republic, 1964–6. Knighted, 1958.

Morris, Frederick G. C. (1883–1959).
Admitted a solicitor, 1907. Partner, Morris, Veasey & Company. Director, Société Générale des Huiles de Pétrole (APOC's French subsidiary), 1921. Appointed Deputy Director, Continental Distribution, APOC, 1927; Managing Director, AIOC, 1946–52.

Morrison, Herbert Stanley (1888–1965).
MP, 1923–4, 1929–31, 1935–59. Leader of London County Council, 1934–40. Home Secretary, 1940–5. Lord President of the Council and Leader of the House of Commons, 1945–51. Foreign Secretary, March–October 1951. Created Baron Morrison of Lambeth, 1959.

Muhammad, Riza Shah Pahlavi (1919–80).
Educated Switzerland and Tehran. Ruled as Shah 1941–79. Died of cancer in Egypt where he had gone for medical treatment from his exiled home in Panama.

Munro, Gordon Richard (1895–1967).
Commissioned, 4th Dragoon Guards, 1914. Invalided, 1923. Joined Herbert, Wagg & Company, bankers, 1923; Managing Director, 1934–46. Financial Adviser to UK High Commissioner, Canada, 1941–6. Treasury Representative, USA, 1946–9. Minister, Washington, 1946–9. Government Director, AIOC/BP, 1954–6. Knighted, 1946.

Musaddiq, Muhammad (1882–1967).
Supporter of Iranian constitutional revolution in 1900s. Governor of Fars, 1920–1. Minister of Finance, 1921. Governor of Azerbaijan, 1922. Minister of Foreign Affairs, 1923. Deputy in

Majlis, 1924–8. Opposed Riza Shah's foundation of the Pahlavi dynasty. Exiled from politics and imprisoned for some of the time during Riza Shah's rule. Returned to active politics after Riza Shah's abdication in 1941. Elected to Majlis, 1944. Led National Front, formed in 1949. Prime Minister, 1951–3 (except for 17–20 July 1952). Deposed in US- and British-assisted coup, 1953, and sentenced to three years solitary confinement. After his release he retired to his country estate.

Northcroft, Ernest G. D. (1896–1976).

APOC/AIOC 1919–51. Chief Representative in Tehran, 1945–51.

Packe, Edward Hussey (1878–1946).

Private Secretary to successive First Lords of the Admiralty, 1916–19. Government Director, APOC/AIOC, 1919–46. Knighted, 1920.

Pattinson, John Mellor (b. 1899).

Assistant, Fields Organisation, APOC, 1922. Assistant General Manager, (Technical), Abadan, 1934. General Manager, AIOC, Iran and Iraq, 1937–45; Deputy Director, Production Department, London, 1949; Director, 1952–65; Deputy Chairman, 1960–5.

Pirnia, Husayn (1914–   ).

Iranian politician. English degree in administration and oil refining. Joined Finance Ministry and became Director-General, Department of Oil Affairs, then Under-Secretary, 1950. Elected to Seventeenth, Eighteenth and Nineteenth Majlis. Professor, Tehran University. Principal, High School of Mathematics and Management, Karaj.

Qavam al-Saltana (1873–1955).

Minister of Justice, 1909. Minister of War, 1910. Minister of Interior, 1911, 1917–18. Minister of Finance, 1914. Governor of Khurasan, 1918–22. Prime Minister, 1921–3, 1942–3, 1946–7 and briefly in 1952.

Ra'is, Muhsin (b. 1896).

Iranian politician. Law graduate. Held several posts in Foreign Ministry, serving in Switzerland, India, France (for the Vichy Government). Foreign Minister. Ambassador, France; Britain. Governor-General, Tehran, 1964. Senator, 1964 and 1975.

Razmara, Ali (d. 1951).

Military education, Iran and St Cyr. Commander of Kirmanshah mixed regiment, 1927. Chief of Staff, 1942, 1944, 1946–50. Prime Minister, June 1950. Assassinated in office, 7 March 1951.

Riza Shah Pahlavi (1878–1944).

Joint leader of bloodless coup d'état, February 1921. Minister for War and Commander-in-Chief of the army, 1921. Prime Minister, 1923. Succeeded to Persian throne, December 1925. Abdicated 1941 and died in Johannesburg.

Robinson, Frederick Percival (1887–1949).

Treasury, 1911–37. Financial Secretary to the King, 1937–41. Permanent Secretary, Ministry of Works, 1943–6. Government Director, AIOC, 1946. Knighted, 1944.

Ruhani, Fu'ad (b. 1902).

Independent oil consultant and lawyer. MA in Law, London University. English translator at Iranian Foreign Ministry 1921–7. Joined APOC as translator, 1928. Became legal adviser to AIOC and then in 1951 to National Iranian Oil Company. Deputy Chairman, National Iranian Oil Company. First Secretary General, OPEC, 1961–4.

Sa'id, Muhammad (b. 1885).

Born Azerbaijan. Consular posts at Baku, Tiflis and Batum. Head of Russian Department in Foreign Ministry, 1933. Ambassador, USSR, 1938. Foreign Minister under Qavam (1942) and Suhaili (1943). Prime Minister and Foreign Minister, March–April 1944. Prime Minister, 1944, 1948–50.

al-Sa'id, Nuri (1888–1958).

Leading Iraqi politician of his generation. Captured by the British in World War I, he joined the Arab Revolt with Faisal and Lawrence. Minister of Defence several times in 1922–8 and 1953, Foreign Minister several times in 1932–6 and Prime Minister 17 times from 1930. He saw Iraq into the League of Nations and it was his initiatives which led to the founding of the Arab League. He was anti-Axis, anti-Communist and pro-British. He was killed in the uprising of 14 July 1958.

Seddon, Norman Richard (1911–89).

Joined APOC in 1933 and held various posts in Iran and Iraq. Assistant General Manager, Abadan, 1948. AIOC's Chief Representative in Tehran at the time of the nationalisation in 1951. Afterwards returned to London, becoming General Manager in Distribution Department. Managing Director and Deputy Chairman of BP Australia, 1957–67.

Shepherd, Francis Michie (1893–1962).

Acting Consul-General, Barcelona, 1938. Consul, Dresden, 1938–9. Wartime service, Congo; Iceland; Poland. British Poli-

tical Representative, Finland, 1944–7. Consul-General, Netherlands East Indies; 1947–9. Ambassador, Iran, 1950–2; Poland,
1952–4. Knighted, 1948.

Shinwell, Emanuel (1884–1986).

MP, 1922–70. Financial Secretary, War Office, 1929–30. Parliamentary Secretary of Mines, 1930–1. Minister of Fuel and Power,
1945–7. Secretary of State for War, 1947–50. Minister of Defence,
1950–1. Created Baron Shinwell of Easington, 1970.

Simon, John Allsebrook (1873–1954).

Liberal statesman and lawyer. Called to Bar, 1899. MP, 1906–18,
1922–40. Solicitor-General, 1910–13. Attorney-General, 1913–15.
Home Secretary, 1915–16. Foreign Secretary, 1931–5. Home
Secretary, 1935–7. Chancellor of Exchequer, 1937–40; Lord
Chancellor, 1940–5. Created Viscount Simon of Stackpole
Elidor, 1940.

Slade, Edmond John Warre (1859–1928).

Director, Intelligence Division of the Admiralty, 1907–9. Commander-in-Chief, East Indies, 1909–13. Head of the Admiralty
Commission to Persia, 1913–14. Government Director, APOC,
1914–16; Vice-Chairman, 1916–28. Knighted, 1911.

Smith, Desmond Abel (1916–74).

Military positions and directorships of Borax Consolidated,
National Westminster Bank, Equitable Life Assurance Society.
Director, AIOC/BP, 1950–62.

Smith, Evan Cadogan Eric (1894–1950).

9th Lancers, 1913–19. Invalided out of army, 1919. Chairman,
Rolls-Royce. Chairman, National Provincial Bank. Director,
AIOC, 1950.

Snow, Harold Ernest (1897–1971).

APOC, 1921. Distribution Department. Secretary, Petroleum
Board, 1939. Deputy Director, Distribution Department, 1946.
Director, AIOC/BP, 1952–4 and 1955–6. Managing Director,
Consortium, 1954–5; Deputy Chairman, 1957–62. Knighted,
1961.

Stevens, Roger Bentham (1906–80).

Consular Service, 1928. Foreign Office, 1946. Assistant Under-
Secretary of State, 1948–51. Ambassador, Sweden, 1951–4; Iran,
1954–8. Deputy Under-Secretary of State, 1958–63. Vice-Chan-
cellor, Leeds University, 1963–70. Director, British Bank of the
Middle East, 1964–77. Knighted, 1954.

Stokes, Richard Rapier (1897–1957).

MP, 1938–57. Minister of Works, 1950. Lord Privy Seal, 1951.

Leader of negotiations with the Iranian Government, August 1951.

Strang, William (1893–1978).

Foreign Office, 1919. Counsellor, Moscow, 1932. Assistant Under-Secretary of State, 1939–43. British Representative European Advisory Commission, 1943–5. Political Adviser to C in C, British Forces of Occupation in Germany, 1945–7. Permanent Under-Secretary, German Section, Foreign Office, 1947–9; Permanent Under-Secretary of State, 1949–53. Knighted, 1943. Created 1st Baron Strang of Stonesfield, 1954.

Suhaili, Ali (b. 1890).

Iranian politician. Graduate of School of Political Science. Under-Secretary, Ministry of Roads and Communications. Under-Secretary, Foreign Ministry, under Riza Shah. Foreign Minister until dismissed in 1938 over a trivial incident. Governor, Kirman, 1939. Minister of Interior under Mansur. Foreign Minister under Furughi. Prime Minister, March–July 1942, February 1943–March 1944.

Taqizadeh, Sa'id Hassan Khan (1877–1970).

Member First Majlis, 1906. Foreign Minister, 1926. Minister, London, 1929–30, 1941–4; Ambassador, 1944–7. Minister of Finance, 1930–3. Minister, Paris, 1933–4. Deputy for Tabriz, 1947–9. Senator, 1949.

Teagle, Walter Clark (1878–1962).

Worked for his father's oil business for two years before it was sold to Rockefeller's Standard Oil Trust. Employed by Standard first in domestic and later in European marketing. Became a Director of Standard Oil's company registered in New Jersey, 1909 and was responsible for production in Peru. After dismemberment of the Standard Trust, became President, Standard Oil (NJ), 1917–37; Chairman, 1937–42.

Tiarks, Frank Cyril (1874–1952).

Partner, J. Henry Schroder & Company, 1902. Director, Bank of England, 1912–45. Director, APOC/AIOC, 1917–49.

Timurtash, Abdul Husayn Khan (1888–1933).

Deputy for Khurasan, 1914, 1920–3. Governor of Gilan, 1919. Minister of Justice, 1922. Minister of Public Works, 1923. Military Governor several provinces. Minister of Court, 1926–32. Imprisoned by Riza Shah, 1932–3 and died in prison.

Vansittart, Robert Gilbert (1881–1957).

Foreign Office, 1902. Assistant Under-Secretary of State and

Principal Private Secretary to the Prime Minister, 1928–30. Permanent Under-Secretary of State to the Foreign Office, 1930–8. Chief Diplomatic Adviser to the Foreign Secretary, 1938–41. Knighted, 1929. Created 1st Baron Vansittart of Denham, 1941.

Watson, Robert Irving (1878–1948).

Burmah Oil Company, London office, 1901–2 and 1912 onwards. Assistant, Finlay, Fleming & Company, Rangoon, 1902–12. Director, Burmah Oil Company, 1918–47; Managing Director, 1920–47; Chairman, 1943. Director, Shell Transport and Trading Company, 1929–47. Director, APOC/AIOC, 1918–47.

Whigham, Gilbert Campbell (1877–1950).

Joined Finlay, Fleming & Company, Rangoon, 1904. Burmah Oil Company's General Manager in India, 1912–14; Director, Burmah Oil Company, 1920–46. Director, APOC/AIOC 1925–46.

Wright, Denis Arthur Hepworth (1911–).

Vice-Consul, Romania, 1931–41. In charge of Consulate, Turkey, 1941–5. First Secretary (Commercial) Belgrade, 1946–8. Head of Economic Relations Department, Foreign Office, 1951–3. Chargé d'Affaires, Iran, 1953–4. Counsellor, Tehran, 1954–5. Assistant Under-Secretary of State, Foreign Office, 1955–9; 1962. Ambassador, Ethiopia, 1959–62; Iran, 1963–71. Director, Shell Transport and Trading, Standard Chartered Bank, Mitchell Cotts Group, 1971–81. Governor, Oversea Service College, Farnham Castle, 1972–86. Knighted, 1961.

Wynne, Trevredyn Rashleigh (1853–1942).

President, Railway Board of India, 1908–14. Imperial Legislative Council of India, 1908–14. Director, APOC/AIOC, 1915–42. Knighted, 1909.

Zahidi, Fazlullah (1890–1963).

Various military posts under Riza Shah. Chief of Police, 1949. Minister of Interior, 1951. Prime Minister, 1953–5. Retired to Switzerland and appointed to represent Iran at UN, 1958.

Bibliography

ARCHIVES

The records deposited with the BP Archives and the BP Corporate Records Office are catalogued on a computer database which is highly flexible and efficient, but does not lend itself to classifying the records by classes, groups or series in the same manner as the Government archives listed below. A list of the individual BP records used for this volume would be impracticably long for inclusion in the bibliography. However, full references to sources are given in the notes to each chapter.

Owing to internal reorganisation, some of the records at BP Sunbury Technical Records (STR) may no longer be extant.

A.   Government archives

Public Records Office (PRO), Kew, London.
   Cabinet Office:
      CAB 23   Minutes, 1916–1939
      CAB 27   Committees: General Series, 1915–1939
      CAB 50   Committee of Imperial Defence: The Oil Board,
                 1925–1939
      CAB 64   Minister for the Co-ordination of Defence:
                 Registered files, 1918–1959
      CAB 67   War Cabinet Memoranda, 1939–1941
      CAB 134   Committees: General Series, 1945–1956
   Treasury:
      T 161   Supply files

T 236   Overseas Finance Division, 1948–1960
T 273   Bridges' Papers, 1927–1956
Foreign Office:
  FO 248   Embassy and Consular Archives – Persia:
            Correspondence, 1807–1955
  FO 371   General Correspondence: Political, 1906–1954
  FO 416   Confidential: Persia, 1899–1957
Ministry of Fuel and Power:
  POWE 33 Petroleum Division: Correspondence and Papers,
            1916–1964
Board of Trade:
  BT 230   Import Licensing Branch, 1939–1956
Ministry of Labour:
  LAB 13   Overseas Department, 1923–1961
US National Archives, Washington.
  SD 888   Records of the State Department in Record Group 59

B.   Business archives

BP Archives, University of Warwick, Coventry
BP Corporate Records Office, Islington, London
BP Sunbury Technical Records (STR), Sunbury, Berkshire
Iraq Petroleum Company Archives (IPC), c/o BP Archives, University
  of Warwick, Coventry

OFFICIAL PUBLICATIONS

Committee of Imperial Defence, *Report of Sub-committee on Oil
  from Coal* (Cmd 5665, 1938).
*Correspondence between His Majesty's Government in the United
  Kingdom and the Persian Government, and related documents con-
  cerning the oil industry in Persia, February to September 1951* (Cmd
  8425, 1951).
*Correspondence between Her Majesty's Government in the United
  Kingdom and the Iranian Government, and Related Documents,
  concerning the Joint Anglo-American Proposals for Settlement of
  the Oil Dispute, August 1952 to October 1952* (Cmd 8677, 1952).
Prices of Petroleum Products (Cmd 3296, 1929).
International Labour Organisation, *Labour Conditions in the Oil
  Industry in Iran* (Geneva, 1950).
League of Nations, *Official Journal*, vol. 13 (1932).

Parliamentary Debates (Hansard), House of Commons, 1928–1954.

82nd Congress, 2nd Session, Senate Small Business Committee, Staff report of the Federal Trade Commission (FTC), *The International Petroleum Cartel* (1952).

UNPUBLISHED PAPERS AND DISSERTATIONS

Greaves, Rose L., 'The record of the British Petroleum Company Ltd' (BP internal record, vol. 3, 1970).

Lockhart, L. and Rose L. Greaves, 'The record of the British Petroleum Company Ltd' (BP internal record, vol. 1, 1968).

Kittner, Nance F., 'Issues in Anglo-Persian diplomatic relations, 1921–1933' (PhD thesis, School of Oriental and African Studies, University of London, 1980).

NEWSPAPERS AND JOURNALS

Autocar
British Medical Journal
BP Shield
Daily Telegraph
Financial Times
Ittila'at
Los Angeles Times
New York Times
Oil and Gas Journal
The Naft (BP magazine)
The Times
Wall Street Journal

BOOKS AND ARTICLES

Acheson, D., *Present at the Creation: My Years in the State Department* (New York, 1969).

Ady, P. H. and G. D. N. Worswick (eds.), *The British Economy, 1945–1950* (Oxford, 1952).

Alexander, Y. and A. Nanes (eds.), *The United States and Iran: A Documentary History* (Frederick, Maryland, 1982).

Amirsadeghi, H. (ed.), *Twentieth-Century Iran* (London, 1977).

Anderson, I. H., *Aramco, the United States and Saudi Arabia: A Study of the Dynamics of Foreign Oil Policy, 1933–1950* (Princeton, 1981).

Banani, A., *The Modernisation of Iran 1921–41* (Stanford, 1961).

Bharier, J., *Economic Development in Iran 1900–1970* (Oxford, 1971).

Bill, J. A., *The Eagle and the Lion: The Tragedy of American-Iranian Relations* (New Haven, 1988).

Bill, J. A. and W. R. Louis (eds.), *Musaddiq, Iranian Nationalism and Oil* (Austin, Texas, 1988).

Birch, S. F. *et al.*, 'Saturated high octane fuels without hydrogenation', *Journal of the Institute of Petroleum Technologists*, vol. 24 (1938).

Blair, J. M., *The Control of Oil* (New York, 1976).

Bostock, Frances and G. Jones, *Planning and Power in Iran: Ebtehaj and Economic Development under the Shah* (London, 1989).

Bowden, J. A., '"That's the Spirit": Russian Oil Products Ltd (ROP) and the British oil market, 1924–39', *Journal of European Economic History*, vol. 17, no. 3 (1989).

British Petroleum Company, *Our Industry*, 5th edn (London, 1977).

Bullard, Sir Reader, *Britain and the Middle East* (London, 1951).

Cadman, Basil (Second Baron) and J. Rowland, *Ambassador for Oil: The Life of John, First Baron Cadman* (London, 1960).

Cairncross, Sir Alec, *Years of Recovery: British Economic Policy, 1945–1951* (London, 1985).

Carpenter, R. E. H. and R. Stansfield, 'The strobophonometer', *Journal of the Institute of Petroleum Technologists*, vol. 18 (1932).

Chandler, A. D., *Strategy and Structure: Chapters in the History of the Industrial Enterprise* (Cambridge, Mass., 1962).

Chatham House Study Group, *British Interests in the Mediterranean and Middle East* (London, 1958).

Chisholm, A. H. T., *The First Kuwait Oil Concession Agreement: A Record of the Negotiations 1911–1934* (London, 1975).

Clark, J. G., *The Political Economy of World Energy: A Twentieth-Century Perspective* (Hemel Hempstead, 1990).

Coleman, D. C., *Courtaulds: An Economic and Social History*, 3 vols. (Oxford, 1969 and 1980).

Comins, D., 'Gas saturation pressure of crude under reservoir conditions as a factor in the efficient operation of oilfields', *Proceedings of First World Petroleum Congress*, vol. 1 (1933).

Corley, T. A. B., *A History of the Burmah Oil Company, 1886–1966*, 2 vols. (London, 1983 and 1988).

DeGolyer and MacNaughton, *Twentieth-Century Petroleum Statistics* (Dallas, 1990).

Diba, F., *Mohammad Mossadegh, A Political Biography* (London, 1986).

Dow, J. C. R., *The Management of the British Economy, 1945–1960* (Cambridge, 1964).

Dunstan, A. E. (principal ed.), *The Science of Petroleum*, 4 vols. (Oxford, 1938).

Eden, A., *Full Circle: The Memoirs of Anthony Eden* (London, 1960).

Egloff, G. and G. Hulla, *The Alkylation of Alkanes – Patents* (New York, 1948).

Elm, M., *Oil, Power and Principle: Iran's Oil Nationalization and its Aftermath* (New York, 1992).

Feinstein, C. H., *National Income, Expenditure and Output in the United Kingdom, 1855–1965* (Cambridge, 1972).

Ferrier, R. W., *The History of the British Petroleum Company: Volume 1, The Developing Years 1901–1932* (Cambridge, 1982).

Ferrier, R. W. and A. Fursenko (eds.), *Oil in the World Economy* (London, 1989).

Fesharaki, F., *Development of the Iranian Oil Industry* (New York, 1976).

Floor, W., *Industrialization in Iran, 1900–1941*, University of Durham, Centre for Middle Eastern and Islamic Studies, Occasional Papers, series 2, no. 23 (1984).

Labour union, law and conditions in Iran (1900–1941), University of Durham, Centre for Middle Eastern and Islamic Studies, Occasional Papers, series 2, no. 26 (1985).

Ford, A. W., *The Anglo-Iranian Oil Dispute of 1951–1952* (Berkeley, 1954).

Fursenko, A. and R. W. Ferrier (eds.), *Oil in the World Economy* (London, 1989).

Gardner, R. N., *Sterling-Dollar Diplomacy: Anglo-American Collaboration in the Reconstruction of Multilateral Trade* (Oxford, 1956).

Ghods, M. Reza, *Iran in the Twentieth Century, A Political History* (London, 1989).

Gibb, G. S. and Evelyn H. Knowlton, *History of Standard Oil Company (New Jersey): The Resurgent Years, 1911–1927* (New York, 1956).

Gibson, H. S., 'The production of oil from the fields of south-western Iran', *Journal of the Institute of Petroleum Technologists*, vol. 34 (1948).

Hannah, L., *The Rise of the Corporate Economy* (London, 1976).

'The rationalization movement – a revolution in business policy', *Journal of Business Policy*, vol. 1 (1970–1).

Hannah, L. (ed.), *Management Strategy and Business Development* (London, 1976).

Herbert, T., *A Discription of the Persian Monarchy* (London, 1634).

Hidayat, Mihdi Quli Khan, *Khatirat va Khatarat* (Recollections and Adversity) (Tehran, 1950).

Hulla, G. and G. Egloff, *The Alkylation of Alkanes – Patents* (New York, 1948).

Ilersic, A. R., *Government Finance and Fiscal Policy in Post-War Britain* (London, 1955).

Iraq Petroleum Company (IPC), *An Account of the Construction of the Pipeline of the IPC Ltd* (London, 1934).

Jenkins, G., *Oil Economists' Handbook* (5th edn, London, 1989).

Jones, D. T., 'The surface tension and specific gravity of crude oil under reservoir conditions', *Proceedings of First World Petroleum Congress*, vol. 1 (1933).

Jones, G., *The State and the Emergence of the British Oil Industry* (London, 1981).

Jones, G. and Frances Bostock, *Power and Planning in Iran: Ebtehaj and Economic Development under the Shah* (London, 1989).

Jones, J. H., 'A seismic method of prospecting', *Proceedings of First World Petroleum Congress*, vol. 1 (1933).

Katouzian, H., *The Political Economy of Modern Iran, 1926–1979* (New York, 1981).

Musaddiq and the Struggle for Power in Iran (London, 1990).

Keddie, N. R. (ed.), *Religion and Politics in Iran* (New Haven, 1983).

Keddie, N. R. and M. E. Bonine (eds.), *Continuity and Change in Modern Iran* (New York, 1981).

Kirk, G. E., *A Short History of the Middle East*, 3rd edn (London, 1955).

Knowlton, Evelyn H. and G. S. Gibb, *History of Standard Oil Company (New Jersey): The Resurgent Years, 1911–1927* (New York, 1956).

Ladjevardi, H., *Labour Unions and Autocracy in Iran* (New York, 1985).

Laird, A. and C. J. May, 'The efficiency of flowing wells', *Journal of the Institute of Petroleum Technologists*, vol. 20 (1934).

Larson, Henrietta M., Evelyn H. Knowlton and C. S. Popple, *History of Standard Oil Company (New Jersey): New Horizons, 1927–1950* (New York, 1971).

Lenczowski, G., *Russia and the West in Iran, 1918–1948: A Study in Big-Power Rivalry* (New York, 1949).

The Middle East in World Affairs, 4th edn (New York, 1980).

Lenczowski, G. (ed.), *Iran under the Pahlavis* (Stanford, 1978).

Longhurst, H., *Adventure in Oil: The Story of British Petroleum* (London, 1959).

Longrigg, S. H., *Iraq, 1900 to 1950: A Political, Social and Economic History* (London, 1953).

Oil in the Middle East: Its Discovery and Development, 3rd edn (London, 1968).

Louis, W. R., *The British Empire in the Middle East, 1945–1951* (Oxford, 1984).

Louis, W. R. and J. A. Bill (eds.), *Musaddiq, Iranian Nationalism and Oil* (Austin, Texas, 1988).

Lucas, A. F., *Industrial Reconstruction and the Control of Competition* (London, 1937).

McBeth, B. S., *British Oil Policy, 1919–1939* (London, 1985).

McGhee, G., *Envoy to the Middle World: Adventures in Diplomacy* (New York, 1983).

MacGregor, D. H. *et al.*, 'Problems of rationalisation', *Economic Journal*, vol. 40 (1930).

May, C. J. and A. Laird, 'The efficiency of flowing wells', *Journal of the Institute of Petroleum Technologists*, vol. 20 (1934).

Miller, A. D., *Search for Security: Saudi Arabian Oil and American Foreign Policy, 1939–1949* (Chapel Hill, 1980).

Milward, A. S., *The Reconstruction of Western Europe, 1945–1951* (London, 1984).

Mond, A., *Industry and Politics* (London, 1927).

Murray, J., *Iran Today: An Economic and Descriptive Survey* (Tehran, 1950).

Nanes, A. and Y. Alexander (eds.), *The United States and Iran: A Documentary History* (Frederick, Maryland, 1982).

Nash, G. D., *United States Oil Policy, 1890–1964: Business and Government in Twentieth-Century America* (Pittsburgh, 1968).

Overseas Consultants Inc., *Report on Seven Year Development Plan for the Plan Organization of the Imperial Government of Iran*, 5 vols. (New York, 1949).

Pahlavi, Muhammad Reza, *Mission for My Country* (London, 1961).

Painter, D. S., 'Oil and the Marshall Plan', *Business History Review*, vol. 58, no. 3 (1984).

Payton-Smith, D. J., *Oil: A Study of Wartime Policy and Administration* (London, 1971).

Pearton, M., *Oil and the Romanian State* (Oxford, 1971).

Penrose, Edith and E. F., *Iraq: International Relations and National Development* (London, 1978).

Pimlott, B., *Hugh Dalton* (London, 1985).

Pirnia, Husayn, *Dah Sal Kushish dar Rah-i Hifz va Bast-i Huquq-i Iran dar Naft* (Ten Years Struggle to Preserve and Expand Iranian Rights in Oil) (Tehran, 1952).

Pym, L. A., 'The measurement of gas-oil ratios and saturation pressures and their interpretation', *Proceedings of First World Petroleum Congress*, vol. 1 (1933).

Ramazani, R. K., *The Foreign Policy of Iran, 1500–1941: A Developing Nation in World Affairs* (Charlottesville, Virginia, 1966).

Iran's Foreign Policy, 1941–1973 (Charlottesville, Virginia, 1975).

Reader, W. J., *Imperial Chemical Industries, A History, 1870–1952*, 2 vols. (London, 1970 and 1975).

Roosevelt, K., *Countercoup: The Struggle for Control of Iran* (New York, 1979).

Rowland, J. and Basil (Second Baron) Cadman, *Ambassador for Oil: The Life of John, First Baron Cadman* (London, 1960).

Royal Institute for International Affairs, *Survey of International Affairs: The Middle East in the War* (Oxford, 1954).

Survey of International Affairs: The Middle East 1945–1950 (Oxford, 1954).

Ruhani, F., *Tarikh-i Milli Shudan-i San'at-i Naft-i Iran* (A History of the Nationalisation of Oil in Iran) (Tehran, 1973).

Sampson, A., *The Seven Sisters: The Great Oil Companies and the World They Made* (London, 1975).

Seamark, M. C., 'The drilling and control of high pressure wells', *Proceedings of First World Petroleum Congress*, vol. 1 (1933).

Shwadran, B., *The Middle East, Oil and the Great Powers*, 2nd edn (New York, 1959).

Skrine, Sir Clarmont, *World War in Iran* (London, 1962).

Slim, Field Marshal Sir William, *Unofficial History* (London, 1959).

Southwell, C. A. P., 'Scientific unit control', *Proceedings of First World Petroleum Congress*, vol. 1 (1933).

Stamp, J., *Criticism and Other Addresses* (London, 1931).

Stansfield, R. and R. E. H. Carpenter, 'The strobophonometer', *Journal of the Institute of Petroleum Technologists*, vol. 18 (1932).

Stocking, G. W., *Middle East Oil: A Study in Political and Economic Controversy* (London, 1970).

Stoff, M. B., *Oil, War and American Security: The Search for a National Policy on Foreign Oil, 1941–1947* (New Haven, 1980).

'The Anglo-American oil agreement and the wartime search for foreign oil policy', *Business History Review*, vol. 55, no. 1 (1981).

Strong, M. W., 'The significance of underground temperatures', *Proceedings of First World Petroleum Congress*, vol. 1 (1933).

Supple, B. J. (ed.), *Essays in British Business History* (Oxford, 1977).

Urwick, L., *The Meaning of Rationalisation* (London, 1929).

Ward, T. E., *Negotiations for Oil Concessions in Bahrain, El Hasa, The Neutral Zone, Qatar and Kuwait* (printed for private circulation, 1965).

Wilber, D. N., *Riza Shah Pahlavi: Resurrection and Reconstruction of Iran* (New York, 1975).

Woodhouse, C. M., *Something Ventured* (London, 1982).

Worswick, G. D. N. and P. H. Ady (eds.), *The British Economy, 1945–1950* (Oxford, 1952).

Yergin, D., *The Prize: The Epic Quest for Oil, Money and Power* (New York, 1991).

Index

For EU product safety concerns, contact us at Calle de José Abascal, 56–1°,
28003 Madrid, Spain or eugpsr@cambridge.org.